Quantification and Mitigation Strategies to Reduce Greenhouse Gas Emissions from Livestock Production Systems

Quantification and Mitigation Strategies to Reduce Greenhouse Gas Emissions from Livestock Production Systems

Special Issue Editors

Mizeck Chagunda
Peter Løvendahl

MDPI • Basel • Beijing • Wuhan • Barcelona • Belgrade

Special Issue Editors
Mizeck Chagunda
University of Hohenheim
Germany

Peter Løvendahl
Aarhus University
Denmark

Editorial Office
MDPI
St. Alban-Anlage 66
4052 Basel, Switzerland

This is a reprint of articles from the Special Issue published online in the open access journal *Animals* (ISSN 2076-2615) from 2018 to 2019 (available at: https://www.mdpi.com/journal/animals/special_issues/Quantification_and_Mitigation_Strategies_to_Reduce_Greenhouse_Gas_Emissions_from_Livestock_Production_Systems).

For citation purposes, cite each article independently as indicated on the article page online and as indicated below:

LastName, A.A.; LastName, B.B.; LastName, C.C. Article Title. *Journal Name* **Year**, *Article Number*, Page Range.

ISBN 978-3-03928-530-3 (Hbk)
ISBN 978-3-03928-531-0 (PDF)

Cover image courtesy of Mizeck Chagunda.

Contents

About the Special Issue Editors

Mizeck Chagunda is a Professor and Chair of Animal Breeding and Husbandry in the Tropics and Subtropics at the University of Hohenheim, Germany. Mizeck earned his doctorate degree in Animal Breeding and Genetics from the University of Göttingen, Germany, after obtaining his MSc and BSc from the University of Malawi. He has previously worked at the University of Malawi, Aarhus University in Denmark, and SRUC (Scotland's Rural College) in Scotland. Mizeck's research interests are in improving biological and economic efficiency in livestock production systems. He achieves this through investigating novel phenotypes and difficult-to-measure traits, the environmental impact of livestock, and the use of technologies and data-driven decision-support systems, optimising animal breeding strategies and breeding goals. This includes methods for quantifying greenhouse gases (GHGs), applying life cycle analysis (LCA) and techniques for measuring methane from breath in ruminants. As an example of direct industry-relevant research, Mizeck was part of the BIOSENS team (Aarhus University) that developed and tested models that ended up as embedded software for HerdNavigator®.

Peter Løvendahl is a Senior Scientist working at Aarhus University (AU-Foulum, Denmark), Center for Quantitative Genetics and Genomics. Peter earned his MSc and later PhD at the Royal Veterinary and Agricultural University in Copenhagen, now a part of Copenhagen University. Peter has studied phenotypic and genetic variation in a wide range of traits in dairy cattle and pigs. The range of traits has included milk production, endocrine regulation, behaviour, milking behaviour, maternal behaviour and aggression in sows, fertility traits in dairy cows, feed efficiency, methane emission in cows, and metagenomic aspects of ruminant digestion. Peter has focused on improved phenotyping as a means of improving genetic progress in dairy cattle and pig production, especially focusing on methods for automated recording of new traits of possible use in genetic selection. Peter has been involved in numerous collaborative projects with industry partners, including the development of new management tools for dairy farming.

Preface to "Quantification and Mitigation Strategies to Reduce Greenhouse Gas Emissions from Livestock Production Systems"

Mizeck Chagunda [1] and Peter Løvendahl [2]

[1] Animal Breeding and Husbandry in the Tropics and Subtropics, University of Hohenheim, 70599 Stuttgart, Germany

[2] Department of Molecular Biology and Genetics, Faculty of Science and Technology, Aarhus University, PO Box 50, DK-8830 Tjele, Denmark

In recent years, the evidence of change in the global climate system has been unequivocal, and climate change has become a growing international concern. Further, it is well-established that the release of greenhouse gases (GHGs) predominantly derived from human activities is a major contributing factor to most of the observed climate change.

As the largest land-use system on earth, the livestock sector occupies 30% of the world's ice-free surface (Herrero et al., 2013). On the one hand, livestock supply chains are estimated to account for 14.5% of total human-induced GHG emissions (Gerber et al., 2013). In livestock value chains, GHG emissions arise from processes both on and off the farm, and these emissions include methane (CH_4), nitrous oxide (N_2O) and carbon dioxide (CO_2). On the other hand, livestock supply chains contribute to 40% of global agricultural gross domestic product, providing income for more than 1.3 billion people and nourishment for at least 800 million food-insecure people, all the while using vast areas of rangelands, one-third of the freshwater, and one-third of global cropland as feed. Different initiatives are being taken to reduce GHG emissions. For example, the European Union has committed itself to reducing its GHG emissions by 20% by the year 2020, relative to 1990 levels. Different countries and groups of countries have different targets. However, it is important to undertake initiatives that can reduce GHG emissions without compromising livestock productivity. Finding a balance between improving productivity and reducing GHG emissions in livestock systems is crucial for maintaining sustainability in the future. Although primarily focused on ruminant production (Henderson et al., 2017) reported that promising practices for reducing enteric CH_4 emissions and for sequestering soil C in grazing lands could abate up to 379 $MtCO_2$-eq yr^{-1} of emissions. This is equivalent to 11% of annual global ruminant GHG emissions. (Henderson et al., 2017). Efficient agricultural practices are key to reducing greenhouse gas emissions. These practices can be achieved through several aspects of livestock production. For example, livestock genetic improvement, changes in feeding strategies, nutritional improvement, disease control and animal health improvement, and improvement in animal welfare and general husbandry. However, care ought to be taken as a focus on reducing emissions in one particular part of a system may result in an inherent increase elsewhere in the system.

Technical solutions to reduce GHG emissions have been and continue to be extensively researched. Globally, different research groups are investigating different components in this regard on an ongoing basis. Although some of this information has been previously reported elsewhere, new knowledge is being generated and more effective strategies are being developed. Accompanying these efforts to reduce GHG emissions should be efforts to develop methods and procedures for quantifying GHG emissions in different livestock systems. Although metabolic calorimetric chambers have been considered the gold standard for quantifying GHG in ruminants for some time, they are expensive, require highly skilled technicians, and are not normally compatible for use in practical animal production environments where the majority of animals are found. This Special Issue aimed to put together contributions in these two broad but related areas: 1) measurement techniques and protocols, use of proxies, and methodological opportunities and challenges including uncertainty in quantification of GHG emissions from livestock systems; and 2) methods, techniques, and strategies for reducing GHG emissions from livestock production systems. These contributions were sought as both reviews and original research. After manuscripts went through the normal peer review process, where some were accepted and others rejected, a total of 10 papers ended up in the Special Issue. Two papers were on quantification while eight were on mitigation strategies for GHGs. In terms of species (and subspecies), papers covered dairy cattle and dairy goats, beef cattle, goats, and sheep.

Papers on the quantification of GHG focused on novel high-throughput non-invasive methods that have the potential for not only generating the much-needed data upon which evidence-based interventions can be developed, but also methods that can be deployed in more extensive production systems. By comparing these novel techniques to traditional techniques, both papers argue that there is sufficient correlation in the data from the different methods tested. These include methods such as the portable and innovative Laser Methane DetectorTM (Chagunda et al., 2009), and the automatic breath sampler (Lassen et al., 2010). Data from these different measuring methods can be combined for either international genetic improvement studies or for providing the much-needed framework for combining data that would inform some mitigation strategies (Garnsworthy et al., 2019 and Jagoba et al., 2019). However, they warn that the joint use of different GHG measuring methods should be considered only if sources of disagreement—which result from different between-subject and within-subject variabilities—are identified and corrected for (Jagoba et al., 2019).

The first paper, on methane emissions mitigation, lays the foundation by initially considering the effect of and environmental stresses that are common factors which negatively influence rumen function and enteric methane (CH_4) emission (Pragna et al., 2018). This is demonstrated using the goat. Further, Pragna et al. (2018) highlight the three mechanisms by which enteric CH_4 can be reduced: targeting the endproduct of digestion to propionate, providing an alternate hydrogen sink, and selectively inactivating rumen methanogens. The strategies that can be implemented to mitigate enteric CH_4 include nutritional interventions, management strategies, and application of advanced biotechnological tools (Pragna et al., 2018). The next four papers on mitigation strategies demonstrate some results from experiments on how some specific nutritional interventions may reduce methane emissions from ruminates. First, is the use of dietary supplements of Moringa oleifera in lactating dairy cows. The second paper investigates the effects of tea saponin in crossbred dorper ewes while the third is on the use of desmanthus in beef cattle (Dong et al., 2019; Liu et al., 2019; Suybeng et al., 2019). The fourth discusses the effect of encapsulated nitrate and microencapsulated blend of essential oils in beef steers (Alemu et al., 2019).

The next group of two papers deals with GHG emissions reduction through management interventions. The first paper in this group deals with the effect of dietary forage proportion in Holstein heifers at various growth stages (Dong et al., 2019) while the other deals with the effect of changes in grazed farmlets for dairy cattle (Van der Weerden et al., 2018). These two papers highlight the fact that improved livestock management systems are a key driver not only to reducing methane emissions but also reducing nitrogen leaching which in itself also contributes to GHG emissions. The data generated from these management systems can be used to develop regional and national emission inventories and mitigation approaches (Dong et al., 2019; Van der Weerden et al., 2018).

Last, but by far not the least, this Special Issue closes with a paper discussing the abatement potential and cost of different GHG mitigation strategies in dairy goat farming systems (Sintori et al., 2019).

References

1. Alemu, A.W.; Romero-Pérez, A.; Araujo, R.C.; Beauchemin, K.A. Effect of Encapsulated Nitrate and Microencapsulated Blend of Essential Oils on Growth Performance and Methane Emissions from Beef Steers Fed Backgrounding Diets. *Animals* **2019**, *9*, 21.
2. Chagunda, M.G.G.; Ross, D.; Roberts, D.J. On the Use of a Laser Methane Detector in Dairy Cows. *Comput. Electron. Agric.* **2009**, *68*, 157–160.
3. Dong, L.; Li, B.; Diao, Q. Effects of Dietary Forage Proportion on Feed Intake, Growth Performance, Nutrient Digestibility, and Enteric Methane Emissions of Holstein Heifers at Various Growth Stages. *Animals* **2019**, *9*, 725.
4. Dong, L.; Zhang, T.; Diao, Q. Effect of Dietary Supplementation of Moringa oleifera on the Production Performance and Fecal Methanogenic Community of Lactating Dairy Cows. *Animals* **2019**, *9*, 262.
5. Garnsworthy, P.C.; Difford, G.F.; Bell, M.J.; Bayat, A.R.; Huhtanen, P.; Kuhla, B.; Lassen, J.; Peiren, N.; Pszczola, M.; Sorg, D.; Visker, M.H.; Yan, T. Comparison of Methods to Measure Methane for Use in Genetic Evaluation of Dairy Cattle. *Animals* **2019**, *9*, 837.
6. Gerber, P.J., Steinfeld, H., Henderson, B., Mottet, A., Opio, C., Dijkman, J., Falcucci, A. and Tempio, G., 2013. Tackling Climate Change through Livestock: A Global Assessment of Emissions and Mitigation Opportunities. Food and Agriculture Organization of the United Nations (FAO).
7. Henderson, B., Falcucci, A., Mottet, A., Early, L., Werner, B., Steinfeld, H. and Gerber, P., Marginal Costs of Abating Greenhouse Gases in the Global Ruminant Livestock Sector. *Mitig. Adapt. Strateg. Glob. Chang.* **2017**, *22*, 199–224.
8. Herrero, M., Havlík, P., Valin, H., Notenbaert, A., Rufino, M.C., Thornton, P.K., Blümmel, M., Weiss, F., Grace, D. and Obersteiner, M. Biomass Use, Production, Feed Efficiencies, and Greenhouse Gas Emissions from Global Livestock Systems. *Proc. Natl. Acad. Sci. USA* **2013**, *110*, 20888–20893.
9. Lassen, J.; Løvendahl, P.; Madsen, J. Accuracy of Noninvasive Breath Methane Measurements using Fourier Transform Infrared Methods on Individual Cows. *J. Dairy Sci.* **2012**, *95*, 890–898.
10. Liu, Y.; Ma, T.; Chen, D.; Zhang, N.; Si, B.; Deng, K.; Tu, Y.; Diao, Q. Effects of Tea Saponin Supplementation on Nutrient Digestibility, Methanogenesis, and Ruminal Microbial Flora in Dorper Crossbred Ewe. *Animals* **2019**, *9*, 29.
11. Pragna, P.; Chauhan, S.S.; Sejian, V.; Leury, B.J.; Dunshea, F.R. Climate Change and Goat Production: Enteric Methane Emission and Its Mitigation. *Animals* **2018**, *8*, 235.
12. Rey, J.; Atxaerandio, R.; Ruiz, R.; Ugarte, E.; González-Recio, O.; Garcia-Rodriguez, A.; Goiri, I. Comparison between Non-Invasive Methane Measurement Techniques in Cattle. *Animals* **2019**, *9*, 563.
13. Sintori, A.; Tzouramani, I.; Liontakis, A. Greenhouse Gas Emissions in Dairy Goat Farming Systems: Abatement Potential and Cost. *Animals* **2019**, *9*, 945.
14. Suybeng, B.; Charmley, E.; Gardiner, C.P.; Malau-Aduli, B.S.; Malau-Aduli, A.E.O. Methane Emissions and the Use of Desmanthus in Beef Cattle Production in Northern Australia. *Animals* **2019**, *9*, 542.
15. Van der Weerden, T.; Beukes, P.; De Klein, C.; Hutchinson, K.; Farrell, L.; Stormink, T.; Romera, A.; Dalley, D.; Monaghan, R.; Chapman, D.; Macdonald, K.; Dynes, R. The Effects of System Changes in Grazed Dairy Farmlet Trials on Greenhouse Gas Emissions. *Animals* **2018**, *8*, 234.

Mizeck Chagunda, Peter Løvendahl

Special Issue Editors

animals

MDPI

Article

Comparison Between Non-Invasive Methane Measurement Techniques in Cattle

Jagoba Rey [1,*], Raquel Atxaerandio [1], Roberto Ruiz [1], Eva Ugarte [1], Oscar González-Recio [2,3], Aser Garcia-Rodriguez [1] and Idoia Goiri [1,*]

1 Department of Animal Production, NEIKER—Tecnalia, Granja Modelo de Arkaute, Apdo. 46, 01080 Vitoria-Gasteiz, Spain
2 Departamento de Producción Agraria, Escuela Técnica Superior de Ingeniería Agronómica, Alimentaria y de Biosistemas, Universidad Politécnica de Madrid, Ciudad Universitaria s/n, 28040 Madrid, Spain
3 Departamento de Mejora Genética Animal, Instituto Nacional de Investigación y Tecnología Agraria y Alimentaria, 28040 Madrid, Spain
* Correspondence: jrey@neiker.eus (J.R.); igoiri@neiker.eus (G.I.); Tel.: +34-945-121-313 (J.R.)

Received: 28 June 2019; Accepted: 8 August 2019; Published: 15 August 2019

Simple Summary: Enteric methane emissions pose a serious issue to ruminant production and environmental sustainability. To mitigate methane emissions, combined research efforts have been put into animal handling, feeding and genetic improvement strategies. For all research efforts, it is necessary to record methane emissions from individual cows on a large scale under farming conditions. The objective of this trial was to compare two large-scale, non-invasive methods of measuring methane (non-dispersive infrared methane analyzer (NDIR) and laser), in order to see if they can be used interchangeably. For this, paired measurements were taken with both devices on a herd of dairy cows and compared. Significant sources of disagreement were identified between the methods, such that it would not be possible to use both methods interchangeably without first correcting the sources of disagreement.

Abstract: The aim of this trial was to study the agreement between the non-dispersive infrared methane analyzer (NDIR) method and the hand held laser methane detector (LMD). Methane (CH_4) was measured simultaneously with the two devices totaling 164 paired measurements. The repeatability of the CH_4 concentration was greater with the NDIR (0.42) than for the LMD (0.23). However, for the number of peaks, repeatability of the LMD was greater (0.20 vs. 0.14, respectively). Correlation was moderately high and positive for CH_4 concentration (0.73 and 0.74, respectively) and number of peaks (0.72 and 0.72, respectively), and the repeated measures correlation and the individual-level correlation were high (0.98 and 0.94, respectively). A moderate concordance correlation coefficient was observed for the CH_4 concentration (0.62) and for the number of peaks (0.66). A moderate-high coefficient of individual agreement for the CH_4 concentration (0.83) and the number of peaks (0.77) were observed. However, CH_4 concentrations population means and all variance components differed between instruments. In conclusion, methane concentration measurements obtained by means of NDIR and LMD cannot be used interchangeably. The joint use of both methods could be considered for genetic selection purposes or for mitigation strategies only if sources of disagreement, which result in different between-subject and within-subject variabilities, are identified and corrected for.

Keywords: NDIR; laser; agreement; enteric emissions; interchangeability

1. Introduction

The livestock sector plays an important role in climate change, representing 14.5% of all anthropogenic emissions. Within livestock, one of the most relevant sources of greenhouse gases is enteric methane (CH_4) representing 39% of sector emissions [1]. Methane is a potent greenhouse gas with a high global warming potential, calculated as 25 times greater than that of carbon dioxide (CO_2) [2]. For these reasons, enteric CH_4 has become a great concern worldwide. In fact, agreements like "The Paris Agreement" in 2015 have pointed out the importance of the reduction of CH_4 emission by cattle.

Moreover, enteric CH_4 production is associated with considerable dietary energy losses for ruminants, ranging from 1.7% to 14.9% of gross energy intake for lactating cows [3] that could lead to decreased energy gain and productivity. Feed is typically the main production cost item in mixed and intensive systems. Wasting part of the feed energy in the form of CH_4 is not only a climate change issue but also a production problem. In this context, the reduction of livestock greenhouse gases emissions, and in particular enteric CH_4, is posed as a top issue in the agribusiness sector.

The reduction of these emissions can be addressed from the combination of handling, feeding and genetic improvement strategies. However, in order to implement any of these strategies and especially to breed animals with lower CH_4 production, it is necessary to have individual CH_4 data on a large number of animals in commercial conditions.

Methane concentration is heritable [4], and accurate and reliable individual measurements are necessary to include this trait in the breeding programs. These measurements must be made on a large number of animals under commercial conditions [5]. Several methods to record CH_4 emissions from individual animals have been proposed, each with specific advantages, scope of application and also flaws [6–8]. Rapid and non-labor-extensive methods of measuring CH_4 are required in order to implement national genetic evaluations. The non–dispersive infrared analyzer CH_4 analyzer (NDIR) and the hand-held laser CH_4 detector (LMD) are two alternative methodologies that measure CH_4 concentrations in the breath of dairy cattle. Both technologies are non-invasive and allow high throughput in commercial conditions. Both methods have been proved to reliably quantify individually CH_4 concentrations in exhaled air at farm conditions [9–12].

Garnsworthy et al. [13] used for the first time a NDIR CH_4 detector to quantify CH_4 emissions from individual cows on a farm, by sampling air released by eructation during milking. With this technique, a sampling point was installed inside the feed bin of automated milking systems (AMS) and CH_4 concentration on exhaled air was continuously measured in each cow's visits to the AMS. The LMD is a hand-held gas detector for remote measurements of column density for CH_4 containing gases that was originally developed for the detection of gas leaks. However, Chagunda, et al. [14] demonstrated its ability to quantify CH_4 concentration in exhaled air and estimate enteric CH_4 output in dairy cows. Their conclusions were later corroborated by other authors [15,16].

Generating large databases of CH_4 emissions that help addressing reduction strategies has been lately proposed. Exchanging data from different measurements devices in different countries for more accurate studies or genetic evaluations would be an advantage. Thus, harmonizing measurement methodologies is a main concern in recent years, and assessing the equivalence or the statistical agreement between measuring methods is crucial to propose strategies to combine data from different devices. It is important to establish the agreement prior to measuring large numbers of individuals. One approach to analyze this agreement, usually used in medical field, is to record simultaneous repeated records on different subjects using the two methods [17] in order to analyze different sources of disagreement like differences on population means, on between-subject or on within-subject variabilities, as well as the calculation of agreement indices, such as concordance correlation coefficient (CCC) and the coefficient of individual agreement (CIA).

Up to date, few comparative studies have been conducted on these two instruments to determine their equivalence or lack thereof [18], but no study has compared their agreement on paired measurements on the same breath air samples, up to the best of our knowledge. Therefore, the aim of this study was to analyze the agreement and concordance of CH_4 emissions through LMD and the NDIR on paired measurements under field conditions.

2. Materials and Methods

2.1. Experimental Setup and CH_4 Concentration Determination

The LMD is a hand held open path laser measuring device. The model used in this study was LaserMethaneMini (Tokyo Gas Engineering Co., Ltd. Anritsu Devices Co., Ltd., Tokyo, Japan). The principle of the LMD measuring technology was described previously by Chagunda et al. [12–14]. Briefly, this device is based on infrared absorption spectroscopy using a semiconductor laser for CH_4 detection. The device must be pointed towards the nostrils of the cow from a fixed distance. Then, the LMD measures the density of the air column between the device and the animal's nostrils. The reflected laser beam is detected by the device, and its signal is processed and converted to the cumulative CH_4 concentration along the laser path in ppm-m. The LMD was connected to a tablet (Samsung Galaxy Tab A6, New Jersey, USA) running GasViewer app (Tokyo Gas Engineering Solutions, Tokyo, Japan) via Bluetooth connection for exporting and storing the data in real time at 0.5 s intervals. The effect of atmospheric ambient CH_4 concentration from the measurements was discounted using the offset function of the LMD.

The NDIR (Guardian NG Edinburg Instruments Ltd., Livinstong, UK) is one of the so-called sniffer methods that measure CH_4 concentration (ppm) in breath or exhaled air. These methods have been previously used by Garnsworthy et al. [10] to assess the CH_4 production of dairy cows at commercial farms. Briefly, a gas sampling tube from the front of a cow's head to a gas analyzer to continuously measure CH_4 concentration in the cow's breath is used. Then, air is drawn through the instrument by an integral pump between the gas inlet port and analyzer. The device used in this study had a range of 0 to 10,000 ppm. For this study, air was sampled continuously at a rate of 1 L/min through an 8 mm polyamide tube, using approximately 2 m of tube from the analyzer to cow´s nostrils. Methane concentration was recorded at 1 s intervals and stored in a datalogger (Data Recorder SRD-99; Simex Sp. z o.o, Gdańsk, Poland). Baseline or ambient CH_4 concentration was calculated as mean CH_4 concentration before starting the measurements and subtracted from the measured data. Each day before starting measurements, the NDIR analyzer was verified using standard mixtures of CH_4 in nitrogen (0.0%, 0.25%, 0.50%, 0.75% and 1.0%; MESA International Technologies INC, Santa Ana, CA, USA).

For this trial, records were measured in 29 Holstein (11) and Brown Swiss (18) dairy cows in the Fraisoro Agricultural School (Zizurkil, Spain). All animal experiments were carried out in accordance with EU Directive 2010/63/EU for animal experiments and approved by the ethics committee (NEIKER-OEBA-2017-004). Cows were 18.5% of the 1st parity, 52% of the 2nd parity, 18.5% of the 3rd parity and 11% of the 4th parity. Average days in milk at the beginning of the experiment was 102 ± 88 d. Cows were offered a partial mixed ration consisting of corn silage, grass silage and straw ad libitum, and a mean of 5 kg of concentrate supplied in the AMS.

Breath CH_4 concentration was measured on six different days during four months, except for two cows that died during the experimental period. Measurements were performed after morning unifeed distribution between 10:00 and 14:00. Animals were restrained and CH_4 was measured simultaneously with the two devices during a 5 min sampling period, obtaining a total of 164 paired measurements. An operator pointed the LMD at a cow's nostril at a fixed distance of 1 m and trying to maintain the angle from which the LMD was pointed to the cow. Data was recorded every 0.5 s in a tablet. Another operator simultaneously placed NDIR sampling tube on the cow's nostrils in order to measure CH_4 concentration in exhaled air with the NDIR and data was recorded every 1 s in a datalogger.

2.2. Calculations and Statistical Analysis

The average CH_4 concentration from LMD and NDIR was calculated as the arithmetic mean of all concentration values within each 5 min profile measured with the LMD and the NDIR analyzer respectively. The number of eructation events or peaks in each 5 min profile was also calculated for both devices as described in [18].

In order to perform a concordance analysis, phenotypes with the same units are required. Some assumptions must be taken into account in order to convert mean LMD values in ppm-m into ppm. Data obtained with LMD at a distance of one meter (D) was transformed considering an exhaled air bubble of 10 cm (X) from the cow's nostrils, and 2 ppm CH_4 concentration in the environment air (A). The resulting phenotype was calculated as:

$$CH_4(\text{ppm-m}) = X(m) \times CH_4(\text{ppm}) + D(\text{m}) \times A(\text{ppm}) \quad (1)$$

Data was analyzed using a linear mixed effects model with Kronecker product covariance structure in a doubly multivariate set-up [19] using the MIXED procedure of SAS with longitudinal repeated measures (SAS® Institute INC, Cary, NC, EEUU, Version 7.15, 2017). A Kenward-Roger correction was used to compute the correct denominator degrees of freedom of fixed effects in the presence of repeated measures [20]. The linear mixed effects model for the ith records can be written as:

$$y_i = \mu + b_1(S_{i1}) + b_2(S_{i2}) + b_3D + b_4B + b_{i1}(Z_{i1}) + b_{i2}(Z_{i2}) + e_i \quad (2)$$

where every method replication response is denoted by superscript i and y_i is the method replication response (CH_4 concentration, ppm). Terms S_{i1} and S_{i2} indicate the instrument each response belongs to (LMD and NDIR) and will take a value of either zero or one to link y_i to the corresponding instrument. The regression coefficients b_1 and b_2 are the respective fixed effects of instruments, b_3 denotes the fixed effect of day of measurement and b_5 is the fixed effect of the breed. Terms b_{i1} and b_{i2} are random effect parameters for each cow with each of the two devices, with bi ~ ND (0,G), and Z_{i1} and Z_{i2} relate responses y_i to the respective methods. Term e_i is the random residuals from each instrument response with e_i ~ ND (0,R_i).

The method described in detail by Roy [19] was employed to formally test the significance of differences for the between-subject and within-subject variances of the two instruments by means of a log likelihood ratio test between models differing for various combinations of structured and unstructured variance–covariance matrices G and R_i. Using the variance components of the previous model, CIA and CCC were calculated.

Repeatability for each method was calculated as the between-cow (σ_B^2) variance divided by the between-cow and within-cow (σ_W^2) variance.

$$\text{REP} = \frac{\sigma_B^2}{\sigma_B^2 + \sigma_W^2} \quad (3)$$

Additionally, the repeated measures correlation (r_p) was calculated as:

$$r_p = \frac{\sigma_{B1B2}^2}{\sqrt{\sigma_{B1}^2 \times \sigma_{B2}^2}} \quad (4)$$

with ${\sigma_{B1B2}}^2$ being the covariance between cows for the LMD and NDIR, and ${\sigma_{B1}}^2$ and ${\sigma_{B2}}^2$ being the between-subject variance for method 1 and 2, respectively.

Individual level of correlation was calculated using variance components applying the statistical model described previously but including the stage of lactation and lactation number as fixed effects [21].

A Pearson correlation coefficient and Spearman´s rank correlation test were used to assess the association between the overall mean CH_4 concentrations across all paired individual cow records [22].

3. Results and Discussion

Strategies to address CH_4 abatement strategies need large amounts of CH_4 measurements. Combining measurements from different countries has been proposed as a possible strategy and is being studied elsewhere. However, different research groups often utilize different types of device to record CH_4. In this context, a common goal is to decide if the measurement systems agree suitably to each other, and thus can be used interchangeably. The current study is the first directly comparing spot-sampling methods based on breath analysis called the NDIR and LMD simultaneously in the same breath air samples.

An example of a CH_4 profile of a cow as measured with the NDIR sensor and the LMD can be seen in Figure 1. Five different eructation events or CH_4 peaks can be seen with both devices. Mean number of peaks differed between instruments (Table 1), being the LMD capable of detecting higher number of peaks (4.69 vs. 4.24, $p < 0.001$), this could be related with the higher sensitivity of the LMD to detect small variations on CH_4 concentration, even away from the animal [11] and with the time of recording (every 0.5 s vs. 1 s with LMD and NDIR, respectively). However, these differences on the mean number of peaks are of small biological relevance. Moreover, the between subjects variance for the number of peaks did not differ between instruments and a large and positive repeated measures correlation and individual-level correlation were observed (Table 1). Thus, it can be said that both methods are capable to detect the eructation events or peaks in a similar way.

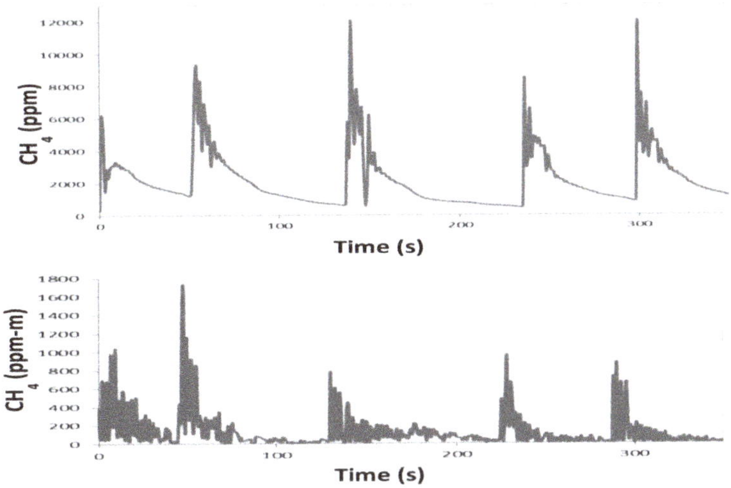

Figure 1. Representation of CH_4 concentration for a cow measured with a non-dispersive infrared (NDIR) sensor (above), and laser methane detector (LMD; below) with different belching events.

In our study, a mean CH_4 concentration of 97 ppm-m with a minimum of 21 ppm-m and a maximum of 303 ppm-m were recorded with the LMD. With the NDIR sensor a mean CH_4 concentration of 1268 ppm with a minimum of 58 ppm and a maximum of 3575 ppm was recorded. The values observed with the LMD agree with those observed in other studies; for instance, Chagunda et al. [12–14] found average CH_4 concentrations between 107–369 ppm-m, depending on the animal activity. The values observed with the NDIR sensor are in line with those observed in other studies [23].

Before considering any method for comparison one needs to ensure that its replication error is acceptable. Repeatability is relevant to the study of method comparison because the repeatabilities of the two methods limit the amount of agreement, which is possible. If one method has poor repeatability the agreement between the two methods is likely to be poor too. As shown in Table 1, the repeatability for the concentration of CH_4 was greater with the NDIR sensor (0.42) than with the LMD (0.23). However, for the number of peaks, the repeatability with the LMD (0.20) was greater than with the NDIR sensor (0.14). The repeatability values obtained in the current trial with the NDIR sensor agree with those reported by Lassen et al. [24] made in the AMS, and with Negussie et al. [25] who reported a repeatability value of 0.36, measuring CH_4 based on CH_4/CO_2 ratio, but were lower than those reported by Sorg et al. [18] who obtained a repeatability of 0.77 (0.13). However, the latter authors processed the NDIR data using head lifting algorithms and modeled it in a Fourier series approach on the time of day of measurement to account for diurnal patterns of CH_4 emission as described by Difford et al. [26]. Repeatability from measurements obtained with LMD was lower than that obtained with the NDIR but similar to those found by other authors [18,27]. The LMD measures were less repeatable than those of NDIR probably because the LMD measurement took place in the open space of the barn, while the NDIR sampled air in the nostrils where environmental influences are likely less variable, e.g., air movement due to ventilation and wind speed.

Though discouraged in comparison studies with repeated measures per subject, we provide the Pearson's and Spearman's correlation coefficients, which were moderately high and positive between instruments for CH_4 concentration (0.73 and 0.74, respectively) and the number of peaks (0.72 and 0.72, respectively). These correlation coefficients could be shrunk because repeated measurements per subject inflate the residual error variance. In these cases, the repeated measures correlation is more appropriate because the effect of repeated measurements per subject are accounted for and prevents a downward bias of correlations due to imprecise measurements. In this sense, repeated measures correlation and individual-level correlation between methods were high (0.98 and 0.94, respectively) for CH_4 concentration (Table 1), which was higher than that observed by Sorg et al. [18]. Individual-level correlations have been used as proxies for genetic correlations in difficult or expensive to measure traits, being the genetic correlation the most informative correlation metric for assessing how to incorporate a method in a selection index [21].

We did not observe a significant breed effect. The CH_4 concentrations population means and all variance components (between-subject and within-subject) differed ($p < 0.05$) between instruments (Table 1). It must be pointed out that this study aimed to assess the agreement or lack thereof between LMD and NDIR. Neither methods are considered as the gold standard, but they allow high throughput recording in a whole cattle population, which is of main interest for the dairy industry. This method comparison study aimed to evaluate whether these methods can be used interchangeably in commercial farms. Traditionally, if there is no reference method, assessing agreement has been based on the intraclass correlation coefficient and CCC. In this study a moderate CCC was observed between instruments for both CH_4 concentration (0.62) and for the number of peaks (0.66; Table 1). This CCC value was higher than that observed by Sorg et al. [18] probably due to the high variance estimates in the present study. This coefficient with fixed within-subject variability increase as between-subject variability increases and some authors question whether this coefficient is adequate in assessing agreement at the individual level, being necessary to evaluate agreement using both CCC and CIA in order to overcome the inherent shortcomings of both metrics [28,29]. The CIA evaluates agreement relative to imprecision in such a way that a good individual agreement means that the individual difference between readings from different methods is close to the difference between replicated readings within a method, and could therefore be inflated by low repeatability values. We observed higher values of CIA between instruments for both CH_4 concentration (0.83) and number of peaks (0.77) than CCC values, which can be explained by the observed moderate repeatability values (Table 1). However, CIA values were moderately high and based on the threshold of "good" interchangeability (0.455) defined by Barnhart et al. [17], CH_4 concentration measurements of both instruments might be used interchangeably.

Nevertheless, based on the criteria for the statistical agreement of the equivalence of means and total variability suggested by Roy [19], the observed differences in the mean populations, between-subject variabilities and within-subject variabilities (Table 1), CH_4 concentration obtained by means of NDIR and LMD could not be considered interchangeable. When two methods are compared without a gold standard, neither provides an unequivocally correct measurement, but CH_4 measures obtained with both devices can be used interchangeably provided that the sources of disagreement are identified and corrected. In this sense, Difford et al. [26] in a study comparing two sniffer methods with simultaneous repeated measurements proposed a process of calibration and standardization to overcome these sources of disagreement with positive results.

Table 1. Sources of (dis)agreement for the CH_4 concentration and number of peaks.

Item	CH_4 Concentration in (ppm)		Number of Peaks	
	NDIR	LMD_cal	NDIR	LMD_cal
LSMean (SE)	1280 (88.3) [a]	991 (57.3) [b]	4.24 (0.1) [a]	4.69 (0.1) [b]
Between-cow variation	149081 [a]	54556 [b]	0.18	0.19
Within-cow variation	352398 [a]	187742 [b]	1.16 [a]	0.76 [b]
Repeatability	0.42	0.23	0.14	0.20
Repeat measures correlation	0.98		1.00	
Individual level correlation	0.94		1.00	
Pearson correlation	0.73		0.72	
Spearman correlation	0.74		0.72	
Concordance correlation coefficient	0.62		0.66	
Coefficient of individual agreement	0.83		0.77	

NDIR: Non-dispersive infrared CH_4 sensor; LMD_cal: Laser CH_4 detector transformed to ppm; SE: Standard error; estimates with subscripts differ ($p < 0.05$).

4. Conclusions

Methane concentration measurements obtained by means of NDIR and LMD instruments cannot be directly used interchangeably under commercial conditions. The joint use of both methods could be considered to establish classifications of individuals, in relation to their CH_4 emissions, in studies for genetic selection purposes or to evaluate CH_4 emissions reduction strategies only if sources of disagreement, which result in different between-subject and within-subject variabilities, are identified and corrected for.

Author Contributions: J.R., R.A., I.G. and A.G.-R. designed the study and methodology. J.R., I.G., R.A. and A.G.-R. have done the field work. J.R., I.G. and A.G.-R., have done the statistical analysis of the data and writing the original manuscript, subsequently supervised by O.G.-R., E.U., R.R. and R.A. Funding acquisition, A.G.-R., R.R., E.U. and O.G.-R.

Funding: This research was financed by RTA2015-00022-C03-02 (METALGEN) project from the national plan of research, development and innovation 2013-2020 and the Department of Economic Development and Competitiveness (Basque Government, Vitoria-Gasteiz, Spain). The first author of this paper was granted a scholarship from the Department of Agriculture of the Basque Country.

Acknowledgments: The authors wish to thank the staff from Escuela Agraria de Fraisoro (Basque Department of Environment, Rural Development, Agriculture and Fishery) for assistance during sampling.

Conflicts of Interest: The authors declare no conflict of interest.

References

1. Gerber, P.J.; Steinfeld, H.; Henderson, B.; Mottet, A.; Opio, C.; Dijkman, J.; Falcucci, A.; Tempio, G. *Tackling Climate Change Through Livestock—A Global Assessment of Emissions and Mitigation Opportunities*; Food and Agriculture Organization of the United Nations (FAO): Rome, Italy, 2013.
2. IPCC (International Panel on Climate Change). *Synthesis Report: Contribution of Working Groups I, II, and III to the Fifth Assessment Report of the Intergovernmental Panel on Climate Change*; Core Writing Team, Pachauri, R.K., Meyer, L.A., Eds.; IPCC: Geneva, Swictzerland, 2014; p. 155.
3. Holter, J.B.; Young, A.J. Methane Prediction in Dry and Lactating Holstein Cows. *J. Dairy Sci.* **1992**, *75*, 2165–2175. [CrossRef]
4. Van Engelen, S.; Bovenhuis, H.; van der Tol, P.P.J.; Visker, M.H.P.W. Genetic background of methane emission by Dutch Holstein Friesian cows measured with infrared sensors in automatic milking systems. *J. Dairy Sci.* **2018**, *101*, 2226–2234. [CrossRef] [PubMed]

5. Negussie, E.; de Haas, Y.; Dehareng, F.; Dewhurst, R.J.; Dijkstra, J.; Gengler, N.; Morgavi, D.P.; Soyeurt, H.; van Gastelen, S.; Yan, T.; et al. Large-scale indirect measurements for enteric methane emissions in dairy cattle: A review of proxies and their potential for use in management and breeding decisions. *J.Dairy. Sci.* **2016**, *100*, 2433–2453. [CrossRef] [PubMed]
6. Storm, I.M.L.D.; Hellwing, A.L.F.; Nielsen, N.I.; Madsen, J. Methods for measuring and estimating methane emission from ruminants. *Animals* **2012**, *2*, 160–183. [CrossRef] [PubMed]
7. Hill, J.; McSweeney, C.; Wright, A.-D.G.; Bishop-Hurley, G.; Kalantar-Zadeh, K. Measuring methane production from ruminants. *Trends Biotechnol.* **2016**, *34*, 26–35. [CrossRef] [PubMed]
8. Patra, A.K. Recent advances in measurements and dietary mitigation of enteric methane emission in ruminants. *Front. Vet. Sci.* **2016**, *3*, 1–17. [CrossRef] [PubMed]
9. Bell, M.J.; Potterton, S.L.; Craigon, J.; Saunders, N.; Wilcox, R.H.; Hunter, M.; Goodman, J.R.; Garnsworthy, P.C. Variation in enteric methane emissions among cows on commercial dairy farms. *Animal.* **2014**, *9*, 1540–1646. [CrossRef]
10. Garnsworthy, P.C.; Craigon, J.; Hernandez-Medrano, J.H.; Saunders, N. Variation among individual dairy cows in methane measurements made on farm during milking. *J.Dairy Sci.* **2012**, *95*, 3181–3189. [CrossRef]
11. Ricci, P.; Chagunda, M.G.G.; Rooke, J.; Houdijk, J.G.M.; Duthie, C.A.; Hyslop, J.; Roeche, R.; Waterhouse, A. Evaluation of the laser methane detector to estimate methane emissions from ewes and steers. *J. Anim. Sci.* **2014**, *92*, 5239–5250. [CrossRef]
12. Chagunda, M.G.G. Opportunities and challenges in the use of the Laser Methane Detector to monitor enteric methane emissions from ruminants. *Animal* **2013**, *7*, 394–400. [CrossRef]
13. Garnsworthy, P.C.; Craigon, J.; Hernandez-Medrano, J.H.; Saunders, N. On-farm methane measurements during milking correlate with total methane production by individual dairy cows. *J. Dairy Sci.* **2012**, *95*, 3166–3180. [CrossRef] [PubMed]
14. Chagunda, M.G.G.; Ross, D.; Roberts, D.J. On the use of a laser methane detector in dairy cows. *Comput. Electron. Agric.* **2009**, *68*, 157–160. [CrossRef]
15. Chagunda, M.G.G.; Yan, T. Do methane measurements from a laser detector and an indirect open-circuit respiation calorimetric chamber agree sufficiently closely? *Anim. Feed Sci. Tecnol.* **2011**, *165*, 8–14. [CrossRef]
16. Sorg, D.; Mühlbach, S.; Rosner, F.; Kuhla, B.; Dermo, M.; Meese, S.; Schwarm, A.; Kreuzer, M.; Swalve, H. The agreement between two next-generation laser methane detectors and respiration chamber facilities in recording methane concentrations in the spent air produced by dairy cows. *Comput. Electron. Agric.* **2017**, *143*, 262–272. [CrossRef]
17. Barnhart, H.X.; Haber, M.; Kosinski, A.S. Assessing individual agreement. *J. Biopharm. Stat.* **2017**, *17*, 697–719. [CrossRef] [PubMed]
18. Sorg, D.; Difford, G.F.; Mühlbach, S.; Kuhla, B.; Swalve, H.; Lassen, J.; Strabel, T.; Pszczola, M. Comparison of a laser methane detector with the GreenFeed and two breath analysers for on-farm measurements of methane emissions from dairy cows. *Comput. Electron. Agric.* **2018**, *153*, 285–294. [CrossRef]
19. Roy, A. An application of linear mixed effects model to assess the agreement between two methods with replicated observations. *J. Biopharm. Stat.* **2009**, *19*, 150–173. [CrossRef]
20. Kenward, M.G.; Roger, J.H. Small sample inference for fixed effects from restricted maximum likelihood. *Biometrics.* **1997**, *53*, 982–997. [CrossRef]
21. Difford, G.F.; Olijhoek, D.W.; Hellwing, A.L.F.; Lund, P.; Bjerring, M.A.; de Haas, Y.; Lassen, J.; Løvendahl, P. Ranking cow's methane emissions under commercial conditions with sniffers versus respiration chambers. *Acta Agric. Scand. A Anim. Sci.* **2019**, *68*, 25–32.
22. SAS. *The little SAS Enterprise Guide Book*; SAS: Cary, NC, USA, 2017.
23. Huhtanen, P.; Cabezas-Garcia, E.H.; Utsumi, S.; Zimmerman, S. Comparison of methods to determine methane emissions from dairy cows in farm conditions. *J. Dairy Sci.* **2015**, *98*, 2294–3409. [CrossRef]
24. Lassen, J.; Lovendahl, P. Heritability estimates for enteric methane emissions from Holstein cattle measured using noninvasive methods. *J. Dairy Sci.* **2015**, *99*, 1959–1967. [CrossRef] [PubMed]
25. Negussie, E.; Liinamo, A.-E.; Mäntysaari, P.; Mäntysaari, E.; Lidauer, M. Between and within-individual variation in methane output measurements in dairy cows. In *Book of Abstracts of the 63rd Annual Meeting of the European Association of Animal Science: Bratislava, Slovakia, 27–31 August 2012*; Wageningen Academic Publishers: Wageningen, The Netherlands, 2012; p. 170.

26. Difford, G.F.; Lassen, J.; Lovendahl, P. Interchangeability between methane measurements in dairy cows assessed by comparing precision and agreement of two non-invasive infrared methods. *Comput. Electron. Agric.* **2016**, *124*, 220–226. [CrossRef]
27. Lassen, J.; Lovendahl, P.; Madsen, J. Accuracy of noninvasive breath methane measurements using Fourier transform infrared methods on individual cows. *J. Dairy Sci.* **2012**, *95*, 890–898. [CrossRef] [PubMed]
28. Barnhart, H.X.; Haber, M.; Lokhnygina, Y.; Kosinski, A.S. Comparison of concordance correlation coefficient and coefficient of individual agreement in assesing agreement. *J. Biopharm. Stat.* **2017**, *17*, 721–738. [CrossRef] [PubMed]
29. Atkinson, G.; Nevill, A. Comment on the use of concordance correlation to assess the agreement between two variables. *Int. Biometr. Soc.* **2016**, *53*, 775–777.

animals

MDPI

Article

Comparison of Methods to Measure Methane for Use in Genetic Evaluation of Dairy Cattle

Philip C. Garnsworthy [1,*], Gareth F. Difford [2,3], Matthew J. Bell [1], Ali R. Bayat [4], Pekka Huhtanen [5], Björn Kuhla [6], Jan Lassen [2], Nico Peiren [7], Marcin Pszczola [8], Diana. Sorg [9,10], Marleen H.P.W. Visker [3] and Tianhai Yan [11]

1 School of Biosciences, University of Nottingham, Sutton Bonington Campus, Loughborough LE12 5RD, UK; Matthew.Bell@nottingham.ac.uk

2 Department of Molecular Biology and Genetics—Center for Quantitative Genetics and Genomics, Aarhus University, Blichers Allé 20, 8830 Tjele, Denmark; garethdifford@gmail.com (G.F.D.); jan.lassen@mbg.au.dk (J.L.)

3 Animal Breeding and Genomics, Wageningen University & Research, P.O. Box 338, 6700 AH Wageningen, The Netherlands; marleen.visker@frieslandcampina.com

4 Milk Production, Production Systems, Natural Resources Institute Finland (Luke), FI 31600 Jokioinen, Finland; alireza.bayat@luke.fi

5 Department of Agricultural Research for Northern Sweden, Swedish University of Agricultural Sciences, SE-901 83 Umeå, Sweden; Pekka.Huhtanen@slu.se

6 Institute of Nutritional Physiology "Oskar Kellner", Leibniz Institute for Farm Animal Biology (FBN), Wilhelm-Stahl-Allee 2, 18196 Dummerstorf, Germany; b.kuhla@fbn-dummerstorf.de

7 Flanders Research Institute for Agriculture, Fisheries and Food, Animal Sciences Unit, Scheldeweg 68, 9090 Melle, Belgium; Nico.Peiren@ilvo.vlaanderen.be

8 Department of Genetics and Animal Breeding, Poznan University of Life Sciences, Wolynska 33, 60-637 Poznan, Poland; mbee@up.poznan.pl

9 Institute of Agricultural and Nutritional Sciences, Martin Luther University Halle-Wittenberg, Animal Breeding, Theodor-Lieser-Str. 11, 06120 Halle, Germany; dianasorg@yahoo.de

10 German Environment Agency (Umweltbundesamt), Wörlitzer Platz 1, 06844 Dessau-Roßlau, Germany

11 Agri-Food and Biosciences Institute, Hillsborough, Co. Down BT26 6DR, UK; Tianhai.Yan@afbini.gov.uk

* Correspondence: phil.garnsworthy@nottingham.ac.uk; Tel.: +44-115-951-6100

Received: 30 August 2019; Accepted: 15 October 2019; Published: 21 October 2019

Simple Summary: Methane is a greenhouse gas with a global warming potential 28 times that of CO_2. Enteric methane accounts for 17% of global methane emissions and 3.3% of total global greenhouse gas emissions from human activities. There is, therefore, significant research interest in finding ways to reduce enteric methane emissions by ruminants. Partners in Expert Working Group 2 (WG2) of the European Cooperation in Science and Technology (COST) Action METHAGENE have used several methods for measuring methane output by individual dairy cattle under various environmental conditions. Methods included respiration chambers, the sulphur hexafluoride (SF_6) tracer technique, breath sampling during milking or feeding, the GreenFeed system, and the laser methane detector. Respiration chambers are considered the 'gold standard', but are unsuitable for large-scale measurements of methane emissions, which are needed for genetic evaluations. In this study, the suitability of methods for large-scale studies was reviewed and compared. All methods showed high correlations with respiration chambers, but comparisons among alternative methods generally had lower correlations. Results confirm, however, that there is sufficient correlation between methods for measurements from all methods to be combined, with appropriate weightings, for use in international genetic studies. This will pave the way for breeding cattle with lower methane emissions.

Abstract: Partners in Expert Working Group WG2 of the COST Action METHAGENE have used several methods for measuring methane output by individual dairy cattle under various environmental conditions. Methods included respiration chambers, the sulphur hexafluoride (SF_6) tracer technique, breath sampling during milking or feeding, the GreenFeed system, and the laser methane detector.

The aim of the current study was to review and compare the suitability of methods for large-scale measurements of methane output by individual animals, which may be combined with other databases for genetic evaluations. Accuracy, precision and correlation between methods were assessed. Accuracy and precision are important, but data from different sources can be weighted or adjusted when combined if they are suitably correlated with the 'true' value. All methods showed high correlations with respiration chambers. Comparisons among alternative methods generally had lower correlations than comparisons with respiration chambers, despite higher numbers of animals and in most cases simultaneous repeated measures per cow per method. Lower correlations could be due to increased variability and imprecision of alternative methods, or maybe different aspects of methane emission are captured using different methods. Results confirm that there is sufficient correlation between methods for measurements from all methods to be combined for international genetic studies and provide a much-needed framework for comparing genetic correlations between methods should these become available.

Keywords: methane; dairy cows; genetic evaluation; greenhouse gases; environment

---

1. Introduction

Methane is a greenhouse gas with a global warming potential 28 times that of CO_2 [1]. Methane from ruminant livestock is generated during microbial fermentation in the rumen and hindgut (enteric methane), and from decomposition of manure. Enteric methane contributes 80% of methane emissions by ruminants, and manure decomposition contributes 20%. Enteric methane accounts for 17% of global methane emissions and 3.3% of total global greenhouse gas emissions from human activities [2]. There is, therefore, a significant research interest in finding ways to reduce enteric methane emissions by ruminants.

Ruminant animals have evolved with a digestive system to digest plant materials efficiently. Like most mammals, ruminants lack the cellulase enzyme required to break the beta-glucose linkages in cellulose, but they play host to diverse populations of rumen microbes that can digest cellulose and other plant constituents. When rumen bacteria, protozoa and fungi ferment carbohydrates and proteins in plant materials, they produce volatile fatty acids, principally acetate, propionate and butyrate. High-fibre diets favour acetate synthesis. Synthesis of acetate and butyrate are accompanied by release of metabolic hydrogen, which, if allowed to accumulate in rumen fluid, has negative effects on microbial growth, and feed digestibility [3]. Rumen Archaea are microorganisms that combine metabolic hydrogen with CO_2 to produce methane and water. Archaea play a vital role, therefore, in protecting the rumen from excess metabolic hydrogen, and the methane they produce is an inevitable product of rumen fermentation.

The amount of methane produced by a ruminant animal is related to the amount of organic matter digested in the rumen, particularly the fibre fraction, and hence the amount of acetate and metabolic hydrogen produced. The main determinants of daily methane production, therefore, are dry matter intake and diet composition: the more feed consumed, and/or the greater the fibre content of the diet, the more methane is produced per day. Nutritional approaches for methane mitigation include reducing the forage to concentrate ratio of diets, increasing dietary oil content, and dietary inclusion of rumen modifiers and methane inhibitors [2,4,5]. Some of these mitigation strategies act through reductions in forage digestibility or feed intake, which can have negative consequences for feed efficiency and methane output per kg of product. Methane output per kg of product is affected mainly by milk yield or growth rate per animal, and by herd-level factors, such as fertility, disease incidence and replacement rate, which affect not only the milk yield of the herd, but also methane produced by replacement animals [6]. Even when all these influences have been taken into account, methane output varies considerably between individual animals. During development of the Metabolisable Energy system, thousands of determinations of methane production by sheep and cattle were performed at the Rowett Research Institute, Aberdeen, UK, using animals fed a restricted amount of feed in closed-circuit respiration chambers. Under these carefully controlled conditions, for animals fed on the same feed, the between-animals coefficient of variation (CV) in methane was 8.1% [7]. Analysis of 1335 records of methane production by cattle fed on a variety of diets in respiration chambers showed a strong relationship (r^2 = 0.91) between methane production and dry matter intake [8]. At the average dry matter intake, however, there was a twofold difference in methane production between the lowest and highest emitters, and at the average methane production rate there was a twofold difference in dry mater intake. These two studies illustrate the range in variation among individual animals that may be encountered under research conditions.

Researchers view individual variation in different ways, according to the aims of their research. When comparing treatments, or evaluating the nutritional value of diets, investigators want to minimise between-animal variation so as to maximise the power of their study, minimise the number of animals required, and increase the chance of detecting a significant difference between treatments. When evaluating populations for genetic studies, on the other hand, variation between animals is of interest, so the aim is to quantify variation, which can then be partitioned into genetic and environmental components. These components are used to determine the amount of variation that is heritable, and genetic correlations between traits currently used in breeding and a possible new trait like methane production. This enables the breeder to decide if there is any merit to adding the new trait to the breeding goal. In terms of methane mitigation, nutritional and management approaches provide greater short-term reductions, but genetic approaches provide greater long-term reductions because genetic improvements are cumulative and permanent. All disciplines require a measurement to be suitably accurate and precise to conduct hypothesis testing and draw reasonable inferences with a given level of certainty. However, several factors influence the choice of the most suitable measurement method such as cost, level of accuracy, precision, scope of application, and scale, which vary across disciplines [9]. For instance, genetic selection programs require methane measurements on thousands of related individuals under the environmental conditions in which the animals are expected to perform [10]. This can be challenging because dairy cattle perform in a wide range of conditions (e.g., grazing vs indoor housing).

Respiration chambers are calibrated to be accurate and precise, and are the gold standard for benchmarking new methods. Where an alternative method may be cheaper, less invasive, easier to implement, or have a wider scope of applications, it is of value to assess the relative accuracy, precision and correlation with the gold standard to assess the relative worth of the alternative method [11]. All methods measure methane with some level of error, so the 'true value' of an individual is not known. However, when the level of measurement error increases, so too does the imprecision. When comparing two methods where one or both methods has high imprecision a phenomenon known as 'attenuation of errors' occurs [12]. The increased measurement error biases the correlation between the two methods downwards and reduces the efficacy of detecting significant differences in accuracy [13]. Or in terms of linear regression terms, when the observed CV of an alternative method is higher than that of the gold standard method, the slope of regression between the methods is decreased and the intercept is biased upwards [14,15].

A variety of technologies are being developed and employed to measure methane emissions by individual dairy cattle under various environmental conditions, as is evidenced by frequent reviews (e.g., [9,16,17]). The aim of the current study was to review and compare the suitability of methods for large-scale measurements of methane output by individual animals, which may be combined with other databases for genetic evaluations. Comparisons included assessing the accuracy, precision and correlation between methods. Combining datasets from different countries and research centres could be a successful strategy for making genetic progress in a trait that is difficult to measure, such as methane emissions if the methods are correlated [17]. Potential for combining large-scale data is of particular interest because data sharing could lead to powerful international collaborations and efficient sharing of resources. Accuracy and precision of methods are important, but data from different sources can be appropriately weighted or adjusted when combined, so any methods can be combined if they are suitably correlated with the 'true' value. The objective of the current study, therefore, was to examine correlations among results obtained by different methods, ultimately leading to an estimate of confidence limits for selecting individual animals that are high or low emitters.

2. Methods for Measuring Methane

Partners in Expert Working Group 2 (WG2) of the European Cooperation in Science and Technology (COST) Action METHAGENE (www.methagene.eu) have used a variety of methods for measuring methane output by individual animals. Methods include respiration chambers, the sulphur hexafluoride (SF_6) tracer technique, breath sampling during milking or feeding, the GreenFeed system, and the laser methane detector. Each method measures different components of methane output. Only respiration chambers measure total emissions from the animal via the oral, nasal and anal routes; all other methods ignore emissions via the anus and only measure methane emitted in breath. Breath measurements are justified because 99% of methane is emitted from the mouth and nostrils, and only 1% via the anus [18]. The SF_6 technique samples breath over 24 h, whereas other techniques use spot samples of breath over periods of minutes throughout the day, so diurnal variation has to be considered. The majority of methane (87%) is released by eructation [18,19], which provides a clear signal for sample processing.

The main features of methods for measuring methane output by individual animals are summarised in Table 1. Values for each feature are based on the experience of experts in METHAGENE WG2 who have used the methods. All values are relative, and somewhat subjective, because absolute values will depend on installation and implementation of each method at different research centres. Each method is described and discussed in more detail in the next five sub-sections.

Table 1. Summary of the main features of methods for measuring methane output by individual animals [1].

Method	Purchase Cost [2]	Running Costs [2]	Labour [2]	Repeatability	Behaviour Alteration [3]	Throughput
Respiration chamber	High	High	High	High	High	Low
SF_6 technique	Medium	High	High	Medium	Medium	Medium
Breath sampling during milking and feeding	Low [4]	Low	Low	Medium	None	High
GreenFeed	Medium	Medium	Low	Medium	Low	Medium
Laser methane detector	Low	Low	High	Low	Low-Medium	Medium

[1] Consensus views based on experiences of METHAGENE WG2 members. [2] Per measuring unit or group of animals. [3] Compared to no methane recording: low = measuring in situ; medium = some handling, training or change in routine; high = confinement. [4] Medium if using FTIR analyser.

2.1. Respiration Chamber

Respiration chambers for open- or closed-circuit indirect calorimetry are considered the 'Gold Standard', and were used extensively in nutrition studies when establishing the Metabolisable Energy system [7]. A single animal (or occasionally more) is confined in a chamber for between 2 and 7 days. Concentration of methane (and other gases if required) is measured at the air inlet and outlet vents of the chamber. The difference between outlet and inlet concentrations is multiplied by airflow to indicate methane emissions rate. In most installations, a single gas analyser is used to measure both inlet and outlet concentrations, often for two or more chambers. This involves switching the analyser between sampling points at set intervals, so concentrations are actually measured for only a fraction of the day.

Respiration chambers vary in construction materials, size of chamber, gas analysis equipment and airflow rate, all of which can influence results. Validation of 22 chambers at six UK research sites revealed an uncertainty of 25.7% between facilities, which was reduced to 2.1% when correction factors were applied to trace each facility to the international standard for methane [20]. The main sources of uncertainty were stability and measurement of airflow, which are crucial for measuring methane emission rate. The authors concluded, however, that chambers were accurate for comparing animals measured at the same site. It is an added challenge, when benchmarking alternative methods against respiration chambers, that the respiration chambers themselves have not been benchmarked against respiration chambers at other facilities.

For large-scale evaluation of methane emissions by individual animals, respiration chambers are challenging, with only a single study in growing Angus steers and heifers exceeding 1000 animals, which found methane production to be moderately heritable $h^2 = 0.27 \pm 0.07$ [21]. Installation costs and running costs are high, and only one animal can be measured at a time. If the monitoring time is three days per animal, and chambers are run continuously, then maximum throughput would be approximately 100 animals per chamber per year. In practice, throughput is likely to be 30 to 50 animals per year. Cows are social animals, and confinement in a chamber may ultimately influence their feeding behaviour, resulting in less feed being consumed and in a different meal pattern compared with farm conditions. Altered feeding patterns or levels is not a problem for metabolic studies evaluating feeds, but can be a problem when evaluating individual animals. Furthermore, the representativeness of respiration chambers to grazing systems has been called into question [22]. However, promising developments have led to more animal friendly respiration chambers constructed from cheaper, transparent materials. These lower the cost and reduce the stress of confinement with minimal disruptions to accuracy, precision and no drop in feed intake of the cows [23].

2.2. The SF_6 Technique

The SF_6 tracer gas technique was developed in an attempt to measure methane emissions by animals without confinement in respiration chambers [24]. Air is sampled near the animal's nostrils through a tube attached to a halter and connected to an evacuated canister worn around the animal's neck or on its back. A capillary tube or orifice plate is used to restrict airflow through the tube so that the canister is between 50 and 70% full after approximately 24 h. A permeation tube containing SF_6 is placed into the rumen of each animal. The pre-determined release rate of SF_6 is multiplied by the ratio of methane to SF_6 concentrations in the canister to calculate methane emission rate.

Many research centres have used the SF_6 technique with variations in design of sampling and collection equipment, permeation tubes, and gas analysis [25]. Reliable results depend on following standard protocols, with greatest variation coming from accuracy of determining SF_6 release rate from permeation tubes and control of sampling rate. With capillary tubes, sampling rate decreases as pressure in the canister increases, whereas an orifice plate gives a steadier sampling rate over 24 h [26]. A source of error that has not been evaluated is that animals might interact and share methane emissions when the sampling tube of one animal is near the head of another animal. There is good agreement between methane emissions measured by the SF_6 technique and respiration chambers, although results from the SF_6 technique are more variable [27,28].

For large-scale evaluation of methane emissions by individual animals, the SF_6 technique is more useful than respiration chambers. Animal behaviour and intake might be affected by wearing the apparatus, and by daily handling to exchange canisters, but the technique is considerably less intrusive than respiration chambers, because cows remain in the herd. Labour and monetary costs for changing canisters each day and for lab analysis are high. Throughput is limited by the number of sets of apparatus available, handling facilities, labour, and the capacity of the lab for gas analysis. Animals need to be measured for 5 to 7 days, and it is recommended that group size should be less than 15 animals [25], so maximum throughput would be about 750 animals per year. Heritability has been estimated for methane production in grazing Holstein cows at $h^2 = 0.33 \pm 0.15$ [29].

2.3. Breath Sampling during Milking and Feeding

Several research groups have developed methods to measure methane concentration in breath of cows during milking and/or feeding. These are often referred to as 'sniffer methods' because they use devices originally designed to detect dangerous gas leaks. Air is sampled near the animal's nostrils through a tube fixed in a feed bin and connected directly to a gas analyser. The feed bin might be in an automatic milking station [14,30–33] or in a concentrate feeding station [34]. Different research centres use different gas analysers (Nondispersive Infrared (NDIR), Fourier-transform infrared (FTIR) or photoacoustic infrared (PAIR)) and different sampling intervals (1, 5, 20 or 90–120 s). Methane concentration during a sampling visit of typically between 3 and 10 min may be specified as the overall mean, or the mean of eructation peaks. Some centres use CO_2 as a tracer gas and calculate daily methane output according to ratio of methane to CO_2 and daily CO_2 output predicted from performance of the cow [35]. Repeatability and rank correlations were higher for eructation peaks than for mean concentrations, and were higher for eructation peaks than for methane to CO_2 ratio [36]. However, all methods show good repeatability.

For large-scale evaluation of methane emissions by individual animals, breath-sampling methods have significant advantages compared with other methods. Breath-sampling methods are non-invasive because, once installed, animals are unaware of the equipment and are in their normal environment. Animals follow their normal routine, which includes milking and feeding, so no training of animals, handling, or change of diet is required. Equipment is relatively cheap, although more expensive gas analysers are available, and running costs are negligible.

The compromise for non-invasiveness of breath-sampling is that concentrations of gasses in the sampled air are influenced by cow head position relative to the sampling tube [15]. The use of head position sensors and data filtering algorithms can remove the effects when the cow's head is completely out of the feed bin [37], but not within the feed bin. Consequently, sniffer measurements are more variable than flux methods, with factors like variable air flow in the barn increasing measurement error (imprecision), and head position, a highly repeatable characteristic, inflating between-cow variability.

Using CO_2 as a tracer gas partly addresses the issue but, because CO_2 arises from metabolism as well as rumen fermentation, variability of CO_2 emissions has to be considered. A further consideration is diurnal variation in breath concentrations of methane and CO_2 because animals are spot-sampled at different times of day and night. Diurnal variation can be accounted for either by fitting a model derived from the whole group of animals, or by including time of measurement in the statistical model [30].

The number of observations per analyser is limited only by number of cows assigned to one automatic milking station or concentrate feeding station and length of time equipment is installed. Typically, each analyser will record 40 to 70 animals 2 to 7 times per day for 7 to 10 days, although the number of sampling stations per analyser can be increased by using an automatic switching system [32]. Throughput per analyser is likely to be 2000 to 3000 animals per year. Estimates of heritability for methane production measured using this method range from $h^2 = 0.12$ to 0.45 over multiple studies [38,39].

2.4. GreenFeed

GreenFeed (C-Lock Inc., Rapid City, SD, USA) is a sophisticated sniffer system where breath samples are provided when animals visit a bait station [15]. As with other sniffer systems, GreenFeed samples breath from individual animals several times per day for short periods (3 to 7 min). GreenFeed is a portable standalone system used in barn and pasture applications, and incorporates an extractor fan to ensure active airflow and head position sensing for representative breath sampling [9]. Measurements are pre-processed by the manufacturer, and data are available in real time through a web-based data management system [40]. As GreenFeed captures a high proportion of emitted air and measures airflow, which can be calibrated using a tracer gas, methane emission is estimated as a flux at each visit. Providing visits occur throughout the 24 h, methane emission can be estimated directly as g/day [15,40]).

A limitation of the GreenFeed system is that animals require training to use the system, although animals which have been trained to use the system will readily use it again [41]. However, some animals will not use the system or will use it infrequently, and frequency of visits is affected by diet [42]. This can be a challenge when screening commercial herds for methane emission under genetic evaluation.

The manufacturer recommends 15 to 25 animals per GreenFeed unit, and recordings are made typically for 7 days. If all animals visit the unit adequately, throughput per unit is likely to be 750 to 1250 animals per year.

2.5. Laser Methane Detector

The laser methane detector (LMD) is a highly responsive, hand-held device that is pointed at an animal's nostrils and measures methane column density along the length of the laser beam (ppm.m). In the first implementation of LMD on a farm, measurements for each cow were taken over periods of 15 to 25 s between eructation events, and could detect methane emitted each time the animal breathed out [43]. In a later study with sheep and beef cattle, monitoring periods of 2 to 4 min allowed authors to separate breathing cycles from eructation events [44]. Typically, animals are restrained either manually or in head yokes at a feed fence for the required length of time. The operator has to stand at the same distance (1 to 3 m) from each animal every time and must be careful to keep the laser pointed at the animal's nostrils throughout the measurement period.

The LMD can be used in the animal's normal environment, although for consistency restraint is required during measurement. Because the LMD measures methane in the plume originating from the animal's nostrils, results can be affected by factors such as: distance from the animal; pointing angle; animal's head orientation and head movement; air movement and temperature in the barn; adjacent animals; and operator variation [45]. Operator variation is likely to be one of the biggest factors, because the operator controls distance and pointing angle, and is responsible for ensuring that the laser remains on target. The structure of the barn and the resulting ventilation conditions and wind speed at the location of the measurement are also considerable sources of variation in recorded methane.

Assuming operator fatigue does not limit measurements, each LMD could record up to 10 animals per hour. If each animal is recorded 3 times (on 3 consecutive days, for example, as in [46]), throughput is likely to be up to 1000 animals per year.

3. Agreement between Methods

In method comparison studies, simultaneous repeated measures per cow with two or more methods are required in order to assess systematic differences between methods (means) and random differences (precision) and correlation between methods free of residual error. Furthermore, short time differences between repeated measures per subject are needed to ensure that the underlying biology of the cow has not changed. Not all methods can be recorded simultaneously, and the methane emission of cows changes both throughout the day and over the lactation period. In such instances, either cross-over designs are needed, or else matched-pair repeated measures designs. Members of METHAGENE WG2 provided data from studies in which two or more methods had been used to measure methane output (g/day) by individual dairy cows. Methods had been applied to each cow either concurrently or consecutively within a short timeframe.

Seven main methods were represented: respiration chambers; SF_6; GreenFeed; LMD; and three breath-sampling systems based on different gas analysers. Gas analysers incorporated different technologies to measure methane: NDIR (e.g., Guardian Plus, Edinburgh Instruments, Edinburgh, UK), FTIR (e.g., Gasmet 4030, Gasmet Technologies Oy, Helsinki, Finland), and PAIR (e.g., F10, Gasera Ltd., Turku, Finland). In the contributing studies, NDIR and FTIR were used in automatic milking stations, and PAIR was used in concentrate feeding stations. One NDIR study and all FTIR and PAIR studies used CO_2 as a tracer gas, with daily CO_2 output calculated either from milk yield, live weight and days pregnant (t1) or from metabolisable energy intake (t2). Two NDIR studies were based on methane concentration in eructation peaks rather than mean methane concentration, so were treated as separate methods. By separating NDIR studies, a total of 8 distinct methods were available, giving a matrix of 28 potential combinations for comparisons. Data were available for 13 method combinations.

Method comparisons were conducted using bivariate models (repeatability animal models) to obtain correlations between 'true values', also known as repeated measures correlations or individual level correlations [33,47]. Variance components, including between-cow variation and within-cow variation (precision) and means (accuracy), were used in the calculation of between-cow coefficient of variation (CV, %) and total CV and repeatability (Table 2). Where single measurements were available for each method Pearson's correlation was reported and where repeated measures per subject were available repeated measures correlation was reported. Lin's concordance correlation coefficient [48] was calculated for each method comparison.

Table 2. Comparisons of methods for recording methane emission in dairy cattle.

Method	N Cows	N Obs	Mean S.E.	Rep S.E.	Between-Cow CV [2]	Total CV	Correlation [3] (S.E.)	CCC [4] (S.E.)
Alternate Methods [1] Versus Respiration Chambers								
SF_6	33	97	471 (14.3)	0.44 (0.13)	11.6	17.4	0.87 (0.08)	0.30 (0.17)
Respiration Chambers	33	97	437 (10.7)	0.36 (0.08)	8.4	14.0		
GreenFeed	27	63	433 (8.7)	0.64 (0.08)	12.8	15.9	0.81 (0.10)	0.41 (0.12)
Respiration Chambers	27	63	459 (6.5)	0.51 (0.09)	8.1	11.3		
NDIR Peaks	12	12	376 (12.1)	N/A	N/A	11.1	0.89 (0.07)	0.88 (0.10)
Respiration Chambers	12	12	377 (10.7)	N/A	N/A	9.4		
NDIR CO_2 t1	20	60	573 (16.8)	0.58 (0.11)	10.1	13.1	0.72 (0.11)	0.38 (0.21)
Respiration Chambers	20	60	521 (13.7)	0.61 (0.12)	9.1	11.7		
PAIR CO_2 t2	21	21	555 (21.3)	N/A	N/A	11.3	0.80 (0.08)	0.70 (N/A)
Respiration Chambers	21	21	585 (14.1)	N/A	N/A	17.1		
Alternate methods [1] versus Alternate methods								
SF_6	48	144	405 (22.5)	N/A	N/A	38.5	0.40 (0.18)	0.34 (N/A)
GreenFeed	48	144	373 (13.9)	N/A	N/A	25.8		
LMD	11	88	432 (24.8)	0.21 (0.11)	19.4	42.7	0.77 (0.23)	0.18 (0.23)
GreenFeed	11	88	423 (18.5)	0.49 (0.12)	11.4	16.8		
NDIR CO_2 t1	27	63	586 (19.4)	0.59 (0.13)	13.2	17.2	0.64 (0.18)	0.14 (0.19)
GreenFeed	27	63	453 (9.8)	0.75 (0.08)	9.7	11.2		
NDIR CO_2 t1	39	118	365 (8.3)	0.66 (0.11)	13.9	17.1	0.60 (0.11)	0.18 (0.19)
LMD	39	118	363 (10.3)	0.14 (0.09)	7.5	19.6		
FTIR CO_2 t2	34	68	315 (12.3)	0.77 (0.13)	21.3	24.3	0.57 (0.25)	0.20 (0.22)
LMD	34	68	299 (6.1)	0.27 (0.15)	7.5	14.5		
NDIR CO_2 t1	45	90	383 (8.7)	0.85 (0.04)	14.0	15.2	0.58 (0.15)	0.14 (0.19)
NDIR Peaks	45	90	393 (8.1)	0.59 (0.09)	10.7	13.9		
FTIR CO_2 t1	43	103	392 (8.1)	0.81 (0.05)	14.1	15.3	0.97 (0.02)	0.79 (0.12)
NDIR CO_2 t1	43	103	382 (8.9)	0.86 (0.04)	12.2	13.6		
FTIR CO_2 t1	45	90	392 (7.9)	0.81 (0.05)	12.2	13.6	0.53 (0.17)	0.15 (0.19)
NDIR Peaks	45	90	382 (8.2)	0.60 (0.09)	10.8	14.0		

[1] SF_6 = Sulphur hexafluoride tracer gas technique, LMD = Laser methane detector; NDIR = Nondispersive Infrared; FTIR = Fourier Transform Infrared; PAIR = Photoacoustic Infrared. CO_2 t1 method uses CO_2 predicted from milk yield, live weight and days pregnant; CO_2 t2 method uses CO_2 predicted from metabolisable energy intake. [2] Coefficient of variation (%). [3] When repeated measures per cow were made the repeated measures correlation was reported, when single measures per cow were made Pearson's correlation was reported, N/A not available, due to single measurements. [4] Lin's concordance correlation coefficient [48].

Respiration chambers were the most precise method, as can be seen by the smaller between-cow CV% and total CV compared to alternative methods, and respiration chambers are by definition the most accurate (Table 2). All methods tested showed high correlations with respiration chambers but none of the correlations exceeded 0.90. This is in part due to the increased imprecision of alternative methods, as even the most accurate and precise method will compare poorly to a less precise method. These correlations are also likely to be underestimated because none of the methods could be recorded simultaneously with respiration chambers and had to be recorded in cross-over designs. Consequently, the true value for each cow may have changed due to changes in the underlying biology of the cow over time between measurements.

For the methods with repeated measures per cow, the two mass flux methods, SF_6 and GreenFeed, had the highest repeated measures correlations (0.87 ± 0.08 and 0.81 ± 0.10), which outperformed the concentration-based NDIR method using CO_2 tracer gas method t1. Of the two concentration methods evaluated against respiration chambers using single measurements, NDIR Peaks had a higher correlation (0.89 ± 0.07) than the PAIR CO_2 tracer gas method t2 (0.80 ± 0.10).

Comparisons among alternative methods generally had lower correlations than comparisons with respiration chambers, despite having relatively higher numbers of animals, and in most cases simultaneous or near simultaneous repeated measures per cow per method. This could be due to the increased variability and imprecision of alternative methods, as seen by the increased CVs or due to the possibility that different aspects of methane emission are captured using different methods. The study of [49] comparing SF_6 and GreenFeed reported a low Pearson correlation of 0.40, despite having a large number of animals with repeated measures per method, the authors appear not to have estimated a repeated measures correlation, which could be larger. Estimating a repeated measures correlation between these two mass flux methods is a priority as it would clarify the inexplicable disagreement between two methods which both correlate highly with the gold standard method. With the exception of the aforementioned study, the imprecision was low in the mass flux measure comparisons as compared to the concentration-based methods [50]. Two of the sniffer methods evaluated, FTIR CO_2t1 and NDIR CO_2t1, correlated close to unity (0.97), most likely due to the shared prediction equation for CO_2 tracer gas. Nevertheless, all correlations derived from actual data were positive. This suggests that combination of datasets obtained with different methods is a realistic proposition for genetic studies. Calculation of adjustment or weighting factors for bias, accuracy and precision is beyond the scope of the current study, but would improve the value of combined datasets.

4. Conclusions

Measuring methane on large numbers of cows is a challenge. The high costs and low throughput of respiration chambers restrict their use to research studies measuring methane emissions on small numbers of individual animals. Respiration chambers remain the gold standard method, but benchmarking alternative methods against respiration chambers is challenging, because simultaneous replicate measures per cow are not feasible. Methods like SF_6 and GreenFeed require lower capital investment and running costs than Respiration Chambers, and have higher throughput and potential for use in extensive and grazing situations, but costs are still prohibitive for recording large numbers of animals. Methods based on concentration are less precise and accurate than flux methods, but they are viable for large-scale measurement, which is a prerequisite of genetic evaluations. Further development is needed to increase the accuracy and precision of concentration methods. Several reviews of methods for measuring methane have made qualitative judgments based on individual comparison studies without expanding scope to genetic evaluations and considering repeated measure correlations between methods as proxies for genetic correlations. Results confirm that there is sufficient correlation between methods for measurements from all methods to be combined for international genetic studies and provide a much-needed framework for comparing genetic correlations between methods should these be made available.

Author Contributions: Conceptualization, P.C.G., G.F.D., M.J.B., A.R.B., N.P., D.S. and M.H.P.W.V.; methodology, G.F.D.; software, G.F.D.; validation, G.F.D., M.J.B. and P.C.G.; formal analysis, G.F.D.; investigation, P.C.G., G.F.D., M.J.B., A.R.B., P.H., B.K., J.L., N.P., M.P., D.S., M.H.P.W.V. and T.Y.; resources, P.C.G., G.F.D., M.J.B., A.R.B., P.H., B.K., J.L., N.P., M.P., D.S., M.H.P.W.V. and T.Y.; data curation, G.F.D. and M.J.B.; writing—original draft preparation, P.C.G., G.F.D. and M.J.B.; writing—review and editing, P.C.G., G.F.D., M.J.B., A.R.B., P.H., B.K., J.L., N.P., M.P., D.S., M.H.P.W.V. and T.Y.; visualization, G.F.D. and P.C.G.; supervision, P.C.G.; project administration, P.C.G. and M.J.B.; funding acquisition, J.L. and P.C.G.

Funding: This research was funded by European Cooperation in Science and Technology (COST), grant number FA1302.

Conflicts of Interest: The authors declare no conflict of interest. The funders had no role in the design of the study; in the collection, analyses, or interpretation of data; in the writing of the manuscript, or in the decision to publish the results.

References

1. Myhre, G.; Shindell, D.; Bréon, F.; Collins, W.; Fuglestvedt, J.; Huang, J.; Koch, D.; Lamarque, J.; Lee, D.; Mendoza, B.; et al. *Anthropogenic and Natural Radiative Forcing*; Stocker, T.F., Qin, D., Plattner, G.-K., Tignor, M., Allen, S.K., Boschung, J., Nauels, A., Xia, Y., Bex, V., Midgley, P.M., Eds.; Cambridge University Press: Cambridge, UK; New York, NY, USA, 2013.
2. Knapp, J.R.; Laur, G.L.; Vadas, P.A.; Weiss, W.P.; Tricarico, J.M. Invited review: Enteric methane in dairy cattle production: Quantifying the opportunities and impact of reducing emissions. *J. Dairy Sci.* **2014**, *97*, 3231–3261. [CrossRef] [PubMed]
3. Janssen, P.H. Influence of hydrogen on rumen methane formation and fermentation balances through microbial growth kinetics and fermentation thermodynamics. *Anim. Feed Sci. Technol.* **2010**, *160*, 1–22. [CrossRef]
4. Beauchemin, K.A.; Kreuzer, M.; O'Mara, F.; McAllister, T.A. Nutritional management for enteric methane abatement: A review. *Aust. J. Exp. Agric.* **2008**, *48*, 21–27. [CrossRef]
5. Cottle, D.J.; Nolan, J.V.; Wiedemann, S.G. Ruminant enteric methane mitigation: A review. *Anim. Prod. Sci.* **2011**, *51*, 491–514. [CrossRef]
6. Garnsworthy, P.C. The environmental impact of fertility in dairy cows: A modelling approach to predict methane and ammonia emissions. *Anim. Feed Sci. Technol.* **2004**, *112*, 211–223. [CrossRef]
7. Blaxter, K.L.; Clapperton, J.L. Prediction of the amount of methane produced by ruminants. *Br. J. Nutr.* **1965**, *19*, 511–522. [CrossRef]
8. Mills, J.A.N.; Crompton, L.A.; Reynolds, C.K. Ruminant Nutrition Regimes to Reduce Methane and Nitrogen Emissions—A Meta-Analysis of Current Databases. Final Project Report MDC 07/04/A. 2008. Available online: https://dairy.ahdb.org.uk/non_umbraco/download.aspx?media=5903 (accessed on 18 October 2019).
9. Hammond, K.J.; Crompton, L.A.; Bannink, A.; Dijkstra, J.; Yáñez-Ruiz, D.R.; O'Kiely, P.; Kebreab, E.; Eugenè, M.A.; Yu, Z.; Shingfield, K.J.; et al. Review of current in vivo measurement techniques for quantifying enteric methane emission from ruminants. *Anim. Feed Sci. Technol.* **2016**, *219*, 13–30. [CrossRef]
10. Falconer, D.; Mackay, T. *Introduction to Quantitative Genetics*, 4th ed.; Longman: Harlow, UK, 1996; p. 480.
11. Barnhart, H.X.; Kosinski, A.S.; Haber, M.J. Assessing Individual Agreement. *J. Biopharm. Stat.* **2007**, *17*, 697–719. [CrossRef]
12. Spearman, C. The Proof and Measurement of Association between Two Things. *Am. J. Psychol.* **1904**, *15*, 72–101. [CrossRef]
13. Adolph, S.C.; Hardin, J.S. Estimating phenotypic correlations: Correcting for bias due to intraindividual variability. *Funct. Ecol.* **2007**, *21*, 178–184. [CrossRef]
14. Garnsworthy, P.C.; Craigon, J.; Hernandez-Medrano, J.H.; Saunders, N. On-farm methane measurements during milking correlate with total methane production by individual dairy cows. *J. Dairy Sci.* **2012**, *95*, 3166–3180. [CrossRef] [PubMed]
15. Huhtanen, P.; Cabezas-Garcia, E.H.; Utsumi, S.; Zimmerman, S. Comparison of methods to determine methane emissions from dairy cows in farm conditions. *J. Dairy Sci.* **2015**, *98*, 3394–3409. [CrossRef] [PubMed]
16. Storm, I.M.L.D.; Hellwing, A.L.F.; Nielson, N.I.; Madsen, J. Methods for Measuring and Estimating Methane Emission from ruminants. *Animals* **2012**, *2*, 160–183. [CrossRef] [PubMed]

17. de Haas, Y.; Pszczola, M.; Soyeurt, H.; Wall, E.; Lassen, J. Phenotypes to genetically reduce greenhouse gas emissions in dairying. *J. Dairy Sci.* **2017**, *100*, 855–870. [CrossRef] [PubMed]
18. Murray, R.M.; Bryant, A.M.; Leng, R.A. Rates of production of methane in the rumen and large intestine of sheep. *Br. J. Nutr.* **1976**, *36*, 1–14. [CrossRef]
19. Blaxter, K.L.; Joyce, J.P. The accuracy and ease with which measurements of respiratory metabolism can be made with tracheostomized sheep. *Br. J. Nutr.* **1963**, *17*, 523–537. [CrossRef]
20. Gardiner, T.D.; Coleman, M.D.; Innocenti, F.; Tompkins, J.; Connor, A.; Garnsworthy, P.C.; Moorby, J.M.; Reynolds, C.K.; Waterhouse, A.; Wills, D. Determination of the absolute accuracy of UK chamber facilities used in measuring methane emissions from livestock. *Measurement* **2015**, *66*, 272–279. [CrossRef]
21. Donoghue, K.A.; Bird-Gardiner, T.; Arthur, P.F.; Herd, R.M.; Hegarty, R.F. Genetic and phenotypic variance and covariance components for methane emission and postweaning traits in Angus cattle. *J. Anim. Sci.* **2016**, *94*, 1438–1445. [CrossRef]
22. Pinares-Patiño, C.S.; Hickey, S.M.; Young, E.A.; Dodds, K.G.; MacLean, S.; Molano, G.; Sandoval, E.; Kjestrup, H.; Harland, R.; Hunt, C.; et al. Heritability estimates of methane emissions from sheep. *Animal* **2013**, *7* (Suppl. 2), 316–321. [CrossRef]
23. Hellwing, A.L.F.; Lund, P.; Weisbjerg, M.R.; Brask, M.; Hvelplund, T. Technical note: Test of a low-cost and animal-friendly system for measuring methane emissions from dairy cows. *J. Dairy Sci.* **2012**, *95*, 6077–6085. [CrossRef]
24. Johnson, K.A.; Huyler, M.; Westberg, H.H.; Lamb, B.; Zimmerman, P.R. Measurement of methane emissions from ruminant livestock using a sulfur hexafluoride tracer technique. *Environ. Sci. Technol.* **1994**, *28*, 359–362. [CrossRef] [PubMed]
25. Berndt, A.; Boland, T.M.; Deighton, M.H.; Gere, J.I.; Grainger, C.; Hegarty, R.S.; Iwaasa, A.D.; Koolaard, J.P.; Lassey, K.R.; Luo, D.; et al. *Guidelines for Use of Sulphur Hexafluoride (SF_6) Tracer Technique to Measure Enteric Methane Emissions from Ruminants*; Lambert, M.G., Ed.; New Zealand Agricultural Greenhouse Gas Research Centre: Palmerston North, New Zealand, 2014; p. 166.
26. Deighton, M.H.; Williams, S.R.O.; Hannah, M.C.; Eckard, R.J.; Boland, T.M.; Wales, W.J.; Moat, P.J. A modified sulphur hexafluoride tracer technique enables accurate determination of enteric methane emissions from ruminants. *Anim. Feed Sci. Technol.* **2014**, *197*, 47–63. [CrossRef]
27. Grainger, C.; Clarke, T.; McGinn, S.M.; Auldist, M.J.; Beauchemin, K.A.; Hannah, M.C.; Waghorn, G.C.; Clark, H.; Eckard, R.J. Methane emissions from dairy cows measured using the sulfur hexafluoride (SF6) tracer and chamber techniques. *J. Dairy Sci.* **2007**, *90*, 2755–2766. [CrossRef] [PubMed]
28. Muñoz, C.; Yan, T.; Wills, D.A.; Murray, S.; Gordon, A.W. Comparison of the sulfur hexafluoride tracer and respiration chamber techniques for estimating methane emissions and correction for rectum methane output from dairy cows. *J. Dairy Sci.* **2012**, *95*, 3139–3148. [CrossRef] [PubMed]
29. Breider, I.S.; Wall, E.; Garnsworthy, P.C.; Pryce, J.E. Genetic relationships between methane emission and milk yield, live weight and dry matter intake. In Proceedings of the World Congress on Genetics Applied to Livestock Production, Abstract No. 134, Auckland, New Zealand, 11–16 February 2018.
30. Lassen, J.; Løvendahl, P.; Madsen, J. Accuracy of noninvasive breath methane measurements using Fourier transform infrared methods on individual cows. *J. Dairy Sci.* **2012**, *95*, 890–898. [CrossRef]
31. Bell, M.J.; Potterton, S.L.; Craigon, J.; Saunders, N.; Wilcox, R.H.; Hunter, M.; Goodman, J.R.; Garnsworthy, P.C. Variation in enteric methane emissions among cows on commercial dairy farms. *Animal* **2014**, *8*, 1540–1546. [CrossRef]
32. Pszczola, M.; Rzewuska, K.; Mucha, S.T. Heritability of methane emissions from dairy cows over a lactation measured on commercial farms. *J. Anim. Sci.* **2017**, *95*, 4813–4819. [CrossRef]
33. Difford, G.F.; Olijhoek, D.W.; Hellwing, A.L.F.; Lund, P.; Bjerring, M.A.; de Haas, Y.; Lassen, J.; Løvendahl, P. Ranking cows' methane emissions under commercial conditions with sniffers versus respiration chambers. *Acta. Agric. Scand. A Anim. Sci.* **2018**, *68*, 25–32. [CrossRef]
34. Negussie, E.; Lehtinen, J.; Mäntysaari, P.; Bayat, A.R.; Liinamo, A.-E.; Mäntysaari, E.A.; Lidauer, M.H. Non-invasive individual methane measurement in dairy cows. *Animal* **2017**, *11*, 890–899. [CrossRef]
35. Madsen, J.; Bjerg, B.S.; Hvelplund, T.; Weisbjerg, M.R.; Lund, P. Methane and carbon dioxide ratio in excreted air for quantification of methane production in ruminants. *Livest. Sci.* **2010**, *129*, 223–227. [CrossRef]

36. Bell, M.J.; Saunders, N.; Wilcox, R.H.; Homer, E.M.; Goodman, J.R.; Craigon, J.; Garnsworthy, P.C. Methane emissions among individual dairy cows during milking quantified by eructation peaks or ratio with carbon dioxide. *J. Dairy Sci.* **2014**, *97*, 6536–6546. [CrossRef] [PubMed]
37. Difford, G.F.; Lassen, J.; Løvendahl, P. Interchangeability between methane measurements in dairy cows assessed by comparing precision and agreement of two non-invasive infrared methods. *Comput. Electron. Agric.* **2016**, *124*, 220–226. [CrossRef]
38. Lassen, J.; Løvendahl, P. Heritability estimates for enteric methane emissions from Holstein cattle measured using noninvasive methods. *J. Dairy Sci.* **2016**, *99*, 1959–1967. [CrossRef] [PubMed]
39. Breider, I.S.; Wall, E.; Garnsworthy, P.C. Heritability of methane production and genetic correlations with milk yield and body weight in Holstein-Friesian dairy cows. *J. Dairy Sci.* **2019**, *102*, 7277–7281. [CrossRef]
40. Hammond, J.K.; Humphries, D.J.; Crompton, L.A.; Green, C.; Reynolds, C.K. Methane emissions from cattle: Estimates from short-term measurements using a GreenFeed system compared with measurements obtained using respiration chambers or sulphur hexafluoride tracer. *Anim. Feed Sci. Technol.* **2015**, *203*, 41–52. [CrossRef]
41. Velazco, J.I.; Cottle, D.J.; Hegarty, R.S. Methane emissions and feeding behaviour of feedlot cattle supplemented with nitrate or urea. *Anim. Prod. Sci.* **2014**, *54*, 1737–1740. [CrossRef]
42. Hammond, K.J.; Jones, A.K.; Humphries, D.J.; Crompton, L.A.; Reynolds, C.K. Effects of diet forage source and neutral detergent fiber content on milk production of dairy cattle and methane emissions determined using GreenFeed and respiration chamber techniques. *J. Dairy Sci.* **2016**, *99*, 7904–7917. [CrossRef]
43. Chagunda, M.G.G.; Ross, D.; Roberts, D.J. On the use of a laser methane detector in dairy cows. *Comput. Electron. Agric.* **2009**, *68*, 157–160. [CrossRef]
44. Ricci, P.; Chagunda, M.G.G.; Rooke, J.; Houdijk, J.G.M.; Duthie, C.-A.; Hyslop, J.; Roehe, R.; Waterhouse, A. Evaluation of the laser methane detector to estimate methane emissions from ewes and steers. *J. Anim. Sci.* **2014**, *92*, 5239–5250. [CrossRef]
45. Sorg, D.; Mühlbach, S.; Rosner, F.; Kuhla, B.; Derno, M.; Meese, S.; Schwarm, A.; Kreuzerc, M.; Swalve, H. The agreement between two next-generation laser methane detectors and respiration chamber facilities in recording methane concentrations in the spent air produced by dairy cows. *Comput. Electron. Agric.* **2017**, *143*, 262–272. [CrossRef]
46. Mühlbach, S.; Sorg, D.; Rosner, F.; Kecman, J.; Swalve, H.H. Genetic analyses for CH_4 concentrations in the breath of dairy cows measured on-farm with the Laser Methane Detector. In Proceedings of the World Congress on Genetics Applied to Livestock Production, Abstract No. 186, Auckland, New Zealand, 11–16 February 2018.
47. Bakdash, J.Z.; Marusich, L.R. Repeated measures correlation. *Front. Psychol.* **2017**, *8*, 456. [CrossRef] [PubMed]
48. Lin, L.; Torbeck, L.D. Coefficient of accuracy and concordance correlation coefficient: New statistics for methods comparison. *J. Pharm. Sci. Technol.* **1998**, *52*, 55–59.
49. Hristov, A.N.; Oh, J.; Giallongo, F.; Frederick, T.; Harper, M.T.; Weeks, H.; Branco, A.F.; Price, W.J.; Moate, P.J.; Deighton, M.H.; et al. Short communication: Comparison of the GreenFeed system with the sulfur hexafluoride tracer technique for measuring enteric methane emissions from dairy cows. *J. Dairy Sci.* **2016**, 5461–5465. [CrossRef] [PubMed]
50. Sorg, D.; Difford, G.F.; Mühlbach, S.; Kuhla, B.; Swalve, H.H.; Lassen, J.; Strabel, T.; Pszczola, M. Comparison of a laser methane detector with the GreenFeed and two breath analysers for on-farm measurements of methane emissions from dairy cows. *Comput. Electron. Agric.* **2018**, *153*, 285–294. [CrossRef]

animals

MDPI

Review

Climate Change and Goat Production: Enteric Methane Emission and Its Mitigation

Pratap Pragna [1,2], Surinder S. Chauhan [3], Veerasamy Sejian [1,*], Brian J. Leury [3] and Frank R. Dunshea [3]

1 ICAR—National Institute of Animal Nutrition and Physiology, Adugodi, Bangalore 560030, India; pragnaprathap@gmail.com
2 Academy of Climate Change Education and Research, Kerala Agricultural University, Vellanikkara 680656, India
3 Faculty of Veterinary and Agricultural Sciences, The University of Melbourne, Parkville, VIC 3010, Australia; ss.chauhan@unimelb.edu.au (S.S.C.); brianjl@unimelb.edu.au (B.J.L.); fdunshea@unimelb.edu.au (F.R.D.)
* Correspondence: drsejian@gmail.com

Received: 7 November 2018; Accepted: 5 December 2018; Published: 7 December 2018

Simple Summary: Given that goats are considered more climate resilient than other ruminant species, research efforts are therefore needed to understand goat productivity during exposure to high ambient temperatures. Heat stress can affect the digestion and rumen fermentation pattern of goats, which contributes to the reduction in production performance in goats. Diet composition, breed and environmental stresses are common factors which negatively influence rumen function and enteric methane (CH_4) emission. There are three mechanisms by which enteric CH_4 can be reduced: targeting end product of digestion to propionate, providing alternate hydrogen sink and selectively inactivating rumen methanogens. The various strategies that can be implemented to mitigate enteric CH_4 include nutritional interventions, management strategies and application of advanced biotechnological tools.

Abstract: The ability of an animal to cope and adapt itself to the changing climate virtually depends on the function of rumen and rumen inhabitants such as bacteria, protozoa, fungi, virus and archaea. Elevated ambient temperature during the summer months can have a significant influence on the basic physiology of the rumen, thereby affecting the nutritional status of the animals. Rumen volatile fatty acid (VFA) production decreases under conditions of extreme heat. Growing recent evidence suggests there are genetic variations among breeds of goats in the impact of heat stress on rumen fermentation pattern and VFA production. Most of the effects of heat stress on rumen fermentation and enteric methane (CH_4) emission are attributed to differences in the rumen microbial population. Heat stress-induced rumen function impairment is mainly associated with an increase in *Streptococcus* genus bacteria and with a decrease in the bacteria of *Fibrobactor* genus. Apart from its major role in global warming and greenhouse effect, enteric CH_4 is also considered as a dietary energy loss in goats. These effects warrant mitigating against CH_4 production to ensure optimum economic return from goat farming as well as to reduce the impact on global warming as CH_4 is one of the more potent greenhouse gases (GHG). The various strategies that can be implemented to mitigate enteric CH_4 emission include nutritional interventions, different management strategies and applying advanced biotechnological tools to find solution to reduce CH_4 production. Through these advanced technologies, it is possible to identify genetically superior animals with less CH_4 production per unit feed intake. These efforts can help the farming community to sustain goat production in the changing climate scenario.

Keywords: climate change; heat stress; goat; immunization; methane; volatile fatty acids

1. Introduction

Morphologically versatile goat species with unique browsing potential adapt to a changing climate more readily than other ruminant species and consequently they continue to be an important source of income and nutrition to many poor and marginal farmers around the world [1]. Goats are also the major means of employment and income for women, children and aged people in tropical and subtropical regions [2]. The important sources of income from the sector include milk, meat, manure, wool and skin [3]. Small ruminant, and in particular goat, farming is very important because of the relatively low input requirements and the corresponding high expected output [4]. Furthermore, goats emit less enteric methane (CH_4) than all other domestic ruminant animals per unit body weight [5].

A changing climate scenario for extensive grazing systems exposes the animals to various types of stressors that may affect their production, health and survival [6]. Among these, heat stress seems to be the major stressor which negatively influences the animal performance [7]. Furthermore, heat stress can also affect the digestion and rumen fermentation pattern of goats which contributes to the reduction in production performance [8]. The ability of an animal to cope and adapt itself to a changing climate depends on maintaining appropriate functioning of the rumen and ruminal microbes [9]. Elevated ambient temperature may prove detrimental to these processes and may ultimately result in influencing the level of CH_4 production particularly with respect to the intensity of its production in goat and this will require appropriate mitigation strategies to curtail such emissions to sustain goat production in the changing climate scenario [8]. Given that goats are considered more climate resilient than other ruminant species, research efforts are therefore needed to understand goat productivity during exposure to high ambient temperatures. This review is therefore an attempt to collate and synthesize existing knowledge and recent research pertaining to the effects of heat stress on rumen fermentation, enteric CH_4 emissions, and the various mechanisms associated with CH_4 production and its mitigation in goats.

2. Goat as Ideal Climate Model Animal

Small ruminants, in particular goats, are considered an important source of income and nutrition for poor and marginal farmers around the world [5]. Low initial investment and high turnover rate for goat production are the primary reasons behind the promotion of the goat industry in developing countries [10]. Goats are often referred to as village banks in some rural areas where the villagers invest their money on purchasing and feeding goats and consider it as an appropriate way to save money for the future [11]. Globally, there are estimated to be over 860 million goats [12] and recent trends show an increased demand for dairy products from goats, particularly in developing countries where they act as a substitute for dairy products from large ruminants for human dietary needs [13].

Goats are versatile animals that adapt to a changing climate more readily than the other ruminant species and are well suited to small farming systems [1]. Much of the global goat population is concentrated in the arid and semi-arid agro-ecological zones that have frequent droughts and famines [14]. However, these species are reported to be less affected by the harsh climate compared with other ruminants that are highly sensitive to subtle changes in the surrounding environmental fluctuations [15]. Hence, goat rearing is a major source of human nutrition and also the means of economic stability for many small and marginal farmers, providing meat and manure as two major sources of income [14].

Because of their browsing habit and the anatomical advantage of the upper lips, goats can thrive well with limited feedstuffs, especially in arid and semi-arid regions [16]. In addition, goats also have a physiological advantage because they efficiently utilize poor quality feedstuffs and produce appreciably good output in terms of milk, meat and manure [17]. During feed scarcity, goats can reduce their metabolic processes to conserve energy resources [8]. Table 1 describes the advantageous characteristics in goats over other livestock species to survive harsh climatic conditions.

Table 1. Advantageous characteristics associated with goats over other livestock species to survive in harsh climatic conditions.

Criteria	Special Characteristics of Goats	References
Adaptability	Goats are better adapted to broad environmental conditions ranging from arid dry to cold arid to hot humid. Goats in the tropical warm climate are more or less dwarf and have less body weight, while goats in colder climates have bigger size and more fur growth. Due to their lesser body size, their metabolic requirements are considerably low, they have the ability to reduce their metabolism and their loose skin aids in easy dissipation of body heat.	[18]
Thermo-tolerance	Goats are more thermo-tolerant than all other ruminant species. They possess the ability to survive in different agro-ecological zones.	[19]
Drought tolerance	Goats possess the ability to thrive well in drought prone areas because of reduced water requirement in comparison to sheep and other domestic ruminants. Goats have better water conservation ability than other ruminant animals because of their browse diet. Further, the gut, especially the rumen, acts as a water reservoir during the periods of dehydration.	[19]
Ability to thrive well on low pasture	Efficient utilizers of poor quality and a wide range of pastures. Goats have improved digestibility compared to all other rumen and animals and, moreover, because the small-sized feed consumption is also low, these factors together favour less CH_4 production.	[19]
Low enteric methane emission	Goats produce less enteric methane compared to sheep and other ruminants.	[20]
More demand for goat meat	Goat meat possesses less fat content and has no religious taboo; hence, it is relished by all. The lower saturated fat content in the goat meat improves the blood cholesterol level and stabilizes the heart rhythm of consumer. Goat meat contains vitamin B, B12 and omega-3 fatty acids. Further goat meat is lower in calories and cholesterol than the meat from other animals.	[21]
Milk with more nutrition	Goat milk is more nutritious than the milk from other species of livestock, easily digestible due to the presence of some beneficial fatty acids and contains fats and proteins in a finer state. Goat milk contains vitamin A, niacin, thiamin, ribofavin and pantotheanate.	[22]
Digestibility and feed conversion efficiency	Increased efficiency to convert feed into milk and meat than all other domestic ruminants, they can even digest poor quality feed. Goats have less proportion of gut in relation their total body weight, which enables the rapid movement of digesta from the rumen and the entire gastrointestinal tract.	[19]
Less initial investment	Minimum investment compared to large ruminants due to lower price. It is possible to get more animals at the cost of one cow. Less quantity of feed is required for goats compared to other domesticated livestock species.	[1]
Women entrepreneurship	Because of their small size, goats are easy to herd by women. They can let the animals graze on common property resources and private fallow lands. As they move as a herd, it is easy to track them.	[19]
Suitable for landless farmers	Small area is required to rear goats because of their small size, they require less feed and they can be easily integrated into other farming systems.	[14]

3. Impact of Heat Stress on Rumen Function

Elevated ambient temperature during the summer months can have a significant influence on the basic physiology of rumen function, thereby affecting the nutritional status of the animals [23]. Rumen volatile fatty acid (VFA) production is altered during the conditions of extreme temperature, while feed digestibility is increased with increasing ambient temperature because of a reduction in feed intake and passage rate, which allows more time for the microbes and enzymes to digest feed [24]. Table 2 describes the various impacts of heat stress on rumen function.

Table 2. Different impacts of heat stress on the rumen function in goats.

Type of Heat Stress	Effect on Rumen Fermentation Pattern	Reference
Summer heat stress	Altered basic physiology of rumen function	[23]
Extreme temperature stress	Reduced VFA production	[24]
Summer heat stress	Decreased rumen pH and acidosis	[25]
Heat stress	Reduction in ruminal pH; reduced rumen fermentation	[26]
Heat stress	Decreased rumen pH	[27]
Summer heat stress	Decreased VFA production; Reduced production of acetate	[28,29]
Heat stress	Decrease in acetate and acetate to propionate ratio and an increase in butyrate	[30]
Heat stress	Increase of *Streptococcus* genus bacteria and a decrease in the bacteria of *Fibrobactor* genus	[31]
Heat stress	Decrease in the *Streptococcus* genus and increase in *Clostridium coccoides–Eubacterium* genus	[32]
Increased temperature and RH	Decline in the concentrations of amylolytic and cellulolytic bacteria; decreased diet digestibility	[9]
Late summer	Increase in enteric CH_4 emissions	[33]
Late summer season	Increase in enteric CH_4 emissions	[34]
Summer heat stress	increase in CH_4 emission	[35]

Note: RH: Relative humidity; VFA: Volatile fatty acid; CH_4: Methane.

3.1. Rumen Fermentation Pattern

Environmental factors such as temperature and relative humidity (RH) can have significant role in the feed consumption of animals. An increase in temperature and RH decreases the dry matter intake of the animals and rumination as a result of increased amount of buffering agents entering the rumen and this could be attributed to the reduced chewing activity [7]. Additionally, blood flow is redirected from the gastrointestinal tract to the periphery for heat dissipation, which further decreases the digestibility [8]. Furthermore, an increased respiration rate during summer season increases expired CO_2 output leading to decreased blood and rumen pH and acidosis [25]. Likewise, Castro-Costa et al. [26] reported a reduction in ruminal pH in heat exposed Murciano-Granadina dairy goats and attributed this to the reduced rumen fermentation during heat stress. Similarly, Yan-fen et al. [27] also reported a decreased rumen pH and NH_3-N concentration in dairy goats exposed to heat stress.

3.2. Volatile Fatty Acid Production

There are reports showing a decrease in VFA production during the periods of heat stress [28,29]. Similarly, Tajima et al. [30] reported a decrease in acetate and acetate to propionate ratio and an increase in butyrate level in heat stressed animals, which they attributed to alterations in the number of rumen microbiota during the periods of heat stress. Likewise, Hirayama et al. [24] reported a reduction in plasma acetate and VFA concentrations in heat exposed (35 °C) Saanen goats compared to Saanen goats kept under thermoneutral conditions (20 °C). They attributed these changes to reduced feed intake and rumen microbial diversity. Further, in a study conducted in indigenous goat breeds, we [28] reported a reduced production of acetate concentrations in heat exposed Osmanabadi and Malabari goats, whereas the Salem goats did not exhibit any change. In the same experiment, we also observed an increase in propionate concentration in the Salem black goats and a decline in the propionate production in Malabari goats. These variations in the heat stress response could be explained by the differences in the adaptive capability among the breeds, suggesting Salem black as the superior adaptive breed in the climate change scenario. Further, Chaidanya et al. [29] reported a reduction in VFA concentrations in rumen of goats exposed to high ambient temperature coupled with high relative humidity. The reduction in the VFA concentration could be attributed to the increased rumen temperature during the heat stress periods.

3.3. *Rumen Microbial Population*

Heat stress induced rumen function impairment is mainly associated with an increase of *Streptococcus* genus bacteria and a decrease in the bacteria of *Fibrobactor* genus [31]. Further, Tajima et al. [30] also reported these changes along with altered rumen bacterial diversity with a decrease in uncultivated Cluster E group sequences during heat stress. Similarly, Uyeno et al. [32] observed a decrease in the *Streptococcus* genus and an increase in both *Streptococcus* spp. and *Clostridium coccoides–Eubacterium* genus in the rumen. Changes in the rumen microbial ecosystem due to heat exposure can influence feed digestibility and composition of the end products by altering the rumen fermentation pattern [32]. Further, Bernabucci et al. [9] observed a decline in the concentrations of amylolytic and cellulolytic bacteria in animals exposed to ambient conditions having a temperature humidity index (THI) 85. The decreased dry matter intake and passage rate in heat stressed animals could reduce the bacterial diversity ultimately culminating in decreased diet digestibility [9]. There are few research reports available on how high ambient temperature selectively affects microbial population. However, this impact could be attributed to the sensitivity of certain rumen microbes to increased temperature exposure.

3.4. *Enteric Methane Emission*

Environmental temperature is a key factor that determines CH_4 production, since feed intake and digestibility differ with ambient temperature. Mbanzamihigo et al. [33] reported an increase in enteric CH_4 emissions during late summer (August–September) compared to early summer (June–July) in the Northern Hemisphere. Similarly, in another experiment conducted in young wethers grazing a moist hilly island pasture, a perennial rye grass/white clover dominant pasture and a late summer season pasture showed CH_4 yields of 4.1%, 3.9% and 5.3%, respectively. Increased CH_4 yield in wethers grazing late summer season pastures is attributed to the quality deterioration (poor dry matter digestibility, lower protein and soluble carbohydrate content and increased cell wall content) of the pastures during the summer season [34]. This study revealed the indirect effect of elevated ambient temperature on the CH_4 production through altered pasture characteristics. Further, Ulyatt et al. [35] reported an increase in CH_4 emission during grazing of summer grassland compared to Kikuyu grassland. Figure 1 shows the impact of heat stress on various rumen functions in goat.

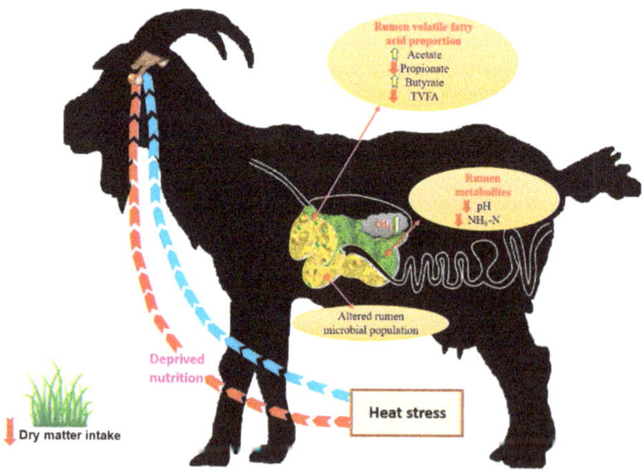

Figure 1. Impact of heat stress on various rumen functions in goat (these concepts were adopted from References [28,29]). TVFA: Total Volatile Fatty Acid.

3.5. Factors Influencing Enteric Methane Emission in Goats

Various factors affect the enteric methane production in goats and these are broadly classified as weather associated factors such season and increased ambient temperature; feed associated factors such as diet composition, time after feeding, and feed additives; and animal associated factors that include inflow of saliva, types of microbial population, and breed [36].

Composition of feed is the primary factor that determines the rumen fermentation pattern and enteric methane emissions [37]. Further, the propionate to acetate ratio also influences the rumen fermentation pattern and is determined by the concentrate to forage content of the diets [38]. In comparison with roughage feed, concentrates contain less structural carbohydrates, so the intake of concentrates may increase the production of propionate and decrease the production of acetate, ultimately resulting in reduced CH_4 production. An increase in concentrate intake is associated with increased propionate production and this may reduce the number of H_2 atoms available to the methanogenic bacteria, again resulting in reduced methane production. However, the higher level of concentrate feeding can cause sub-acute acidosis, both sub-clinical and clinical, which may adversely impact normal ruminal fermentation processes through both alteration of the functions of essential rumen microbes and impaired VFA absorption due to low ruminal pH [39].

In recent years, the usage of microbial feed additives has increased to improve growth performance of meat animals. In addition, some microbial feed additives have been used to reduce CH_4 production in ruminant animals. Malik et al. [40] used acetogens as a feed additive to replace prominent CH_4-producing methanogenic bacteria to reduce enteric methane production by acting as alternate hydrogen. The prominent CH_4-producing methanogenic bacteria have a low H_2 threshold level, thus do not allow the naturally resident acetogens to utilize hydrogen. Other feed additives such as fat and oil supplements have also been reported to have an effect on the rumen fermentation profile, thereby reducing rumen protozoan population and CH_4 reduction [41]. However, high fat diets can alter the rumen microbial population and ultimately it can hamper the fibre digestibility by specifically inactivating the rumen microbes that are associated with fibre digestion [41]. Plant bioactives, including saponins and tannins, can reduce CH_4 production in ruminants [42].

Breed is another important factor that determines enteric CH_4 production [43]. These breed-to-breed differences in enteric CH_4 production could be attributed to their variation in body size, adaptation, rumen volume and the variation in the feed intake [43]. Rumen associated factors such as rumen pH, type of volatile fatty acids fermented, type of substrates fermented, rate of fermentation, absorption capacity of rumen wall, and rumen protozoa concentration determine the level of CH_4 production [44]. Rumen methanogens remove H_2 molecules that are synthesized during the organic matter fermentation produced during fermentation of organic matter in the hind gut and rumen and produce CH_4 [45]. Further, the increased production of propionate decreases the CH_4 production by consuming H_2 molecules [46].

Geographic location and climate are known to be the most crucial factors significantly affecting CH_4 production and this could be due to ambient temperature differences as well as difference in feed resources available [44]. Animals reared in arid and semi-arid regions have been reported to produce less CH_4 production compared with animals in temperate regions, and this could be due to the differences in the type or amount of feed consumed in different locations [44]. Among the climate variables temperature, humidity, solar radiation and wind velocity are the important variables that influences CH_4 production. Increased ambient temperature coupled with high relative humidity (RH) directly affects CH_4 production by altering the rumen fermentation profile and indirectly by altering the quality of pasture or forage [46]. Although heat stress may reduce the feed intake, the increased methane emission could still be attributed to the heat stress associated negative impact on feed digestibility by inhibiting the rumen microbial populations that are essential for the normal digestion process. The various factors influencing enteric methane production from goats are summarized in Figure 2.

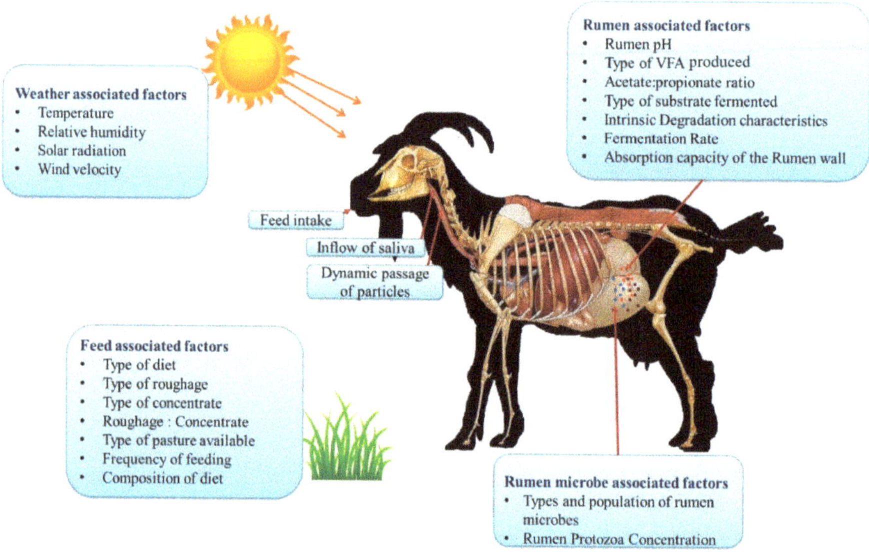

Figure 2. Various factors influencing enteric methane emission in goats (these concepts were adopted from References [8,28,29]).

3.6. Enteric Methane Mitigation Strategies in Goats

Apart from its major role in global warming and the greenhouse effect, enteric CH_4 is also considered as a dietary energy loss of around 2–12% in ruminants. Consequently, the global scientific community is targeting the development of suitable CH_4 mitigation strategies to reduce both global warming and dietary energy loss. The various strategies that can be implemented to mitigate enteric CH_4 include feeding feed sources containing plant secondary metabolites, ration manipulation, fat and oil supplementation, bacteriocin supplementation, rumen modification, etc. [29,45–47]. Various mechanisms to reduce enteric methane production in goats are summarised in Figure 3.

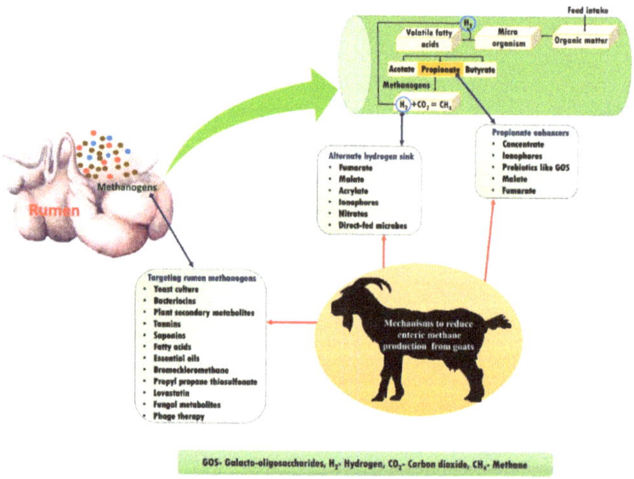

Figure 3. Various mechanisms to reduce enteric methane emission in goats.

3.6.1. Nutritional Intervention to Reduce Enteric Methane Production in Goats

Among the various CH_4 mitigation strategies, nutritional intervention or dietary manipulation is the most effective and commonly used strategy to mitigate enteric CH_4 emission in ruminant livestock [48,49]. It is well known that increasing the ratio of concentrate to forage in the diet can reduce the amount of energy loss as enteric CH_4 and this is mainly due to change of fermented substrate from fibre to starch [48]. In an experiment conducted on Murciano-Granadina goats in late lactation, Ibáñez et al. [50] observed a lower CH_4 production in goats fed with concentrate (ground corn) diet than beet pulp fed goats. However, concentrate feeding beyond a certain limit is not appreciable as it can cause severe damage to the animal itself and to its production performance. In addition, grains that may be used for concentrates are more valuable for human feeds in arid and semi-arid regions where much of the global goat production is located.

Supplementing the feed with more lipids and fatty acids was reported to reduce the dietary energy loss in goats [51]. However, the effectiveness of lipid supplementation relies on the source, inclusion rate, fatty acid profile and the composition of the rest of the diet [48,52]. Reduction in enteric CH_4 emission to the tune of around 40% is possible using high quality lipid feed supplements [46]. By differing the mode of action, lipid feed additives may reduce the methanogen and ciliated protozoan population in the rumen. Further, lipid supplementation reduces fibre and organic matter degradability and decreases the fermentable substrate availability and thereby minimising CH_4 production [53]. Abubakr et al. [54] conducted an experiment in Boer X Catcang crossbred goats where they found that adding decanter cake and palm kernel cake at up to 80% inclusion decreases methanogenesis by reducing rumen protozoa in goats. Further, Zhou et al. [55] reported the ability of lauric acid to reduce CH_4 production by reducing the viability of *Methanobrevibacter ruminantium*. Likewise, Kong et al. [56] reported a significant reduction in the methanogenesis without affecting the quantity of rumen methanogenic archaea after flaxseed supplementation.

Ionophore supplementation is another extensively researched CH_4 abatement strategy. Ionophores cause a shift in the rumen fermentation pattern from acetate and butyrate production to propionate by increasing the gram-positive bacteria population, resulting in decreasing the production of CH_4 [57]. Monensin is the most studied ionophore and routinely used as an animal nutrition supplement [58]. Saanen goats supplemented with oils with sodium bicarbonate and monensin showed a shift in the production of molar concentrations of acetate to propionate, thereby reducing the production of CH_4 [59]. Furthermore, up to a 75% reduction in CH_4 production was observed on addition of 10% encapsulated fumarate to the diet without any negative effect on animal growth [58].

Although several anti-methonogenic compounds are well proved in terms of their CH_4 reduction potential, certain individual components have antinutritional properties that inhibit their commercial usage. However, data obtained from anti-methonogenic supplementation studies are good models and they can pave a way towards effective CH_4 mitigation strategies [60,61]. Abecia et al. [45] conducted an experiment in Murciano-Granadina lactating goats to evaluate the potential of bromochloromethane (BCM) complex to reduce enteric CH_4 production and they observed 32% reduction in BCM fed goats as compared to the control group. In another experiment conducted in Murciano-Granadina goats, Martínez-Fernández et al. [62] observed 33% and 64% methane reduction per kg of dry matter intake with propyl propane thiosulfinate (PTS) and BCM supplementation, respectively. Further, Murciano–Granadina goats supplemented with PTS and BCM decreased CH_4 production by 48% and 98%, respectively, which was attributed to the redirection of H_2 from CH_4 production to propionate metabolic pathways [63]. Similarly, Mitsumori et al. [64] also reported 71% and 91% reductions in CH_4 production in Shiba Japanese goats supplemented with 2 g/100 kg Live Weight and 5 g/100 kg LW of BCM, respectively. Candyrine et al. [65] conducted a study on Saanen goats with three levels of lovastatin (naturally produced from fermentation of palm kernel with *Aspergillus terreus*) supplementation and the authors observed 7.8%, 20% and 21% CH_4 reduction for low (2 mg lovastatin/kg BW/day), medium (4 mg lovastatin/kg BW/day) and high (6 mg lovastatin/kg BW/day) treatment groups, respectively. Further, Azlan et al. [66] reported 32% reduction in enteric

CH_4 production when supplementing Boer crossbred goats with 14 mg/kg BW of lovastatin produced from rice straw treated with *Aspergillus terreus*.

Microbial feed additives are another important nutritional intervention in the CH_4 mitigation studies. Apart from the effects on CH_4 mitigation, probiotic feeding can improve the growth performance of meat animals and it can also reduce the incidence of diarrhoea [67]. However, studies proving the efficiency of direct fed probiotics to reduce the production of enteric CH_4 are few [68]. The same authors also reported that nitrate as feed additive can reduce rumen methanogenesis in different ruminant species and production conditions [68]. Chaucheyras-Durand et al. [69] showed that yeast cells can reduce the production of enteric CH_4 by deviating hydrogen atoms from methanogens to acetogenic strains of ruminal bacteria to enhance the production of acetate. Yeasts such as *S. cerevisiae* and the lactic acid utilizing bacteria *Propionibacterium* spp. and *Megasphaera elsdenii* can decrease rumen methanogenesis when included in the diet as supplements [60]. Wang et al. [61] found that replacing ordinary rice feed with red yeast rice, which is a traditional Chinese culinary and medicinal product, resulted in a 13% reduction in CH_4/DM intake in Boer crossbred goats.

Organic acids such as malic acid and fumerate have the potential to reduce CH_4 production in the ruminant by serving as an alternative hydrogen sink. Organic acid administration has been proven to reduce methane production in a dose-dependent manner in several in vitro studies [70]. In an experiment conducted in Xinong Saanen dairy goats, Li et al. [71] reported a significant reduction in CH_4 production in goats supplemented with fumaric acid. Further, in the same study, along with fumeric acid supplementation, the authors also altered the particle size of concentrate and forage feed and observed 32% and 18% CH_4 reduction in low forage and concentrate particle size diet and high forage and concentrate particle size diet, respectively [71].

Phenolic monomers, condensed tannins and other plant secondary metabolites in dose-dependent manner can reduce enteric CH_4 emission from the ruminants because of their ability to reduce methanogenesis. Puchala et al. [72] reported 57% reduction in CH_4 in terms of g/kg DMI in condensed tannin containing *Lespedeza cuneata* fed Angora goats compared to Angora goats fed a combination of *Festucaarundinacea* and *Digitariaischaemum*. Dietary tannins can directly hinder CH_4 production as well as indirectly limit methanogenesis through reducing the availability of hydrogen atoms. In a meta-analysis using 30 experiments comprising 171 treatments to evaluate the extent of dietary tannins to reduce the CH_4 emission, Jayanegara et al. [73] found a negative correlation between enteric CH_4 production and tannin supplementation. Furthermore, Wina et al. [74] reported a reduction in methanogens in methanol extract saponin containing *Sapindus rarak* fed animals. Similarly, Mao et al. [75] reported a 27% reduction in enteric CH_4 production with tea saponin supplementation. Further, in an experiment conducted in goats fed with natural tannin containing *Mimosa* spp., Bhatta et al. [46] reported a CH_4 reduction after *Mimosa* spp. supplementation even at low concentrations (2–8 g/kg DM of the diet). In a study conducted in Nanjiang Yellow goats, Dong et al. [76] reported a reduction in enteric methane production on *Artemisiae annuae* extract and herbal medicines mixture supplementation to different diets. Further, under in vitro condition, Denman et al. [77] reported 91% reduction in methane production using bromochloromethane at 5 g/100 kg LW in Japanese native goats.

3.6.2. Management Strategies to Reduce CH_4 Production from Goats

Improving management strategies not only reduces enteric methane emission but also helps to improve animal productivity [78]. Reduction or culling of unproductive animals from the herd has the potential to simultaneously improve the productivity and to reduce CH_4 emission [79,80]. In subsistence production systems, reduction in the herd size allows distribution of adequate amount of feed and proper veterinary care to all animals. Additionally, selective culling can reduce CH_4 production both per unit of animal product and for the total herd [81]. However, in some subsistence farming systems, there may be insufficient high breeding value animals to allow selective culling. Slaughter weight of goats can be advanced at a young age through early finishing approaches. This can potentially reduce the lifetime net CH_4 emissions, thus making available proportionally few CH_4 producing animals [79].

Reductions in enteric CH_4 production can be achieved through efficient pasture management practices in goats. Feeding animals high quality fodder can reduce the wastage of dietary energy. Improving quality of the forage also increases feed intake and reduces the retention time of digesta in the rumen, thereby stimulating energetically more efficient post-ruminal digestion and decreases the percentage of energy transformed to CH_4 [79]. Sejian et al. [82] reported a reduced CH_4 emission in animals fed with high quality fodder as compared to animals consuming low quality fodder. Reductions in enteric CH_4 production can also be achieved through feeding high quality fodder with higher soluble carbohydrates and lower fibre or through grazing on less-matured pastures [48,79]. Harvesting or grazing of forage at early stages of maturity also reduces the plant cell wall lignification, thereby increasing digestibility and reducing the CH_4 emission per unit of digestible dry matter [83]. Similarly, Pinares-Patiño et al. [84] conducted a grazing experiment in timothy pasture at four different vegetative phases, namely, early vegetative stage, heading, flowering and senescence, and they observed lower CH_4 production only at heading stage, which confirms the significance of growth stages of forage in CH_4 production. Waghorn and Hegarty [85] calculated that animals grazing on high quality pasture (20% higher ME value) may show a reduction in enteric CH_4 production of approximately 50%. Likewise, animals consuming certain high quality tropical and temperate legumes show reduced enteric CH_4 production, as the legumes contain condensed tannins that are toxic to methanogenic archaea, ciliate protozoa, and fibre degrading bacteria [86]. Further, the grasses with high concentrations of water-soluble carbohydrates have been investigated as suitable tool to reduce enteric CH_4 emission from the ruminant livestock [87]. De Ramus et al. [88] reported 22% reduction in enteric CH_4 production annually through the efficient use of grazed forage crops through management-intensive grazing. Furthermore, around 5% reduction in enteric CH_4 production is possible through improving total tract NDF digestibility [53]. Archimede et al. [89] reported 17% more CH_4 production from animals fed with C4 grasses than the animals fed with C3 grasses.

In the majority of regions around the globe, goats are raised under continuous grazing systems, where animals have ad libitum access to pasture. However, unrestricted access to the pasture can result in the elimination of edible pasture and the domination of less edible pasture due to the uncontrolled selective grazing [31]. Hence, adoption of controlled grazing is a reliable strategy to reduce enteric CH_4 and to improve productivity. In these systems, grazing land is divided into different paddocks that are alternatively grazed and rested until the pasture restores its quality. A continuous supply of uniform quality feed throughout the year enables animals to increase their production and to decrease CH_4 production per kilogram of weight gain [20].

Size of the forage has profound effect on the CH_4 emissions. Animals deviate considerable amount of their energy to the chewing process [90]. Particle size reduction of fodder by mechanical means helps to enhance digestibility through bringing more microbial access to the substrate, decreasing energy expenses, CH_4 production and increasing the passage rate of digesta and animal productivity [91].

Selection of genetically superior animals with less CH_4 production per unit feed intake is another management strategy that can be employed to reduce CH_4 production from the ruminants [92]. The direct selection of low CH_4 producing animals is practically impossible because of high cost for measuring CH_4. However, selection is possible through the indirect means such as rumen digesta retention time and feed intake [80]. Genetic selection of goats with higher feed conversion efficiency generates a reduced amount of CH_4. Further, genetic selection for the less CH_4 production indirectly helps the farmer to increase their profits without any extra carbon credits by increasing the feed conversion efficiency and growth rate per animal [92]. A 3–10% reduction in CH_4 production can be achieved through improving the feed use efficiency by 10% [93].

3.6.3. Advanced Biotechnological Tools for Methane Mitigation

The inhibition of enteric CH_4 emission in ruminant animals is possible though biotechnological interventions. One of the possible future strategies to reduce enteric CH_4 production is to immunize the animals against their own methanogens. In an experiment conducted in Australia using vaccines against three selected methanogens, Wright et al. [94] reported 8% reduction in CH_4 production. However, another experiment conducted in different geographical zone with vaccines prepared using different set of bacteria could not elicit any positive response [94]. The reasons for the immunization failures could be due to the variation in rumen methanogenic diversity present in the animals raised in different conditions and the replacement of the biological niche left by the targeted species by another methanogen [95]. CH_4 inhibition was also attempted through oral supplementation IgY as a feed additive [96]. Zhang et al. [97] conducted an experiment in Boer goats to evaluate the efficiency of a candidate vaccine protein (EhaF) on the rumen methanogens and microbes but did not find any changes in CH_4 production among control and vaccinated goats. However, vaccination influenced the composition of rumen bacteria.

Use of bacteriocins offers another possible strategy to reduce CH_4 emission from ruminant animals. Bacteriocins are the proteins produced by bacteria that can obstruct certain microbial species in the rumen [98]. An in vitro study conducted by Lee et al. [99] using bovicin HC5 (a bacteriocin produced by *Streptococcus* spp.) showed 50% reduction in CH_4 production without inducing methanogen adaptation. Likewise, Santoso et al. [100] reported a 10% reduction in CH_4 emission in an in vivo study that used nisin, a bacteriocin produced by *Lactobacillus lactis* subsp.

The lytic potential and genes of the bacteriophages makes them potential tools to mitigate enteric CH_4 emission [101]. Certain bacteriophages may inhabit the rumen wall to maintain the homeostasis of the rumen micro fauna. Due to their host specific nature, they lyse certain microbes such as methanogens and *Streptococcus bovis* or pathogens such as *salmonella* and *E. coli* O157:H7 [47]. McAllister and Newbold [70] reported that siphophages (*Siphoviridae* phage) can infect methanogens such as *Methano brevibacter*, *Methanobacterium* and *Methanococcus* spp.; however, siphophages have yet to be isolated from the rumen. However, there are few available data relating to the genetic functionality and blueprint of the archaeal methanogenic phages and, to date, no bacteriophages from rumen have been isolated [47].

Another plausible method of biological control of methanogens is the use of CH_4 oxidizers. The CH_4 oxidizing bacteria have already been isolated from the rumen [102]. However, in vitro studies conducted using carbon isotopes reveal that only 0.3–8% of CH_4 oxidation to CO_2 happens in the rumen [103]. Valdez et al. [104] reported a reduction in CH_4 production by adding CH_4 oxidizing bacterium isolated from the gut of young pigs. However, detailed in vivo studies are needed to establish the level of CH_4 reduction. Another novel approach for enteric CH_4 reduction is through the genetic modification of fermentation characteristics of rumen bacteria. However, research is still in the preliminary stages and very little progress has been made pertaining to applying the molecular techniques to characterize and quantify the microbial populations [86].

4. Conclusions and Future Perspectives

Goats undoubtedly need to be the priority focus for livestock industries due to their advantages over other ruminant animals from a climate resilience point of view. Elevated ambient temperature during the summer months can have a significant influence on the basic physiology of rumen, thereby affecting the rumen fermentation pattern, VFA and other rumen metabolites production. Furthermore, growing evidence suggests that heat stress influences the rumen microbial population, resulting in alterations in ruminal digestion process in goats. In addition, heat stress has also been shown to increase the production of enteric CH_4 emission resulting in dietary energy loss. Thus, the productive performances of the animals are compromised. Nutritional interventions and other management strategies are traditional ways by which enteric CH_4 emission is reduced in goats. More recently, several researchers have targeted reducing enteric CH_4 through advanced biotechnological tools such immunization therapy, using bacteriocins, etc. but without much success. Further refinements in these technologies are essential before these technologies are implemented at field level. In the near future, these technologies offer scope for identifying genetically superior animals with less CH_4 production per unit feed intake. However, further research efforts are needed to elucidate the mechanisms associated with enteric CH_4 emission during heat stress exposure by establishing the relationships among the rumen microbes through metagenomics approaches in goats in the changing climate scenario. Such efforts may help to develop more focussed mitigation strategies for reducing enteric CH_4 emission in goats. This may help to sustain goat production in the changing climate scenario by preventing the dietary energy loss incurred during the process of enteric CH_4 emission.

Author Contributions: Conceptualization, P.P. and V.S.; Preparation of Manuscript, P.P., V.S., S.S.C., B.J.L. and F.R.D.; Figure Preparation, P.P., V.S. and S.S.C.; Table Preparation, P.P., S.S.C. and V.S.; and Scientific Editing, F.R.D., B.J.L. and S.S.C.

Funding: This research received no external funding.

Acknowledgments: The authors profusely thank Frank R. Dunshea for supporting this manuscript by bearing the publication charge.

Conflicts of Interest: The authors declare no conflict of interest.

References

1. Feleke, F.B.; Berhe, M.; Gebru, G.; Hoag, D. Determinants of adaptation choices to climate change by sheep and goat farmers in Northern Ethiopia: The case of Southern and Central Tigray, Ethiopia. *SpringerPlus* **2016**, *5*, 1692. [CrossRef] [PubMed]
2. Bezabih, M.Y.; Berhane, G. Livestock Production Systems Analysis. *Am. Int. J. Contemp. Sci. Res.* **2014**, *1*, 16–51.
3. Thornton, P.K. Livestock production: Recent trends, future prospects. *Philos. Trans. R. Soc. Lond. B Biol. Sci.* **2010**, *365*, 2853–2867. [CrossRef] [PubMed]
4. Brahmi, A.; Khaldi, R.; Jaouad, M.; Hicheri, A.; Touati, I.; Rkhissi, A.; Ifaoui, H.; Khaldi, G. Impacts of climate change on the small ruminants farming systems in north western Tunisia and adaptation tools. In *New Approaches for Grassland Research in a Context of Climate and Socio-Economic Changes*; Acar, Z., López-Francos, A., Porqueddu, C., Eds.; Options Méditerranéennessérie; CIHEAM: Zaragoza, Spain, 2012; pp. 427–431.
5. Koluman, N.; Silanikove, N.; Koluman, A. Climate Change and Goat Agriculture Interactions in the Mediterranean Region. In *Sustainable Goat Production in Adverse Environments*; Simões, J., Gutiérrez, C., Eds.; Springer: Cham, Switzerland, 2017; pp. 393–405.
6. Sejian, V.; Bhatta, R.; Gaughan, J.B.; Dunshea, F.R.; Lacetera, N. Review: Adaptation of animals to heat stress. *Animal* **2018**. [CrossRef] [PubMed]
7. Bernabucci, U.; Lacetera, N.; Baumgard, L.H.; Rhoads, R.P.; Ronchi, B.; Nardone, A. Metabolic and hormonal acclimation to heat stress in domesticated ruminants. *Animal* **2010**, *4*, 1167–1183. [CrossRef] [PubMed]
8. Yadav, B.; Gynendra, S.; Verma, A.K.; Dutta, N.; Sejian, V. Impact of heat stress on rumen functions. *Vet. World* **2013**, *6*, 992–996. [CrossRef]
9. Bernabucci, U.; Lacetera, N.; Danieli, P.P.; Bani, P.; Nardone, A.; Ronchi, B. Influence of different periods of exposure to hot environment on rumen function and diet digestibility in sheep. *Int. J. Biometeorol.* **2009**, *53*, 387–395. [CrossRef]
10. Pollott, G.; Wilson, R.T. *Sheep and Goats for Diverse Products and Profits*; FAO Diversification Booklet No. 9; FAO Library: Rome, Italy, 2009; 42p.
11. Oluwatayo, I.B.; Oluwatayo, T.B. *Small Ruminants as a Source of Financial Security, a Case Study of Women in Rural Southwest Nigeria*; Working Paper 1; Institute for Money, Technology and Financial Inclusion (IMTFI): Irvine, CA, USA, 2012; 21p.
12. Aziz, M.A. Present status of the world goat populations and their productivity. *World* **2010**, *861*, 1.
13. Lérias, J.R.; Hernández-Castellano, L.E.; Suárez-Trujillo, A.; Castro, N.; Pourlis, A.; Almeida, A.M. The mammary gland in small ruminants: Major morphological and functional events underlying milk production—A review. *J. Dairy Res.* **2014**, *81*, 304–318. [CrossRef]

14. Kumar, S.; Roy, M.M. Small Ruminant's Role in Sustaining Rural Livelihoods in Arid and Semiarid Regions and their Potential for Commercialization. In *New Paradigms in Livestock Production from Traditional to Commercial Farming and Beyond*; Agrotech Publishing Academy: Udaipur, India, 2013; pp. 57–80.
15. Agossou, D.J.; Dougba, T.D.; Koluman, N. Recent Developments in Goat Farming and Perspectives for a Sustainable Production in Western Africa. *Int. J. Environ. Agric. Biotechnol.* **2017**, *2*, 2047–2051. [CrossRef]
16. Yami, A.; Merkel, R.C. Sheep and Goat Production Handbook for Ethiopia. Ethiopia Sheep and Goat Productivity Improvement Program. 2008. Available online: https://archive.org/details/SheepGoatHandbook (accessed on 15 October 2018).
17. Kosgey, I.S.; Rowlands, G.J.; Van Arendonk, J.A.; Baker, R.L. Small ruminant production in smallholder and pastoral/extensive farming systems in Kenya. *Small Rumin. Res.* **2008**, *77*, 11–24. [CrossRef]
18. Amankwah, K.; Klerkx, L.; Oosting, S.J.; Sakyi-Dawson, O.; Van der Zijpp, A.J.; Millar, D. Diagnosing constraints to market participation of small ruminant producers in northern Ghana: An innovation systems analysis. *NJAS-Wageningen J. Life Sci.* **2012**, *60*, 37–47. [CrossRef]
19. Silanikove, N. The physiological basis of adaptation in goats to harsh environments. *Small Rumin. Res.* **2000**, *35*, 181–193. [CrossRef]
20. Soren, N.M.; Sejian, V.; Terhuja, M.; Dominic, G. Enteric methane emission in sheep: Process description and factors influencing production. In *Sheep Production Adapting to Climate Change*; Sejian, V., Bhatta, R., Gaughan, J., Malik, P.K., Naqvi, S.M.K., Lal, R., Eds.; Springer: Singapore, 2017; pp. 209–233.
21. Archana, P.R.; Sejian, V.; Ruban, W.; Bagath, M.; Krishnan, G.; Aleena, J.; Manjunathareddy, G.B.; Beena, V.; Bhatta, R. Comparative assessment of heat stress induced changes in carcass traits, plasma leptin profile and skeletal muscle myostatin and HSP70 gene expression patterns between indigenous Osmanabadi and Salem Black goat breeds. *Meat Sci.* **2018**, *141*, 66–80. [CrossRef] [PubMed]
22. Getaneh, G.; Mebrat, A.; Wubie, A.; Kendie, H. Review on Goat Milk Composition and its Nutritive Value. *J. Nutr. Health Sci.* **2016**, *3*, 1–10.
23. Baumgard, L.H.; Rhoads, R.P. Ruminant Nutrition Symposium: Ruminant production and metabolic responses to heat stress. *J. Anim. Sci.* **2012**, *90*, 1855–1865. [CrossRef] [PubMed]
24. Hirayama, T.; Katoh, K.; Obara, Y. Effects of heat exposure on nutrient digestibility, rumen contraction and hormone secretion in goats. *Anim. Sci. J.* **2004**, *75*, 237–243. [CrossRef]
25. Choubey, M.; Kumar, A. Nutritional interventions to combat heat stress in dairy animals. *Vetscan* **2012**, *7*, 19–27.
26. Castro-Costa, A.; Salama, A.A.; Moll, X.; Aguiló, J.; Caja, G. Using wireless rumen sensors for evaluating the effects of diet and ambient temperature in non-lactating dairy goats. *J. Dairy Sci.* **2015**, *98*, 4646–4658. [CrossRef]
27. Ma, Y.-F.; Du, R.-P.; Gao, M. Effect of Heat Stress on Dairy Goat Performance and Rumen Epithelial Cell Morphology. *Sci. Agric. Sin.* **2013**, *46*, 4486–4495.
28. Pragna, P.; Sejian, V.; Soren, N.M.; Bagath, M.; Krishnan, G.; Beena, V.; Devi, P.I.; Bhatta, R. Summer season induced rhythmic alterations in metabolic activities to adapt to heat stress in three indigenous (Osmanabadi, Malabari and Salem Black) goat breeds. *Biol. Rhythm Res.* **2018**, *49*, 551–565. [CrossRef]
29. Chaidanya, K.; Soren, N.M.; Sejian, V.; Bagath, M.; Manjunathareddy, G.B.; Kurien, K.E.; Varma, G.; Bhatta, R. Impact of heat stress, nutritional stress and combined (heat and nutritional) stresses on rumen associated fermentation characteristics, histopathology and HSP70 gene expression in goats. *J. Anim. Behav. Biometeorol.* **2017**, *5*, 36–48. [CrossRef]
30. Tajima, K.; Nonaka, I.; Higuchi, K.; Takusari, N.; Kurihara, M.; Takenaka, A.; Mitsumori, M.; Kajikawa, H.; Aminov, R.I. Influence of high temperature and humidity on rumen bacterial diversity in Holstein heifers. *Anaerobe* **2007**, *13*, 57–64. [CrossRef] [PubMed]
31. Freestone, P.; Lyte, M. Stress and microbial endocrinology: Prospects for ruminant nutrition. *Animal* **2010**, *4*, 1248–1257. [CrossRef] [PubMed]
32. Uyeno, Y.; Sekiguchi, Y.; Tajima, K.; Takenaka, A.; Kurihara, M.; Kamagata, Y. An rRNA-based analysis for evaluating the effect of heat stress on the rumen microbial composition of Holstein heifers. *Anaerobe* **2010**, *16*, 27–33. [CrossRef] [PubMed]
33. Mbanzamihigo, L.; Fievez, V.; da Costa Gomez, C.; Piattoni, F.; Carlier, L.; Demeyer, D. Methane emissions from the rumen of sheep fed a mixed grass-clover pasture at two fertilisation rates in early and late season. *Can. J. Anim. Sci.* **2002**, *82*, 69–77. [CrossRef]

34. Ulyatt, M.J.; Lassey, K.R.; Shelton, I.D.; Walker, C.F. Methane emission from sheep grazing four pastures in late summer in New Zealand. *N. Z. J. Agric. Res.* **2005**, *48*, 385–390. [CrossRef]
35. Ulyatt, M.J.; Lassey, K.R.; Shelton, I.D.; Walker, C.F. Methane emission from dairy cows and wether sheep fed subtropical grass-dominant pastures in midsummer in New Zealand. *N. Z. J. Agric. Res.* **2002**, *45*, 227–234. [CrossRef]
36. Aluwong, T.; Wuyep, P.A.; Allam, L. Livestock-environment interactions: Methane emissions from ruminants. *Afr. J. Biotechnol.* **2011**, *10*, 1265–1269.
37. Brouček, J. Methane yield from cattle, sheep, and goats housing with emphasis on emission factors: A review. *Slovak J. Anim. Sci.* **2015**, *48*, 122–139.
38. Bhatta, R.; Malik, P.K.; Sejian, V. Enteric methane emission and reduction strategies in sheep. In *Sheep Production Adapting to Climate Change*; Sejian, V., Bhatta, R., Gaughan, J., Malik, P.K., Naqvi, S.M.K., Lal, R., Eds.; Springer: Singapore, 2017; pp. 291–305.
39. Castillo-González, A.R.; Burrola-Barraza, M.E.; Domínguez-Viveros, J.; Chávez-Martínez, A. Rumen microorganisms and fermentation. *Arch. Med. Vet.* **2014**, *46*, 349–361. [CrossRef]
40. Malik, P.K.; Thulasi, A.; Soren, N.M.; Jose, L.; Prasad, K.S.; Prasad, C.S. Phylogenic diversity analysis of rumen acetogens in adult sheep fed on conventional roughage diet. *Indian J. Anim. Sci.* **2015**, *85*, 1104–1107.
41. Patra, A.K. A meta-analysis of the effect of dietary fat on enteric methane production, digestibility and rumen fermentation in sheep, and a comparison of these responses between cattle and sheep. *Livest. Sci.* **2014**, *162*, 97–103. [CrossRef]
42. Rochfort, S.; Parker, A.; Dunshea, F.R. Plant bioactives for ruminant health and productivity—A review. *Phytochemistry* **2008**, *69*, 299–322. [CrossRef] [PubMed]
43. Fraser, M.D.; Fleming, H.R.; Moorby, J.M. Traditional vs Modern: Role of Breed Type in Determining Enteric Methane Emissions from Cattle Grazing as Part of Contrasting Grassland-Based Systems. *PLoS ONE* **2014**, *9*, e107861. [CrossRef] [PubMed]
44. Cottle, D.J.; Nolan, J.V.; Wiedemann, S.G. Ruminant enteric methane mitigation: A review. *Anim. Prod. Sci.* **2011**, *51*, 491–514. [CrossRef]
45. Pragna, P.; Sejian, V.; Bagath, M.; Krishnan, G.; Devaraj, C.; Afsal, A.; Aleena, J.; Archana, P.R.; Bhatta, R. Metabolic adaptation of livestock in heat stress challenges. In *National Initiative on Climate Resilient Agriculture Workshop Compendium on "Livestock and Climate Change"*; Hemanth, G.K., Sejian, V., Giridhar, K., Naveen, K.G.S., Eds.; Veterinary College, Karnataka Veterinary Animal and Fishery Sciences University: Hassan, Karnataka, India; National Initiatives on Climate Resilient Agriculture, Indian Council of Agriculture Research and Hyderabad: India, 2018; pp. 33–36.
46. Chaidanya, K.; Soren, N.M.; Sejian, V.; Bagath, M.; Kurien, E.K.; Varma, G.; Bhatta, R. Impact of heat stress, nutritional stress and combined (heat and Nutritional) stresses on rumen enzymes and fermentation metabolites in Osmanabadi bucks. In Proceedings of the Interenational Conference Compendium on Climate Change Adaptation and Biodiversity: Ecological Sustainability and Resource Management for Livelihood Security, Portblair, Andaman and Nicobar Island, 8–10 December 2016; Andaman Science Association at ICAR-Central Island Agricultural Research Institute: Portblair, Andaman and Nicobar Island.
47. Baumgard, L.H.; Rhoads, R.P.; Rhoads, M.L.; Gabler, N.K.; Ross, J.W.; Keating, A.F.; Boddicker, R.L.; Lenka, S.; Sejian, V. Impact of climate change on livestock production. In *Environmental Stress and Amelioration in Livestock Production*; Sejian, V., Naqvi, S.M.K., Ezeji, T., Lakritz, J., Lal, R., Eds.; Springer-Verlag GMbH Publisher: Germany, 2012; pp. 413–468.
48. Beauchemin, K.A.; Kreuzer, M.; O'mara, F.; McAllister, T.A. Nutritional management for enteric methane abatement: A review. *Aust. J. Exp. Agric.* **2008**, *48*, 21–27. [CrossRef]
49. Sejian, V.; Lal, R.; Lakritz, J.; Ezeji, T. Measurement and prediction of enteric methane emission. *Int. J. Biometeorol.* **2011**, *55*, 1–6. [CrossRef]
50. Ibáñez, C.; López, M.C.; Criscioni, P.; Fernández, C. Effect of replacing dietary corn with beet pulp on energy partitioning, substrate oxidation and methane production in lactating dairy goats. *Anim. Prod. Sci.* **2015**, *55*, 56–63. [CrossRef]
51. Lascano, C.E.; Carulla, J.E.; Vargas, J.J. Strategies for reducing methane emissions from ruminants. *Rev. Bras. Geogr. Fís.* **2011**, *6*, 1315–1335.

52. Knapp, J.R.; Laur, G.L.; Vadas, P.A.; Weiss, W.P.; Tricarico, J.M. Invited review: Enteric methane in dairy cattle production: Quantifying the opportunities and impact of reducing emissions. *J. Dairy Sci.* **2014**, *97*, 3231–3261. [CrossRef] [PubMed]
53. McGinn, S.M.; Beauchemin, K.A.; Coates, T.; Colombatto, D. Methane emissions from beef cattle: Effects of monensin, sunflower oil, enzymes, yeast, and fumaric acid. *J. Anim. Sci.* **2004**, *82*, 3346–3356. [CrossRef] [PubMed]
54. Abubakr, A.R.; Alimon, A.R.; Yaakub, H.; Abdullah, N.; Ivan, M. Digestibility, rumen protozoa, and ruminal fermentation in goats receiving dietary palm oil by-products. *J. Saudi Soc. Agric. Sci.* **2013**, *12*, 147–154. [CrossRef]
55. Zhou, X.; Meile, L.; Kreuzer, M.; Zeitz, J.O. The effect of saturated fatty acids on methanogenesis and cell viability of *Methanobrevibacter ruminantium*. *Archaea* **2013**, *2013*, 106916. [CrossRef] [PubMed]
56. Kong, Y.; He, M.; McAlister, T.; Seviour, R.; Forster, R. Quantitative fluorescence in situ hybridization of microbial communities in the rumens of cattle fed different diets. *Appl. Environ. Microbiol.* **2010**, *76*, 6933–6938. [CrossRef] [PubMed]
57. Kataria, R.P. Use of feed additives for reducing greenhouse gas emissions from dairy farms. *Microbiol. Res.* **2016**, *6*, 19–25. [CrossRef]
58. Wallace, R.J.; Wood, T.A.; Rowe, A.; Price, J.; Yanez, D.R.; Williams, S.P.; Newbold, C.J. Encapsulated fumaric acid as a means of decreasing ruminal methane emissions. *Int. Congr. Ser.* **2006**, *1293*, 148–151. [CrossRef]
59. Li, X.Z.; Yan, C.G.; Long, R.J.; Jin, G.L.; Khuu, J.S.; Ji, B.J.; Choi, S.H.; Lee, H.G.; Song, M.K. Conjugated linoleic acid in rumen fluid and milk fat, and methane emission of lactating goats fed a soybean oil-based diet supplemented with sodium bicarbonate and monensin. *Asian-Aust. J. Anim. Sci.* **2009**, *22*, 1521–1530. [CrossRef]
60. Seo, J.K.; Kim, S.W.; Kim, M.H.; Upadhaya, S.D.; Kam, D.K.; Ha, J.K. Direct-fed microbials for ruminant animals. *Asian-Australas. J. Anim. Sci.* **2010**, *23*, 1657–1667. [CrossRef]
61. Wang, L.Z.; Zhou, M.L.; Wang, J.W.; Wu, D.; Yan, T. The effect of dietary replacement of ordinary rice with red yeast rice on nutrient utilization, enteric methane emission and rumen archaeal diversity in goats. *PLoS ONE* **2016**, *11*, e0160198. [CrossRef]
62. Martínez-Fernández, G.; Abecia, L.; Martín-García, A.I.; Ramos-Morales, E.; Hervás, G.; Molina-Alcaide, E.; Yáñez-Ruiz, D.R. In vitro–in vivo study on the effects of plant compounds on rumen fermentation, microbial abundances and methane emissions in goats. *Animal* **2013**, *7*, 1925–1934. [CrossRef] [PubMed]
63. Martínez-Fernández, G.; Abecia, L.; Martín-García, A.I.; Ramos-Morales, E.; Denman, S.E.; Newbold, C.J.; Molina-Alcaide, E.; Yáñez-Ruiz, D.R. Response of the rumen archaeal and bacterial populations to anti-methanogenic organosulphur compounds in continuous-culture fermenters. *FEMS Microbiol. Ecol.* **2015**, *91*. [CrossRef] [PubMed]
64. Mitsumori, M.; Shinkai, T.; Takenaka, A.; Enishi, O.; Higuchi, K.; Kobayashi, Y.; Nonaka, I.; Asanuma, N.; Denman, S.E.; McSweeney, C.S. Responses in digestion, rumen fermentation and microbial populations to inhibition of methane formation by a halogenated methane analogue. *Br. J. Nutr.* **2012**, *108*, 482–491. [CrossRef] [PubMed]
65. Candyrine, S.C.; Mahadzir, M.F.; Garba, S.; Jahromi, M.F.; Ebrahimi, M.; Goh, Y.M.; Samsudin, A.A.; Sazili, A.Q.; Chen, W.L.; Ganesh, S.; et al. Effects of naturally-produced lovastatin on feed digestibility, rumen fermentation, microbiota and methane emissions in goats over a 12-week treatment period. *PLoS ONE* **2018**, *13*, e0199840. [CrossRef] [PubMed]
66. Azlan, P.M.; Jahromi, M.F.; Ariff, M.O.; Ebrahimi, M.; Candyrine, S.C.; Liang, J.B. Aspergillus terreus treated rice straw suppresses methane production and enhances feed digestibility in goats. *Trop. Anim. Health Prod.* **2018**, *50*, 565–571. [CrossRef] [PubMed]
67. Soren, N.M.; Tripathi, M.K.; Bhatt, R.S.; Karim, S.A. Effect of yeast supplementation on the growth performance of Malpura lambs. *Trop. Anim. Health Prod.* **2013**, *45*, 547–554. [CrossRef] [PubMed]
68. Jeyanathan, J.; Martin, C.; Morgavi, D.P. The use of direct-fed microbials for mitigation of ruminant methane emissions: A review. *Animal* **2014**, *8*, 250–261. [CrossRef] [PubMed]
69. Chaucheyras-Durand, F.; Walker, N.D.; Bach, A. Effects of active dry yeasts on the rumen microbial ecosystem: Past, present and future. *Anim. Feed Sci. Technol.* **2008**, *145*, 5–26. [CrossRef]
70. McAllister, T.A.; Newbold, C.J. Redirecting rumen fermentation to reduce methanogenesis. *Aust. J. Exp. Agric.* **2008**, *48*, 7–13. [CrossRef]

71. Li, Z.; Liu, N.; Cao, Y.; Jin, C.; Li, F.; Cai, C.; Yao, J. Effects of fumaric acid supplementation on methane production and rumen fermentation in goats fed diets varying in forage and concentrate particle size. *J. Anim. Sci. Biotechnol.* **2018**, *9*, 21. [CrossRef] [PubMed]
72. Puchala, R.; Min, B.R.; Goetsch, A.L.; Sahlu, T. The effect of a condensed tannin-containing forage on methane emission by goats. *J. Anim. Sci.* **2005**, *83*, 182–186. [CrossRef] [PubMed]
73. Jayanegara, A.; Leiber, F.; Kreuzer, M. Meta-analysis of the relationship between dietary tannin level and methane formation in ruminants from in vivo and in vitro experiments. *J. Anim. Physiol. Anim. Nutr.* **2012**, *96*, 365–375. [CrossRef] [PubMed]
74. Wina, E.; Muetzel, S.; Hoffmann, E.; Makkar, H.P.; Becker, K. Saponins containing methanol extract of Sapindus rarak affect microbial fermentation, microbial activity and microbial community structure in vitro. *Anim. Feed Sci. Technol.* **2005**, *121*, 159–174. [CrossRef]
75. Mao, H.L.; Wang, J.K.; Zhou, Y.Y.; Liu, J.X. Effects of addition of tea saponins and soybean oil on methane production, fermentation and microbial population in the rumen of growing lambs. *Livest. Sci.* **2010**, *129*, 56–62. [CrossRef]
76. Dong, G.Z.; Wang, X.J.; Liu, Z.B.; Wang, F. Effects of phytogenic products on in vitro rumen fermentation and methane emission in goats. *J. Anim. Feed Sci.* **2010**, *19*, 218–229. [CrossRef]
77. Denman, S.E.; Martinez-Fernandez, G.; Shinkai, T.; Mitsumori, M.; McSweeney, C.S. Metagenomic analysis of the rumen microbial community following inhibition of methane formation by a halogenated methane analog. *Front. Microbiol.* **2015**, *6*, 1087. [CrossRef] [PubMed]
78. Naqvi, S.M.; Sejian, V. Global climate change: Role of livestock. *Asian J. Agric. Sci.* **2011**, *3*, 19–25.
79. Eckard, R.J.; Grainger, C.; De Klein, C.A. Options for the abatement of methane and nitrous oxide from ruminant production: A review. *Livest. Sci.* **2010**, *130*, 47–56. [CrossRef]
80. Hegarty, R.S.; McEwan, J.C. Genetic opportunities to reduce enteric methane emissions from ruminant livestock. In Proceedings of the 9th World Congress on Genetics Applied to Livestock Production, Leipzig, Germany, 1–6 August 2010; pp. 1–6.
81. Tarawali, S.; Herrero, M.; Descheemaeker, K.; Grings, E.; Blümmel, M. Pathways for sustainable development of mixed crop livestock systems: Taking a livestock and pro-poor approach. *Livest. Sci.* **2011**, *139*, 11–21. [CrossRef]
82. Sejian, V.; Lakritz, J.; Ezeji, T.; Lal, R. Forage and Flax seed impact on enteric methane emission in dairy cows. *Res. J. Vet. Sci.* **2011**, *4*, 1–8. [CrossRef]
83. Boadi, D.A.; Wittenberg, K.M. Methane production from dairy and beef heifers fed forages differing in nutrient density using the sulphur hexafluoride (SF6) tracer gas technique. *Can. J. Anim. Sci.* **2002**, *82*, 201–206. [CrossRef]
84. Pinares-Patiño, C.S.; Baumont, R.; Martin, C. Methane emissions by Charolais cows grazing a monospecific pasture of timothy at four stages of maturity. *Can. J. Anim. Sci.* **2003**, *83*, 769–777. [CrossRef]
85. Waghorn, G.C.; Hegarty, R.S. Lowering ruminant methane emissions through improved feed conversion efficiency. *Anim. Feed Sci. Technol.* **2011**, *166*, 291–301. [CrossRef]
86. Lascano, C.E.; Cárdenas, E. Alternatives for methane emission mitigation in livestock systems. *R. Bras. Zootec.* **2010**, *39*, 175–182. [CrossRef]
87. Ellis, J.L.; Dijkstra, J.; France, J.; Parsons, A.J.; Edwards, G.R.; Rasmussen, S.; Kebreab, E.; Bannink, A. Effect of high-sugar grasses on methane emissions simulated using a dynamic model. *J. Dairy Sci.* **2012**, *95*, 272–285. [CrossRef] [PubMed]
88. De Ramus, H.A.; Clement, T.C.; Giampola, D.D.; Dickison, P.C. Methane emissions of beef cattle on forages. *J. Environ. Qual.* **2003**, *32*, 269–277. [CrossRef]
89. Archimède, H.; Eugène, M.; Magdeleine, C.M.; Boval, M.; Martin, C.; Morgavi, D.P.; Lecomte, P.; Doreau, M. Comparison of methane production between C3 and C4 grasses and legumes. *Anim. Feed Sci. Technol.* **2011**, *166*, 59–64. [CrossRef]
90. Gerber, P.J.; Hristov, A.N.; Henderson, B.; Makkar, H.; Oh, J.; Lee, C.; Meinen, R.; Montes, F.; Ott, T.; Firkins, J.; et al. Technical options for the mitigation of direct methane and nitrous oxide emissions from livestock: A review. *Animal* **2013**, *7*, 220–234. [CrossRef]
91. Hristov, A.N.; Oh, J.; Firkins, J.L.; Dijkstra, J.; Kebreab, E.; Waghorn, G.; Makkar, H.P.; Adesogan, A.T.; Yang, W.; Lee, C.; et al. Special topics—Mitigation of methane and nitrous oxide emissions from animal operations: I. A review of enteric methane mitigation options 1. *J. Anim. Sci.* **2013**, *91*, 5045–5069. [CrossRef]

92. Pinares-Patiño, C.S.; Hickey, S.M.; Young, E.A.; Dodds, K.G.; MacLean, S.; Molano, G.; Sandoval, E.; Kjestrup, H.; Harland, R.; Hunt, C.; et al. Heritability estimates of methane emissions from sheep. *Animal* **2013**, *7*, 316–321. [CrossRef]
93. Sudmeyer, R. Carbon Farming: Selective Breeding of Sheep for Reduced Methane Emissions, Climate Change, Agriculture and Food, Department of Primary Industries and Regional Development, Government of Western Australia. 2018. Available online: https://www.agric.wa.gov.au/climate-change/carbon-farming-selective-breeding-sheep-reduced-methane-emissions (accessed on 15 October 2018).
94. Wright, A.D.; Kennedy, P.; O'Neill, C.J.; Toovey, A.F.; Popovski, S.; Rea, S.M.; Pimm, C.L.; Klein, L. Reducing methane emissions in sheep by immunization against rumen methanogens. *Vaccine* **2004**, *22*, 3976–3985. [CrossRef]
95. Williams, Y.J.; Popovski, S.; Rea, S.M.; Skillman, L.C.; Toovey, A.F.; Northwood, K.S.; Wright, A.D. A vaccine against rumen methanogens can alter the composition of archaeal populations. *Appl. Environ. Microbiol.* **2009**, *75*, 1860–1866. [CrossRef]
96. Cook, S.R.; Maiti, P.K.; Chaves, A.V.; Benchaar, C.; Beauchemin, K.A.; McAllister, T.A. Avian (IgY) anti-methanogen antibodies for reducing ruminal methane production: In vitro assessment of their effects. *Aust. J. Exp. Agric.* **2008**, *48*, 260–264. [CrossRef]
97. Zhang, L.; Huang, X.; Xue, B.; Peng, Q.; Wang, Z.; Yan, T.; Wang, L. Immunization against rumen methanogenesis by vaccination with a new recombinant protein. *PLoS ONE* **2015**, *10*, e0140086. [CrossRef]
98. Patra, A.; Park, T.; Kim, M.; Yu, Z. Rumen methanogens and mitigation of methane emission by anti-methanogenic compounds and substances. *J. Anim. Sci. Biotechnol.* **2017**, *8*, 13. [CrossRef]
99. Lee, S.S.; Hsu, J.T.; Mantovani, H.C.; Russell, J.B. The effect of bovicin hc5, a bacteriocin from *Streptococcus bovis* hc5, on ruminal methane production in vitro. *FEMS Microbiol. Lett.* **2002**, *217*, 51–55. [CrossRef]
100. Santoso, B.; Mwenya, B.; Sar, C.; Gamo, Y.; Kobayashi, T.; Morikawa, R.; Kimura, K.; Mizukoshi, H.; Takahashi, J. Effects of supplementing galacto-oligosaccharides, Yucca schidigera or nisin on rumen methanogenesis, nitrogen and energy metabolism in sheep. *Livest. Prod. Sci.* **2004**, *91*, 209–217. [CrossRef]
101. Ackermann, H.W. 5500 Phages examined in the electron microscope. *Arch. Virol.* **2007**, *152*, 227–243. [CrossRef]
102. Moss, A.R.; Jouany, J.P.; Newbold, J. Methane production by ruminants: Its contribution to global warming. *Ann. Zootech.* **2000**, *49*, 231–253. [CrossRef]
103. Boadi, D.; Benchaar, C.; Chiquette, J.; Massé, D. Mitigation strategies to reduce enteric methane emissions from dairy cows: Update review. *Can. J. Anim. Sci.* **2004**, *84*, 319–335. [CrossRef]
104. Valdez, C.; Newbold, C.J.; Hillman, K.; Wallace, R.J. Evidence for methane oxidation in rumen fluid in vitro. *Ann. Zootech.* **1996**, *45*, 351. [CrossRef]

animals

Article

Effect of Dietary Supplementation of *Moringa Oleifera* on the Production Performance and Fecal Methanogenic Community of Lactating Dairy Cows

Lifeng Dong, Tingting Zhang and Qiyu Diao *

Beijing Key Laboratory for Dairy Cow Nutrition, Sino-US Joint Lab on Nutrition and Metabolism of Ruminants, Feed Research Institute, Chinese Academy of Agricultural Sciences, Key Laboratory of Feed Biotechnology of the Ministry of Agriculture, Beijing 100081, China; donglifeng@caas.cn (L.D.); tingting5_2_0@163.com (T.Z.)
* Correspondence: diaoqiyu@caas.cn; Tel.: +86-010-82106055

Received: 25 February 2019; Accepted: 20 May 2019; Published: 22 May 2019

Simple Summary: High-quality forages such as protein-rich ingredients are essential to maximize production performance in dairy production. However, enteric methane produced by methanogenesis represents a substantial waste of feed energy for ruminants. Thus, it is important to evaluate the environmental effect when such feed ingredients are used to provide necessary nutrients. The aim of the present study was to examine the effects of dietary supplementation of *Moringa oleifera* on the production performance and fecal methanogenic community in lactating cows. The study's main results suggest that inclusion of *Moringa oleifera* improved milk fat content and changed the composition and diversity of methanogens. This study indicates that secondary metabolites from *Moringa oleifera* may regulate fermentation conditions and associations between some methanogens and other microbes. These findings provide basic information on the utilization of alternative forage resources for dairy cows and can help to better understand the regulation of microbial metabolic function and methane emissions.

Abstract: Development of alternative forage resources is of great importance to provide necessary nutrients and minimize greenhouse gas emissions in ruminant production. The aim of this study was to examine the effects of dietary supplementation of *Moringa oleifera* on the production performance and fecal methanogenic community in dairy cows using methyl-coenzyme M reductase α-subunit gene. Sixty-four cows were allocated to one of four treatments: basal diet without *M. oleifera* (control) or low (3% *w/w*, M3), medium (6%, M6), or high (9%, M9) supplementation with *M. oleifera*. This study demonstrated that different supplementation levels of *Moringa oleifera* in the diet achieved similar feed intake and milk production, but adding 6% of *Moringa oleifera* improved milk fat content. Two families, two phyla, three genera, and three species in total were identified among the four treatments. The fecal archaeal community in the control treatment was predominated by *Methanobrevibacter* (39.1% of the total sequence reads) followed by *Methanosphaera* and *Methanocorpusculum* at the genus level. The increased abundance of the *Methanosphaera* genus and *Methanosphaera* sp. ISO3-F5 species was induced by secondary metabolites of *Moringa oleifera* in the diet. Results indicated that *Moringa oleifera* supplementation not only improved dairy product quality but could also potentially reduce methane emissions.

Keywords: *Moringa oleifera*; fecal methanogenic community; dairy cows; *mcr*A gene sequencing technique

1. Introduction

Forage source and nutrient composition hold significant importance for dairy production systems due to the constant cost of commercial concentrates. This cost is a serious constraint for smallholder farms when the dietary protein sources are restricted or the cost is unaffordable. In the past few decades, many efforts have been made to explore less-expensive ingredients such as agricultural by-products, tree foliage, and plant leaves to supply adequate nutrients [1].

As an indigenous native tree in the Himalayas, *Moringa oleifera* (*M. oleifera*) is a perennial leafy tree that produces a high biomass in a short period and is widely distributed in tropical and subtropical areas around the world [2]. Recently, it has been increasingly considered as an alternative ingredient for animal feed because of its high content of protein, vitamins, and minerals. The average crude protein (CP) contents of *M. oleifera* range from 180 to 270 g CP/kg DM, similar to that of sesame meal (260 g CP/kg DM) [1]. In addition, saponins, tannins, and polysaccharides in *M. oleifera* demonstrate beneficial anti-inflammatory, antioxidant, and antimicrobial activities and can increase milk yield when dairy cows are offered dried or fresh leaves and soft twigs [3,4]. A recent study [5] found that alfalfa hay and maize silage can be partially replaced by *M. oleifera* silage without negative effects on nutrient digestibility and milk yield. This enhanced production performance was due to the considerable amounts of secondary metabolites in *M. oleifera*, which were also used as a potential feed additive to reduce greenhouse gas (GHG) emission of ruminants.

Methane (CH_4) emission from ruminants is a major contributor to atmospheric CH_4 accumulation. Reduction of ruminal CH_4 emissions would thus improve energy utilization efficiency and alleviate environmental issues within the dairy industry. For example, ruminal CH_4 emissions were significantly decreased when *M. oleifera* leaves were used in an in vitro experiment [6]. Similarly, supplementation of pomegranate pulp in the diet decreased enteric CH_4 emissions in dairy cows [7]. These reductions in CH_4 emissions were attributed to the direct reduction of ruminal methanogenesis by active polyphenolic compounds either in *M. oleifera* leaves or pomegranate pulp. However, the variation and responses of fecal methanogenic community to *M. oleifera* supplementation in the diet of dairy cows have not been elucidated.

Archaeal methanogens are obligate anaerobes that use methanogenesis pathways to facilitate fiber digestion by converting hydrogen into CH_4. Daquiado et al. [8] found the predominant species in rumen fluid and manure was *Methanobrevibacter ruminantium* (63.6% and 62.4%, respectively), whereas *Methanocorpusculum labreanum* was most abundant in rectal dung for beef cattle (53.2%). In addition, the community structures of fecal microbiota reflect not only animal productivity but also health and food safety. Jin et al. [9] found that dietary supplementation of active dried yeast significantly increased the relative abundance of *Methanocorpuschulum* and *Thermoplasma* species but decreased *Methanobrevibacter* in the feces. Mohammadzadeh et al. [10] observed a decrease in the fecal methanogenic archaea abundance in goats when the diet changed from alfalfa hay to a combination of alfalfa hay and oats. Determining the fecal methanogen composition would help us understand the effect of dietary supplementation on methanogenesis and CH_4 emissions. Therefore, the objective of the present study was to determine the effect of dietary supplementation of *M. oleifera* on the production performance and on the population and diversity of the fecal methanogenic community in lactating Holsteins dairy cows.

2. Materials and Methods

This study was conducted at the Guoxiu dairy farm located in Boading, China (latitude: 38°45′54″ and longitude: 115°08′06″) in 2017. The experiment design and animal care and handling procedures were evaluated and approved by the Animal Ethics Committee of the Chinese Academy of Agricultural Sciences (protocol number 023-2017) prior to the commencement of the experiment.

2.1. Animals and Experimental Design

Sixty-four multiparous lactating Holstein dairy cows (120 ± 8.0 days in milk; 31.9 ± 1.20 kg/day of milk yield at the beginning of the trial) were used in this experiment. Animals were randomly assigned to one of four treatments: (1) control, basal diet without *M. oleifera*; (2) a low supplementation of *M. oleifera* (3% *w/w*; M3); (3) medium supplementation of *M. oleifera* (6% *w/w*; M6); (4) high supplementation of *M. oleifera* (9% *w/w*; M9). The basal diet was formulated to be isoenergetic and isonitrogenous to meet the nutrient requirements of lactating dairy cows (NY/T 34-2004, Table 1). The treatments were balanced for milk yield, body weight, and lactation period. Rachises and twigs of *M. oleifera* at 56 days of age were harvested in Guangdong, China (23°8′ N, 113°17′ E); these materials were then chopped and dried on plastic sheets for 3 days for further preparation of a total mixed ration (TMR) diet. The CP, neutral detergent fiber (NDF) and acid detergent fiber (ADF) content of *M. oleifera* were 71.5, 743 and 552 g/kg DM, respectively, while the ether extract (EE) and ash content was 48.6 and 78.3 552 g/kg DM, respectively. The TMR diet was offered ad libitum in amounts resulting in 5% refusals. The whole experiment (77 days) consisted of 14 days for adaptation to the diet and 63 days for feeding period, with fecal samples collected in the last 5 days of the feeding period. Animals were fed with TMR at 07:00 and 19:00 and milked twice at 06:00 and 16:00 on a daily basis. Cows were housed in individual tie-stalls in a barn with good ventilation and had continuous access to water throughout the experiment. Artificial light was provided by suspended bulbs, and the floor was cleaned twice daily for good hygiene.

Table 1. Ingredients and nutrient composition of the experimental diets.

Items	Dietary *Moringa Oleifera* Content			
	0	3%	6%	9%
Ingredients, % of DM				
Ground corn	21.2	21.7	22.1	22.7
Soybean meal	10.5	11.6	12.7	13.8
DDGS	8.4	8.4	8.4	8.4
Cottonseed meal	7.5	7.5	7.5	7.5
Palm fat	7.5	7.5	7.5	7.5
Beet pulp	4.7	4.7	4.7	4.7
Alfalfa hay	20.5	17.1	13.7	10.2
Corn silage	16.8	15.7	14.5	13.4
Moringa oleifera	0.0	3.0	6.0	9.0
Premix [1]	2.0	2.0	2.0	2.0
Sodium hydrogen carbonate	0.6	0.6	0.6	0.6
Calcium hydrogen phosphate	0.2	0.2	0.2	0.2
Sodium chloride	0.3	0.3	0.3	0.3
Nutrient composition [2]				
CP	178.70	179.00	179.30	179.60
EE	44.80	46.90	48.80	50.90
Ash	76.80	84.60	73.50	79.20
NDF	437.00	434.30	431.70	428.60
ADF	211.30	214.30	217.20	219.30
NE_L, MJ/kg	6.60	6.61	6.62	6.63

[1] One kilogram of premix contained the following: 100,000 IU VA; 40,000 IU VD; 1000 IU VE; 330 mg Fe; 250 mg Cu; 400 mg Mn; 500 mg Zn; 10 mg Se; 10 mg I; 5 mg Co; [2] All values were measured from the monthly total mixed ration (TMR) samples, while NE_L (net energy for lactation) was calculated based on Ministry of Agriculture (MOA) of P.R. China individual feedstuffs recommendations based on chemical composition (MOA, 2004).

2.2. Sample Collection and Measurements

Samples of *M. oleifera*, TMR, and refusals were collected daily. All samples were composited and analyzed for DM (65 °C in a forced-air oven to a constant weight), CP (method 990.03; AOAC International, 2016), ADF and NDF (Ankom200 Fiber Analyzer, Ankom Technology, Macedon, NY, USA). Ash and EE concentration was determined with method 942.05 (AOAC International, 2016) [11]. Milk samples were recorded in the last 5 days of the feeding period and treated with 2-bromo-2-nitropropane-1-2-diol for determination of milk protein, fat, lactose, and total solids.

Energy-corrected milk yield (ECM), standardized to 4.0% fat and 3.3% protein, was calculated using the equation below:

$$\text{ECM (kg/cow per day)} = \textit{mikl yield}\ (\text{kg/d}) \times \frac{376 \times \textit{fat}\ \% + 209 \times \textit{protein}\ \% + 948}{3138}. \quad (1)$$

Rectal fecal samples (200 g per sampling) were collected four times a day (08:00, 12:00, 16:00, and 20:00) and bulked per animal according to institutional animal care guidelines. Fecal samples were stored at −20 °C for later DNA extraction, high-throughput sequencing, and bioinformatics analysis.

2.3. DNA Extraction and High-Throughput Sequencing using McrA Gene

Fecal samples were freeze-dried, and total DNA was extracted from 200 mg samples with a QIAamp Fast DNA Stool Mini Kit (QIAGEN, CA, USA) according to the manufacturer's instructions. DNA extracts were dissolved in 200 μL elution buffer and the quality and quantity of the extracted DNA were determined using a NanoDrop ND-1000 Spectrophotometer (Nyxor Biotech, Paris, France). The PCR primers used to amplify the *mcr*A fragments were from those of Luton et al. [12]: 5′-GGTGGTGTMGGATTCACACARTAYGCWACAGC-3′ (forward) and 5′-TTCATTGCRTAGTTWGGRTAGTT-3′ (reverse). The PCR was performed using the TaKaRa rTaq DNA Polymerase system and 2 μL of 10× buffer, 2 μL of deoxynucleotide triphosphate (dNTPs) mixture (2.5 mmol/L), 0.2 μL of rTaq polymerase, and 0.8 μL of each primer (forward and reverse). This reaction mixture (25 μL) used the following program: 95 °C for 3 min, followed by 30 cycles of 95 °C for 30 s, 55 °C for 30 s, and 72 °C for 45 s, and a final extension of 72 °C for 10 min. PCR products were electrophoresed in 1% agarose in Tris-acetate-EDTA buffer and visualized with ethidium bromide staining. Amplicons were extracted from 1% agarose gels and purified using the AxyPrep DNA Gel Extraction Kit (Axygen Biosciences, CA, USA) according to the manufacturer's instructions and quantified using QuantiFluor-ST (Promega, Madison, WI, USA). Purified amplicons were then paired-end sequenced (2 × 300) on an Illumina MiSeq platform according to standard protocols.

2.4. Bioinformatics Analysis of the Sequence Data

In the present study, raw FASTQ files were demultiplexed and quality-filtered with the following criteria: (i) The 300-bp reads were truncated at any site that had an average quality score < 20 over a 50-bp sliding window, and truncated reads < 50 bp were discarded; (ii) exact barcode matching was required, and any 2-nucleotide mismatch in primer matching and reads containing ambiguous characters were removed; and (iii) only sequences that overlapped by more than 10 bp were assembled according to their overlap sequences. The length of over 89% of the total valid sequence was between 421 and 440 bp, while around 10% of sequences were ranged between 441 and 460 bp. Operational taxonomic units (OTUs) were clustered with 97% similarity cutoff using Uparse algorithm (version 7.1, http://drive5.com/uparse/), and chimeric sequences were identified and removed using UCHIME. The taxonomy of each *mcr*A gene sequence was analyzed by the RDP Classifier against the FunGene Database using a confidence threshold of 70% [13]. Good's coverage and rarefaction curves were determined to estimate the coverage and sampling effort. QIIME (version 1.17) was also used to calculate the archaeal population diversity (Simpson's diversity index), evenness (Shannon's diversity index), and richness (Chao1 and Ace index). Venn diagram was constructed according to Oliveros [14] to show the shared and unique OTUs among samples. Heatmap analysis and identification of significant features were also used to determine changes among samples. The data obtained in the present study have been submitted to NCBI (submission ID: SUB 3898364, BioProject ID: PRJNA449795).

2.5. Statistical Analyses

Data were analyzed using SPSS software (version 22.0 for Windows, SPSS Inc., Chicago, IL, USA). Dietary DM intake, milk yield, ECM and milk composition were analyzed the using one-way ANOVA package. The model used was as follows:

$$Y_{ij} = \mu + A_i + T_j + e_{ij} \quad (2)$$

where Y_i = the observations for the dependent variable, μ = overall mean, A_i = the random animal effect, T_j = the fixed effect of the jth *Moringa oleifera* amount (treatment, j = 3, 6, or 9% *w/w*), and e_{ij} = the random residual assumed to be normally distributed with mean zero.

Relative abundance data of fecal methanogenic archaea are presented as percentages/proportions. These data were analyzed using Welch's t-test package, and the model was as follows:

$$Y_{ij} = \mu + S_i + T_j + e_{ij} \quad (3)$$

where Y_i = the observations for the dependent variable, μ = overall mean, S_i = the random effect of sampling, T_j = the fixed effect of treatment j, and e_{ij} = the random residual assumed to be normally distributed with mean zero.

Treatment means were considered statistically different at $p < 0.05$, and SEM values are presented with p values.

3. Results

3.1. Feed Intake and Milk Yield and Composition

The DM intake, milk yield, and ECM were similar among the four treatments (Table 2). For the milk composition, fat content (38.2 g/kg) was highest in the M6 treatment ($p < 0.05$) compared with the other three treatments. Protein, lactose, and total solid content did not differ significantly among the 4 treatments.

Table 2. Effects of supplementation of *Moringa oleifera* rachises and twigs on the dry matter intake, milk yield and composition of Holstein dairy cows.

Item [1]	Treatment [2]				SEM [3]	*p*-Value
	Control	M3	M6	M9		
DM intake, kg/day	20.6	20.9	20.7	19.3	0.53	0.09
Milk yield, kg/day	29.6	29.7	31.0	28.6	0.46	0.46
ECM kg/day	31.7	32.2	34.5	31.4	0.57	0.69
Milk composition, %						
Fat	3.54 ^{b}	3.62 ^{b}	3.82 ^{a}	3.68 ^{ab}	0.07	0.04
Protein	3.65	3.70	3.66	3.70	0.03	0.49
Lactose	5.10	5.08	5.08	5.05	0.03	0.14
Total solid	13.0	13.1	13.2	13.1	0.08	0.13

[1] DM = dry matter; [2] M3, M6, and M9 = dietary *Moringa oleifera* supplementation of 3%, 6%, and 9% *w/w*; Means within rows lacking common superscript differ ($p < 0.05$); [3] SEM = standard error of means.

3.2. Composition and Dynamics of Fecal Methanogen Community

A total of 450,500 high-quality sequences with an average length of 439.5 bp were obtained. The Good's coverage indices obtained from each treatment were all above 0.999, indicating a high-quality of sampling and sequencing. The richness and diversity indices obtained for fecal samples of cows fed different levels of *M. oleifera* are presented in Table 3. Richness indices after the 4 treatments did not differ significantly, although the control group had relatively higher values of Ace and Chao (21.05 and 20.63). By contrast, Shannon's index was highest for the M9 group (1.783; $p < 0.05$) and similar among the control, M3, and M6 groups (1.653, 1.438, and 1.628, respectively). The Simpson index was lowest for the M9 group ($p < 0.05$) and did not differ significantly among the control, M3, and M6 groups.

Table 3. Effects of supplementation of *Moringa oleifera* rachises and twigs on the diversity indices of fecal methanogenic archaea based on *mcr*A gene sequences of lactating Holstein dairy cows.

Item [1]	**Treatment** [2]				**SEM** [3]	***p*-Value**
	Control	**M3**	**M6**	**M9**		
Ace	21.05	19.78	15.30	14.35	1.788	0.510
Chao	20.63	19.25	19.75	18.25	0.569	0.561
Shannon	1.653^{b}	1.438^{b}	1.628^{b}	1.783^{a}	0.0570	0.019
Simpson	0.310^{a}	0.393^{a}	0.328^{a}	0.265^{b}	0.0199	0.014

[1] Indices of Ace, Chao, Shannon, and Simpson were calculated to measure alpha diversity of the methanogens in the sample; [2] M3, M6, and M9 = dietary *Moringa oleifera* supplementation of 3%, 6%, and 9% *w/w*; Means within rows lacking common superscript differ ($p < 0.05$); [3] SEM = standard error of means.

A Venn diagram constructed using the OTUs for the sequences from the fecal samples is presented in Figure 1. Shared and unique OTUs were represented at a 97% similarity level among the 4 treatments. A total of 51, 50, 48, and 50 OTUs were found for the control, M3, M6, and M9 treatments, respectively, and 38 OTUs were common to all 4 treatments. In pairwise comparisons of treatments, 43 OTUs were shared between the M3 and M6 treatments, 41 between the M6 and M9 treatments, and 41 between the M3 and M9 treatments. Between the control and treatment M3, M6, or M9, respectively, 47, 44, and 44 OTUs were shared. In addition, a heatmap was constructed to determine the relationship between OTUs and experimental treatments based on the log-transformed sequence abundance and is presented based on the species level at the 97% similarity level (Figure 2). The heatmap showed a change in the abundance of *Methanosphaera* sp. ISO3-F5, *Methanobrevibacter ruminantium*, and *Methanobrevibacter* SM9, indicating that supplementation of *M. oleifera* in the diet resulted in archaeal populations distinct from the control, the same as shown in the abundance analysis.

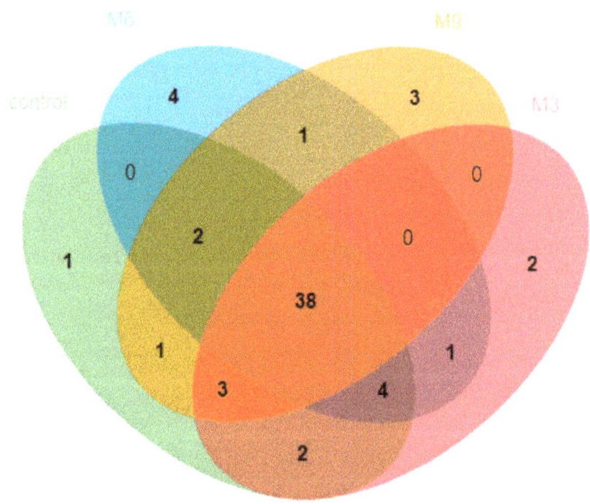

Figure 1. Venn diagram representation of the shared and exclusive Operational taxonomic units (OTUs) at 97% similarity level of four groups: Control (without *Moringa oleifera* supplementation), M3 (a low dose of *Moringa oleifera* supplementation, 3% *w/w*), M6 (a medium dose of *Moringa oleifera* supplementation, 6% *w/w*), M9 (a high dose of *Moringa oleifera* supplementation, 9% *w/w*) in lactating Holstein dairy cows.

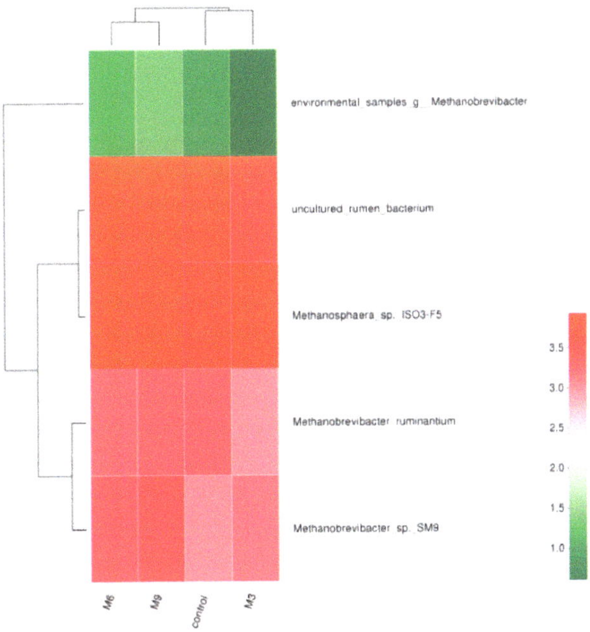

Figure 2. Heatmap under the species level of the four treatments of the lactating Holstein dairy cows fecal sample (control: without *Moringa oleifera* supplementation; M3: a low dose of *Moringa oleifera* supplementation, 3% *w/w*; M6: a medium dose of *Moringa oleifera* supplementation, 6% *w/w*; M9: a high dose of *Moringa oleifera* supplementation, 9% *w/w*).

The relative abundance of fecal methanogenic archaea in the dairy cows with different supplements of *M. oleifera* based on the *mcr*A gene sequencing is listed by taxonomic level in Table 4; three phyla, four orders, four families, five genera, and seven species were determined. The predominant archaeal family across the four treatments in the order *Methanobacteriales* was *Methanobacteriaceae*, with species *Methanosphaera* sp. ISO3-F5, *Methanobrevibacter* sp. SM9, and *Methanobrevibacter ruminantium*.

Specifically, at the order level, no significant difference was observed among the four treatments in terms of *Methanobacteriales* and *Methanomicrobiales*. However, *Methanobacteriales* was relatively more abundant in the M3 treatment (0.677) when compared with the control (0.582), M6 (0.615), and M9 (0.606) treatments. At the genus level, the *Methanocorpusculum* was the third most abundant genus with an average value of 15.0%, 6.5%, 14.2%, and 9.9% for the control, M3, M6, and M9 treatments, respectively. Similar values were obtained for *Methanobrevibacter* among the 4 treatments, whereas the abundance of *Methanosphaera* was significantly higher in the M3 and M9 treatments (0.297 and 0.293, respectively, $p < 0.05$). There was no significant difference between the control and M6 group or between M3 and M9 groups. The species abundance of *Methanosphaera* sp. ISO3-F5 was significantly higher in the M3 and M9 groups (0.297 and 0.291, respectively) than in the control and M6 groups (0.191 and 0.215, respectively, $p < 0.05$), but no significant difference was found between the control and M6 groups or between the M3 and M9 groups. The abundance of *Methanobrevibacter ruminantium* was significantly higher in the control than in the other three groups ($p < 0.05$), but there was no difference among the M3, M6 and M9 group. In addition, values for *Methanobrevibacter SM9* were similar across the four groups but accounted for less than 10% of the total species in each group.

Table 4. Effects of supplementation of *Moringa oleifera* rachises and twigs on the relative abundance of fecal methanogenic archaea based on *mcr*A gene sequences of lactating Holstein dairy cows [1].

Item	Treatment				SEM [2]	*p*-Value
	Control	M3	M6	M9		
Phylum						
Euryarchaeota	0.734	0.743	0.757	0.705	0.022	0.883
Uncultured rumen archaea	0.216	0.150	0.213	0.272	0.016	0.456
Unclassified	0.051ab	0.108a	0.03ab	0.023b	0.015	0.172
Order						
Methanobacteriales	0.582	0.677	0.615	0.606	0.023	0.561
Methanomicrobiales	0.150	0.065	0.142	0.099	0.026	0.676
Uncultured rumen archaea	0.216	0.150	0.213	0.272	0.016	0.456
Family						
Methanobacteriaceae	0.582	0.677	0.615	0.606	0.023	0.561
Methanocorpusculaceae	0.150	0.065	0.142	0.099	0.026	0.676
Uncultured rumen archaea	0.216	0.150	0.213	0.272	0.016	0.456
Genus						
Methanobrevibacter	0.391	0.380	0.400	0.315	0.026	0.698
Methanosphaera	0.191b	0.297a	0.215b	0.291a	0.016	0.016
Methanocorpusculum	0.150	0.065	0.142	0.099	0.026	0.676
Uncultured rumen archaea	0.216	0.150	0.213	0.272	0.016	0.456
Species						
Unclassified *Methanobrevibacter*	0.274	0.279	0.281	0.172	0.031	0.569
Methanosphaera sp. ISO3-F5	0.191b	0.297a	0.215b	0.291a	0.016	0.016
Unclassified *Methanocorpusculum*	0.150	0.065	0.142	0.099	0.026	0.676
Uncultured rumen archaea	0.216	0.150	0.213	0.272	0.026	0.456
Methanobrevibacter sp. SM9	0.038	0.060	0.065	0.080	0.007	0.258
Methanobrevibacter ruminantium	0.078a	0.041b	0.054b	0.062b	0.014	0.035
Unclassified	0.051	0.108	0.030	0.023	0.015	0.172

[1] M3, M6, and M9 = dietary *Moringa oleifera* supplementation of 3%, 6%, and 9% *w/w*; [2] SEM = standard error of means.

4. Discussion

4.1. Feed Intake, Milk Yield and Composition

Application of agricultural by-products in the ruminant production system has been extensively investigated because of their relatively high biomass yield and low cost. In this research, the production, and fecal methanogenic archaea were examined when lactating dairy cows were subjected to different levels of supplementation of *M. oleifera* in the diet. The milk fat content in the M6 treatment was significantly higher than that in the control and M3 treatments, although M6 treatment had relatively higher values of milk yield and ECM. Several consistent results were reported previously that supplementation of *M. oleifera* enhanced milk yield and milk composition. For example, Cohen-Zinder et al. [15] found a significant increase in milk yield, milk fat and protein content in lactating cows offered an *M. oleifera* diet. Azzaz et al. [3] reported that milk and total solid yield increased by 11.3 and 17.7% in lactating ewes fed a supplement of 15 g/kg DM *M. oleifera*. This positive effect of *M. oleifera* on production performance can be attributed to improved feed intake, apparent nutrient digestibility, and ruminal fermentation conditions [4,15]. Moderate concentrations of phenolics and tannins in *M. oleifera* exhibited antioxidant and antimicrobial properties, which have beneficial effects in productive ruminants [16]. Aerts et al. [17] reported that ruminal methanogenesis would be inhibited by phenolics and tannins, which leading to repartition of consumed energy in CH_4 and milk production. This was in agreement with the results of Shaani et al. [7] that improved fat yield and ECM production efficiency resulted from inhibition of CH_4 production by ruminal methanogenic bacteria.

4.2. Fecal Methanogenic Composition and Dynamics

The composition and function of ruminal methanogens have been studied in great detail, whereas less is known in the lower gastrointestinal tract (GIT) of ruminants [18]. Previous research showed relationships between methanogens in feces and those present in the pregastric compartments [19,20]. In the present study, the functional *mcr*A gene sequencing technique was used, and the results revealed the presence of *Methanobrevibacter*, *Methanosphaera*, and *Methanocorpusculum* in the feces of lactating dairy cows. These results were consistent with previous studies that detected two phyla and six genera in the feces of multiparous dairy cows, whereas fecal *mcr*A sequences had the closest similarity to *Methanocorpusculum*, *Methanobacterium*, and *Methanobrevibacter* species [9]. Jin et al. [9] found that fecal archaeal community was predominated by *Methanobrevibacter* (86.9% of the total sequence reads) and *Methanocorpusculum* (10.4%). A bTEFAP pyrosequencing study reported the dominant methanogens of *Methanobrevibacter*, *Methanophaera*, and *Methanobacteriaceae* in the hindgut of goats [21]. In addition, Mohammadzadeh et al. [10] suggested that the fermentation characteristics as digesta pass from the rumen into the small intestine and out of the animal would affect methanogen diversity. The sequencing technique may also influence the results, although the *mcr*A gene-based approach was thought to be comparable to the 16S rRNA gene for phylogenetic studies.

Different from previous studies, however, the present study demonstrated that *Methanobrevibacter* sequence made up approximately 35% of the total. Guzman et al. [22] hardly detected *Methanobrevibacter* in the feces of calves in the first 3 days after birth. In mature cows, *Methanobrevibacter* represented 62% of the rumen archaea, and they were among the most important and dominant archaea in the rumen fluid. A higher percentage was also found in the hindgut of goats with *Methanobrevibacter* accounting for 74.8% of the total sequenced reads, while Jin et al. [9] reported that *Methanobrevibacter* was the dominant phylotype at the genus level accounting for over 86% of the total sequence reads. These results indicated that the presence and abundance of *Methanobrevibacter* may be influenced by dietary composition, enteric fermentation, and even environmental factors.

At the genus level, the relative abundance of *Methanosphaera* and *Methanocorpusculum* was generally around 20% and 15% in the fecal sample of dairy cows, higher than previously reported from the rumen and fecal samples of ruminants [19]. For example, Liu et al. [19] did not detect *Methanocorpusculum* in the rumen but made up only 2% of the archaeal community in the feces of sheep. Jin et al. [9] reported a very low content of *Methanosphaera* in the feces of lactating cows (0.8%). The great diversity of the fecal microbiota can be attributed to various factors such as animal breeds, diet sources, and composition.

4.3. Effect of Moringa Oleifera Supplementation on Fecal Methanogenic Archaea

A range of studies reported that the fecal microbial relative abundance and composition were affected by types of diet or different dietary supplementations [23–25]. For example, the fecal microbial community structure was significantly changed as cattle were fed either high-grain diets or high-forage diets [24]. In our *mcr*A gene-based sequencing study of the methanogenic archaeal community in the feces of lactating cows, the richness indexes remained similar when *M. oleifera* was added at different levels to the diet. However, the Simpson diversity index was significantly lower compared to the control treatment. Changes in the fecal methanogen diversity might be dependent on nutrient contents and the fermentation profile of the fecal samples as a result of secondary metabolites from *M. oleifera*, similar to the alteration of the ruminal environment and microbial activity when *M. oleifera* was fed to ruminants [1,9]. For example, previous results showed that feeding *M. oleifera* plant improved nutrient digestibility and increased SCFA concentration in the rumen of goat, which resulted in the growth of propionate-producing bacterial species and inhibition of CH_4-producing archaea [26]. The high protein (241–277 g/kg DM) and polyphenol content make *M. oleifera* a high-quality feed resource [27]. Bioactive products such as saponins (80 g/kg) and tannins (12 g/kg) in *M. oleifera* leaves have an antimicrobial function and play a key role in improving nutrient digestibility and fermentation efficiency [28,29]. When steers were supplemented with up to 30 g/day tea saponin, daily CH_4 emission (g/day) was reduced by 18%, and yield (CH_4/DM intake, g/kg) was reduced by 22% [30]. In the present study, as demonstrated in our previous experiment, the calculated saponin intake was 144 g/day when 9% of *M. oleifera* was included in the diet. We thus assumed that the composition and distribution of methanogenic archaea changed along the gastrointestinal tract, and methanogenesis and CH_4 emissions would be inhibited by such a large amount of saponin intake. However, feeding saponin-containing *Yucca schidigera* and *Quillaja saponaria* powder (10 g/kg DM) differed little in CH_4 emission (g/day) and yield (CH_4-E/GE intake) from the basal diet [31]. This discrepancy of CH_4 reduction may be attributed to the source and the actual saponin content. Further research is needed to compare responses when similar saponin sources or supplementation levels are used.

When *M. oleifera* was added to the diet, *Methanosphaera* and *Methanosphaera* sp. ISO3-F5 increased in abundance but *Methanobrevibacter ruminantium* decreased compared with the control group. *Methanobrevibacter ruminantium* is a strict anaerobe that can produce CH_4 from H_2, CO_2, and formate and has a close syntrophic association with protozoa [32]. Secondary metabolites from *M. oleifera* such as saponins and tannins have antiprotozoal properties that affect cell membrane integrity [33]. Soliva et al. [34] found that approximately 30% of ciliate protozoa concentration was reduced when extracted *M. oleifera* was added in vitro experiment. This result was in accordance with the inhibitory effect of saponin on ruminal ciliate protozoa population in cattle or sheep [35]. In line with a range of in vivo and in vitro experiments adding different sources and levels of secondary metabolites from *M. oleifera*, *Methanosphaera* sp. ISO3-F5 increased as levels of *M. oleifera* increased. As one of the main methylotrophic methanogens, *Methanosphaera* sp. ISO3-F5 was found to be associated with different bacteria including members of *Lachnospiraceae* [36]. Thus, it would be interesting to examine the pectin content of *M. oleifera* to see its influence on *Methanosphaera* sp. ISO3-F5 abundance. In addition, more future work will be needed to investigate the interaction between some specific methanogens and ruminal fermentation conditions, which may help for a better understanding of rumen microbial metabolic function and development of CH_4 mitigation approaches.

5. Conclusions

This study demonstrated that different supplementation levels of *Moringa oleifera* in the diet achieved similar feed intake, milk production, but adding 6% of *Moringa oleifera* improved milk fat content. The fecal methanogenic archaea diversity changed as the increased abundance of the *Methanosphaera* genus and *Methanosphaera* sp. ISO3-F5 species was induced by secondary metabolites of *Moringa oleifera* in the diet. This study provided some basic information on the utilization of *Moringa oleifera* as forage resources for dairy cows, and helped to elucidate the interaction between methanogens and other microbes, regulation of microbial metabolic function and methane emissions.

Author Contributions: Conceptualization, Q.D.; methodology, L.D.; data curation, T.Z.; writing—original draft preparation, L.D.; writing—review and editing, Q.D.; visualization, L.D.; supervision, Q.D; project administration, Q.D.; funding acquisition, L.D. and Q.D.

Funding: This research was funded by the Program of International S and T Cooperation (grant number 2016YFE0109000), the National Science Foundation for Young Scientists of China (grant number 31802085), the National Key Research and Development Program of China (grant number 2017YFF0211702), and Young Scientist Lifting Project (2017–2019).

Acknowledgments: The authors thank colleagues in Feed Research Institute and Century Love International Investment Group Co. Ltd., Beijing for providing kind assistance in animal experiments, sample processing, and data collection. The authors would like to thank reviewers for their suggestions and comments on this work.

Conflicts of Interest: The authors declare no conflict of interest.

References

1. Kholif, A.E.; Gouda, G.A.; Anele, U.Y.; Galyean, M.L. Extract of *Moringa oleifera* leaves improves feed utilization of lactating Nubian goats. *Small Rumin. Res.* **2018**, *158*, 69–75. [CrossRef]
2. Zhang, T.T.; Si, B.W.; Deng, K.D.; Tu, Y.; Zhou, C.L.; Diao, Q.Y. Effects of feeding a *Moringa oleifera* rachis and twig preparation to dairy cows on their milk production and fatty acid composition, and plasma antioxidants. *J. Sci. Food Agric.* **2018**, *98*, 661–666. [CrossRef] [PubMed]
3. Azzaz, H.H.; Farahat, E.S.A.; Morsy, T.A.; Aziz, H.A.; Hadhoud, F.I.; Abd-Alla, M.S. *Moringa oleifera* and *Echinacea purpurae* as supplements for Rhamani lactating ewe's diets and their effect on rumen characteristics, nutrients digestibility, blood parameters, milk production, composition and its fatty acid profile. *Asian J. Anim. Vet. Adv.* **2016**, *11*, 684–692. [CrossRef]
4. Mendieta-Araica, B.; Sporndly, R.; Reyes-Sanchez, N.; Sporndly, E. Moringa (*Moringa oleifera*) leaf meal as a source of protein in locally produced concentrates for dairy cows fed low protein diets in tropical areas. *Livest. Sci.* **2011**, *137*, 10–17. [CrossRef]
5. Zeng, B.; Sun, J.J.; Chen, T.; Sun, B.L.; He, Q.; Chen, X.Y.; Zhang, Y.L.; Xi, Q.Y. Effects of *Moringa oleifera* silage on milk yield, nutrient digestibility and serum biochemical indexes of lactating dairy cows. *J. Anim. Physiol. Anim. Nutr.* **2018**, *1*, 75–81. [CrossRef]
6. Sarkar, S.; Mohini, M.; Nampoothiri, V.M.; Mondal, G.; Pandita, S.; Mahesh, M.S. Effect of supplementation of *Moringa oleifera* leaves on *in vitro* methane emissions and rumen fermentation on roughage based ration. In Proceedings of the XVI Biennial Animal Nutrition Conference on Innovative Approaches for Animal Feeding and Nutritional Research, Karnal, India, 6–8 February 2016.
7. Shaani, Y.; Eliyahu, D.; Mizrahi, I.; Yosef, E.; Ben-Meir, Y.; Nikbachat, M.; Solomon, R.; Mabjeesh, S.J.; Miron, J. Effect of feeding ensiled mixture of pomegranate pulp and drier feeds on digestibility and milk performance in dairy cows. *J. Dairy Res.* **2016**, *1*, 35–41. [CrossRef] [PubMed]
8. Daquiado, A.R.; Cho, K.M.; Kim, T.Y.; Kim, S.C.; Chang, H.H.; Lee, Y.B. Methanogenic archaea diversity in Hanwoo (*Bos taurus coreanae*) rumen fluid, rectal dung, and barn floor manure using a culture-independent method based on *mcr*A gene sequence. *Anaerobe* **2014**, *27*, 77–81. [CrossRef]
9. Jin, D.X.; Kang, K.; Wang, H.Z.; Wang, Z.S.; Xue, B.; Wang, L.Z.; Xu, F.; Peng, Q.H. Effects of dietary supplementation of active dried yeast on fecal methanogenic archaea diversity in dairy cows. *Anaerobe* **2017**, *44*, 78–86. [CrossRef] [PubMed]
10. Mohammadzadeh, H.; Yáñez-Ruiz, D.R.; Martínez-Fernandez, G. Molecular comparative assessment of the microbial ecosystem in rumen and faeces of goats fed alfalfa hay alone or combined with oats. *Anaerobe* **2014**, *29*, 52–58. [CrossRef]
11. AOAC International. *Official Methods of Analysis*, 20nd ed.; AOAC International: Rockville, MD, USA, 2016.
12. Luton, P.E.; Wayne, J.M.; Sharp, R.J.; Riley, P.W. The *mcr*A gene as an alternative to 16S rRNA in the phylogenetic analysis of methanogen populations in landfill. *Microbiol.* **2002**, *148*, 3521–3530. [CrossRef]
13. Amato, K.R.; Yeoman, C.J.; Kent, A.; Righini, N.; Carbonero, F.; Estrada, A.; Gaskins, H.R.; Stumpf, R.M.; Yildirim, S.; Torralba, M.; et al. Habitat degradation impacts black howler monkey (*Alouatta pigra*) gastrointestinal microbiomes. *ISME J.* **2013**, *7*, 1344–1353. [CrossRef] [PubMed]
14. Oliveros, J.C. *VENNY. An Interactive tool for Comparing Lists with Venn Diagrams*; BioinfoGP of CNB-CSIC: Madrid, Spain, 2007.
15. Cohen-Zinder, M.; Leibovich, H.; Vaknin, Y.; Sagi, G.; Shabtay, A.; Ben-MEIR, Y.; Nikbachat, M.; Protnik, Y.; Yishay, M.; Miron, J. Effect of feeding lactating cows with ensiled mixture of *Moringa oleifera*, wheat hay and molasses, on digestibility and efficiency of milk production. *Anim. Feed Sci. Technol.* **2016**, *211*, 75–83. [CrossRef]
16. Verma, A.R.; Vijayakumar, M.; Mathela, C.S.; Rao, C.V. *In vitro* and *in vivo* antioxidant properties of different fractions of *Moringa oleifera* leaves. *Food Chem. Toxicol.* **2009**, *47*, 2110–2196. [CrossRef]
17. Aerts, R.J.; Barry, T.N.; McNabbm, W.C. Polyphenols and agriculture: Beneficial effects of *proanthocyanidins* in forages. *Agric. Ecosyst. Environ.* **1999**, *75*, 1–12. [CrossRef]
18. Faulkner, M.J.; Wenner, B.A.; Solden, L.M.; Weiss, W.P. Source of supplemental dietary copper, zinc, and manganese affects fecal microbial relative abundance in lactating dairy cows. *J. Dairy Sci.* **2017**, *100*, 1037–1044. [CrossRef] [PubMed]

19. Liu, C.; Zhu, Z.P.; Liu, Y.F.; Guo, T.J.; Dong, H.M. Diversity and abundance of the rumen and fecal methanogens in Altay sheep native to Xinjiang and the influence of diversity on methane emissions. *Arch. Microbiol.* **2012**, *194*, 353–361. [CrossRef]
20. Zhou, M.; Chen, Y.H.; Griebel, P.J.; Guan, L.L. Methanogen prevalence throughout the gastrointestinal tract of pre-weaned dairy calves. *Gut Microbes* **2014**, *5*, 628–638. [CrossRef]
21. Min, B.R.; Solaiman, S.; Shange, R.; Eun, J.S. Gastrointestinal bacterial and methanogenic archaea diversity dynamics associated with condensed tannin-Containing pine bark diet in goats using 16S rDNA amplicon pyrosequencing. *Int. J. Microbiol.* **2014**, *4*, 1–11. [CrossRef]
22. Guzman, C.E.; Bereza-Malcolm, L.T.; Groef, B.D.; Franks, A.E. Presence of selected methanogens, fibrolytic bacteria, and proteobacteria in the gastrointestinal tract of neonatal dairy calves from birth to 72 hours. *PLoS ONE* **2015**, *7*, e0133048. [CrossRef]
23. Dowd, S.E.; Callaway, T.R.; Wolcott, R.D.; Sun, Y.; McKeehan, T.; Hagevoort, R.G.; Edrington, T.S. Evaluation of the bacterial diversity in the feces of cattle using 16S rDNA bacterial tag-encoded FLX amplicon pyrosequencing (bTEFAP). *BMC Microbiol.* **2008**, *8*, 125. [CrossRef]
24. Shanks, O.C.; Kelty, C.A.; Archibeque, S.; Jenkins, M.; Newton, R.J.; McLellan, S.L.; Huse, S.M.; Sogin, M.L. Community structures of fecal bacteria in cattle from different animal feeding operations. *Appl. Environ. Microbiol.* **2011**, *77*, 2992–3001. [CrossRef]
25. Durso, L.M.; Wells, J.E.; Harhay, G.P.; Rice, W.C.; Kuehn, L.; Bono, J.L.; Shackelford, S.; Wheeler, T.; Smith, T.P.L. Comparison of bacterial communities in faeces of beef cattle fed diets containing corn and wet distillers' grain with solubles. *Lett. Appl. Microbiol.* **2012**, *55*, 109–114. [CrossRef] [PubMed]
26. Wallace, R.J.; McEwan, N.R.; McIntosh, F.M.; Teferedegne, B.; Newbold, C.J. Natural products as manipulators of rumen fermentation. *Asian Austral. J. Anim. Sci.* **2002**, *15*, 1458–1468. [CrossRef]
27. Elghandour, M.M.Y.; Vallejo, L.H.; Salem, A.Z.M.; Mellado, M.; Camacho, L.M.; Cipriano, M.; Olafadehan, O.A.; Olivares, J.; Rogas, S. *Moringa oleifera* leaf meal as an environmental friendly protein source for ruminants: Biomethane and carbon dioxide production, and fermentation characteristics. *J. Clean. Prod.* **2017**, *165*, 1229–1238. [CrossRef]
28. Ferreira, P.M.P.; Farias, D.F.; de Abreu Oliveira, J.T.; de Fátima Urano Carvalho, A. Moringa oleifera: Bioactive compounds and nutritional potential. *Revista de Nutrição* **2008**, *21*, 431–437. [CrossRef]
29. Teixeira, E.M.B.; Carvalho, M.R.B.; Neves, V.A.; Silva, M.A.; Arantes-Pereira, L. Chemical characteristics and fractionation of proteins from Moringa oleifera Lam. leaves. *Food Chem.* **2014**, *147*, 51–54. [CrossRef]
30. Ramírez-Restrepoa, C.A.; Tan, C.; O'Neill, C.J.; López-Villalobos, N.; Padmanabha, J.; Wang, J.K.; McSweeney, C.S. Methane production, fermentation characteristics, and microbial profiles in the rumen of tropical cattle fed tea seed saponin supplementation. *Anim. Feed. Sci. Technol.* **2016**, *216*, 58–67. [CrossRef]
31. Holtshausen, L.; Chaves, A.V.; Beauchemin, K.A.; McGinn, S.M.; McAllister, T.A.; Odongo, N.E.; Cheeke, P.R.; Benchaar, C. Feeding saponin-containing *Yucca schidigera* and *Quillaja saponaria* to decrease enteric methane production in dairy cows. *J. Dairy Sci.* **2009**, *92*, 2809–2821. [CrossRef] [PubMed]
32. Finlay, B.J.; Esteban, G.; Clarke, K.J.; Williams, A.G.; Embley, T.M.; Hirt, R.P. Some rumen ciliates have endosymbiotic methanogens. *FEMS Microbiol. Lett.* **1994**, *117*, 157–162. [CrossRef]
33. Morgavi, D.P.; Martin, C.; Jouany, J.P.; Ranilla, M.J. Rumen protozoa and methanogenesis: Not a simple cause-effect relationship. *Br. J. Nutr.* **2012**, *107*, 388–397. [CrossRef]
34. Soliva, C.R.; Kreuzer, M.; Foidl, N.; Foidl, G.; Machnüller, A.; Hess, H.D. Feeding value of whole and extracted *Moringa oleifera* leaves for ruminants and their effects on ruminal fermentation *in vitro*. *Anim. Feed Sci. Technol.* **2005**, *118*, 47–62. [CrossRef]
35. Zhou, M.; Hernandez-Sanabria, E.; Guan, L.L. Assessment of the microbial ecology of ruminal methanogens in cattle with different feed efficiency. *Appl. Environ. Microbiol.* **2009**, *75*, 6524–6533. [CrossRef] [PubMed]
36. Henderson, G.; Cox, F.; Ganesh, S.; Jonker, A.; Young, W.; Janssen, P.H. Rumen microbial community composition varies with diet and host, but a core microbiome is found across a wide geographical range. *Sci. Rep.* **2015**, *5*, 14567. [CrossRef] [PubMed]

animals

MDPI

Article

Effects of Tea Saponin Supplementation on Nutrient Digestibility, Methanogenesis, and Ruminal Microbial Flora in Dorper Crossbred Ewe

Yunlong Liu [1], Tao Ma [1], Dandan Chen [1], Naifeng Zhang [1], Bingwen Si [1], Kaidong Deng [2], Yan Tu [1,*] and Qiyu Diao [1,*]

[1] Key Laboratory of Feed Biotechnology of the Ministry of Agriculture, Feed Research Institute, Chinese Academy of Agricultural Sciences, Beijing 100081, China; liuyunlong9438@163.com (Y.L.); matao@caas.cn (T.M.); chendancd@126.com (D.C.); zhangnaifeng@caas.cn (N.Z.); sibingwen@caas.cn (B.S.)

[2] College of Animal Science, Jinling Institute of Technology, Nanjing 210000, China; kdeng3@yeah.net

* Correspondence: tuyan@caas.cn (Y.T.); diaoqiyu@caas.cn (Q.D.); Tel.: +86-10-8210-6055 (Y.T.); +86-10-8210-6090 (Q.D.)

Received: 18 December 2018; Accepted: 17 January 2019; Published: 21 January 2019

Simple Summary: Greenhouse gas emissions are a serious cause of global warming and climate change, and have become a common focus for all countries. Methane has been proven the second most commonly occurring greenhouse gas. Ruminants have been blamed for substantially contributing to methane emissions. Supplementation with tea saponin (TS) effectively decreased methane emissions and nitrogen emissions. It is not only beneficial for environmental protection, but also has potential economic benefits.

Abstract: Two experiments were conducted using Dorper × thin-tailed Han crossbred ewes. In experiment 1, eighteen ewes were randomly assigned to two dietary treatments (a basal diet, or the same basal diet supplemented with 2.0 g tea saponin (TS)/head/day) to investigate the effects of TS supplementation on nutrient digestibility and methane emissions. In experiment 2, six ewes with ruminal cannulae were assigned to the same two dietary treatments as in experiment 1 to investigate the effects of TS supplementation on rumen fermentation and microbial flora. TS supplementation increased the apparent digestibility of organic matter (OM) ($p = 0.001$), nitrogen (N) ($p = 0.036$), neutral detergent fibre (NDF) ($p = 0.001$), and acid detergent fibre (ADF) ($p < 0.001$). Urinary N ($p = 0.001$) and fecal N ($p = 0.036$) output were reduced, and N retention ($p = 0.001$) and nitrogen retention/nitrogen intake ($p = 0.001$) were increased. Supplementary TS did not decrease absolute methane emissions ($p = 0.519$) but decreased methane emissions scaled to metabolic bodyweight by 8.80% ($p = 0.006$). Ammonia levels decreased ($p < 0.001$) and total volatile fatty acid levels increased ($p = 0.018$) in response to TS supplementation. The molar proportion of propionate increased ($p = 0.007$), whereas the acetate:propionate ratio decreased ($p = 0.035$). Supplementation with TS increased the population of *Fibrobacter succinogenes* ($p = 0.019$), but the population of protozoans tended to decrease ($p = 0.054$). Supplementation with TS effectively enhanced the apparent digestibility of OM, N, NDF, and ADF, and decreased methane emissions scaled to metabolic bodyweight.

Keywords: methane emissions; nitrogen balance; reduction strategy; rumen fermentation; microbial flora; tea saponins

1. Introduction

Greenhouse gas emissions are a serious cause of global warming and climate change, and have become a common focus for all countries, according to the Intergovernmental Panel on Climate Change [1]. Methane is the second most important anthropogenic greenhouse gas, which has 21 times the global warming potential of carbon dioxide [2]. Agriculture accounts for approximately 40% of the total methane emissions from anthropogenic sources, with 25% coming from enteric fermentation in livestock [3]. In ruminants, approximately 95.5% of the methane is produced by feed fermentation in the rumen [4] and is exhaled through the nose and mouth; it represents a loss of 2–12% of the feed energy, depending on the diet [5]. Consequently, numerous efforts are underway to manipulate rumen fermentation and the rumen microbial ecosystem to reduce methane emissions. Limiting the methane emissions from ruminants is not only beneficial for environmental protection, but also has potential economic benefits [6].

Many chemical feed additives have been used to inhibit methane emissions, but these additives are either toxic to the hosts or only have a transient effect on methanogenesis [7]. By contrast, plant extracts are attractive as additives for animal feeds and animal health agents, as they are considered natural, safe, and efficient, and have no hormonal consequences or negative side effects [8]. One promising plant compound is tea saponin (TS), which is a class of pentacyclic triterpenoid glucoside compounds found in a variety of tea plants (Camelliaceae). The basic structure consists of ligands, sugars, and organic acids. Tea saponin has been reported to have an inhibitory effect on protozoa by affecting cell membrane integrity. As protozoa are known to be positively correlated with methanogenesis, tea saponin's biological properties can be used to suppress methane production [9], reduce rumen protozoan counts, and modulate rumen fermentation patterns [10,11].

The rumen is a fermentation chamber where a large number of microbes, including bacteria, protozoa, and fungi, coexist and conduct complicated fermentation processes. Previous studies have been carried out on the effect of TS on methane emissions [12–14]; however, most trials were carried out in vitro [9,15], so the results do not necessarily reflect the situation in vivo [16]. Consequently, the mechanism of action of TS remains unclear. The aim of the present study was to investigate the effects of dietary TS supplementation on ruminal fermentation characteristics, digestibility, methanogenesis, and the ruminal microbial flora using sheep as in vivo model. We hypothesized that TS supplementation could reduce methane emissions by inhibiting the growth of ruminal methanogens and protozoa, and may have different effects on cellulolytic bacteria.

2. Materials and Methods

This study was conducted from March to May 2013 at the Experimental Station of the Chinese Academy of Agricultural Science (CAAS), Beijing, China. The experimental procedures were approved by the Animal Ethics Committee of the CAAS, and humane animal care and handling procedures were followed throughout the experiment (protocol number: AEC-CAAS-2013-01).

2.1. Animals, Diets, and Experimental Design

2.1.1. Experiment 1

Eighteen primiparous Dorper × thin-tailed Han crossbred ewes (60.0 ± 1.73 kg body weight (BW)), 12 months of age, were randomly divided, according to the principle of uniform weight, into two dietary treatment groups: a basal diet, or the same basal diet supplemented with TS at 2.0 g/head/day (TS was extracted from Camellia seeds, Xi'an Feida Bio-Tech Co., Ltd., Shanxi, China). The basal diets included pelleted total mixed rations (concentrate) and Chinese wildrye hay (Table 1). For the experimental diet, the TS was mixed with the pelleted concentrate. The ewes were fed 1500 g pelleted concentrate at 800 h and 200 g of Chinese wild rye hay at 1200 h, daily. This feeding level fulfilled the maintenance and growth requirements of yearling ewes (60 kg BW) according to the NRC [17]. All animals were housed in individual pens, had free access to fresh water throughout the experimental period.

All ewes were moved into metabolism crates after a 14-day adaptation to the diets and another 7-day adaptation to the metabolism crates. The amounts of feed offered, ort, and produced feces were weighed daily and homogenized. A 10% sample was collected during an 8-day collection period, as described by Ma et al. [18]. Urine was collected daily in buckets containing 100 mL of 10% (v/v) H_2SO_4. The volume was measured and a sample (10 mL/L of total volume) was collected and stored at −20 °C until analysis. Samples of feed, ort, feces, and urine were pooled to form a composite sample for each ewe.

Ruminal methane production was measured using an open-circuit respirometry system (Sable Systems International, Las Vegas, NV, USA) with three metabolism cages, each fitted with a polycarbonate head box. Measurements of methane production were staggered because only three measurement units were available. On days 0, 2, 4, and 6 of each 8-day collection period, the ewes were moved in sequence from their own metabolism cages to metabolism cages equipped with head boxes for digestibility assays and methane output assessments. After a 24-h adaption period, individual methane production was measured over a 24-h period, as described by Deng et al. [19]. All ewes had been previously trained for confinement in head boxes attached to metabolism cages.

The ewes were weighed when entering and leaving the gas metabolism cages and the average body weight was used as the basis for calculating the metabolic body weight.

Table 1. Ingredients and chemical compositions of experimental diets (% of dry matter (DM)).

Item [a]	Total Mixed Ration	Chinese Wildrye Hay
Ingredient, % of DM		
Corn	17.0	
Soybean meal	12.0	
Chinese wildrye hay	68.7	
$CaHPO_4$	1.35	
Limestone	0.25	
NaCl	0.50	
Premix [b]	0.24	
Chemical composition (determined)		
DM, (% as fed)	88.6	91.4
OM	80.8	90.6
GE, MJ/kg of DM	17.2	17.6
CP	12.2	8.50
NDF	41.4	70.7
ADF	21.8	38.1

[a] DM: dry matter; OM: organic matter; GE: gross energy; CP: crude protein; NDF: neutral detergent fibre; ADF: acid detergent fibre. [b] Manufactured by Precision Animal Nutrition Research Centre, Beijing, China. The premix contained (per kg): 22.1 g Fe, 2.25 g Cu, 9.82 g Mn, 27.0 g Zn, 0.19 g Se, 0.54 g I, 0.09 g Co, 3.2 g Vitamin A, 0.8 g Vitamin D3, and 0.4 g Vitamin E.

2.1.2. Experiment 2

Six ruminally cannulated Dorper × thin-tailed Han crossbred ewes (65.2 ± 2.0 kg BW) were divided into two groups of three, according to a crossover design, and fed one of the following diets: basal diet, or basal diet supplemented with TS (2.0 g/head/day). The composition of the basal diets and the experimental regime were the same as described for Experiment 1. The experiment lasted for 42 days and consisted of two periods lasting 21 days, including 7 days of adaptation. On days 16 and 37, two 50 mL samples of ruminal digesta were collected from the rumen cannula using a syringe attached to a plastic tube (20 mm internal diameter). Samples were collected at 0, 1, 3, 6, and 9 h after the morning feeding for the measurements of ruminal fermentation parameters and microbial flora populations. The pH was measured immediately using a pH meter (Model PB-10, Sartorius Co, Goettingen, Germany). All samples were frozen in liquid nitrogen within 5 min and stored at −80 °C until analysis.

2.2. Analytical Procedures

Dry matter (DM) content was measured by drying samples in a forced-air oven at 135 °C for 2 h (method 930.15; AOAC) [20]. Ash content was measured by placing samples into a muffle furnace at 550 °C for 5 h (method 938.08; AOAC) [20]. Organic matter (OM) was measured as the difference between DM and the ash content. Nitrogen (N) was measured according to the methods of Kjeldahl, using Se as a catalyst. Crude protein (CP) was calculated as 6.25 × N. Gross energy (GE) was measured using a bomb calorimeter (C^{200}, IKA Works Inc., Staufen, Germany). Ether extracts (EE) were measured by the weight loss of the DM following extraction with diethyl ether in a Soxhlet extraction apparatus for 8 h (method 920.85; AOAC) [20]. Neutral detergent fibre (NDF) and acid detergent fibre (ADF) were measured according to [21,22]. NDF was measured without a heat stable amylase and expressed inclusive of residual ash. Ruminal volatile fatty acid (VFA) was measured according to the procedure described by Ma et al. [23], and ammonia (NH_3) was assessed according to Broderick and Kang [24].

The frozen samples were thawed at room temperature, and the total DNA from rumen fluid was extracted using the bead-beating method described by Zhang et al. [25]. The microbial cells were resuspended in a lysis buffer in tubes containing zirconium beads and were bead-beaten at 4600 rpm for 3 min in a mini-bead beater (MM400, Retsch, Hann, Germany), followed by phenol-chloroform extraction [26]. After centrifugation of the sample at 14,000× *g* for 15 min at 4 °C, the supernatant was mixed with a glass milk kit (Gene Clean II Kit, ZZBio Co., Ltd., Shanghai, China) and washed before a final elution step to release the DNA from the glass milk.

Table 2 shows the amplifying primers used for quantitative polymerase chain reaction (qPCR) analysis for microbial flora [27], including total bacteria, methanogens, protozoans, *Fibrobacter succinogenes*, *Ruminococcus flavefaciens*, *Ruminococcus albus*, and *Butyrivibrio fibrisolvens*. All primers were verified using a sequencing and melting curve analysis with a C^{1000TM} Thermal Cycler and bundled software CFX^{96} Manager^{TM} software version 2.1 (Bio-Rad Laboratories, Hercules, CA, USA). The PCR products were purified by gel extraction and ligated into the pGM-T vector (Promega, Madison, WI, USA). The recombinant plasmids were extracted using a plasmid minikit (Omega, Norcross, GA, USA) according to the manufacturer's instructions and quantified by A260 measurements. Standard curves for microbes were generated with 101–107 copies of recombinant plasmids per μL. The qPCR was performed using SsoFast EvaGreen Supermix (Bio-Rad) on a C^{1000TM} thermal cycler qPCR detection system, with genomic DNA as the template. All PCR amplifications used the following thermal cycling: 95 °C for 10 min, followed by 40 cycles of 94 °C for 20 s, 60 °C for annealing, extension, and collection of fluorescent signals. All samples were prepared from the ewes and each sample was assayed in triplicate.

Table 2. Primers for qPCR assay.

Target Species	Primer Sequence (5′→3′) [a]	Amplicon
Total bacteria	F: CGGTGAATACGTTCYCGG R: GGWTACCTTGTTACGACTT	123
Methanogens	F: TTCGGTGGATCDCARAGRGC R: GBARGTCGWAWCCGTAGAATCC	140
Protozoans	F: GCTTTCGWTGGTAGTGTATT R: CTTGCCCTCYAATCGTWCT	223
Fibrobacter succinogenes	F: GTTCGGAATTACTGGGCGTAAA R: CGCCTGCCCCTGAACTATC	121
Ruminococcus flavefaciens	F: GATGCCGCGTGGAGGAAGAAG R: CATTTCACCGCTACACCAGGAA	286
Ruminococcus albus	F: GTTTTAGGATTGTAAACCTCTGTCTT R: CCTAATATCTACGCATTTCACCGC	270
Butyrivibrio fibrisolvens	F: TAACATGAGAGTTTGATCCTGGCTC R: CGTTACTCACCCGTCCGC	135

[a] Primers were designed according to Denman and McSweeney [27].

2.3. Statistical Analyses

The data on digestibility and nitrogen balance were analysed using one-way ANOVA. Data for ruminal fermentation parameters and microbial flora measured at each sampling time were analysed using Repeated Measures and Multivariate of General Linear Model. Statistical analyses were performed by using SPSS (SPSS Inc., Chicago, IL, USA). Group differences were considered significant when $p < 0.05$ and tendencies were discussed when $0.05 < p < 0.10$.

3. Results

3.1. Nutrient Digestibility

The intake of DM, total tract apparent digestibility of nutrients, and N balance are shown in Table 3. Supplementation with TS increased the apparent digestibility of OM ($p = 0.001$), N ($p = 0.036$), NDF ($p = 0.001$), and ADF ($p < 0.001$) (Table 3). Daily fecal N output decreased from 10.7 to 9.90 g ($p = 0.036$), urinary N decreased from 14.9 to 12.5 g ($p = 0.001$). Overall, the N retention and the ratio of N retention/N intake increased ($p = 0.001$).

Table 3. Effects of tea saponin (TS) supplementation on the apparent digestibility of nutrients and nitrogen balance in ewes.

Item [a]	Treatments [b]		SEM	*p*-Value
	CON	TS		
Apparent digestibility, %				
OM	60.3	66.1	0.99	0.001
N	66.6	69.2	0.63	0.036
NDF	37.9	48.5	1.79	0.001
ADF	35.0	48.3	2.07	<0.001
Fecal N, g/d	10.7	9.90	0.20	0.036
Urinary N, g/d	14.9	12.5	0.42	0.001
N retention, g/d	6.54	9.78	0.56	0.001
N retention/N intake, %	20.3	30.4	1.74	0.001

[a] DM: dry matter; OM: organic matter; GE: gross energy; CP: crude protein; NDF: neutral detergent fibre; ADF: acid detergent fibre. [b] Control (CON) ewes were fed a basal diet; TS ewes were fed the same basal diet supplemented with tea saponin (TS).

3.2. Ruminal Fermentation and Methanogenesis

The methane production, ruminal pH, and ruminal concentrations of ammonia and VFA are shown in Table 4. Supplementation of TS did not affect daily methane production by the ewes ($p > 0.05$), but the methane output, scaled to $BW^{0.75}$, decreased from 2.84 to 2.59 ($p = 0.006$) (Table 4). Ruminal pH was similar between the two treatments ($p = 0.912$). TS supplementation decreased ammonia production from 10.7 to 8.3 mmol/L ($p < 0.001$), while total VFA increased from 101.6 to 118.1 mmol/L ($p = 0.018$). The molar proportions of propionate ($p = 0.007$), isobutyrate ($p = 0.001$), butyrate ($p = 0.002$), and isovalerate ($p = 0.001$) were increased by TS supplementation. No difference was observed in the molar proportion of acetate ($p = 0.171$) and valerate ($p = 0.107$). The molar proportion of the ratio of acetate to propionate decreased from 5.23% to 4.50% ($p = 0.035$).

Table 4. Effects of tea saponin (TS) supplementation on daily methane production and ruminal fermentation in ewes.

Item [a]	Treatments [b]		SEM	p-Value
	CON	TS		
DM intake, g/d	1512.5	1512.6	0.04	0.593
$BW^{0.75}$, kg	21.4	24.2	0.52	<0.001
Methane production				
L	61.1	62.2	0.77	0.519
L/kg $BW^{0.75}$	2.86	2.57	0.05	0.001
L/kg DMI	40.4	41.1	0.51	0.519
pH	5.98	5.96	0.05	0.912
Ammonia, mmol/L	10.7	8.30	0.33	<0.001
Total VFA, mmol/L	101.6	118.1	3.56	0.018
Molar proportions, %				
Acetate	74.0	80.0	2.17	0.171
Propionate	14.4	18.7	0.82	0.007
Isobutyrate	1.34	2.04	0.12	0.001
Butyrate	9.62	14.2	0.76	0.002
Isovalerate	1.39	2.16	0.13	0.001
Valerate	0.89	1.06	0.05	0.107
Acetate:propionate	5.23	4.50	0.17	0.035

[a] BW: bodyweight; DMI: dry matter intake; VFA: volatile fatty acids. [b] Control (CON) ewes were fed a basal diet; TS ewes were fed the same basal diet supplemented with tea saponin (TS).

3.3. Ruminal Microbial Flora

The effect of TS supplementation on ruminal microbial population is shown in Table 5. Supplementation of TS tended to decrease the population of protozoans (p = 0.054) and increased the populations of *F. succinogenes* (p = 0.019), whereas population of total bacteria, methanogen, *R. flavefaciens*, *R. albus*, and *B. fibrisolvens* did not change.

Table 5. Effects of tea saponin (TS) supplementation on ruminal microbial population.

Microbial Population, per mL of Ruminal Fluid	Treatments [a]		SEM	p-Value
	CON	TS		
Total bacteria, $\times 10^9$	7.77	8.23	0.40	0.569
Protozoans, $\times 10^7$	5.44	4.59	0.22	0.054
Methanogens, $\times 10^7$	7.09	6.18	0.45	0.318
F. succinogenes, $\times 10^5$	4.36	5.41	0.23	0.019
R. flavefaciens, $\times 10^8$	4.06	4.40	0.19	0.372
R. albus, $\times 10^7$	5.30	5.02	0.16	0.385
B. fibrisolvens, $\times 10^8$	6.31	6.49	0.12	0.476

[a] Control (CON) ewes were fed a basal diet; TS ewes fed the same basal diet supplemented with tea saponin (TS).

4. Discussion

4.1. Effect of Tea Saponin on Apparent Digestibility and Nitrogen Balance

The effect of TS on nutrient digestibility has been poorly studied. In the present experiment, supplementation with TS increased the apparent digestibility of OM, N, NDF, and ADF. Tea saponins have an important effect on nutrient digestibility [5], whereas saponins from other sources, such as *Quillaja saponaria* or *Yucca schidigera*, are reported to have no effect on diet digestibility [10,28]. The lower N outputs in the urine and feces in the TS group in the present study are consistent with the higher apparent digestibility of dietary N, and may reflect the reduction in protozoan numbers [29]. A related study reported that supplementation with TS improved in vitro OM digestibility [8].

Similarly, Guyader et al. [14] reported a numerical increase in NDF and ADF digestibility in dairy cows; this outcome may be related to the numerically lower dry matter intake (DMI) of lactating dairy cows fed this plant extract, given that a reduction of DMI can be associated with lower rumen filling and greater fiber digestibility. In the present experiment, DMI was not modified in ewes supplemented with TS in DM. Overall, the increase in nutrient digestibility is not related to DMI; it could be explained by the increase in the population of *F. succinogenes*, generally considered the primary organisms responsible for the degradation of plant cell walls in the rumen [30], and by the decrease in the protozoan population.

Nitrogen retention in ruminants has significant benefits for ruminant survival, health, production, and ruminal protection [31]. In the present study, TS supplementation decreased fecal N and urinary N outputs, resulting in a significant N retention. The decrease in urinary N output could be attributed to the decrease in the protozoan population, as previously confirmed by Van Soest [32]. Protozoa contribute to 10–40% of the total rumen nitrogen, so a reduction in this population would mean less predation and lysis of bacteria and, consequently, a lower release of the products of protein breakdown. Jouany [29] assumed that urinary N always decreases with defaunation, due to both the decreased ammonia concentration in the ruminal fluid and the increased capture of urea N for microbial protein synthesis prior to its delivery to the large intestine for recycling in the blood. Koening et al. [33] also reported increases in microbial protein entering the post-digestive tract from the rumen and a promotion of nitrogen utilization.

4.2. *Effect of Tea Saponin on Methane Production*

In the present study, TS supplementation resulted in an 8.8% decrease in the daily methane emissions, scaled to metabolic BW. TS has been reported to reduce methane production by inhibiting the proliferation of rumen protozoa and perhaps by inhibiting interspecies hydrogen transfer between the protozoa and methanogens, although inhibitory effects on hydrogen-producing bacteria are also possible [34]. Similar results were reported in other studies using TS as a plant extract additive to reduce methane emissions. For example, Guo et al. [12] determined that the mechanism of TS inhibition of methane production involved inhibition of the expression of *mcrA*, a key gene encoding the methyl-coenzyme M reductase enzyme involved in methane synthesis. Hess et al. [35] showed that TS can act directly on methanogens to reduce methane production to levels consistent with those reported by Whitelaw et al. [36] and Dohme et al. [37]. In our study, supplementation with TS had no significant effect on the population of methanogens, but it decreased the population of protozoans. A similar observation was made by Hess et al. [35], who reported a 54% decrease in protozoan counts and a 20% decline in in vitro methane emissions, with no effect on methanogens. These researchers suggested that defaunation reduced methane emissions because of the lower H_2 supply, which reduced the activity per methanogen. In the present study, TS supplementation increased the molar proportion of propionate (1 mole H_2 consumed per mole propionate) and decreased the acetate:propionate ratio, indicating a transformation of the rumen from acetate fermentation to propionate fermentation. This switch would lead to a reduction in H_2 availability for methanogenic archaea [38]. Overall, the observed reduction in methane output, scaled to $BW^{0.75}$, may be related to the size of the protozoan population and the VFA patterns.

4.3. *Effect of Tea Saponin on Ruminal Fermentation*

Supplementation with TS modified the fermentation patterns, resulting in changes in rumen pH, ammonia release, and total VFA content. Ruminal pH is an important index of normal rumen function, and the rumen pH values (pH 5.96–5.98) in the present study were within the normal range for efficient rumen function [39]. Supplementation with TS significantly decreased the ruminal concentration of ammonia but increased the levels of total VFA. Wina et al. [40] suggested that decreases in rumen ammonia concentration were an indirect result of the decreased protozoan numbers caused by addition of TS. Similar to our results, most studies have shown that TS supplementation increased the molar proportion of propionate [13,41,42], although Ramírez-Restrepo et al. [43] reported that butyrate concentration increased and propionate concentration decreased at their highest TS supplementation level. Guyader et al. [44] reported that supplementation with TS decreased the acetate:propionate ratio in an in vitro experiment, in agreement with the results of Hu et al. [9,15] and Guo et al. [12] but a subsequent in vivo experiment in lactating cows showed an increase in the acetate:propionate ratio. Overall, the change in the molar proportions of propionate and butyrate and in the acetate:propionate ratio suggested that TS supplementation could modify rumen fermentation profiles by changing the microbial population or the rate of passage of digesta through the rumen.

4.4. Effect of Tea Saponin on Microbial Flora

Supplementation with TS tended to decrease the population of protozoans in the present study, in agreement with the findings of Mao et al. [45] and Zhou et al. [13], who also showed that the rumen protozoan numbers were lower in sheep supplemented with TS. A toxic effect of TS towards protozoa has also been reported previously [9,12,44] in vitro. Wallace et al. [31] indicated that TS might kill or damage protozoa by forming complexes with sterols on the protozoan membrane surface, leading to membrane impairment and eventual disintegration. However, several reports have shown no effect of saponin on protozoa, and some have showed an increase in protozoan numbers [41]. These differences may reflect differences in the experimental diets and the TS dosages. Methanogenic archaea have been observed on the exterior surfaces of rumen protozoa [46]. About 10% to 20% of methanogens live in association with protozoa [47], so a reduction in protozoan numbers would also be expected to reduce methanogen numbers, and thereby decrease methane emissions. However, in vitro [12] and in vivo [45] experiments have shown that TS addition has little effect on the methanogen population, which is consistent with the present findings. Similarly, previous studies indicated that the relative abundance of methanogens in sheep was unaffected by TS supplementation [13,45]. However, the activity of the methanogens could be reduced, as Guo et al. [12] found that TS supplementation inhibited the expression of the *mcrA* gene.

In the present study, we also used qPCR to quantify four main cellulolytic bacteria and found a selective effect of TS supplementation on rumen bacteria. Unlike the case for *F. succinogenes*, the populations of *R. flavefaciens*, *R. albus*, and *B. fibrisolvens* were unchanged by TS supplementation. Several studies have examined the effects of TS on ruminal microbial flora, but the results have been inconsistent. For example, Guo et al. [12] reported that number of *F. succinogenes* increased significantly with the addition of TS in vitro, in agreement with our results. Conversely, in vivo TS supplementation has been reported to have no effect on the populations of *R. flavefaciens* or *F. succinogenes* [45]. Zhou et al. [13] also reported no changes in the population of *F. succinogenes*. The rumen is a complex system where billions of microbes live, so the effects of TS supplementation on ruminal microbial populations deserve further study.

In order to avoid the adverse effects of stress on entering and leaving the gas metabolism cage, the ewes were weighed twice, before and after the start of the experiment, the average body weight was taken as the basis for the calculation of metabolic body weight. Of course, we considered that body weight may be related to methane emissions, so we used metabolic body weight to eliminate this factor. Converting the average body weight into metabolic body weight is equivalent to unifying the body weight and eliminating the influence of body weight. The relative (per kg metabolic BW) methane emissions are more indicative of differences in methane emissions between different diets.

5. Conclusions

In the present study, dietary TS supplementation effectively enhanced OM, N, NDF, and ADF digestibility and reduced daily methane emissions (L/kg $BW^{0.75}$) in ewes. These effects were probably due to decreases in the population of ruminal protozoans and modifications in the VFA profile in response to TS. Further investigation is necessary to explain the mechanisms by which TS exerts these effects on methanogenesis and ruminal microbial flora.

Author Contributions: Conceptualization, T.M. and Q.D.; methodology, Y.T.; software, Y.L.; validation, Y.T., N.Z. and Q.D.; formal analysis, T.M.; investigation, D.C.; resources, Q.D.; data curation, D.C.; writing—original draft preparation, Y.L.; writing—review and editing, T.M. and K.D.; visualization, T.M.; supervision, B.S.; project administration, B.S.; funding acquisition, Q.D.

Funding: This work was supported by the [China Agriculture Research System] under Grant [CARS-38] and [National Natural Science Foundation of China] under Grant [No: 41705129]. All authors participated in writing the final draft of the manuscript and agreed on the final format.

Conflicts of Interest: We certify that there are no conflicts of interest with any financial organization regarding the material discussed in the manuscript.

References

1. IPCC (Intergovernmental Panel on Climate Change). Climate change: Synthesis report. In *Contribution of Working Groups I, II and III to the Fifth Assessment Report of the Intergovernmental Panel on Climate Change*; Pachauri, R.K., Meyer, L.A., Eds.; IPCC: Geneva, Switzerland, 2014; p. 151.
2. UNFCC (United Nations Framework Convention on Climate Change). *Greenhouse Gas Inventory Data*; UNFCC: Bonn, Germany, 2006.
3. Olivier, J.G.J.; Aardenne, J.A.; Dentener, F.; Ganzeveld, L.; Peters, J.A.H.W. Recent trends in global greenhouse gas emissions: Regional trends and spatial distribution of key sources. In Proceedings of the Non-CO_2 Greenhouse Gases (NCGG-4), Utrecht, The Netherlands, 4–6 July 2005; pp. 325–330.
4. AGO (Australian Greenhouse Office). *National Greenhouse Gas Inventory 2001 with Methodology Supplements*; AGO: Canberra, Australia, 2003.
5. Johnson, K.A.; Johnson, D.E. Methane emissions from cattle. *J. Anim. Sci.* **1995**, *73*, 2483–2492. [CrossRef] [PubMed]
6. Alford, A.R.; Hegarty, R.S.; Parnell, P.F.; Cacho, O.J.; Herd, R.M.; Griffith, G.R. The impact of breeding to reduce residual feed intake on enteric methane emissions from the Australian beef industry. *Aust. J. Exp. Agric.* **2006**, *46*, 813–820. [CrossRef]
7. Moss, A.R.; Jouany, J.P.; Newbold, C.J. Methane production by ruminants: Its contribution to global warming. *Ann. Zootech.* **2000**, *49*, 231–235. [CrossRef]
8. Wei, M.L.; Ren, L.P.; Zhou, Z.M.; Meng, Q.X. Effect of addition of three plant extracts on gas production, ruminal fermentation, methane production and ruminal digestibility based on an in vitro technique. *J. Anim. Vet. Adv.* **2012**, *11*, 4304–4309.
9. Hu, W.L.; Wu, Y.M.; Liu, J.X.; Guo, Y.Q.; Ye, J.A. Tea saponins affect in vitro fermentation and methanogenesis in faunated and defaunated rumen fluid. *J. ZheJiang Univ. Sci. B* **2005**, *6*, 787–792. [CrossRef]
10. Lila, Z.A.; Mohammed, N.; Kanda, S.; Kamada, T.; Itabashi, H. Effect of sarsaponin on ruminal fermentation with particular reference to methane production in vitro. *J. Dairy Sci.* **2003**, *86*, 3330–3336. [CrossRef]
11. Wang, C.J.; Wang, S.P.; Zhou, H. Influences of flavomycin, ropadiar, and saponin on nutrient digestibility, rumen fermentation, and methane emission from sheep. *Anim. Feed Sci. Technol.* **2009**, *148*, 157–166. [CrossRef]
12. Guo, Y.Q.; Liu, J.X.; Lu, Y.; Zhu, W.Y.; Denman, S.E.; McSweeney, C.S. Effect of tea saponin on methanogenesis, microbial community structure and expression of *mcrA* gene, in cultures of rumen microorganisms. *Lett. Appl. Microbiol.* **2008**, *47*, 421–426. [CrossRef]
13. Zhou, Y.Y.; Mao, H.L.; Jiang, F.; Wang, J.K.; Liu, J.X.; McSweeney, C.S. Inhibition of rumen methanogenesis by tea saponins with reference to fermentation pattern and microbial communities in Hu sheep. *Anim. Feed Sci. Technol.* **2011**, 166–167. [CrossRef]
14. Guyader, J.; Eugène, M.; Doreau, M.; Morgavi, D.P.; Gérard, C.; Loncke, C.; Martin, C. Nitrate but not tea saponin feed additives decreased enteric methane emissions in nonlactating cows. *J. Anim. Sci.* **2015**, *93*, 5367–5377. [CrossRef]
15. Hu, W.L.; Wu, Y.M.; Liu, J.X.; Guo, Y.Q.; Ye, J.A. Effect of tea saponins on in vitro ruminal fermentation and growth performance in growing Boer goat. *Arch. Anim. Nutr.* **2006**, *60*, 89–97. [CrossRef]
16. Macome, F.M.; Pellikaan, W.F.; Hendriks, W.H.; Warner, D.; Schonewille, J.T.; Cone, J.W. In vitro gas and methane production in rumen fluid from dairy cows fed grass silages differing in plant maturity, compared to in vivo data. *J. Anim. Physiol. Anim. Nutr.* **2018**, *102*, 843–852. [CrossRef] [PubMed]

17. NRC (Nutrient Requirements of Small Ruminants). *Sheep, Goats, Cervids and New World Camelids*; National Academy Press: Washington, DC, USA, 2007.
18. Ma, T.; Chen, D.D.; Tu, Y.; Zhang, N.F.; Si, B.W.; Deng, K.D.; Diao, Q.Y. Effect of dietary supplementation with resveratrol on nutrient digestibility, methanogenesis and ruminal microbial flora in sheep. *J. Anim. Physiol. Anim. Nutr.* **2015**, *99*, 676–683. [CrossRef] [PubMed]
19. Deng, K.D.; Jiang, C.G.; Tu, Y.; Zhang, N.F.; Liu, J.; Ma, T.; Zhao, Y.G.; Xu, G.S.; Diao, Q.Y. Energy requirements of Dorper crossbred ewe lambs. *J. Anim. Sci.* **2014**, *92*, 2161–2169. [CrossRef] [PubMed]
20. AOAC (Association of Official Analytical Chemists). *Official Methods of Analysis*, 15th ed.; AOAC: Washington, DC, USA, 1990.
21. Van Soest, P.J.; Robertson, J.B.; Lewis, B.A. Methods for dietary fiber, neutral detergent fiber and non-starch polysaccharides in relation to animal nutrition. *J. Dairy Sci.* **1991**, *74*, 3583–3597. [CrossRef]
22. Goering, H.G.; Van Soest, J.P. Forage fiber analysis. In *Agricultural Handbook*; UPSDA: Washington, DC, USA, 1970; Volume 379.
23. Ma, T.; Deng, K.D.; Tu, Y.; Zhang, N.F.; Jiang, C.G.; Liu, J.; Zhao, Y.G.; Diao, Q.Y. Effect of dietary forage-to-concentrate ratios on urinary excretion of purine derivatives and microbial nitrogen yields in the rumen of Dorper crossbred sheep. *Livest. Sci.* **2014**, *160*, 37–44. [CrossRef]
24. Broderick, G.A.; Kang, J.H. Automated simultaneous determination of ammonia and total amino acids in ruminal fluid and in vitro media. *J. Dairy Sci.* **1980**, *63*, 64–75. [CrossRef]
25. Zhang, C.M.; Guo, Y.Q.; Yuan, Z.P.; Wu, Y.M.; Wang, J.K.; Liu, J.X.; Zhu, W.Y. Effect of octadeca carbon fatty acids on microbial fermentation, methanogenesis and microbial flora in vitro. *Anim. Feed Sci. Technol.* **2008**, *146*, 259–269. [CrossRef]
26. Zoetendal, E.G.; Akkermans, A.D.; De Vos, W.M. Temperature gradient gel electrophoresis analysis of 16S rRNA from human fecal samples reveals stable and host-specific communities of active bacteria. *Appl. Environ. Microb.* **1998**, *64*, 3854–3859.
27. Denman, S.E.; McSweeney, C.S. Development of a real-time *PCR* assay for monitoring anaerobic fungal and cellulolytic bacterial populations within the rumen. *FEMS Microbiol. Ecol.* **2006**, *58*, 572–582. [CrossRef]
28. Holtshausen, L.; Chaves, A.V.; Beauchemin, K.A.; McGinn, S.M.; McAllister, T.A.; Odongo, N.E.; Cheeke, P.R.; Benchaar, C. Feeding saponin-containing Yucca schidigera and Quillaja saponaria to decrease enteric methane production in dairy cows. *J. Dairy Sci.* **2009**, *92*, 2809–2821. [CrossRef] [PubMed]
29. Jouany, J.P. Effect of rumen protozoa on nitrogen utilization by ruminants. *J. Nutr.* **1996**, *126*, 1335S–1346S. [CrossRef] [PubMed]
30. Michalet-Doreau, B.; Fernandez, I.; Peyron, C.; Millet, L.; Fonty, G. Fibrolytic activities and cellulolytic bacterial community structure in the solid and liquid phases of rumen contents. *Reprod. Nutr. Dev.* **2001**, *41*, 187–194. [CrossRef] [PubMed]
31. Wallace, R.J.; McEwan, N.R.; McIntosh, F.M.; Teferedegne, B.; Newbold, C.J. Natural products as manipulators rumen fermentation. *Asian Australas. J. Anim.* **2002**, *15*, 1458–1468. [CrossRef]
32. Van Soest, P.J. *Nutritional Ecology of the Ruminant*, 2nd ed.; Cornell University Press: New York, NY, USA, 1994; p. 470.
33. Koening, K.M.; Newbold, C.J.; Mcintosh, F.M.; Rode, L.M. Effects of protozoa on bacterial nitrogen recycling in the rumen. *J. Dairy Sci.* **2000**, *78*, 2431–2445.
34. Attwood, G.; McSweeney, C.S. Methanogen genomics to discover targets for methane mitigation technologies and options for alternative H_2 utilisation in the rumen. *Aust. J. Exp. Agric.* **2008**, *48*, 28–37. [CrossRef]
35. Hess, H.D.; Kreuzer, M.; Diaz, T.E.; Lascano, C.E.; Carulla, J.E.; Soliva, C.R.; Machmuller, A. Saponin rich tropical fruits affect fermentation and methanogenesis in faunated and defaunated rumen fluid. *Anim. Feed Sci. Technol.* **2003**, *109*, 79–94. [CrossRef]
36. Whitelaw, F.G.; Eadie, J.M.; Bruce, L.A.; Shand, W.J. Methane formation in faunated and ciliate-free cattle and its relationship with rumen volatile fatty acid proportions. *Br. J. Nutr.* **1984**, *52*, 261–275. [CrossRef]
37. Dohme, F.; Machmüller, A.; Estermann, B.L.; Pfister, P.; Wasserfallen, A.; Kreuzer, M. The role of the rumen ciliateprotozoa for methane suppression caused by coconut oil. *Lett. Appl. Microbiol.* **1999**, *29*, 187–192. [CrossRef]
38. Guyader, J.; Eugène, M.; Meunier, B.; Doreau, M.; Morgavi, D.P.; Silberberg, M.; Rochette, Y.; Gérard, C.; Loncke, C.; Martin, C. Additive methane-mitigating effect between linseed oil and nitrate fed to cattle. *J. Anim. Sci.* **2015**, *93*, 3564–3577. [CrossRef]

39. Yoon, I.K.; Stern, M.D. Effect of Saccharomyces cerevisiae and Aspergillus oryzae cultures on ruminal fermentation in dairy cows. *J. Dairy Sci.* **1996**, *79*, 411–417. [CrossRef]
40. Wina, E.; Muetzel, S.; Becker, K. The impact of saponins or saponin-containing plant materials on ruminant production-areview. *J. Agric. Food Chem.* **2005**, *53*, 8093–8105. [CrossRef]
41. Santoso, B.; Kilmaskossu, A.; Sambodo, P. Effect of saponin from Biophytum petersianum Klotzsch on ruminal fermentation: Microbial protein synthesis and nitrogen utilization in goats. *Anim. Feed Sci. Technol.* **2007**, *137*, 58–68. [CrossRef]
42. Poungchompu, O.; Wanapat, M.; Wachirapakorn, C.; Wanapat, S.; Cherdthong, A. Manipulation of ruminal fermentation and methane production by dietary saponins and tannins from mangosteen peel and soapberry fruit. *Arch. Anim. Nutr.* **2009**, *63*, 389–400. [CrossRef] [PubMed]
43. Ramírez-Restrepo, C.A.; Tan, C.; O'Neil, C.J.; López-Villalobos, N.; Padmanabha, J.; Wang, J.K.; McSweeney, C.S. Methane production, fermentation characteristics, and microbial profiles in the rumen of tropical cattle fed tea saponin supplementation. *Anim. Feed Sci. Technol.* **2016**, *216*, 58–67.
44. Guyader, J.; Eugène, M.; Doreau, M.; Morgavi, D.P.; Gérard, C.; Loncke, C.; Martin, C. Tea saponin reduced methanogenesis in vitro but increased methane yield in lactating dairy cows. *J. Dairy Sci.* **2017**, *100*, 1845–1855. [CrossRef] [PubMed]
45. Mao, H.L.; Wang, J.K.; Zhou, Y.Y.; Liu, J.X. Effects of addition of tea saponins and soybean oil on methane production, fermentation and microbial population in the rumen of growing lambs. *Livest. Sci.* **2010**, *129*, 56–62. [CrossRef]
46. Vogels, G.D.; Hoppe, W.F.; Stumm, C.K. Association of methanogenic bacteria with rumen ciliates. *Appl. Environ. Microbiol.* **1980**, *40*, 608–612.
47. Tokur, M.; Chagan, I.; Ushida, K.; Kojima, Y. Phylogenetic study of methanogens associated with rumen ciliates. *Curr. Microbiol.* **1999**, *39*, 123–128. [CrossRef]

animals

MDPI

Review

Methane Emissions and the Use of *Desmanthus* in Beef Cattle Production in Northern Australia

Bénédicte Suybeng [1], Edward Charmley [2], Christopher P. Gardiner [1], Bunmi S. Malau-Aduli [3] and Aduli E. O. Malau-Aduli [1,*]

[1] Animal Genetics and Nutrition, Veterinary Sciences Discipline, College of Public Health, Medical and Veterinary Sciences, Division of Tropical Health and Medicine, James Cook University, Townsville, QLD 4811, Australia

[2] CSIRO Agriculture and Food, Private Mail Bag Aitkenvale, Australian Tropical Sciences and Innovation Precinct, James Cook University, Townsville, QLD 4811, Australia

[3] College of Medicine and Dentistry, Division of Tropical Health and Medicine, James Cook University, Townsville, QLD 4811, Australia

* Correspondence: aduli.malauaduli@jcu.edu.au; Tel.: +61-747-815-339

Received: 10 May 2019; Accepted: 6 August 2019; Published: 9 August 2019

Simple Summary: An in-depth review of Australia's tropical beef cattle production system is presented with emphasis on the use of *Desmanthus*, a tropical legume, as a nutritional supplementation strategy for the abatement and mitigation of methane emissions. It also identifies current knowledge gaps in *in vivo* methane emissions research.

Abstract: The Australian beef industry is a major contributor to the economy with an estimated annual revenue generation of over seven billion dollars. The tropical state of Queensland accounted for 48% of Australian beef and veal production in 2018. As the third biggest beef exporter in the world, Australia supplies 3% of the world's beef exports and its agricultural sector accounts for an estimated 13.2% of its total greenhouse gas emissions. About 71% of total agricultural emissions are in the form of methane and nitrous oxide. In this review, an overview of the carbon footprint of the beef cattle production system in northern Australia is presented, with emphasis on the mitigation of greenhouse gases. The review also focuses on the tropical legume, *Desmanthus*, one of the more promising nutritional supplements for methane abatement and improvement of animal growth performance. Among the review's findings is the need to select environmentally well-adapted and vigorous tropical legumes containing tannins that can persistently survive under the harsh northern Australian conditions for driving animal performance, improving meat quality and reducing methane emissions. The paper argues that the use of appropriate legumes such as *Desmanthus*, is a natural and preferred alternative to the use of chemicals for the abatement of methane emanating from tropical beef cattle production systems. It also highlights current gaps in knowledge and new research opportunities for *in vivo* studies on the impact of *Desmanthus* on methane emissions of supplemented tropical beef cattle.

Keywords: methane emission; tropical beef cattle; *Desmanthus*; supplementation; growth performance; ruminant nutrition; legumes

1. Introduction

Global climate change is principally caused by greenhouse gas (GHG) emissions that result in warming of the atmosphere [1]. According to the Australian National Greenhouse Accounts [2], 13.2% of GHG emissions emanate from agriculture, with methane and nitrous oxide accounting for 71% of total agricultural emissions [3]. In 2017, Australia produced an estimated 51,543.56 Gg CO_2-e of CH_4 from enteric fermentation [3].

The world population is predicted to increase from 7.6 to 9.8 billion by 2050 [4]. Consequently, the world has to match the increased demand for food from a larger and more affluent population to its supply in an environmentally sustainable manner [5]. Livestock products constitute an important source of food for global food security by providing 33% and 17% of world protein and kilocalorie consumption, respectively [6]. Climate change constitutes a risk to livestock production due to its impact on the feed quality of crops and forages, animal performance, milk production, water availability, animal reproduction, livestock diseases and biodiversity [6]. Therefore, the challenge is to find ways to increase livestock productivity without compromising household food security while sustainably improving the natural resource base [7].

Beef cattle productivity in north Queensland is beset with climatic and nutritional challenges, due to prolonged drought, high climate variability, inadequate feed resources, low-quality pastures and the poor body condition of cattle [8]. In this seasonally dry, low-elevation, heavy textured soils and inland tropical region of north Queensland, there is an overwhelming need for integrating more productive, nutritious and persistent summer-growing legumes into existing low quality, grass-dominant pastures [8].

Existing cultivars lack the capacity to adapt to seasonally waterlogged duplex soils, infertile light-textured soils, heavy cracking clays and low rainfall conditions [9]. Gardiner [10] evaluated the performance characteristics of *Desmanthus* in contrasting tropical environments and found that it thrived and spred on heavier vertisol soils. Hall and Walker [9] conducted a study over a 15-year period in six different environments in the seasonally dry tropics of north Queensland and found that on cracking clay soils, *Desmanthus* species and *Clitoria ternatea* were the most persistent and productive legumes among 118 legume accessions.

Further evaluation and development of *Desmanthus* under commercial grazing management will be highly beneficial to northern Australian beef cattle graziers for improved productive and reproductive performances, better animal body condition and higher meat quality, particularly in the live cattle trade where northern Australia is the main gateway to this key business export market. The State of Queensland accounts for about 43% of the Australian cattle population [11] and has a CH_4 emission from ruminants that has been estimated to account for 3% of Australia's GHG [3]. Australia has a target to reduce its emissions by 5% below the 2000 level by 2020 and 26–28% below 2005 emissions by 2030. The Government allocated $2.55 billion to the Emissions Reduction Fund (ERF) to help livestock producers use modern farming methods to store carbon in vegetation and soils towards reducing GHG [12]. Research into contemporary, scientific and sustainable ways to produce high quality beef in tropical Northern Australia with low methane emissions are paramount to Australia being competitive in the international market. *Desmanthus* spp. are among the most promising sown legume species for the vastly undeveloped semi-arid clay soil regions across northern Australia.

This review focusses on the carbon footprint of the beef cattle industry in northern Australia, explores mechanisms and methods of enteric methane production and abatement with a focus on *Desmanthus* as a potential pasture legume for mitigating methane emissions. Finally, the current knowledge gaps that could underpin future research are also reviewed.

2. Carbon Footprint from the Beef Industry in Queensland

2.1. The Australian Beef Cattle Market

Australia is the third biggest beef exporter in the world, supplying 3% of the world's beef exports with 1,500,000 tons of carcass weight exported annually. The Australian beef cattle industry accounted for $11.4 billion in 2017–2018 [11]. Furthermore, the beef cattle industry employed 191,800 people in 2016–2017. Therefore, the beef industry plays a central role in the Australian economy, especially in the state of Queensland, where its 11.1 million head of cattle accounted for 48.1% of the Australian beef and veal production in 2017–2018 [11].

2.2. *The Different Sectors Included in the Carbon Footprint of the Beef Industry in Queensland*

The total net emissions attributed to agriculture in Queensland was 18,672.5 Gg CO_2-e in 2017 [3]. The beef industry in Queensland is the largest agricultural industry in the state [13]. Sources of GHG emissions from a typical beef enterprise comprise enteric fermentation in cattle (CH_4 and N_2O), burning of vegetation (intentional or accidental), energy use (electricity and fuel), land clearing, loss of pasture and decline in soil carbon [13,14]. A study conducted by Eady et al. [14] in two beef farms in Queensland showed that the carbon footprint of beef products at the farm gate ranged from 17.5–22.9 kg CO_2-e/kg liveweight at Gympie and 11.6–15.5 kg CO_2-e/kg liveweight in the Arcadia Valley. They also found that enteric fermentation represented about 80% (74% at Arcadia Valley and 85% at Gympie) of the overall 'cradle-to-farm gate' GHG emissions [14]. The last figures can be linked with the 70% (12,995.97 Gg CO_2-e) of agriculture GHG emissions coming from enteric fermentation from grazing beef cattle in Queensland in 2017 [3].

2.3. *The Principal Causes Inducing Enteric Methane Emissions*

2.3.1. Rumen Microbial Fermentation

The rumen is a dynamic and complex ecosystem composed essentially of anaerobic bacteria, protozoa, anaerobic fungi, methanogenic archaea and phages [15]. The microbes interact with each other and have a symbiotic relationship with the host. The breakdown of plant cell wall carbohydrates that are inedible by humans provides energy to the host [16]. Methane is produced exclusively by methanogenic archaea [15] via the hydrogenotrophic pathway using CO_2 as the carbon source and H_2 as the main electron donor, and less so through the utilization of methyl groups (methylotrophic pathway), or even less commonly from acetate (acetoclastic pathway) [15]. The methanogenesis reaction uses H_2 to reduce CO_2 to CH_4: $CO_2 + 4H_2 = CH_4 + 2H_2O$ [17].

The main products of rumen microbial fermentation, as depicted in Figure 1, are volatile fatty acids (VFA) (acetic, propionic and butyric acids), carbon dioxide and methane [18]. In the rumen, the VFA formed are absorbed and used as a source of energy. On the contrary, CO_2 and CH_4 are eliminated by eructation from the rumen. Over 80% of the methane is synthesised in the rumen and the lower digestive tract produces the rest [18]. Northern beef cattle in Australia can generate about 32.2 to 184 g of methane per day [19], which represents an important energy loss to the animal ranging from 2% to 12% of gross energy intake depending on the nature of the diet [20]. Under a high forage diet, these losses are on the average, 7.2% of gross energy intake; 6.3% for an intermediate forage and 3.84% for a low forage (feedlot) [21].

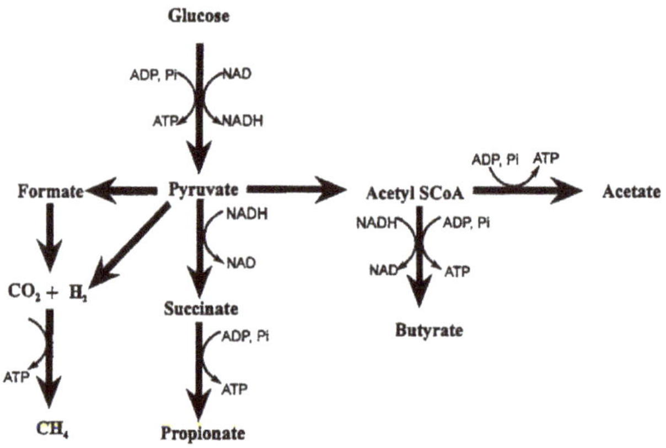

Figure 1. Principal end-products of carbohydrate fermentation in the rumen [18].

2.3.2. Low Animal Performance Increases Methane Production

Less efficient cattle can take longer to reach market weight and might only breed two out of three seasons. The longer an animal takes to reach market weight, the longer that animal is producing methane, with very little beef being marketed in return [20,22]. Arthur et al. [23] estimated genetic and phenotypic parameters for feed intake in Angus bulls and heifers, and showed that the feed conversion ratio defined by the amount of feed consumed divided by live weight gain was correlated genetically (−0.62) and phenotypically (−0.74) with the average daily gain (ADG). For instance, Charmley et al. [22] showed that by maintaining a liveweight (LW) gain of 0.5 kg/day for steers in the northern spear grass region by adding supplements to the pasture diet would reduce the turn-off age of the Japanese Ox market from 4 years (526 kg LW) to 2.3 years (650 kg LW). Gross margin budget and cashflow analyses for a 100-cow herd showed a 61% internal rate of return over a 25-year investment period, despite the higher cost for purchasing efficient bulls. It represents an annual benefit per cow of A$8.76 [24]. Low animal productivity is associated with high methane output per unit of product (methane intensity) and low pasture quality is associated with high methane output per unit of dry matter intake [12]. For that reason, northern Australian beef herds are estimated to produce more methane than the more intensive systems in southern Australia [20]. For instance, Eady [25] showed that the GHG emissions of beef produced from cattle supply chain from Northern Australia to the Indonesian market were higher (26 kg CO_2 equivalent/kg liveweight) than beef produced in Southern Australian systems, where GHG emissions ranged from 5.4 to 14.5 kg CO_2 equivalent/kg liveweight for finished steers. They attributed it to the higher reproduction rate, faster turn-off and lower methane emissions per unit of feed intake permitted by a high pasture quality in the southern systems [25].

2.3.3. Northern Australian Forage Diet Influences Rumen Microbiome and Methane Production

In northern Australia, comprising the Kimberley and Pilbara districts of Western Australia, the Northern Territory and Queensland above the Tropic of Capricorn, the beef industry is dominated by large pastoral properties [10]. This part of Australia is characterised by a vast array of heavy clay or vertosol soils, where the range of available sown pasture legumes has long been regarded as being deficient [26]. There are also vast areas of light textured soils where the legume *Stylosanthes* has been successfully introduced. Pasture production is highly seasonal, with a wet season (November to April) characterised by growth, and a senescent period during the dry season. This induces a marked seasonal pattern of pasture availability and quality [27]. The prevailing pasture species are mainly C4 grasses, which have lower nutritional value than temperate grasses, and result in lower animal productivity than in temperate regions [28,29]. During the wet, hot summers, these pastures grow quickly and persist through the dry winter seasons as mature grasses [30–32]. The low livestock productivity in northern Australia is especially due to low protein content and low digestibility during the dry season [33]. The low digestibility (45% organic matter) and nitrogen content (less than 7g N/kg dry matter (DM)) of these grasses during the dry season results in poor forage intakes and low annual growth rate of young cattle [30–32]. Animals tend to put on weight in the wet season and lose weight in the dry season. In northern Australia, it is not uncommon for 4–6 years old steers to be marketed [34]. Consequently, depending on the time of the year, liveweight gains in Northern Australia are around 70–240 kg/year for native pastures [35] compared to 250–300 kg/year for temperate pastures [36]. Growth rate is directly related to metabolisable energy intake, and can be markedly increased by replacing the feed base or by giving supplements to the animals [36].

Archimede et al. [37] showed that ruminants fed C4 grass produced 17% more methane as L/kg organic matter intake than those fed C3 grass. Likewise, Perry et al. [29] found that steers fed a wet season pasture (crude protein (CP) = 90 g/kg DM) or a high quality hay (CP = 88 g/kg DM) produced 5–10 g CH_4/kg, digested less dry matter intake (DMI) and had about 3% less digestible energy intake than steers fed low quality hay (CP = 25 g/kg DM). They observed shorter rumen retention times in high quality hay fed steers, which decreased methane production per kilogram of DMI compared with low quality hay and the dry season pasture. This phenomenon can be explained with an increased rumen outflow rate [38]. The rise in rumen outflow rates is associated with higher concentrations of dissolved H_2 that increase the growth rate of methanogens. The greater cellulose and hemicellulose content in tropical C4 grasses rather than neutral detergent soluble carbohydrates in grain diets results in higher methane emissions and a shift in rumen fermentation pathways from propionate to acetate [12,29]. The production of methane in the rumen is associated with the production of VFA. The formation of both acetic and butyric acids is accompanied by the production of H_2 and CO_2, whereas propionic acid production requires a net uptake of H_2, which can reduce methanogenesis [38]. The production of propionic acid instead of acetic acid can be realised by replacing structural carbohydrates (forage) with easily fermented carbohydrates [38].

3. Mitigation Techniques against Methane Emission

3.1. The Use of Chemicals for Rumen Manipulation to Reduce Methane Production

3.1.1. The Use of Chemicals to Control Protozoa, the Main Hydrogen Producer

Some techniques such as defaunation and the utilisation of ionophores, have been used to control protozoa, the major producers of H_2 from the rumen [39], so that less H_2 is accessible for CH_4 formation.

Defaunation

Defaunation techniques comprise synthetic chemicals such as copper sulphate, dioctylsodium sulfosuccinate, calcium peroxide, detergents and natural compounds, such as vitamin A, steroidal hormones or non-protein amino acids [40]. Dohme et al. [41] showed that defaunation using coconut oil immediately reduced methane formation by about 40% *in vitro* using non-lactating Brown Swiss cow fed hay. However, like other inhibitors of methanogenesis, numerous defaunation agents are toxic to the animal [42]. Moreover, defaunation techniques on-farm are currently non-existent [40].

Ionophores

Ionophores are classified as antibiotics and are synthetized by soil microorganisms that can modify the movement of cations, such as calcium, potassium and sodium through cell membranes. The ionophores that are particularly used to reduce methane emissions are monensin and lasalocid [43]. Guan et al. [44] showed that supplementing ionophores to 36 Angus yearling steers decreased enteric CH_4 emissions (expressed as litres per kilogram) by 30% for the first two weeks for animals on a highly concentrated diet and by 27% for the first four weeks for animals on high and low-concentrate diets, respectively. They also indicated that alternative feeding of cattle with monensin and lasalocid in comparison to only monensin did not result in further decreases or longer periods of depressed enteric methane emissions. In contrast, McCaughey et al. [45], observed no difference in methane production in pasture-fed steers supplemented with 270 mg/d monensin controlled release capsule. According to Russell and Houlihan [46], the possibility of transmission of antibiotic resistance from animals to man through ionophores in animal feeds is not likely to happen. However, the use of monensin in cattle as a feed additive to increase growth and feed efficiency was phased out by the European Union Council Regulation in January 2006, but it has been re-evaluated and authorized as a feed additive for the control of coccidiosis in poultry [47].

Another technique using probiotics has also been developed. Although the mechanism used to decrease CH_4 production is not yet clear, it may be due to the utilisation of metabolic H_2 by acetogenic bacteria to produce acetate [40] or by decreasing the numbers of rumen ciliate protozoa [48]. Probiotics are microbial feed additives that affect fermentation in the rumen. The most widely used probiotics are yeasts such as *Saccharomyces cerevisiae* and *Lactobacillus sporogenes* [40]. McGinn et al. [49] found that a commercial yeast product (procreatin 7 yeast) fed to growing beef cattle induced a 3% reduction in CH_4 production (g/g DMI). The use of probiotics appears to be an interesting method, but results have been unconvincing or yet to be confirmed *in vivo* [50].

3.1.2. The Use of Chemicals to Control the Methanogen Numbers

Methane inhibitors are chemical compounds with inhibitory effects on rumen archaea [40]. Studies using methane inhibitors such as chloroform, 3-nitrooxypropanol (3-NOP), carbon tetrachloride, methylene chloride, bromoethanesulphonate or bromochloromethane showed significant reductions in CH_4 production [51–54]. For instance, Martinez Fernandez et al. [54] showed that methane production (in g/kg DMI) reduced by 38% in animals supplemented with 3-NOP and by 30% for Brahman steers supplemented with chloroform compared with the control group (*Chloris gayana*). Mathison et al. [42] indicated that methane inhibitors can reduce CH_4 emissions on short-term basis by preventing the accumulation of H_2 in the rumen, but because of microbial adaptation, the effects are rapidly neutralized and feed intake often depressed [42].

Overall, the utilisation of chemicals for rumen manipulation with subsequent mitigation of methane emission appears promising, but requires considerable further development due to inconclusive results (probiotics, ionophores), microbial adaptation (defaunation, methane inhibitors) and prohibited use of antibiotics in some countries [40].

3.2. The Use of Diet Manipulation to Reduce Methane Production

3.2.1. The Use of Concentrates to Reduce Methane Production

Supplements are frequently used in grazing systems when availability and/or quality of pasture is limiting animal performance. To promote good animal health, supplementary feeding should satisfy the animals' needs for protein, energy, roughage and minerals. This can be a regular part of the production cycle during the dry season. The use of supplements depends on the enterprise's production objectives and seasonal conditions [42]. Table 1 sums up the typical tropical supplements for critical seasons used in northern Australia, often chosen for their low cost [55–57].

Table 1. Typical tropical animal supplements for critical seasons [55,56].

Animal Nutrient Needs	Supplement	Critical Season
Energy	Grains, molasses	Dry
Protein	Urea	Dry
Roughage	Silage, hay	Dry and wet
Minerals	Phosphorus	Wet

Purnomoadi et al. [58] found that offering concentrates to Indonesian Ongole crossbred young bulls twice a day significantly reduced methane production (32.76 CH_4 g/kg DMI) compared to other bulls fed concentrate only once a day (36.33 CH_4 g/kg DMI). The same study also showed that increasing the feeding frequency of concentrates resulted in a better feed utilisation (lower feed conversion rate) and increased animal productivity with a higher ADG (0.44 vs. 0.38 kg/day) [58]. This phenomenon can be explained by the change in fermented substrate from fibre to starch and the decline in ruminal pH, inducing a reduction in the proportion of dietary energy converted to CH_4 thereby increasing the level of concentrates in the diet [59]. Although increasing dietary concentrates may sometimes increase total carbon footprint by increasing the amount of emissions associated with total production, the use of pesticides, fertilisers and transportation infrastructure are indirect contributing factors [59].

3.2.2. The Use of Legumes to Reduce Methane Production

Interest in secondary plant compounds as possible methane mitigation strategy is rising, as plant preparations are viewed as natural alternatives to chemical additives, which are prone to negative perception from consumers [50]. The production of methane from rumen fermentation is generally lower with legumes than grass forages, principally due to the lower fibre content inducing a more rapid rate of passage through the rumen [59].

One of the plant extracts used to reduce methane emissions belongs to the tannin families [50].

Tannins are polyphenolic compounds of plant origin. There are two main types: Hydrolysable tannins (HT) (polyesters of gallic acid and various sugars) and condensed tannins (CT) (polymers of flavonoids) as depicted in Figure 2 [60]. Tannins are broadly distributed in the plant kingdom and are known to protect against infection, insects or animal herbivory [40]. Tannins have the ability to form complexes with dietary proteins, minerals and polymers, such as hemicellulose, cellulose and pectin, thus delaying digestion; this confers tannins with their anti-nutritive property [61].

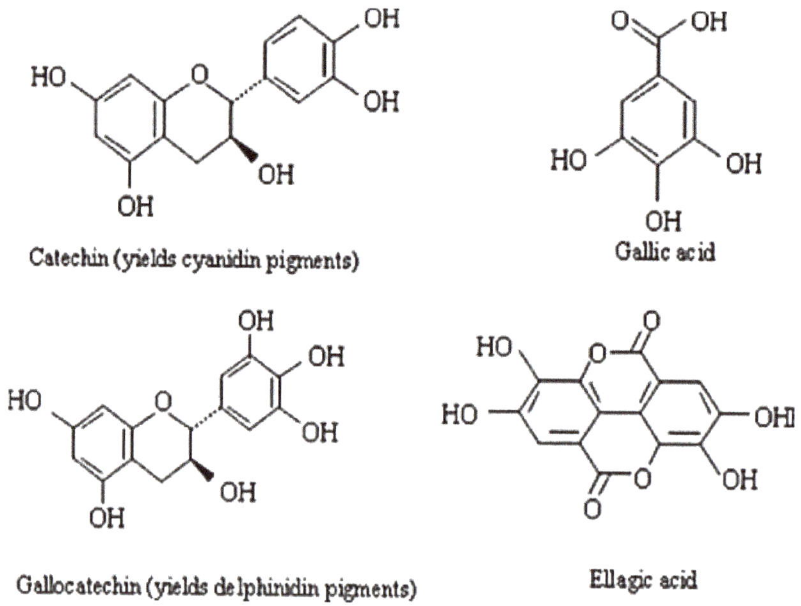

Figure 2. Monomeric units of condensed (catechin and gallocatechin) and hydrolysable tannins (gallic and ellagic acid) [62].

Several legumes have been studied for their methane reduction properties. Hess et al. [63] showed that extracted tannins and legumes with high tannin levels from *Calliandra calothyrsus* induced a reduction in methane emissions, but also reduced the feeding value of the diet. The same observation was made by Tiemann et al. [64], who reported a reduction in CH_4 production by up to 24% when an herbaceous high-quality legume (*Vigna unguiculata*) was replaced with tannin-rich plants (*Calliandra calothyrsus* or *Flemingia macrophylla*). They concluded that this reduction was mainly due to a reduction in fibre digestion and organic matter.

Leucaena leucocephala, a leguminous shrub that is abundant in the tropics, contains a significant amount of CT (33 to 61 g/kg DM) [65] and a high protein content of 200 to 250 g/kg DM [66]. *Leucaena* contains mimosine ranging from 40 to 120 g/kg DM [67], and mimosine is an anti-nutritive compound that can be toxic at high DM intake [65,67]. However, *in vitro* [65,68,69] and *in vivo* [70,71] studies showed that the addition of *Leucaena* in the diet induces methane reduction. Soltan et al. [71] conducted an *in vivo* study with Santa Inês sheep and showed that *Leuceana*, compared to Bermuda grass (*Cynodon dactylon*) in the diet, decreased CH_4 emissions and enhanced intake, body nitrogen retention, faecal nitrogen excretion and the elimination of urinary purine derivatives (a sign of the synthesis and availability of microbial proteins). In order to test the effect of tannins on methane production, they added polyethylene glycol (PEG), a tannin inhibitor, at a ratio of 1:1 PEG:*Leucaena* into the diet and did not see any significant difference in methane reduction with or without PEG. They suggested that there was no clear efficiency of tannins on methane emissions in sheep. Jones and Mangan [72] showed that the interchange reaction of PEG with an already formed tannin-protein complex depends on the quantity of tannins and complex age before PEG addition. They explained that any increase in both factors decreases the exchange. McSweeney et al. [73] showed that PEG addition (10 mg PEG/50 mg plant substrate) to *in vitro* fermentation can be used to analyse the effect of tannins on nitrogen digestibility. Bhatta et al. [74] showed that tannins suppress methanogenesis by reducing methanogenic populations in the rumen by either direct inhibition of methanogens or indirect interference with the protozoal population, resulting in a decrease in the number of methanogens symbiotically associated with the protozoal population. Beauchemin et al. [75] found that supplementing *quebracho* tannin extract linearly decreased the proportion of acetate, resulting in a linear decrease of the acetate to propionate ratio.

The antimethanogenic activity of tannin-containing plants has been credited mostly to the condensed tannin group because hydrolysable tannins are more toxic for the animal [76]. However, a study conducted by Jayanegara et al. [77] showed that HT had a greater effect in reducing CH_4 emissions and had less negative effects on digestibility than CT. They attributed this observation to the lower risk of toxicity of CT than HT [59]. Ruminants consuming forage plants containing a high level of HT (*Terminalia oblongata* and the Indonesian shrub *Clidemia hirta*) showed toxicity symptoms through simple phenolics liberated in the gut [78] beyond the capacity of the liver to detoxify [79]. McMahon et al. [80] reported that high tannin concentrations exceeding 40 to 50 g/kg dry matter in forages may diminish protein and dry matter digestibility in ruminants. Several experiments showed that a level of HT lower than 20 g/kg DM did not cause detrimental effects on production parameters [77]. At low to moderate concentrations, CT raises dietary protein quantity, in particular, the essential amino acids. CT (polyphenolics) are able to form complexes with proteins in the rumen under the near-neutral condition of pH 6.5 and protect them from deamination, thus reducing nitrogen availability to rumen microorganisms [60,72]. However, at pH 2.5 in the abomasum and abomasal end of the duodenum, the complex becomes disrupted and unstable, thereby permitting protein degradation by acidic proteases [72].

In summary, legumes and plant extracts such as tannins, seem to be good alternatives for methane abatement as they are perceived to be more natural than the other methods [50]. However, the addition of plant extracts does not always show conclusive results. For instance, the addition of *Leucaena* can be toxic due to high mimosine content [67], and Calliandra can decrease feed digestibility [64]. Only *Desmanthus*, a tropical legume containing CT, has so far shown promising results in reducing methane emissions [68,81] and improving animal growth performance [82–85].

4. The Use of Legumes to Increase Pasture Quality and Animal Performance in Northern Australia

4.1. The Use of Legumes to Increase Pasture quality

4.1.1. Ability to Fix Nitrogen

Legumes are rich in nitrogen because they have the capacity to biologically fix nitrogen and transform it into leguminous protein [86]. For instance, Wetselaar [87] measured the amount of nitrogen fixed by four legumes: Townsville Lucerne (*Stylosanthes humilis*), guar (*Cyamopsis tetragonoloba*), cowpea (cv. *Poon*) and peanut (cv. *Natal common*) on Tippera clay loam in three growing seasons. They showed that the total amount of N added to the soil-plant system in three seasons by the four legumes was 220, 220, 270 and 125 kg/ha respectively. Another study on Tippera clay loam soil in the Northern Territory displayed a higher nitrogen uptake by 30 kg/ha after the first year, and by 55 kg/ha after the third year of maize crops on a Caribbean stylo (*Stylosanthes hamata* cv. *Verano*) legume ley compared to a grass ley [88]. The presence of *Rhizobium* bacteria-legume symbioses is capable of fixing nitrogen under dry conditions that benefits not only the legumes, but associated grasses also [89].

Northern Australian graziers are concerned about the 'rundown' of buffel grass, which constitutes the dominant sown species in the area. Buffel grass pastures older than 10–20 years since establishment have declined by up to 50% in all districts. This decrease is principally related to the lack of nitrogen in the soil. Economic analysis suggests that the best solution to overcome this 'rundown' is to establish a range of adapted pasture legumes into existing grass-only pastures in order to introduce more nitrogen. Seeding legumes into a predominantly grass pasture can enable a regain of 30–50% of lost production from pasture rundown and improve economic returns [90].

4.1.2. Ability to Extract Moisture and Nutrients from the Soil

Legumes have taproots that allow for moisture and nutrient extraction from deep down the soil profile. This assists with more drought tolerance, greener and productive longevity than grasses [91]. Thus, forage legumes can have significant impacts on the environment, including nitrogen fixation, improvement of soil quality, protection from water and wind erosions [92] and improvement of carbon accumulation [93].

4.2. *The Use of Legumes to Increase Animal Productivity*

Studies have shown that legumes increase animal productivity due to improved crude protein content and feed digestibility [10,94]. For instance, liveweight gains of 190 kg/head/year were observed on improved Townsville *Stylosanthes* legumes compared to 80 kg/head/year on native pastures at a stocking rate of one beast per 2.4 hectares [8]. Bowen et al. [95] conducted a study on 21 sites located in the Fitzroy river catchment (Queensland) across 12 commercial beef cattle properties. They showed that tropical legume forages constituted high quality diets (*Leucaena*-grass (120 and 59), lablab (115 and 59), and butterfly pea-grass (97 and 59), g CP/kg DM) and dry matter digestibility (DMD) in comparison with perennial grass pastures that had 66 g CP/kg DM and 55% DMD. These high quality diets resulted in an annual per ha liveweight gain of 2.6 kg when cattle grazed paddocks containing *Leucaena* and Butterfly peas with perennial C4 grass which was 1.6 times higher than for cattle grazing only perennial grass pastures. Coates et al. [9] found that the introduction of legumes such as stylo pastures improved annual liveweight gains (0.45 kg/day), decreased turn-off age by at least 3–6 months, extended cattle growth into the late wet season and minimised dry season liveweight loss [95].

Thus, it seems the sowing of legumes in grass improves pasture quality and animal performance. Throughout the long annual dry seasons of northern Australia, the semiarid clay soil region has no sown pasture legumes with recognized adaptation and persistence [96]. Therefore, to help meet beef cattle production requirements, farmers use nutritional supplementation strategies [97], agistment or selling of stock to reduce stocking rates [55].

4.3. *Northern Australian Legumes*

Northern Australian legumes such as *Crotalaria* spp., *Cullen* spp., *Glycine* spp., *Indigofera* spp., *Rhynchosia* spp., *Sesbania* spp. *and Vigna* spp. are often described as grazing intolerant [98], toxic and/or unpalatable [10]. Some legumes such as *Stylosanthes* with its cultivars Seca (*S. scabra*) and Verano (*S. hamata*) have been incorporated into native grass pastures on light textured soils such as black spear grass (*Heteropogon contortus*). This legume has been shown to be beneficial in increasing cattle liveweight gains in the range of 30–60 kg/head/year and improving stocking rates [9,10]. In semi-arid northern regions with textured clay soils (vertisols), the stylos are not usually well adapted and few other sown legume species have shown persistence in such environments [10]. *Leucaena* is another notable success in the development of exotic species in northern Australia, especially after the discovery by Raymond Jones that a bacterium (*Synergistes jonesii*) could degrade DHP (3-hydroxy-4(IH) pyridone), a breakdown product of mimosine, the anti-nutritional toxic agent in *Leucaena* [98,99]. The search for legumes broadly adapted to the Australian subtropics had limited success. Twining tropical legumes including *C. pascuorum*, *Clitoria ternatea* (butterfly pea), Sirano (*Macroptilium atropurpureum*) and *Centrosema mole* (centro) did not persist under grazing and could not regenerate from seeds when the first-established plants died [98]. Some other legumes were persistent but suffered from other deficiencies such as limited environmental adaptation to the wide range of the Australian subtropical environment, low palatability and weedy characteristics that reduced their attractiveness [98]. However, *Desmanthus*, a legume native to the Americas has been shown to persist under heavy grazing on clay soils [14]. In the 1990s, various *Desmanthus* accessions persisted for more than two decades in abandoned trial sites across remote northern and central west Queenslands' semi-arid clay soil regions [26]. The Commonwealth Scientific and Industrial Research Organisation and Queensland Department of Primary Industries have introduced numerous accessions of *Desmanthus* over the past 50 years [100].

5. *Desmanthus* as a Potential Pasture Species for Ruminants

5.1. *Performance Characteristics of* Desmanthus

Desmanthus is included in the *Dichrostachys* group of the tribe *Mimoseae* [101]. It can grow on a wide range of soil types from coastal sands to rocky limestone and saline soils. *Desmanthus* spp. are often selected for their persistence on heavy clay such as alkaline soils, but will grow on lighter soils of neutral to alkaline pH [102]. In exotic locations such as Queensland, with its average annual rainfall of 616 mm (1900 to 2015) [103], *Desmanthus* is well adapted and capable of thriving in a 550–1000 mm average rainfall environment [102]. The plant grows better in humid-tropical locations with annual average temperatures ranging from 22 to 28 °C. The legume can be defoliated by heavy frost, but is able to regrow from crowns when the moisture and heat conditions are sufficient [103]. Its deep roots enable it to be grown with stoloniferous grasses such as buffel grass (*Cenchrus ciliaris*), Bambatsi panic (*Panicum coloratum var. makarikariense*) and Queensland bluegrass (*Dichanthium sericeum*). Minor damages in seed crops by psyllid insects (*Accizia* spp.) in northern Australia and by seed-eating bruchid beetle (5 *Acanthoscelides* spp. and *Stator* sp.) have been reported [102]. Jones and Brandon [104] studied the persistence and productivity of eight accessions of *Desmanthus virgatus* under grazing at five levels of presentation yield at the end of the growing season in subtropical and subcoastal Queensland from 1989 to 1996. After surface sowing *Desmanthus* at 4 kg/ha in 1989, they found that the yields averaged 0.7 t/ha at the highest grazing pressure and 4.7 t/ha at the lowest grazing pressure [104]. The best of these varieties has been selected, evaluated, propagated and commercialised by Agrimix Pty Ltd. (James Cook University's commercialisation partner, Virginia, QLD, Australia), as Progardes™ which stands for PROtein, GARdiner and *Desmanthus*; and includes new selections of the species *D. bicornutus*, *D. leptophyllus* and *D. virgatus*. The five selected cultivars are: JCU1 (*D. leptophyllus*), JCU 2, 3, 5 (*D. virgatus*) and JCU 4 (*D. bicornutus*) [10]. The different species give a large collection of early to late maturity types, habits (herbaceous to suffructicose), edaphic and climatic tolerances [104]. Progardes™ seeds have been sown in about 20,000 ha of commercial paddocks across northern New South Wales, Northern Territory and principally Queensland, using several sowing techniques such as aerial seeding, seeding following a blade plough and stick raking [10]. *Desmanthus* has an average crude protein content of 21% [105] with 20.2% crude protein in the leaf, 11.9% in the stem and 17% in the pods of Progardes™ *Desmanthus* [106]. On the contrary, Australian native grasses (bluegrass, spear grass) have average crude protein levels between 10% at the beginning and 5% at the end of the wet season [101]. During the dry or winter season, *Desmanthus* dies back to the base, and each year, when moisture and/or temperature conditions are favourable, new stems sprout [101]. A shallow planting depth (0.5–2.0 cm in at least 50–60 cm depth of good moist soil [103]) and weed control have been shown to be beneficial for *Desmanthus* cultivar Progardes™ establishment, particularly in central and southern Queensland. In general, the end of the dry season/start of the wet season is a good period to sow *Desmanthus* seeds and enable grazing during the summer/autumn in northern Queensland [10]. However, due to unpredictable annual rainfall, it is advisable to plant 3 kg of Progardes™ seeds/ha as a combination of half-hard and half-soft (scarified) seeds. Scarification has been used in the horticultural industry to improve the rate of seed germination by chemically or physically altering the seed coat. The purpose is to increase the diffusion rates of water and gases into the seeds [107]. Scarification of Progardes™ by hot water or with a mechanical abrasive disc for commercial batches enhances germination from 10% to 70–80% (with scarification) [10]. Its seed yield range varies between 400 and 600 kg/ha from direct harvesting [102,108]. The ability of *Desmanthus* to spread and become a potential weed is limited. Late flowering cultivars such as cv. Bayamo produce limited seeds while early flowering cultivars have high seed yields resulting in high soil seed reserves. These reserves lead to a thickening of the planted areas with a slow spread from the original plantings [102,108]. However, hard seeds of leguminous species are known to resist digestion and can be dispersed by ruminants in faeces (endozoochory). Gardiner et al. [109] found that most JCU2 seeds fed to sheep passed through the animals in 48h with only 9% of the fed seeds recovered, with about 60% remaining viable.

Consequently, *Desmanthus* seems to be a promising legume in northern Australia due to its high DM productivity, seed production, tolerance of heavy grazing in alkaline, sodic, saline and heavy clay soils and its persistence in low rainfall environments [102].

5.2. *Desmanthus as a Potential Pasture to Reduce Methane Production*

As depicted in Table 2, Vandermeulen et al. [81] evaluated organic matter degradability (OMD) and methane production via *in vitro* incubation of ruminal fluid from grazing Brahman (*Bos indicus*) steers on Rhodes grass (as control), *Desmanthus bicornutus*, *D. leptophyllus* and *D. virgatus* harvested from Agrimix Pty. Ltd. commercial plots. They showed that *D. leptophyllus* had a significantly lower methane emission per unit of fermented organic matter during winter in comparison to the control and other *Desmanthus* species. For instance, after 72 h of incubation, 29.56 mL CH_4/g OM (organic matter) fermented was emitted in the presence of *D. leptophyllus;* 38.72 mL CH_4/g OM was fermented for the control; and 39.90 and 32.94 mL CH_4/g OM fermented for *D. virgatus* and *D. bicornutus* respectively [81]. They also found a negative correlation between HT concentration in *Desmanthus* forages and CH_4 emission per g of OM fermented. Consequently, they hypothesised a possible anti-methanogenic property of HT [81]. Durmic et al. [68] in their study comparing fermentation parameters and nutritive values between plant species and across seasons, showed that *Desmanthus leptophyllus* produced less methane than *Leucaena*, and had reduced volatile fatty acid concentrations.

5.3. *Desmanthus as a Potential Pasture to Increase Animal Production*

Gardiner and Parker [83] showed that steers grazing a mixed buffel grass-Progardes™ pasture in central Queensland gained an extra 40 kg liveweight over a 90-day period in comparison to steers on a buffel grass-only based diet during the dry season (Table 2). Another study conducted in central Queensland has shown that cattle grazing paddocks containing buffel grass with Progardes™ at a population density of 7 plants/m^2 had an additional gain of 40 kg/head compared to steers grazing only buffel grass [82]. A 56-day feeding trial with 24 growing goats showed that supplementing animals with 40% *D. bicornutus* and alfalfa induced an average daily gain of 60.9 g/day compared to 82.3 g/day on alfalfa only [97]. Rangel and Gardiner [85] showed the potential advantage of providing 30% *Desmanthus* to sheep on a Mitchell grass hay diet. They observed reduced weight loss, higher feed intake and wool growth exceeding 19% over the 6 week experimental duration. Sheep showed a positive nitrogen balance and significantly enhanced weight gains and intakes by supplementing *D. leptophyllus* to a Flinders grass diet [84].

Table 2. Effects of *Desmanthus* on methane production, growth performance and rumen fermentation [a].

Desmanthus Species	Experiment	Dosage	Control Dosage	Effects	References
D. bicornutus, *D. leptophyllus* or *D. virgatus*	*In vitro* (Brahman steers rumen fluid)	1 g *Desmanthus* + 125 mL rumen fluid	1 g *Rhodes grass* forage + 125 mL rumen fluid	↓ ME, VFA	[81]
D. leptophyllus	*In vitro* (sheep rumen fluid)	10 mL of 1:1.3 or 1:1.5 dilution of inoculum:buffer + 0.1 g *Desmanthus*	10 mL of 1:1.3 or 1:1.5 dilution of inoculum:buffer + 0.1 g grass	↓ ME, VFA	[68]

Table 2. *Cont.*

Desmanthus Species	Experiment	Dosage	Control Dosage	Effects	References
Progardes™	Steers	Paddock with buffel grass and Progardes™	Paddock with buffel grass	↑ LW	[83]
Progardes™	Steers	Paddock Progardes™ (7 plants/m^2) and buffel grass	Paddock with buffel grass	↑ LW	[82]
D. bicornutus	Goats	40% *Desmanthus* in the diet + alfalfa	Alfalfa	↓ LW	[110]
D. virgatus, *D. pubescens* or *D. leptophyllus*	Sheep	30% *Desmanthus* + Mitchell grass hay	Mitchell grass	↑ LW, ↑ Intake, ↑ Wool growth	[85]
D. leptophyllus	Sheep	*Ad libitum* flinders grass hay + *D. leptophyllusor* either *D. leptophyllus* or flinders grass hay		↑ LW, ↑ positive N balance with *Desmanthus*	[84]

[a] ME, methane emissions; VFA, volatile fatty acids; LW, liveweight, ↓, decrease; ↑, increase.

6. Implications, Future Research and Conclusions

Australia as the third biggest beef exporter in the world, and particularly the state of Queensland, that produced almost half of Australia's beef and veal in 2017–2018 [11], is heavily reliant on the beef industry. Enteric fermentation in livestock represents three quarters of the agricultural GHG emissions in the form of methane and nitrous oxide, and methane production represents a significant energy loss to the animal (2 to 12% of gross energy) [3,20]. The Australian government allocated $2.55 billion to the Emissions Reduction Fund in 2018 [26]. This was to encourage livestock producers to use innovative methods to store carbon in vegetation and soils for reducing GHG. Queensland is most concerned by enteric fermentation emissions because its beef production is the largest agricultural industry in the state [13]. Its enteric fermentation coming from grazing beef cattle represents 70% of agricultural GHG emissions [3] and also represents about 80% of the overall 'cradle-to-farm gate' GHG emissions [14]. However, prolonged drought, high climate variability, low quality pastures and heavy textured soils in north Queensland constitute a challenge for beef cattle productivity characterised by the poor body condition of cattle [97]. Selection of environmentally well-adapted and vigorous legumes that can persist in the harsh climatic conditions of northern Australia is a good solution for alleviating various nutritional problems faced by livestock in this tropical part of Australia. Legumes enable an increase in animal production due to higher protein content and digestibility in comparison to native tropical grasses [10]. The roots of legumes enable ready access to deep water, introduce nitrogen in the soil and stabilize associated grasses [89]. The tropical legume, *Desmanthus*, seems to be a promising legume, due to its high DM productivity, seed production, tolerance of heavy grazing in alkaline, sodic, saline and heavy clay soils and its persistence in low rainfall environments [102]. For future studies using *Desmanthus*, it is important to keep in mind its establishment limitations on heavy soils due to its small sized seeds that can also constitute a risk for short-term pastures (<3 years) [102]. Furthermore, *Desmanthus* containing condensed tannins, showed promising results in decreasing methane emissions [68,81] and improving animal growth performance [82–85]. The legume also seems to be a good alternative for methane abatement, because it is a better natural alternative to chemical methods and concentrate supplementation [50]. However, no study has been conducted on the impact of *Desmanthus* on *in vivo* methane emissions in northern Australia. Thus, further studies should be conducted *in vivo* to test the effects of *Desmanthus* on methane emissions from supplemented live cattle in northern Australia.

Author Contributions: Conceptualization, A.E.O.M.-A., E.C., C.P.G., B.S.M.-A. and B.S.; methodology, A.E.O.M.-A., E.C., C.P.G., B.S.M.-A. and B.S.; software, A.E.O.M.-A.; validation, A.E.O.M.-A., E.C., C.P.G. and B.S.M.-A.; formal analysis, B.S.; investigation, B.S.; resources, A.E.O.M.-A., E.C., C.P.G. and B.S.M.-A.; data curation, writing—original draft preparation, B.S.; writing—reviewing and editing, A.E.O.M.-A., E.C., C.P.G. and B.S.M.-A.; supervision, A.E.O.M.-A., E.C., C.P.G. and B.S.M.-A.; project administration, A.E.O.M.-A., C.P.G; funding acquisition, A.E.O.M.-A., C.P.G. and E.C.

Funding: This research was funded by the Cooperative Research Centre for Developing Northern Australia (CRC-DNA) Projects [grant number CRC P-58599] from the Australian Government's Department of Industry, Innovation and Science, and a PhD scholarship funded by the College of Public Health, Medical and Veterinary Sciences, James Cook University, Queensland, Australia, awarded to the first named author.

Acknowledgments: The authors gratefully acknowledge James Cook University's (JCU) College of Public Health, Medical and Veterinary Sciences, the Cooperative Research Centre for Developing Northern Australia (CRC-DNA), Meat and Livestock Australia (MLA) and the Commonwealth Scientific and Industrial Research Organisation (CSIRO)-JCU-Agrimix Joint Research Project.

Conflicts of Interest: The authors declare no conflict of interest. The funders had no role in the design of the study; collection, analyses, or interpretation of data; in the writing of the manuscript, or in the decision to publish the results.

References

1. Intergovernmental Panel on Climate Change. Climate change 2007: The physical science basis. *Agenda* **2007**, *6*, 333.
2. Australia's National Greenhouse Accounts. *Quarterly Update of Australia's National Greenhouse Gas Inventory: June 2018*; Australian Government Department of the Environment and Energy: Canberra, Australia, 2018. Available online: http://www.environment.gov.au/climate-change/climate-science-data/greenhouse-gas-measurement/publications/quarterly-update-australias-national-greenhouse-gas-inventory-june-2018 (accessed on 17 March 2019).
3. Australian Greenhouse Emissions Information System. In *National Greenhouse Gas Inventory—Kyoto Protocol Classifications*; Australian Government Department of the Environment and Energy: Canberra, Australia, 2017. Available online: http://ageis.climatechange.gov.au/ (accessed on 26 July 2019).
4. United Nations. World Population Projected to Reach 9.8 Billion in 2050, and 11.2 Billion in 2100. 2017. Available online: https://www.un.org/development/desa/en/news/population/world-population-prospects-2017.html (accessed on 17 March 2019).
5. Godfray, H.C.J.; Beddington, J.R.; Crute, I.R.; Haddad, L.; Lawrence, D.; Muir, J.F.; Toulmin, C. Food security: The challenge of feeding 9 billion people. *Science* **2010**, *327*, 812–818. [CrossRef]
6. Rojas-Downing, M.M.; Nejadhashemi, A.P.; Harrigan, T.; Woznicki, S.A. Climate change and livestock: Impacts, adaptation, and mitigation. *Clim. Risk Manag.* **2017**, *16*, 145–163. [CrossRef]
7. Wright, I.A.; Tarawali, S.; Blümmel, M.; Gerard, B.; Teufel, N.; Herrero, M. Integrating crops and livestock in subtropical agricultural systems. *J. Sci. Food Agric.* **2012**, *92*, 1010–1015. [CrossRef]
8. Winks, L. Townsville stylo research at Swan's Lagoon. *Trop. Grassl.* **1973**, *7*, 201.
9. Hall, T.J.; Walker, R.W. Pasture legume adaptation to six environments of the seasonally dry tropics of north Queensland. *Trop. Grassl.* **2005**, *39*, 182–196.
10. Gardiner, C.P. Developing and commercializing new pasture legumes for clay soils in the semi-arid rangelands of northern Australia: The new Desmanthus cultivars JCU 1–5 and the Progardes story. In *Tropical Forage Legumes: Harnessing the Potential of Desmanthus and Other Genera for Heavy Clay Soils*; Lazier, J.R., Ahmad, N., Eds.; CABI: Wallingford, UK, 2016; pp. 283–304.
11. MLA. Fast Facts Australia's Beef Industry. 2018. Available online: https://www.mla.com.au/globalassets/mla-corporate/prices--markets/documents/trends--analysis/fast-facts--maps/mla_beef-fast-facts-2018.pdf (accessed on 17 March 2019).
12. Commonwealth of Australia. Emissions Reduction Fund-Overview. 2019. Available online: http://www.environment.gov.au/system/files/resources/cef73480-9dea-4436-bb15-4882345ae9a5/files/erf-factsheet-overview.pdf (accessed on 17 March 2019).
13. Bray, S.; Willcocks, J. *Net Carbon Position of the Queensland Beef Industry*; Queensland Department of Employment, Economic Development and Innovation: Brisbane, Australia, 2009.

14. Eady, S.; Viner, J.; MacDonnell, J. On-farm greenhouse gas emissions and water use: Case studies in the Queensland beef industry. *Anim. Prod. Sci.* **2011**, *51*, 667–681. [CrossRef]
15. Morgavi, D.; Forano, E.; Martin, C.; Newbold, C. Microbial ecosystem and methanogenesis in ruminants. *Animal* **2010**, *4*, 1024–1036. [CrossRef] [PubMed]
16. Huws, S.A.; Creevey, C.J.; Oyama, L.B.; Mizrahi, I.; Denman, S.E.; Popova, M.; Hess, M. Addressing global ruminant agricultural challenges through understanding the rumen microbiome: Past, present, and future. *Front. Microbiol* **2018**, *9*, 2161. [CrossRef]
17. Czerkawski, J.W. *An Introduction to Rumen Studies*; Elsevier: Amsterdam, The Netherlands, 2013.
18. Immig, I. The rumen and hindgut as source of ruminant methanogenesis. *Environ. Monit. Assess.* **1996**, *42*, 57–72. [CrossRef] [PubMed]
19. Charmley, E.; Williams, S.R.O.; Moate, P.J.; Hegarty, R.S.; Herd, R.M.; Oddy, V.H.; Hannah, M.C. A universal equation to predict methane production of forage-fed cattle in Australia. *Anim. Prod. Sci.* **2016**, *56*, 169–180. [CrossRef]
20. Johnson, K.A.; Johnson, D.E. Methane emissions from cattle. *J. Anim. Sci.* **1995**, *73*, 2483–2492. [CrossRef] [PubMed]
21. Gavrilova, O.; Leip, A.; Dong, H.; MacDonald, J.D.; Alfredo, C.; Bravo, G.; Widiawati, Y. 2019 Refinement to the 2006 IPCC Guidelines for National Greenhouse Gas Inventories. Available online: https://www.ipcc-nggip.iges.or.jp/public/2019rf/index.html (accessed on 26 July 2019).
22. Charmley, E.; Stephens, M.L.; Kennedy, P.M. Predicting livestock productivity and methane emissions in northern Australia: Development of a bio-economic modelling approach. *Aust. J. Exp. Agric.* **2008**, *48*, 109–113. [CrossRef]
23. Arthur, P.; Archer, J.; Johnston, D.; Herd, R.; Richardson, E.; Parnell, P. Genetic and phenotypic variance and covariance components for feed intake, feed efficiency, and other postweaning traits in Angus cattle. *J. Anim. Sci.* **2001**, *79*, 2805–2811. [CrossRef] [PubMed]
24. Exton, S.; Herd, R.; Davies, L.; Archer, J.; Arthur, P. Commerical Benefits to the Beef Industry from Genetic Improvement in Net Feed Efficiency. *Asian Australas. J. Anim.* **2000**, *13*, 338–341.
25. Eady, S.J. *Undertaking a Life Cycle Assessment for the Livestock Export Trade*; Meat & Livestock Australia Limited: North Sydney, Australia, 2011.
26. Pengelly, B.C.; Conway, M.J. Pastures on cropping soils: Which tropical pasture legume to use? *Trop. Grassl.* **2000**, *34*, 162–168.
27. Tothill, J.C.; Gillies, C. *The Pasture Lands of Northern Australia: Their Condition, Productivity and Sustainability*; Tropical Grasslands Society of Australia: St Lucia, Australia, 1992.
28. Hattersley, P. The distribution of C 3 and C 4 grasses in Australia in relation to climate. *Oecologia* **1983**, *57*, 113–128. [CrossRef] [PubMed]
29. Perry, L.; Al Jassim, R.; Gaughan, J.; Tomkins, N. Effect of feeding forage characteristic of wet-or dry-season tropical C4 grass in northern Australia, on methane production, intake and rumen outflow rates in Bos indicus steers. *Anim. Prod. Sci.* **2017**, *57*, 2033–2041. [CrossRef]
30. Hennessy, D.; Williamson, P.; Nolan, J.; Kempton, T.; Leng, R. The roles of energy-or protein-rich supplements in the subtropics for young cattle consuming basal diets that are low in digestible energy and protein. *J. Agric. Sci.-Camb.* **1983**, *100*, 657–666. [CrossRef]
31. McLennan, S.R. *Developing Profitable Strategies for Increasing Growth Rates of Cattle Grazing Tropical Pastures*; Meat & Livestock Australia Limited: North Sydney, Australia, 1997; Project DAQ, 100.
32. Shaw, N.; Bisset, W. Characteristics of a bunch spear grass (Heteropogon contortus (L). BEAUV.) pasture grazed by cattle in subtropical Queensland. *Aust. J. Agric. Res.* **1955**, *6*, 539–552. [CrossRef]
33. Poppi, D.P.; McLennan, S.R. Nutritional research to meet future challenges. *Anim. Prod. Sci.* **2010**, *50*, 329–338. [CrossRef]
34. Poppi, D.P.; McLennan, S.R. Protein and energy utilization by ruminants at pasture. *J. Anim. Sci.* **1995**, *73*, 278–290. [CrossRef] [PubMed]
35. Bortolussi, G.; McIvor, J.G.; Hodgkinson, J.; Coffey, S.; Holmes, C. The northern Australian beef industry, a snapshot. 3. Annual liveweight gains from pasture based systems. *Aust. J. Agric. Res.* **2005**, *45*, 1093–1108. [CrossRef]

36. Poppi, D.P.; McLennan, S.R. Nutrition R&D: Past, present and future. In Proceedings of the Northern Beef Update Research Conference, Southbank Convention Centre, Townsville, Australia, 21–22 March 2007; Pattie, B., Restall, B., Eds.;
37. Archimède, H.; Eugène, M.; Magdeleine, C.M.; Boval, M.; Martin, C.; Morgavi, D.; Doreau, M. Comparison of methane production between C3 and C4 grasses and legumes. *Anim. Feed Sci. Technol.* **2011**, *166*, 59–64. [CrossRef]
38. Benchaar, C.; Pomar, C.; Chiquette, J. Evaluation of dietary strategies to reduce methane production in ruminants: A modelling approach. *Can. J. Anim. Sci.* **2001**, *81*, 563–574. [CrossRef]
39. Joblin, K. Ruminal acetogens and their potential to lower ruminant methane emissions. *Aust. J. Agric. Res.* **1999**, *50*, 1307–1314. [CrossRef]
40. Broucek, J. Options to methane production abatement in ruminants: A review. *J. Anim. Plant Sci.* **2018**, *28*, 348–364.
41. Dohme, F.; Machmüller, A.; Estermann, B.; Pfister, P.; Wasserfallen, A.; Kreuzer, M. The role of the rumen ciliate protozoa for methane suppression caused by coconut oil. *Lett. Appl. Microbiol.* **1999**, *29*, 187–192. [CrossRef]
42. Mathison, G.; Okine, E.; McAllister, T.; Dong, Y.; Galbraith, J.; Dmytruk, O. Reducing methane emissions from ruminant animals. *J. Appl. Anim. Res.* **1998**, *14*, 1–28. [CrossRef]
43. Iqbal, M.F.; Cheng, Y.-F.; Zhu, W.-Y.; Zeshan, B. Mitigation of ruminant methane production: current strategies, constraints and future options. *World J. Microb. Biot.* **2008**, *24*, 2747–2755. [CrossRef]
44. Guan, H.; Wittenberg, K.; Ominski, K.; Krause, D. Efficacy of ionophores in cattle diets for mitigation of enteric methane. *J. Anim. Sci.* **2006**, *84*, 1896–1906. [CrossRef] [PubMed]
45. McCaughey, W.; Wittenberg, K.; Corrigan, D. Methane production by steers on pasture. *Can. J. Anim. Sci.* **1997**, *77*, 519–524. [CrossRef]
46. Russell, J.B.; Houlihan, A.J. Ionophore resistance of ruminal bacteria and its potential impact on human health. *FEMS Microbiol. Rev.* **2003**, *27*, 65–74. [CrossRef] [PubMed]
47. European Medicines Agency Veterinary Medicines and Inspections. Committee for Medical Products for Veterinary Use. 2007. Available online: https://www.ema.europa.eu/documents/mrl-report/monensin-cattle-including-dairy-cows-summary-report-committee-veterinary-medicinal-products_en.pdf (accessed on 17 March 2019).
48. Newbold, C.; McIntosh, F.; Wallace, R. Changes in the microbial population of a rumen-simulating fermenter in response to yeast culture. *Can. J. Anim. Sci.* **1998**, *78*, 241–244. [CrossRef]
49. McGinn, S.; Beauchemin, K.; Coates, T.; Colombatto, D. Methane emissions from beef cattle: Effects of monensin, sunflower oil, enzymes, yeast, and fumaric acid. *J. Anim. Sci.* **2004**, *82*, 3346–3356. [CrossRef] [PubMed]
50. Martin, C.; Morgavi, D.; Doreau, M. Methane mitigation in ruminants: From microbe to the farm scale. *Animal* **2010**, *4*, 351–365. [CrossRef] [PubMed]
51. Martinez-Fernandez, G.; Denman, S.E.; Yang, C.; Cheung, J.; Mitsumori, M.; McSweeney, C.S. Methane inhibition alters the microbial community, hydrogen flow, and fermentation response in the rumen of cattle. *Front. Microbiol.* **2016**, *7*, 1122. [CrossRef] [PubMed]
52. McCrabb, G.; Berger, K.; Magner, T.; May, C.; Hunter, R. Inhibiting methane production in Brahman cattle by dietary supplementation with a novel compound and the effects on growth. *Aust. J. Agric. Res.* **1997**, *48*, 323–329. [CrossRef]
53. Van Nevel, C.; Demeyer, D. Control of rumen methanogenesis. *Environ. Monit. Assess.* **1996**, *42*, 73–97. [CrossRef]
54. Martinez Fernandez, G.; Duval, S.M.; Kindermann, M.; Schirra, H.J.; Denman, S.E.; McSweeney, C.S. 3-NOP vs. Halogenated compound: Methane production, ruminal fermentation and microbial community response in forage fed cattle. *Front. Microbiol.* **2018**, *9*, 1582. [CrossRef] [PubMed]
55. Supplementary Feeding. 2019. Available online: https://www.mla.com.au/Research-and-development/Feeding-finishing-nutrition/Supplementary-feeding (accessed on 17 March 2019).
56. Department of Agriculture Forestry and Fisheries. Supplementation Feeding: Some Basic Considerations. 2019. Available online: https://www.daf.qld.gov.au/business-priorities/agriculture/disaster-recovery/drought/managing/supplementation-feeding-considerations (accessed on 17 March 2019).

57. McLennan, S.R.; Dunster, P.; O'rourke, P.; Murphy, G. Comparison of dry season urea supplements containing salt, sulfur or molasses for steers grazing native pasture in the dry tropics of northern Queensland. *Aust. J. Exp. Agric.* **1981**, *21*, 457–463. [CrossRef]
58. Purnomoadi, A.; Rianto, E.; Kurihara, M. Reduction of methane production from Ongole Crossbreed cattle in Indonesia by increasing the concentrate feeding frequency. In *Greenhouse Gas Control Technologies*; Elsevier: Amsterdam, The Netherlands, 2005; Volume 7, pp. 2513–2516.
59. Beauchemin, K.A.; Kreuzer, M.; O'mara, F.; McAllister, T.A. Nutritional management for enteric methane abatement: A review. *Aust. J. Exp. Agric.* **2008**, *48*, 21–27. [CrossRef]
60. McSweeney, C.; Palmer, B.; McNeill, D.; Krause, D. Microbial interactions with tannins: Nutritional consequences for ruminants. *Anim. Feed Sci. Techol.* **2001**, *91*, 83–93. [CrossRef]
61. Woodward, S.; Waghorn, G.; Thomson, N. Supplementing dairy cows with oils to improve performance and reduce methane-does it work? In Proceedings of the New Zealand Society of Animal Production, Napier, New Zealand, 26–28 June 2006; Volume 66, pp. 176–181.
62. Patra, A.K.; Saxena, J. Exploitation of dietary tannins to improve rumen metabolism and ruminant nutrition. *J. Sci. Food Agric.* **2011**, *91*, 24–37.
63. Hess, H.D.; Tiemann, T.T.; Noto, F.; Carulla, J.E.; Kreuzer, M. Strategic use of tannins as means to limit methane emission from ruminant livestock. In *International Congress Series*; Elsevier: Amsterdam, The Netherlands, 2006; Volume 1293, pp. 164–167.
64. Tiemann, T.T.; Lascano, C.E.; Wettstein, H.-R.; Mayer, A.C.; Kreuzer, M.; Hess, H.D. Effect of the tropical tannin-rich shrub legumes Calliandra calothyrsus and Flemingia macrophylla on methane emission and nitrogen and energy balance in growing lambs. *Animal* **2008**, *2*, 790–799. [CrossRef]
65. Tan, H.; Sieo, C.; Abdullah, N.; Liang, J.; Huang, X.; Ho, Y. Effects of condensed tannins from Leucaena on methane production, rumen fermentation and populations of methanogens and protozoa *in vitro*. *Anim. Feed Sci. Techol.* **2011**, *169*, 185–193. [CrossRef]
66. Brewbaker, J.L. Leucaena: A multipurpose tree genus for tropical agroforestry. In *Agroforestry: A Decade of Development*; Steppler, H.A., Ramachandran Nair, P.K., Eds.; International Council for Research in Agroforestry: Nairobi, Kenya, 1987; pp. 289–323.
67. Dalzell, S.; Burnett, D.; Dowsett, J.; Forbes, V.; Shelton, H. Prevalence of mimosine and DHP toxicity in cattle grazing Leucaena leucocephala pastures in Queensland, Australia. *Anim. Prod. Sci.* **2012**, *52*, 365–372. [CrossRef]
68. Durmic, Z.; Ramírez-Restrepo, C.A.; Gardiner, C.; O'Neill, C.J.; Hussein, E.; Vercoe, P.E. Differences in the nutrient concentrations, *in vitro* methanogenic potential and other fermentative traits of tropical grasses and legumes for beef production systems in northern Australia. *J. Sci. Food Agric.* **2017**, *97*, 4075–4086. [CrossRef] [PubMed]
69. Soltan, Y.A.; Morsy, A.S.; Sallam, S.M.; Lucas, R.C.; Louvandini, H.; Kreuzer, M.; Abdalla, A.L. Contribution of condensed tannins and mimosine to the methane mitigation caused by feeding *Leucaena leucocephala*. *Arch. Anim. Nutr.* **2013**, *67*, 169–184. [CrossRef] [PubMed]
70. Harrison, M.T.; McSweeney, C.; Tomkins, N.W.; Eckard, R.J. Improving greenhouse gas emissions intensities of subtropical and tropical beef farming systems using *Leucaena leucocephala*. *Agric. Syst.* **2015**, *136*, 138–146. [CrossRef]
71. Soltan, Y.; Morsy, A.; Sallam, S.; Louvandini, H.; Abdalla, A. Comparative *in vitro* evaluation of forage legumes (*prosopis*, *acacia*, *atriplex*, and *leucaena*) on ruminal fermentation and methanogenesis. *J. Anim. Feed Sci.* **2012**, *21*, 759–772. [CrossRef]
72. Jones, W.T.; Mangan, J.L. Complexes of the condensed tannins of sainfoin (*Onobrychis viciifolia* Scop.) with fraction 1 leaf protein and with submaxillary mucoprotein, and their reversal by polyethylene glycol and pH. *J. Sci. Food Agric.* **1977**, *28*, 126–136. [CrossRef]
73. McSweeney, C.; Palmer, B.; Bunch, R.; Krause, D. *In vitro* quality assessment of tannin-containing tropical shrub legumes: Protein and fibre digestion. *Anim. Feed Sci. Techol.* **1999**, *82*, 227–241. [CrossRef]
74. Bhatta, R.; Uyeno, Y.; Tajima, K.; Takenaka, A.; Yabumoto, Y.; Nonaka, I.; Kurihara, M. Difference in the nature of tannins on *in vitro* ruminal methane and volatile fatty acid production and on methanogenic archaea and protozoal populations. *J. Dairy Sci.* **2009**, *92*, 5512–5522. [CrossRef] [PubMed]
75. Beauchemin, K.; McGinn, S.; Martinez, T.; McAllister, T. Use of condensed tannin extract from quebracho trees to reduce methane emissions from cattle. *J. Anim. Sci.* **2007**, *85*, 1990–1996. [CrossRef] [PubMed]

76. Field, J.; Kortekaas, S.; Lettinga, G. The tannin theory of methanogenic toxicity. *Biol. Waste* **1989**, *29*, 241–262. [CrossRef]
77. Jayanegara, A.; Goel, G.; Makkar, H.P.; Becker, K. Divergence between purified hydrolysable and condensed tannin effects on methane emission, rumen fermentation and microbial population *in vitro*. *Anim. Feed Sci. Techol.* **2015**, *209*, 60–68. [CrossRef]
78. Murdiati, T.B.; McSweeney, C.S.; Lowry, J. Complexing of toxic hydrolysable tannins of yellow-wood (*Terminalia oblongata*) and harendong (*Clidemia hirta*) with reactive substances: An approach to preventing toxicity. *J. Appl. Toxicol.* **1991**, *11*, 333–338. [CrossRef]
79. Makkar, H.; Francis, G.; Becker, K. Bioactivity of phytochemicals in some lesser-known plants and their effects and potential applications in livestock and aquaculture production systems. *Animal* **2007**, *1*, 1371–1391. [CrossRef] [PubMed]
80. McMahon, L.R.; McAllister, T.A.; Berg, B.P.; Majak, W.; Acharya, S.N.; Popp, J.D.; Cheng, K.-J. A review of the effects of forage condensed tannins on ruminal fermentation and bloat in grazing cattle. *Can. J. Anim. Sci.* **2000**, *80*, 469–485. [CrossRef]
81. Vandermeulen, S.; Singh, S.; Ramírez-Restrepo, C.A.; Kinley, R.D.; Gardiner, C.P.; Holtum, J.A.; Bindelle, J. *In vitro* assessment of ruminal fermentation, digestibility and methane production of three species of Desmanthus for application in northern Australian grazing systems. *Crop Pasture Sci.* **2018**, *69*, 797–807. [CrossRef]
82. Collins, J.; Gardiner, C.P.; Kempe, N.; Hannah, I. Successful Pasture Development at Cungelella: A grazier, a researcher and a seed company's perspective. In Proceedings of the Northern Beef Research Update Conference, Rockhampton, Australia, 15–18 August 2016; North Australia Beef Research Council: Gympie, Australia, 2016.
83. Gardiner, C.P.; Parker, A. Steer liveweight gains on ProgardesTM/buffel pastures in Qld. In Proceedings of the 2nd Australian and New Zealand Societies of Animal Production Joint Conference, Lincoln University, Christchurch, New Zealand, 2–5 July 2012.
84. Ngo, T.; Parker, A.; Gardiner, C.P. The effects of diet preference on feed intake, digestibility and nitrogen balance of sheep given Flinders grass (*Iseilema* spp.) hay and/or Desmanthus leptophyllus cv. JCU 1 *ad libitum*. In Proceedings of the TropAg2017: International Tropical Agriculture Conference, Brisbane, Australia, 20–22 November 2017.
85. Rangel, J.; Gardiner, C.P. Stimulation of wool growth by *Desmanthus* spp. as a supplement to a diet of Mitchell grass hay. *Trop. Grassl.* **2009**, *43*, 106–111.
86. Gardiner, C.; Kempe, N.; Hannah, I. New pasture legumes for clay soils in dry environments. *North. Muster* **2012**, *30*, 27–28.
87. Wetselaar, R. Estimation of nitrogen fixation by four legumes in a dry monsoonal area of north-western Australia. *Aust. J. Exp. Agric.* **1967**, *7*, 518–522. [CrossRef]
88. Jones, R.; Probert, M.; Dalgliesh, N.; McCown, R. Nitrogen inputs from a pasture legume in rotations with cereals in the semi-arid tropics of northern Australia: Experimentation and modelling on a clay loam soil. *Aust. J. Exp. Agric.* **1996**, *36*, 985–994. [CrossRef]
89. Zahran, H.H. Rhizobium-legume symbiosis and nitrogen fixation under severe conditions and in an arid climate. *Microbiol. Mol. Biol. Rev.* **1999**, *63*, 968–989. [PubMed]
90. Peck, G.; Buck, S.; Hoffmann, A.; Holloway, C.; Johnson, B.; Lawrence, D.; Paton, C. *Review of Productivity Decline in Sown Grass Pastures*; Meat & Livestock Australia Limited: North Sydney, Australia, 2011; p. 86.
91. Sturz, A.; Christie, B.; Matheson, B.; Nowak, J. Biodiversity of endophytic bacteria which colonize red clover nodules, roots, stems and foliage and their influence on host growth. *Biol. Fertil. Soils* **1997**, *25*, 13–19. [CrossRef]
92. Rao, I.M.; Peters, M.; Castro, A.; Schultze-Kraft, R.; White, D.; Fisher, M.; Rudel, T. LivestockPlus: The sustainable intensification of forage-based agricultural systems to improve livelihoods and ecosystem services in the tropics. *Trop. Grassl.* **2015**, *3*, 59–82. [CrossRef]
93. Peters, M.; Herrero, M.; Fisher, M.; Erb, K.-H.; Rao, I.; Subbarao, G.V.; Searchinger, T. Challenges and opportunities for improving eco-efficiency of tropical forage-based systems to mitigate greenhouse gas emissions. *Trop. Grassl.* **2013**, *1*, 156–167.

94. Mero, R.N.; Udén, P. Promising tropical grasses and legumes as feed resources in Central Tanzania V. Effect of supplementing Cenchrus ciliaris hay with leaves from four legumes on intake and digestibility by growing Mpwapwa bulls. *Anim. Feed Sci. Techol.* **1998**, *70*, 111–122. [CrossRef]
95. Bowen, M.; Chudleigh, F.; Buck, S.; Hopkins, K. Productivity and profitability of forage options for beef production in the subtropics of northern Australia. *Anim. Prod. Sci.* **2016**, *58*, 332–342. [CrossRef]
96. Gardiner, C.P.; Swan, S.J. Abandoned pasture legumes offer potential economic and environmental benefits in semiarid clay soil rangelands. In Proceedings of the 15th Biennial Conference Proceedings 'A Climate of Change in the Rangelands', Charters Towers, Australia, 28 September–2 October 2008.
97. CSIRO. *Nutrient Requirements of Domesticated Ruminants*; CSIRO publishing: Clayton, VIC, Australia, 2007.
98. Clements, R.J.; Henzell, E.F. Pasture research and development in northern Australia: An ongoing scientific adventure. *Trop. Grassl.* **2010**, *44*, 221–230.
99. Jones, R.; Lowry, J. Australian goats detoxify the goitrogen 3-hydroxy-4 (1H) pyridone (DHP) after rumen infusion from an Indonesian goat. *Experientia* **1984**, *40*, 1435–1436. [CrossRef]
100. Pengelly, B.C.; Liu, C.J. Genetic relationships and variation in the tropical mimosoid legume Desmanthus assessed by random amplified polymorphic DNA. *Genet. Resour. Crop Evol.* **2001**, *48*, 93–101. [CrossRef]
101. Luckow, M. Monograph of Desmanthus (*leguminosae-mimosoideae*). In *Systematic Botany Monographs*; Anderson, C., Ed.; The American Society of Plant Taxonomists: Laramie, WY, USA, 1993; Volume 38, pp. 1–166.
102. Tropical Forages. Desmanthus Leptophyllus. Available online: http://www.tropicalforages.info/key/forages/Media/Html/entities/desmanthus_leptophyllus.htm (accessed on 26 July 2019).
103. State of the Environment. Average Rainfall. 2017. Available online: https://www.stateoftheenvironment.des.qld.gov.au/2015/climate/climate-observations/average-rainfall (accessed on 26 July 2019).
104. Jones, R.; Brandon, N. Persistence and productivity of eight accessions of *Desmanthus virgatus* under a range of grazing pressures in subtropical Queensland. *Trop. Grassl.* **1998**, *32*, 145–152.
105. Department of Agriculture Forestry and Fisheries. *Pasture Management for the Inland Burnett*, 3rd ed.; DAFF: Brisbane, Australia, 2014.
106. Gardiner, C.P.; Kempe, N.; Hannah, I.; McDonald, J. PROGARDES TM: A legume for tropical/subtropical semi-arid clay soils. *Trop. Grassl.* **2013**, *1*, 78–80. [CrossRef]
107. Pandrangi, S.; Elwell, M.; Anantheswaran, R.; LaBorde, L. Efficacy of sulfuric acid scarification and disinfectant treatments in eliminating Escherichia coli O157: H7 from alfalfa seeds prior to sprouting. *J. Food Sci.* **2003**, *68*, 613–617. [CrossRef]
108. Clem, B. Desmanthus. 2009. Available online: https://keys.lucidcentral.org/keys/v3/pastures/Html/Desmanthus.htm (accessed on 26 July 2019).
109. Gardiner, C.; Wright, C.; Coventry, M. The germination, passage and viability of *Desmanthus virgatus* (L.) Willenow seed through sheep and its implication for dispersal in tropical rangelands. In Proceedings of the 16th Australian Society of Agronomy Conference, Armidale, NSW, Australia, 14–18 October 2012.
110. Kanani, J.; Lukefahr, S.; Stanko, R. Evaluation of tropical forage legumes (*Medicago sativa*, *Dolichos lablab*, *Leucaena leucocephala* and *Desmanthus bicornutus*) for growing goats. *Small Rumin. Res.* **2006**, *65*, 1–7. [CrossRef]

animals MDPI

Article

Effect of Encapsulated Nitrate and Microencapsulated Blend of Essential Oils on Growth Performance and Methane Emissions from Beef Steers Fed Backgrounding Diets

Aklilu W. Alemu [1], Atmir Romero-Pérez [1,†], Rafael C. Araujo [2] and Karen A. Beauchemin [1,*]

1 Lethbridge Research and Development Centre, Agriculture and Agri-Food Canada, Lethbridge, AB T1J 4B1, Canada; aklilu.alemu@canada.ca (A.W.A.); atmir@umam.mx (A.R.-P.)
2 GRASP Ind. & Com. LTDA, Curitiba, Paraná, Brazil 81260-000/EW | Nutrition GmbH, 49429 Visbek, Germany; rafael@grasp.ind.br
* Correspondence: karen.beauchemin@canada.ca; Tel.: +1-403-317-2235
† Current address: Departamento de Nutrición Animal y Bioquímica, Universidad Nacional Autónoma de México, Ciudad de Mexico 04510, Mexico.

Received: 3 December 2018; Accepted: 1 January 2019; Published: 10 January 2019

Simple Summary: The use of supplemental dietary nitrate (NO_3^-) to minimize enteric methane (CH_4) emissions from ruminants is hindered by potential toxicity effects. In the current study, the potential effects of feeding encapsulated NO_3^- (EN), microencapsulated blend of essential oils (MBEO), and their combination on growth performance and enteric CH_4 emissions of beef cattle were evaluated. There was no interaction effect between feeding EN and MBEO on CH_4 emissions and the presence of MBEO did not affect the potential of EN to reduce CH_4. Feeding MBEO increased CH_4 emissions without affecting animal performance. Inclusion of EN as a replacement for urea reduced CH_4 emissions without incurring any adverse effects on cattle health and performance.

Abstract: A long-term study (112 days) was conducted to examine the effect of feeding encapsulated nitrate (NO_3^-), microencapsulated blend of essential oils (EO), and their combination on growth performance, feeding behavior, and enteric methane (CH_4) emissions of beef cattle. A total of 88 crossbred steers were purchased and assigned to one of four treatments: (i) control, backgrounding high-forage diet supplemented with urea (1.17% in dietary DM); (ii) encapsulated NO_3^- (EN), control diet supplemented with 2.5% encapsulated NO_3^- as a replacement for urea (1.785% NO_3^- in the dietary DM); (iii) microencapsulated blend of EO (MBEO), control diet supplemented with 150 mg/kg DM of microencapsulated blend of EO and pepper extract; and (iv) EN + MBEO, control diet supplemented with EN and MBEO. There was no interaction ($p \geq 0.080$) between EN and MBEO on average dry matter intake (DMI), average daily gain (ADG), gain to feed ratio (G:F), feeding behavior, and CH_4 emission (using GreenFeed system), implying independent effects of feeding EN and MBEO. Feeding MBEO increased CH_4 production (165.0 versus 183.2 g/day; $p = 0.005$) and yield (18.9 versus 21.4 g/kg DMI; $p = 0.0002$) but had no effect ($p \geq 0.479$) on average DMI, ADG, G:F, and feeding behavior. However, feeding EN had no effect on ADG and G:F ($p \geq 0.119$) but reduced DMI (8.9 versus 8.4 kg/day; $p = 0.003$) and CH_4 yield (21.5 versus 18.7 g/kg DMI; $p < 0.001$). Feeding EN slowed ($p = 0.001$) the feeding rate (g of DM/min) and increased ($p = 0.002$) meal frequency (events/day). Our results demonstrate that supplementing diets with a blend of EO did not lower CH_4 emissions and there were no advantages of feeding MBEO with EN. Inclusion of EN as a replacement for urea reduced CH_4 emissions but had no positive impact on animal performance.

Keywords: backgrounded cattle; encapsulated nitrate; essential oil; methane

1. Introduction

Over the past decades, livestock research has been focused on developing strategies to reduce the environmental impacts of ruminant animals [1]. As enteric methane (CH_4) emission is the major contributor of total emissions in ruminant farming, different mitigation strategies including feed additives (e.g., inhibitors, ionophores, plant bioactive compounds, electron receptors, dietary lipids), feed (e.g., high starch grains, lipids), and feeding management (e.g., forage quality and management, feed processing, feeding frequency, precision feeding) have been directed towards minimizing enteric CH_4 emissions [1].

Feeding nitrate (NO_3^-) to ruminant animals as a replacement for urea has received attention as a promising methane-mitigating approach, as several studies have shown that feeding NO_3^- can decrease enteric CH_4 [2–7]. Similarly, a recent in vitro experiment [8] and metabolism study using beef heifers [9] at our lab were also in line with the previous reports. Conversely, reduction in enteric CH_4 was not observed from feedlot animals managed outdoors and supplemented with encapsulated NO_3^- at 1.25 and 2.5% on a dry matter (DM) basis [10,11]. Furthermore, despite its positive effects on CH_4 reduction, feeding NO_3^- could pose a potential risk of NO_3^-/nitrite (NO_2^-) toxicity to animals. Nitrate intoxication can occur when the concentration of NO_2^- (reduced form of NO_3^-) accumulates in the rumen and is absorbed into the blood stream, increasing methemoglobin (MetHb) level. When ample hemoglobin (Hb) is converted to MetHb, the animal suffers from oxygen starvation [12]. A slow release form of NO_3^- (encapsulated NO_3^-) was developed to ensure the slow release of NO_3^- to rumen microbes and minimize potential toxicity [7–11].

Bacterial resistance to multiple antibiotics is a worldwide health problem. As such, following the prohibition of the use of growth-promoting antibiotics in animal feeds by the European Union (1831/2003) [13], interest in the use of essential oils (EO) as potential alternatives to antibiotics and studying their effects and mechanisms on ruminal fermentation has been the focus of livestock research [14,15]. Large numbers of in vitro and in vivo studies have investigated the potential effects of EO on modifying rumen function [14–16]. However, the mode of action remain poorly understood [17]. Furthermore, in addition to its impact on rumen function, EO have been shown to have antioxidant, anti-inflammatory, immune modulation, mucolytic, as well as thermoregulation and blood oxygenation properties [14,18,19] and minimize stress in feedlot cattle [20]. These impacts have not been studied in detail yet. Hori et al. [21] reported that capsaicin (an alkaloid from chili pepper) increased peripheral blood flow with positive impact on body thermoregulation. Recently, Silva et al. [19,22] reported that the use of a blend of EO (Activo® Premium) increased milk efficiency, digestible organic matter intake, and O_2 saturation of Hb in dairy cows. The improved oxygenation of blood may be beneficial for cattle fed NO_3^-. Using the same product for sheep, Soltan et al. [23] reported a reduction in CH_4 emissions without affecting dry matter intake (DMI) and nutrient digestibility. Overall, research on the effects of EO in beef cattle diets is fairly limited [17,24].

Therefore, the current study aimed to explore the effects of feeding encapsulated NO_3^- (EN) and a microencapsulated blend of EO (MBEO) alone or in combination on feed consumption and behavior, animal performance, and enteric CH_4 emissions from feedlot beef steers fed a high-forage diet. Because the mode of action of NO_3^- and EO differ, we hypothesized that feeding NO_3^- in combination with EO would improve animal performance and reduce enteric CH_4 production.

2. Materials and Methods

All experimental procedures were reviewed and approved by the Animal Care and Use Committee at the Lethbridge Research and Development Centre (ACC 1626) under the guidelines of the Canadian Council on Animal Care [25] and the Veterinary Drug Directorate of Health Canada (DSTS No. 197834).

2.1. Animals and Experimental Design

A total of 88 crossbred steers (mean arrival BW of 287 ± 19 kg) were purchased from the local auction market. The experiment was conducted as a completely randomized design in a 2 × 2 factorial arrangement of treatments. A total of 22 animals per treatment were assigned and housed in four large adjacent pens (17 × 12.7 m; 10 m^2 per animal). The four treatments (Table 1) were: (i) control, a typical backgrounding high-forage diet (800 g/kg DM corn silage) supplemented with urea (1.17% in dietary DM); (ii) EN, control diet supplemented with 2.5% encapsulated calcium ammonium NO_3^- in dietary DM providing 1.785% NO_3^- in the dietary DM (GRASP Ind. & Com. LTDA, Curitiba, Brazil); (iii) MBEO, control diet supplemented with 150 mg/kg DM of commercial microencapsulated blend of EO and pepper extract (Activo® Premium, GRASP Ind. & Com. LTDA, Curitiba, Brazil); and (iv) EN + MBEO, control diet supplemented with 2.5% encapsulated calcium ammonium NO_3^- in dietary DM and 150 mg/kg in the dietary DM of MBEO. The commercial blend of EO was a blend of natural and identical to natural terpenoids (carvacrol), phenylpropanoids (cinnamaldehyde and eugenol), and alkaloids (capsaicin from capsicum oleoresin) and fed to the animals according to the manufacturer's recommended level. It was mixed with ground barley before feeding, and the blend was fed at the rate of 75 g/day to provide the full dose starting day 1. Encapsulated NO_3^- was added directly into the total mixed ration (TMR) daily and contained 85.6% DM, 17.6% N, 19.6% Ca, and 71.4% NO_3^- on a DM basis. Diets were formulated to be isonitrogenous, although chemical analysis indicated that the TMR containing EN were slightly lower in crude protein (CP) content (13.1 versus 14.3% DM; Table 1). The TMR were offered twice daily at 0900 h and 1600 h. Due to the high amount of Ca in encapsulated NO_3^-, the concentration of limestone was reduced in the EN and EN + MBEO diets to provide a similar Ca level across the diets.

The experiment was conducted over a total of 112 days (28 days adaptation and 84 days of measurement), with the measurement period conducted in three consecutive periods of four weeks. In order to avoid the risk of intoxication, animals that received diets containing encapsulated NO_3^- were acclimatized gradually using a step-up protocol during the first 28 days of adaptation; 0.625%, 1.25%, 1.875%, and 2.5% NO_3^- in dietary DM. Each pen was equipped with five automated feeding stations (GrowSafe System Ltd., Airdrie, AB, Canada) to measure individual daily feed intake and feeding behavior. Animals were fitted with radio-frequency identification (RFID) ear tags to record feeding events of individual animals. Standard feedlot management procedures were implemented. Pens were bedded with straw and animals were implanted with steroids following the Standard Operating Procedure (SOP code: GEN. 1001) at Lethbridge Research and Development Centre. However, ionophores and antibiotics for liver abscess control were not added to the diets.

Table 1. Feed ingredients and chemical composition of the experimental diets with no additives (control, −EN, and −MBEO), or supplemented with encapsulated nitrate (+EN), microencapsulated blend of essential oils (+MBEO), and combination of EN and MBEO (+EN +MBEO).

Item	−EN [1]		+EN	
	−MBEO [1]	+MBEO	−MBEO	+MBEO
Ingredients, % of dry matter (DM)				
Corn silage [2]	80	80	80	80
Barley grain, dry rolled [3]	10	10	10	10
Supplement	10	9.99	7.5	7.48
Canola meal	3.70	3.70	3.70	3.70
Limestone	1.55	1.55	0.43	0.43
Salt (NaCl)	0.11	0.11	0.11	0.11
Urea	1.17	1.17	0.24	0.24
LeRDC beef feedlot premix [4]	0.05	0.05	0.05	0.05
Molasses, dried	0.05	0.05	0.05	0.05
Barley ground	3.31	3.31	2.85	2.85
Canola oil	0.07	0.05	0.07	0.06
EN	0.0	0.0	2.5	2.5
MBEO	0.0	0.015	0.0	0.015

Table 1. *Cont.*

Item	−EN [1]		+EN	
	−MBEO [1]	+MBEO	−MBEO	+MBEO
Chemical composition (% of DM)				
DM (as-is)	44.0	44.0	43.8	43.8
OM	93.9	93.9	92.5	92.5
CP	14.3	14.3	13.1	13.1
NDF	42.0	42.0	41.2	41.1
ADF	28.1	28.1	27.9	27.9
Starch	29.8	29.8	29.3	29.3
NO_3^-	0.12	0.16	1.66	1.68
GE (Mcal/kg DM) [5]	5.39	5.39	5.32	5.29

DM = dry matter; OM = organic matter; CP = crude protein; NDF = neutral detergent fiber; ADF = acid detergent fiber; NO_3^- = nitrate. [1] EN = Encapsulated nitrate (EN) was manufactured by GRASP Ind. & Com. LTDA, Curitiba, Paraná, Brazil; DM, 85.6%; N, 17.6%; Ca, 19.6%; and NO_3^-, 71.4% on a DM basis. The source of nitrate was the double salt of calcium ammonium nitrate decahydrate [$5Ca(NO_3)_2 \bullet NH_4NO_3 \bullet 10H_2O$]; MBEO = Commercial microencapsulated blend of natural and identical to natural terpenoids (carvacrol), phenylpropanoids (cinnamaldehyde and eugenol), and alkaloids (capsaicin from capsicum oleoresin) manufactured by GRASP Ind. & Com. LTDA, Curitiba, Paraná, Brazil. [2] DM, 32% (±1.3 SD) on as-is basis and OM, 96% (±0.1 SD); CP, 8% (±0.2 SD); NDF, 47.3% (±2.1 SD); ADF, 32.8% (±3.8 SD); starch, 28% (±3.6 SD) on a DM basis. [3] OM, 98% (±0.5 SD); CP, 14% (±0.8 SD); NDF, 16.6% (±1.8 SD); starch, 55% (±1.6 SD) on a DM basis. [4] Lethbridge Research and Development Centre (LeRDC) beef feedlot vitamin-mineral premix contained (on a DM basis) $CaCO_3$, 34.83%; $ZnSO_4$, 28.37%; $CuSO_4$, 10.31%; ethylenediamine dihydriodide (80% concentration), 0.15%; selenium 1% (10,000 mg Se/kg, Na_2SeO_3), 5.04%; $CoCO_3$, 0.08%; $MnSO_4$, 14.61%; vitamin A (500,000,000 IU/kg), 1.72%; vitamin D (500,000,000 IU/kg), 0.17%; and vitamin E (500,000 IU/kg), 4.73%. [5] Gross energy (GE, Mcal/kg DM); corn silage, 5.60 (±0.35 SD); dry rolled barley grain, 4.91 (±0.05 SD); control and MBEO supplement, 4.24 (±0.15 SD); and EN supplement, 4.69 (±0.18 SD).

2.2. Sample Collection

Body weight (BW) was measured before feeding (nonfasted BW) on 2 consecutive days at the start and end of the experiment and once at the end of each period (4 weeks) to calculate average daily gain (ADG). Feed ingredients, TMR offered, and orts were sampled weekly and composited by period for further chemical and particle size analyses. Blood samples from all animals were collected before feeding from the jugular vein into two sodium and one lithium heparinized tubes (10-mL) and one K_2EDTA Vacutainer® tube (8-mL; Becton Dickinson Breda, Etten-Leur, The Netherland) on weeks 0 (experimental day 28, which was end of the adaptation period and beginning of Period 1), 4, 8, and 12 to determine acid-base balance, blood gases, total Hb, MetHb, packed cell volume (PCV), and NO_3^- and NO_2^- concentrations. Whole blood was analyzed within 30 min for total Hb and MetHb levels.

2.3. Emission Measurements

Methane and hydrogen were measured using the GreenFeed emission monitoring (GEM) system (C-Lock Inc., Rapid City, SD, USA). The GEM system was placed in one of the pens and animals were moved rotationally (conveyer belt approach) once a week, such that a new pen of cattle could access the system each week. Thus, once the animals were adapted to the GEM system (28 days), each treatment group had access to the system for seven days within a period, totaling three weeks of measurement per treatment group (pen) during the 84-day period. This approach allowed us to eliminate any possible pen effect because all animals spent the same amount of time in each pen.

The GEM system allows free movement of animals (in and out of the system) and gasses are measured only when the animal's head is in the "head chamber" unit as determined by a proximity sensor. The system is equipped with RFID reader to recognize individual animal visits by its electronic ear tag. Upon visiting the system, animals were provided with pellets from the overhead hopper (as bait) to keep them in the unit for sufficient eructation time to achieve a representative measurement. The pellet was composed of ground barley, canola meal, canola oil, dried molasses, and salt (NaCl) with a composition of 14.6% CP, 42.1% starch, 19.6% neutral detergent fiber (NDF), 11.8% acid detergent fiber (ADF), and 4.8 Mcal/kg DM gross energy (GE, DM basis). Maximum daily pellet drops per

Table 2. Body weight, average daily gain, dry matter intake, feed efficiency, and feeding behavior of backgrounding beef steers fed a high-forage diet with no additives (control, −EN −MBEO), or supplemented with encapsulated nitrate (+EN), microencapsulated blend of essential oils (+MBEO), and combination of EN and MBEO (+EN +MBEO).

Item [2]	−EN [1]		+EN [1]		SEM	*p*-Value			
	−MBEO	+MBEO	−MBEO	+MBEO		EN	MBEO	EN × MBEO	Period
Number of animals	22	22	22	22	---	---	---	---	---
Body weight, kg									
Initial	332	331	331	331	5.03	0.939	0.953	0.989	---
day 28	371	363	359	356	5.59	0.088	0.348	0.604	---
day 56	397	395	387	382	6.12	0.060	0.541	0.839	---
day 84	439	426	413	407	6.92	0.002	0.184	0.561	---
Final (d 112)	467	458	446	441	7.60	0.012	0.336	0.835	---
ADG (kg/day)									
day 1 to 28	1.417	1.137	0.986	0.910	0.060	<0.0001	0.004	0.093	---
day 29 to 56	0.934 ^{b}	1.136 ^{a}	1.010 ^{a,b}	0.916 ^{b}	0.050	0.153	0.281	0.004	---
day 57 to 84	1.497 ^{a}	1.110 ^{b}	0.917 ^{b}	0.909 ^{b}	0.055	<0.0001	0.001	0.001	---
day 85 to 112	0.998	1.153	1.190	1.173	0.053	0.049	0.200	0.111	---
DMI (kg/day) [3]									
day 1 to 28	7.71	7.53	7.05	7.12	0.180	0.004	0.768	0.489	---
day 29 to 56	7.85	8.31	7.82	7.69	0.180	0.072	0.359	0.100	---
day 57 to 84	9.35	8.81	8.67	8.47	0.197	0.011	0.063	0.388	---
day 85 to 112	9.47	9.53	8.95	8.73	0.180	0.001	0.673	0.447	---
G:F									
day 1 to 28	0.184	0.150	0.139	0.127	0.007	<0.0001	0.002	0.123	---
day 29 to 56	0.118 ^{b}	0.136 ^{a}	0.129 ^{a,b}	0.118 ^{b}	0.006	0.504	0.571	0.009	---
day 57 to 84	0.160 ^{a}	0.126 ^{b}	0.106 ^{c}	0.105 ^{c}	0.005	<0.0001	0.002	0.002	---
day 85 to 112	0.105	0.121	0.132	0.134	0.005	0.0001	0.088	0.182	---
Average (day 29 to 112)									
DMI (kg/day) [3]	8.93	8.87	8.47	8.28	0.176	0.003	0.479	0.713	<0.0001
ADG (kg/day)	1.065	1.157	1.073	1.018	0.042	0.119	0.657	0.080	0.0001
G:F	0.123	0.128	0.125	0.121	0.004	0.432	0.921	0.240	0.811
Average daily NO_3^- consumed (g/animal) [4]	9.43	12.43	127.20	120.37	5.030	<0.001	0.723	0.370	0.160
Feeding behavior [5]									
Total meal duration, min/day	183.6	183.4	188.2	186.6	4.71	0.410	0.849	0.883	<0.0001
Head down duration per meal, min/meal	9.6	9.3	10.2	9.6	0.68	0.572	0.519	0.793	0.323
Head down duration, min/day	80.9	84.0	97.1	91.6	4.83	0.016	0.810	0.377	<0.0001
Feeding/meal frequency, events/day	9.2	9.6	10.4	10.4	0.30	0.002	0.495	0.440	<0.0001
Feeding rate, g DM/min	46.2	46.1	42.7	41.1	1.29	0.001	0.514	0.553	<0.0001

[1] EN represents the main effect of encapsulated nitrate (−EN −MBEO and −EN +MBEO) versus (+EN −MBEO and +EN +MBEO); MBEO represents the main effects of microencapsulated blend of essential oils (−EN −MBEO and +EN −MBEO) versus (−EN +MBEO and +EN +MBEO); +EN +MBEO represents the interaction between main effects of EN and MBEO. [2] Animals that received diets containing encapsulated NO_3^- were acclimatized gradually using a step-up protocol during the first 28 days of adaptation; 0.625%, 1.25%, 1.875%, and 2.5% NO_3^- in dietary DM. Animals that received MBEO were supplemented with 150 mg/kg DM microencapsulated blend of EO since the beginning of the experiment. [3] Dry matter intake for animals that visited the GreenFeed emission monitoring system included the amount of pellet consumed while visiting the system. [4] NO_3^- consumption was calculated from NO_3^- analysis of TMR and ort samples. [5] A meal was defined as a visit to the bunk, followed by an absence from the bunk for 300 s or greater. Total meal duration = total time spent at feeder, head down duration per meal = meal duration/number of meals per day, feeding rate = DM consumed by time at feeder (DMI/meal duration). [a, b, c] Means within a row for each treatment with different lower case letter are significantly different ($p \leq 0.05$).

Feeding or eating behavior was not affected ($p \geq 0.377$) by MBEO and EN × MBEO (Table 2). However, feeding EN reduced feeding rate (g DM/min; $p = 0.001$), which resulted in longer head down duration (min/day; $p = 0.016$) during meals and greater ($p = 0.002$) meal frequency per day.

Inclusion of EN and MBEO in the diet did not change ($p \geq 0.128$) particle size distribution of TMR or the large (≥18 mm) and medium (8 to 18 mm) particles of orts (Table 3). However, there was an interaction ($p < 0.04$) between EN and MBEO for the small (1.2 to 8 mm) and bottom (fine, <1.2 mm) particles of orts because inclusion of EN induced selective sorting in favor of fine particles to a greater extent when MBEO was not added.

Table 3. Particle size distribution of total mixed ration (TMR) and orts (n = 3) from beef steers fed a backgrounding diet with no additives (control, −EN −MBEO), or supplemented with encapsulated nitrate (+EN), microencapsulated blend of essential oils (+MBEO), and combination of EN and MBEO (+EN +MBEO).

Item	−EN [1]		+EN [1]		SEM	*p*-Value			
	−MBEO	+MBEO	−MBEO	+MBEO		EN	MBEO	EN × MBEO	Period
TMR, % (as-is basis)									
Large (≥18 mm)	2.57	3.93	3.43	3.23	0.331	0.817	0.128	0.056	0.020
Medium (8 to 18 mm)	61.58	66.65	65.19	66.61	1.917	0.462	0.219	0.451	0.413
Small (1.2 to 8 mm)	31.99	28.13	29.63	28.89	1.430	0.706	0.388	0.510	0.707
Bottom (<1.2 mm)	2.92	1.83	1.41	2.02	0.554	0.479	0.766	0.401	0.269
Orts, % (as-is basis)									
Large (≥18 mm)	11.48	14.27	12.65	9.52	3.886	0.732	0.973	0.586	0.115
Medium (8 to 18 mm)	57.26	62.71	61.01	58.86	2.211	0.982	0.483	0.136	0.111
Small (1.2 to 8 mm)	29.65 [a]	24.11 [b]	24.02 [b]	30.70 [a]	0.775	0.564	0.49	0.0001	0.001
Bottom (<1.2 mm)	1.71 [a]	0.79 [a,b]	0.49 [b]	0.78 [a,b]	0.232	0.039	0.225	0.04	0.017
Orts, % of total offered (DM basis)	1.08	1.01	1.36	2.62	0.477	0.096	0.259	0.215	0.037

[1] EN represents the main effect of encapsulated nitrate (−EN −MBEO and −EN +MBEO) versus (+EN −MBEO and +EN +MBEO); MBEO represents the main effects of microencapsulated blend of essential oils (−EN −MBEO and +EN −MBEO) versus (−EN −MBEO and +EN +MBEO); +EN +MBEO represents the interaction between main effects of EN and MBEO. [a, b] Means within a row for each treatment with different lower case letter are significantly different ($p \leq 0.05$).

Animal visits to the GEM system and the impacts of feeding EN and MBEO on enteric CH_4 and hydrogen emissions and yield are presented in Table 4. An interaction between EN and MBEO was observed for the number of animals that visited the GEM system (p = 0.021), but there were no treatment effects ($p \geq 0.089$) on the average number of good visits or visit duration. The effect of EN and MBEO on CH_4 emissions was independent as indicated by the lack of interaction effects ($p \geq 0.174$) for these variables. Feeding EN reduced enteric CH_4 emissions by 17.6% (190.9 versus 157.3 g/day for –EN and +EN, respectively; $p < 0.0001$), whereas feeding MBEO increased CH_4 emissions by 11.0% (165.0 versus 183.2 g/day for –MBEO and +MBEO, respectively; p = 0.005). Similarly, CH_4 yield (emissions corrected for intake) was 13.0% lower for EN (21.5 versus 18.7 g/kg DMI for −EN and +EN, respectively; $p < 0.0001$) but 13.6% higher for MBEO (18.9 versus 21.4 g/kg DMI for −MBEO and +MBEO, respectively; p = 0.0002). When CH_4 emission was expressed in terms of energy loss, feeding EN resulted in a 10.8% lower loss of GE as CH_4 (5.37 versus 4.79% of GE intake for −EN and +EN, respectively; p = 0.001), whereas feeding MBEO resulted in 10.6% more loss of GE as CH_4 (4.83 versus 5.34% GE intake for –MBEO and +MBEO, respectively; p = 0.002). Correspondingly, an interaction (p = 0.02) was observed between additives for daily hydrogen production because addition of EN increased hydrogen production to a greater extent when MBEO was not added. However, when corrected for DMI and expressed as yield, feeding EN increased hydrogen yield by 57.3% (0.052 versus 0.081 g/kg DMI for −EN and +EN, respectively; $p < 0.001$), whereas MBEO and EN × MBEO had no effect ($p \geq 0.12$).

The diurnal pattern of CH_4 production and animal visits to the GEM system by hour are presented in Figure 2. Hourly CH_4 emissions increased after morning (1000 h) and afternoon (1600 h) feeding and production rate for +EN was consistently lower throughout the day relative to −EN. Furthermore, +EN reduced CH_4 emissions consistently throughout the experimental period, implying that the effectiveness of EN did not decline over time (Figure 3). Animals frequented the GEM system to the greatest extent at midnight (0000 h) and the lowest number of visits was observed between 0300 h and 0400 h for all treatments.

Table 4. Visitation to the GreenFeed emissions monitoring (GEM) system, pellet consumption, and emission and yield of CH_4 and hydrogen from backgrounding beef animals fed a high-forage diet with no additives (control, −EN, −MBEO) or supplemented with encapsulated nitrate (+EN), microencapsulated blend of essential oils (+MBEO), and combination of EN and MBEO (+EN +MBEO).

Item	−EN [1]		+EN [1]		SEM	*p*-Value			
	−MBEO	+MBEO	−MBEO	+MBEO		EN	MBEO	EN × MBEO	Period
GEM system visitation									
Number of animals that visited	18 [a]	16 [b]	17 [a,b]	18 [a]	0.5	0.741	0.339	0.021	0.010
Good visits per animal per period [2]	33.0	33.1	33.3	34.3	1.66	0.647	0.743	0.807	0.210
Visit duration (min:s)	4:06	4:07	4:39	4:18	0.12	0.089	0.210	0.389	<0.0001
Pellet consumed, kg DM/day	0.79	0.84	0.82	0.88	0.030	0.262	0.065	0.911	<0.0001
DMI [3], kg/day	9.02	8.89	8.56	8.30	0.207	0.010	0.322	0.713	<0.0001
GE intake, Mcal/day	46.64	47.30	44.83	43.56	1.075	0.009	0.766	0.353	<0.0001
CH_4									
g/day	184.14	197.69	145.89	168.78	6.533	<0.0001	0.005	0.458	<0.0001
g/kg of DMI	20.69	22.35	17.00	20.46	0.683	<0.0001	0.0002	0.174	<0.0001
% of GE intake	5.19	5.55	4.46	5.12	0.169	0.001	0.002	0.350	0.0002
Hydrogen [4]									
g/day	0.428 [c]	0.455 [c]	0.734 [a]	0.639 [b]	0.0265	<0.0001	0.192	0.022	<0.0001
g/kg of DMI	0.050	0.053	0.084	0.078	0.0031	<0.0001	0.571	0.123	<0.0001

[1] EN represents the main effect of encapsulated nitrate (−EN −MBEO and −EN +MBEO) versus (+EN −MBEO and +EN +MBEO); MBEO represents the main effects of microencapsulated blend of essential oils (−EN −MBEO and +EN −MBEO) versus (−EN +MBEO and +EN +MBEO); +EN +MBEO represents the interaction between main effects of EN and MBEO. [2] Good visits were selected based on the distance of the animal's head from the proximity sensor and the duration that the animal's head in the "head chamber". Good visits were used to calculate the average daily CH_4 emissions. [3] DM intake included both TMR and pellet consumption. [4] Hydrogen emission was calculated using the "arithmetic averaging method", a straight-forward averaging of the visit fluxes defined as the sum of the visit fluxes divided by the number of measurements [26]. [a, b, c] Means within a row for each treatment with different lower case letter are significantly different ($p \leq 0.05$).

There was no interaction ($p \geq 0.20$) between EN and MBEO on blood partial pressure of carbon dioxide (pCO_2) and oxygen (pO_2), total concentration of CO_2 (tCO_2), saturation of O_2 ($SatO_2$) and CO_2 ($SatCO_2$), bicarbonate (HCO_3^-), total Hb, base excess (BE), pH, and packed cell volume (PCV) (Table 5). However, an interaction between the additives was observed for blood MetHb ($p = 0.008$) and plasma NO_3^--N ($p = 0.003$) contents, because blood MetHb and plasma NO_3^--N increased to a lesser extent when MBEO was added. None of the animals showed visual signs of methemoglobinemia throughout the experiment, observing a maximum individual MetHb concentration of 4.1% of total Hb for the +EN +MBEO treatment, a level that is not a threat to animal health and wellbeing. Furthermore, feeding EN (+EN) increased ($p = 0.05$) blood HCO_3^- and total CO_2 relative to treatment without EN. Feeding MBEO had no effect on all measured blood parameters. Plasma NO_3^--N concentration for the EN and EN + MBEO treatments was reduced over the experimental period (Figure 4a); however, plasma NO_2^--N concentrations reached maximum at the fourth week of the experimental period for all the treatments and sharply reduced thereafter (Figure 4b).

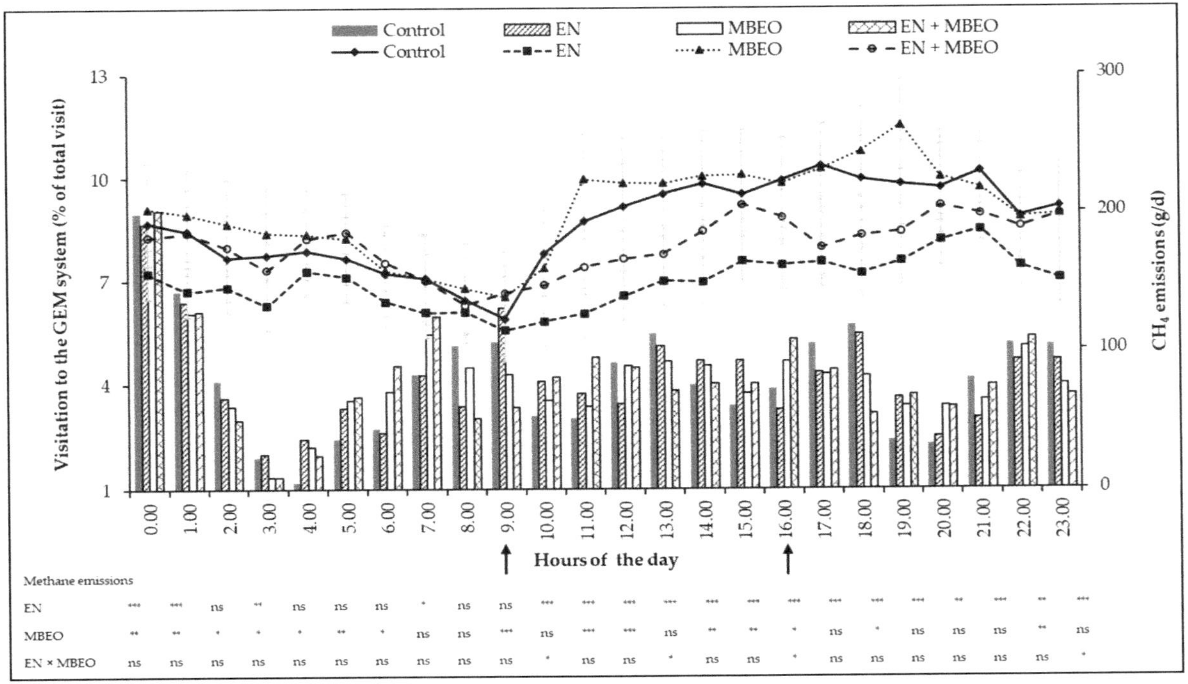

Figure 2. Diurnal pattern of CH_4 emissions (g/day, solid and broken lines) and animal visits to the GreenFeed emission monitoring (GEM) system over 24-h period (% of total visits, bar graphs) for backgrounding beef steers fed a high-forage diet with no additives (control, −EN, −MBEO), or supplemented with encapsulated nitrate (EN), microencapsulated blend of essential oils (MBEO), and combination of EN and MBEO (EN + MBEO). The arrows indicate time of feeding at 0900 h and 1600 h, and 0000 h indicates midnight. There were no significant differences among treatments for the animal visits at individual time points throughout the hours of the day. For CH_4 emissions, error bars indicate standard deviation and ns = $p > 0.05$; * $p \leq 0.05$; ** $p \leq 0.01$; *** $p \leq 0.001$. For the statistical analysis, EN represents the main effect of encapsulated nitrate (−EN −MBEO and −EN +MBEO) versus (+EN −MBEO and +EN +MBEO); MBEO represents the main effects of microencapsulated blend of essential oils (−EN −MBEO and +EN −MBEO) versus (−EN +MBEO and +EN +MBEO); EN + MBEO represents the interaction between main effects of EN and MBEO.

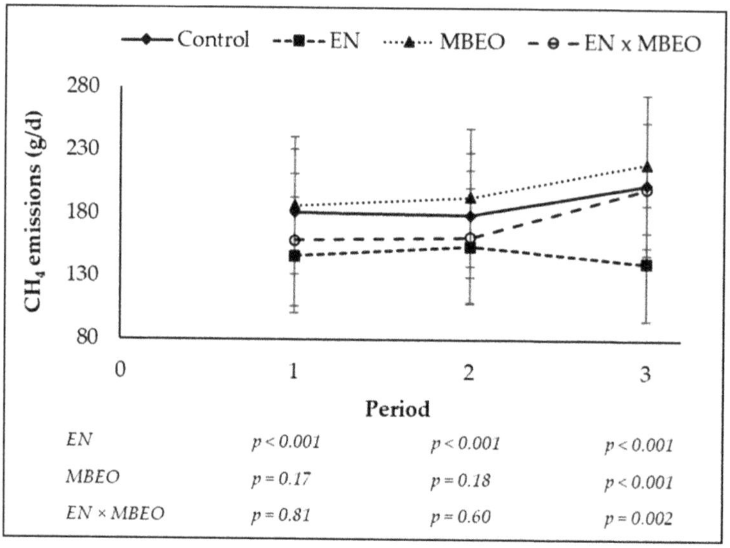

Figure 3. Enteric CH_4 emissions over the experimental period for beef steers consuming a high-forage diet with no additives (control, −EN −MBEO) or supplemented with encapsulated nitrate (EN), microencapsulated blend essential oils (MBEO), and combination of EN and MBEO (EN + MBEO). Error bars indicate standard deviation. For the statistical analysis, EN represents the main effect of encapsulated nitrate (−EN −MBEO and −EN +MBEO) versus (+EN −MBEO and +EN +MBEO); MBEO represents the main effects of microencapsulated blend of essential oils (−EN −MBEO and +EN −MBEO) versus (−EN +MBEO and +EN +MBEO); EN + MBEO represents the interaction between main effects of EN and MBEO.

Table 5. Jugular blood acid-base balance of beef steers (n = 22) fed a high forage backgrounding diet with no additives (control, −EN, −MBEO) or supplemented with encapsulated nitrate (+EN), microencapsulated blend of essential oils (+MBEO), and combination of EN and MBEO (+EN +MBEO).

Item [2]	−EN [1]		+EN [1]		SEM	*p*-Value			
	−MBEO	+MBEO	−MBEO	+MBEO		EN	MBEO	EN × MBEO	Period
pCO_2, mmHg	40.91	39.55	41.39	41.82	0.974	0.165	0.632	0.364	<0.0001
tCO_2, mmol/L [3]	29.30	29.55	30.43	30.49	0.506	0.048	0.762	0.844	<0.0001
HCO_3^-, mmol/L [3]	28.06	28.34	29.15	29.20	0.486	0.051	0.724	0.813	<0.0001
BE, mmol/L [3]	4.82	5.28	5.49	5.65	0.380	0.180	0.424	0.691	0.209
pH	7.48	7.50	7.48	7.49	0.008	0.857	0.334	0.200	0.0001
pO_2, mmHg	42.79	43.03	41.88	44.03	1.127	0.968	0.294	0.402	0.0077
$SatO_2$, % total Hb	78.00	79.15	77.27	78.91	1.223	0.694	0.261	0.844	<0.001
PCV, %	43.03	43.40	43.49	43.83	0.571	0.444	0.539	0.979	0.0001
Total Hb, g/dL	16.15	15.98	16.42	16.29	0.210	0.170	0.479	0.934	<0.0001
MetHb, g/100 g Hb	0.70 [b]	0.87 [b]	1.45 [a]	1.22 [a]	0.053	<0.0001	0.710	0.008	0.0003
Min., g/100 g Hb	0.00	0.25	0.20	0.55	---	---	---	---	---
Max., g/100 g Hb	1.95	2.53	3.40	4.10	---	---	---	---	---
Plasma									
NO_3^--N, mg/L	0.082 [c]	0.076 [c]	1.135 [a]	0.812 [b]	0.052	<0.0001	0.002	0.003	<0.0001
NO_2^--N, µg/L	2.536	2.621	2.178	1.971	0.129	0.0002	0.636	0.260	<0.0001

[1] EN represents the main effect of encapsulated nitrate (−EN −MBEO and −EN +MBEO) versus (+EN −MBEO and +EN +MBEO); MBEO represents the main effects of microencapsulated blend of essential oils (−EN −MBEO and +EN −MBEO) versus (−EN +MBEO and +EN +MBEO); +EN +MBEO represents the interaction between main effects of EN and MBEO. [2] pCO_2, partial pressure of carbon dioxide; tCO_2, total concentration of CO_2; HCO_3^-, bicarbonate; BE, base excess; pO_2, partial pressure of O_2; $SatO_2$, O_2 saturation as percent of oxygen based on total hemoglobin saturation capacity; PCV, packed cell volume; Hb, hemoglobin; MetHb, methemoglobin; NO_3^-, nitrate and NO_2^-, nitrite. [3] These parameters were calculated from parameters measured by VetStat analyzer. [a, b, c] Means within a row for each treatment with different lower case letter are significantly different ($p \leq 0.05$).

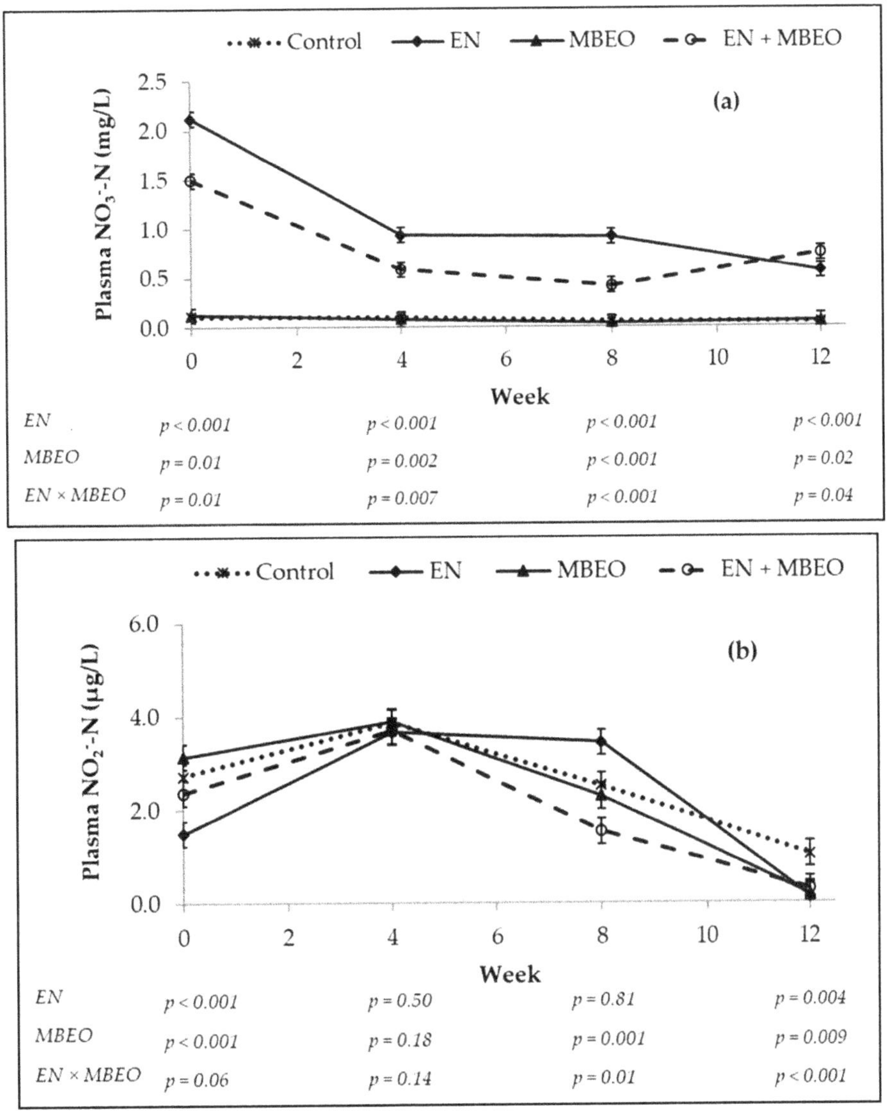

Figure 4. Plasma NO_3^--N (**a**) and NO_2^--N (**b**) concentration before morning feeding in beef steers (n = 22) consuming a high-forage diet with no additives (control, −EN, −MBEO) or supplemented with encapsulated nitrate (EN), microencapsulated blend essential oils (MBEO), and combination of EN and MBEO (EN + MBEO). Week zero indicates the end of adaptation period (experimental day 28). Error bars indicate standard deviation. For the statistical analysis, EN represents the main effect of encapsulated nitrate (−EN −MBEO and −EN +MBEO) versus (+EN −MBEO and +EN +MBEO); MBEO represents the main effects of microencapsulated blend of essential oils (−EN −MBEO and +EN −MBEO) versus (−EN +MBEO and +EN +MBEO); EN + MBEO represents the interaction between main effects of EN and MBEO.

4. Discussion

Supplementation of NO_3^- in ruminant diets has been proposed as an alternative to increase non-protein nitrogen intake while effectively minimizing enteric CH_4 emissions [30]. However, it is also well documented that over-consumption of NO_3^- can be toxic to animals [12]. Encapsulation of NO_3^- has been used [7,29] to ensure slow release of NO_3^- in the rumen and increase the efficiency of microbes to fully reduce NO_3^- to ammonia, thus minimizing the risk of NO_3^-/NO_2^- toxicity.

Essential oils have been shown to favorably affect rumen fermentation in vitro, but the observed responses have not translated into improved production characteristics in the few existing studies with beef cattle [17]. Furthermore, previous studies have reported that the immune modulation, antioxidant, thermoregulation, and blood oxygenation properties of EO may improve animal productivity and energetics [14,18]. Despite the use of NO_3^- and EO in ruminant diets, previous studies have not explored their possible interaction on enteric CH_4 mitigation and animal productivity. The main finding in the current study is that the effects of EN were mostly independent from those of MBEO, as most of the variables examined showed a lack of significant interaction between EN and MBEO, thus considering the responses to these additives as generally independent.

4.1. Nitrate

In the literature, the impact of feeding NO_3^- to ruminants on DMI varies among studies. For example, using unencapsulated NO_3^- (2% in diet DM) Lund et al. [31] reported 11% reduction in DMI for dairy cattle fed a high-forage diet (58% DM). Similarly, Hulshof et al. [2] used unencapsulated NO_3^- (2.2% in dietary DM) and observed a decrease in DMI of 6% in beef cattle fed high-forage diets (60% DM). Encapsulation of NO_3^- ensures not only slow release in the rumen [8] but also has a potential to minimize its negative impact on feed intake caused by its organoleptic properties [32]. This has been the case in some previous reports that used encapsulated NO_3^- [7,10] but not in others [11].

In our study, the lack of effects of EN on ADG and G:F were in agreement with Lee et al. [10] and El-Zaiat et al. [7]. Lee et al. [10] supplemented encapsulated NO_3^- (2.5% in dietary DM) to beef cattle fed a high-forage diet (65% DM corn silage) and reported no effect on DMI and feed efficiency. Whereas with the same inclusion rate in a high concentrate diet (80% DM of barley grain), the same authors observed a 7.5% reduction in DMI and 11% improvement in feed efficiency for finishing beef cattle [11]. Changes in feeding and eating behavior following NO_3^- supplementation may contribute to DMI and feed efficiency responses [32,33]. For example, Lee et al. [10] observed significant sorting of the TMR for large particles, which increased the proportion of small particles and decreased the proportion of large particles in orts, as well as a considerable increase in NO_3^- concentration in orts. Conversely, in our study, feeding EN induced sorting in favor of fine particles but had no effect on either the large and medium particles of orts or the total amount of orts (% of total offered). The reduction in feeding rate (g DM/min) for EN was manifested in longer head down duration during meals and more frequent meals per day. These changes in feeding behavior of cattle fed EN were consistent with Velazco et al. [34], where NO_3^--fed cattle consumed a large number of meals per day and smaller in size when compared to cattle fed a control diet.

Previous studies reported a reduction in enteric CH_4 production in several species and categories of animals due to NO_3^- feeding [1,30]. Reductions in CH_4 yield ranging between 4% (with 1% NO_3^- in diet DM; [9,10]) and 33% (with 2.7% NO_3^- in diet DM; [7]) have been reported for ruminants fed high-forage diets supplemented with encapsulated NO_3^-. The observed reduction in CH_4 yield in our study for +EN (13.0%) was within the range of 12.2% and 18.3% reduction reported for beef heifers fed a forage diet (50% DM barley silage) supplemented with encapsulated NO_3^- at 2% and 3% in diet DM, respectively [9]. Lower rate of reduction (6.2%) was also reported for backgrounding steers fed a high-forage diet (65% DM corn silage) supplemented with encapsulated NO_3^- (2.5% in diet DM [10]). Little is known about the factors that may interfere with the efficiency of NO_3^- reduction in the rumen. Encapsulation of NO_3^- [8], amount of NO_3^- ingested, and intake rate of NO_3^- [32,35], type of diet (e.g., roughage inclusion, N and S concentrations [10,11,35] as well as type of animals [35]) affect ruminal NO_3^- utilization, and consequently, CH_4 reduction. Furthermore, duration of feeding a dietary additive may affect its efficacy in reducing enteric CH_4 production over time [36]; however, there was no decline in the effectiveness of EN over time in the current study (Figure 3).

Multiple in vitro [8] and in vivo studies [4,5,31] have reported an increase in hydrogen production after feeding NO_3^-. Similarly, a significant ($p < 0.001$) increase in hydrogen production and yield was observed for the EN treatment in our study. It is generally believed that NO_3^- reduction is a thermodynamically favorable process relative to methanogenesis in which NO_3^- acts as a hydrogen sink [37]. However, considering the observed increase in hydrogen production following NO_3^- supplementation in previous studies as well as in the current study, the earlier hypothesis needs to be re-examined. Perhaps the direct toxicity of NO_3^- and its reduced intermediate (NO_2^-) on rumen microbes [3] may contribute as an additional mode of action in decreasing enteric CH_4 production. Furthermore, hydrogen is an energy-dense gas (142 kJ/g of hydrogen, [38]) and its emission by animals could partially offset the energy gain by the decrease in CH_4 production. For example, the calculated energy lost in hydrogen production for the +EN treatment was 23.3 kcal per day or 6.9% of the observed CH_4 decrease with the use of EN.

The increased concentration of plasma NO_3^- and NO_2^- following supplementation of NO_3^- in the diets of ruminants implies that NO_3^- is not fully reduced to ammonia. Lee et al. [10,32] observed a dose-response increase in blood NO_3^- and NO_2^- when encapsulated NO_3^- was fed. The observed increase in blood NO_3^- concentration for the EN treatment in our study was comparable to previous reports [10,33]. Furthermore, NO_2^- was present in the blood in a detectable range (2 to 3 µg/L of NO_2^--N) but did not elevate blood MetHb levels (less than 4.1% of total Hb) to the threshold that is considered to cause subclinical methemoglobinemia (30 to 40% [12]). Feeding NO_3^- at 2 to 3% of dietary DM has been widely reported without any toxicity issues [30]. Although NO_3^- consumption was relatively consistent over the experimental periods, blood NO_3^- and NO_2^- concentrations gradually decreased during the study, which could be due to a combination of factors, including a possible gradual improvement in microbial capacity to reduce dietary NO_3^- [39], physiological change of the experimental animals [8], and change in the feeding behavior [8,10,40]. The lower feeding rate (g DM/min) and higher meal frequency per day for EN may have helped to spread out the availability of NO_3^- to rumen microbes over a longer period, thus reducing the size of NO_3^- pulses occurring in the rumen, which in turn would have lowered the concentration of NO_3^- and NO_2^- in rumen fluid and blood.

4.2. Essential Oils

The lack of effect of MBEO on average DMI and feed efficiency observed in our study is consistent with previous in vivo studies that supplemented EO or blend of EO to beef cattle. For example, Beauchemin and McGinn [41] reported that growing beef cattle fed a high-forge diet (75% DM whole-grain barley silage) supplemented with 1 g/day blend of EO (Crina® Ruminants, mixture of thymol, eugenol, vanillin, limonene, and guaiacol) did not show any difference in DMI and feed efficiency. Similarly, using that same product (Crina® Ruminants) at 1 and 2 g/day, Tomkins et al. [39] found no differences in DMI and animal performance for steers fed Rhodes grass hay (ad libitum). For beef animals on finishing diets, a study conducted by Yang et al. [20] tested the effects of 3 doses (0.4, 0.8, 1.6 g/day) of cinnamaldehyde or monensin on feedlot cattle performance, and reported that none of the treatments affected performance variables. Furthermore, Meyer et al. [42] reported no effect on DMI and feed efficiency for feedlot steers over a 115-day finishing period when fed a blend of EO (thymol, eugenol, vanillin, guaiacol, and limonene) at 1 g/animal/day. However, the authors reported improved efficiency for diets containing a blend of EO and tylosin. Furthermore, using higher doses (3.5 and 7 g/animal/day) of a blend of EO (MixOil®, extracts from oregano, garlic, lemon, rosemary, thyme, eucalyptus, and sweet orange), Rivaroli et al. [43] reported no effects on DMI and animal performance parameters for crossbred bulls fed high-grain finishing diets for 120-days. A similar lack of effect has been observed in other animal species, including sheep [44] and dairy cows [18,42]. Overall, the results from our study are consistent with the literature that suggests supplementation of diets with EO has no effects on DMI and performance of beef cattle.

Only few in vivo studies investigated the effect of EO on enteric CH_4 emission with conflicting results [16,17]. Furthermore, due to the variation in type of diets, dose rate, and the range of EO and EO compounds used, it is challenging to make a direct comparison of CH_4 emission outputs among studies. Reduction in enteric CH_4 emissions following supplementation of EO and blend of EO has been reported for sheep [23,44] and buffalo [45], although others reported no effect on CH_4 emissions for beef [41,46] and dairy cattle [47]. However, in the current study, enteric CH_4 emissions and yield were increased by 11.0% and 13.6%, respectively. It is difficult to explain the observed increase in CH_4 production. However, several factors, including the wide range of non-specific antibacterial activity of EO that may favor methanogenesis [15] or positive impacts of EO on ruminal feed degradability [19,48], may play a role.

Information on the effect of EO on the process of methanogenesis in ruminants is ambiguous [49,50]. The impact of EO on CH_4 emissions may be attributed to direct impact on methanogenic archaea (changing community structure or activity of methanogenesis pathway) and indirect impact on microbial metabolic processes contributing to methanogenesis [17]. Essential oils can also affect some protozoa that are symbiotically associated with archaea. Using meta-analysis, Khiaosa-ard and Zebeli [51] reported a dose-response effect of EO on reducing protozoa, whereas Cobellis et al. [17] reported no effect of EO on protozoa in most in vivo studies in ruminants. Furthermore, the antimicrobial activity of EO varies with the quantity used, chemical composition (both components present and their proportion), interaction among EO components, and chemical configurations [52,53]. Additive, antagonistic, and synergistic effects have been observed between components of EO [15,52]. The Activo® Premium used in the MBEO treatment contained carvacol, eugenol, capsaicin, cinnamaldehyde, and pepper extract with diverse antimicrobial activities. For example, carvacrol has shown a negative effect on Gram-negative bacteria [54], whereas eugenol has shown a broad antibacterial activity by affecting both Gram-negative and Gram-positive bacteria [14,55].

Although reduction in DM digestibility following supplementation of EO has been reported in most in vitro studies, in vivo studies have been inconsistent [17]. Yang et al. [48] fed garlic (5 g/cow per day) and juniper berry (2 g/cow per day) to dairy cows consuming a ration containing forage (40% DM) and reported a 12 to 15% increase in rumen DM and OM digestibility. However, total tract digestibility of DM, OM, fiber, and starch were not affected. They suggested that the increased ruminal digestibility was due to an 11% increase in dietary protein digestibility in the rumen compared with control. In another study, Silva et al. [19] fed Activo® Premium at the rate of 150 mg/kg DM to dairy cows consuming a diet containing corn silage (48% DM) and observed an increase in total tract OM digestibility. It has also been reported that the effects of EO and EO blends are rumen pH and diet dependent [14]. Benchaar et al. [56] reported that supplementation of 750 mg Crina® Ruminants per day to dairy cows tended to increase total VFA concentration in the rumen of lactating cows when the diet contained alfalfa silage, but tended to decrease total VFA concentration when the diet contained corn silage. Overall, further long-term in vivo studies are required to determine the potential of using EO in ruminant diets to lower enteric CH_4 production.

Although studies report anti-inflammatory, immune modulation, antioxidant, thermoregulation, and blood oxygenation properties of EO [14,18], blood parameters measured in our study did not differ for the MBEO and EN + MBEO treatments. Recent findings suggest that due to their phenolic nature, some EO are likely less susceptible to microbial degradation in the rumen and exhibit activities post-ruminally by binding to specific receptors expressed in neurons, intestines, and other cells [18,57]. However, Oh et al. [18] stated that these impacts of EO are likely dependent on the type and physiological status of the experimental animals, as well as the type of diets. In dairy cattle, providing EO either in the diet or by direct infusion into the abomasum had an effect on the immune system of the animals, post-ruminal nutrient use, and animal physiology [18]. Silva et al. [19,22] fed

Activo® Premium at 150 mg/kg diet DM to dairy cows in mid-lactation for 8 weeks and reported increased O_2 saturation of Hb and a greater proportion of O_2 transported by blood in relation to total gases for cows fed a blend of EO compared with a control. Furthermore, for feedlot cattle fed a finishing diet containing dry-rolled barley grain (86% DM), Yang et al. [20] reported that supplementation with cinnamaldehyde (0.4 to 1.6 g/day per animals) reduced stress and increased DMI during the early feeding period when stress is greater. In our study, animals were fed a high-forage diet (80% DM) and were likely under minimal stress.

5. Conclusions

The effects of feeding EN as a replacement for urea and MBEO alone or in combination with EN on animal performance and enteric CH_4 emissions from beef steers fed a high-forage diet were investigated. Our results demonstrate that there were no advantages of feeding EN with MBEO. Supplementing diets with MBEO neither improved animal performance nor lowered CH_4 emissions. However, EN reduced CH_4 emissions and altered feeding behavior, whereas it had no impact on animal health and performance. Accordingly, the use of EN could have important implications for the Canadian beef sector in particular and global ruminant agriculture in general. In 2016, CH_4 emissions from enteric fermentation represented 3.5% of the Canadian national greenhouse gas inventory, with beef cattle production accounting for 80% of the emissions. A 30% adoption rate of EN by beef producers, combined with 17.6% reduction of enteric CH_4 emissions following the use of EN, would result in 4% less total enteric emissions and 1.8% less agricultural emissions in Canada.

Author Contributions: Conceptualization, A.W.A., A.R.-P., R.C.A., and K.A.B.; methodology, A.W.A., A.R.-P., R.C.A., and K.A.B.; software, A.W.A.; validation, A.W.A., A.R.-P., and K.A.B.; formal analysis, A.W.A., A.R.-P., and K.A.B.; investigation, A.W.A., A.R.-P., R.C.A., and K.A.B.; resources, A.W.A., A.R.-P., R.C.A., and K.A.B.; data curation, A.W.A., A.R.-P., and K.A.B.; writing—original draft preparation, A.W.A.; writing—review and editing, A.W.A., A.R.-P., R.C.A., and K.A.B.; supervision, K.A.B.; project administration, A.W.A., A.R.-P., and K.A.B.; funding acquisition, K.A.B.

Funding: This research was funded by Emissions Reduction Alberta and Agriculture and Agri-Food Canada.

Acknowledgments: Special thanks to the staff of the Research Feedlot and Controlled Environment Facilities for animal care and lab technicians (K. Andrews, B. Farr, L. Holtshausen, N. Johnson, R. Roth, and D. Vedres) for sample analyses.

Conflicts of Interest: The authors declare no conflict of interest.

References

1. Hristov, A.N.; Oh, J.; Lee, C.; Meinen, R.; Montes, F.; Ott, T.; Firkins, J.; Rotz, A.; Dell, C.; Adesogan, A.; et al. *Mitigation of Greenhouse Gas Emissions in Livestock Production—A Review of Technical Options for non-CO_2 Emissions*; Gerber, P.J., Henderson, B., Makkar, H.P.S., Eds.; FAO Animal Production and Health Paper No. 177; FAO: Rome, Italy, 2013.
2. Hulshof, R.B.A.; Berndt, A.; Gerrits, W.J.J.; Dijkstra, J.; van Zijderveld, S.M.; Newbold, J.R.; Perdok, H.B. Dietary nitrate supplementation reduces methane emission in beef cattle fed sugarcane-based diets. *J. Anim. Sci.* **2012**, *90*, 2317–2323. [CrossRef]
3. Van Zijderveld, S.M.; Gerrits, W.J.J.; Apajalahti, J.A.; Newbold, J.R.; Dijkstra, J.; Leng, R.A.; Perdok, H.B. Nitrate and sulfate: Effective alternative hydrogen sinks for mitigation of ruminal methane production in sheep. *J. Dairy Sci.* **2010**, *93*, 5856–5866. [CrossRef] [PubMed]
4. Van Zijderveld, S.M.; Gerrits, W.J.J.; Dijkstra, J.; Newbold, J.R.; Hulshof, R.B.A.; Perdok, H.B. Persistency of methane mitigation by dietary nitrate supplementation in dairy cows. *J. Dairy Sci.* **2011**, *94*, 4028–4038. [CrossRef] [PubMed]
5. Guyader, J.; Eugene, M.; Meunier, B.; Doreau, M.; Morgavi, D.P.; Silberberg, M.; Rochette, Y.; Gerard, C.; Loncke, C.; Martin, C. Additive methane-mitigating effect between linseed oil and nitrate fed to cattle. *J. Anim. Sci.* **2015**, *93*, 3564–3577. [CrossRef] [PubMed]
6. Nolan, J.V.; Hegarty, R.S.; Hegarty, J.; Godwin, I.R.; Woodgate, R. Effects of dietary nitrate on fermentation, methane production and digesta kinetics in sheep. *Anim. Prod. Sci.* **2010**, *50*, 801–806. [CrossRef]

7. El-Zaiat, H.M.; Araujo, R.C.; Soltan, Y.A.; Morsy, A.S.; Louvandini, H.; Pires, A.V.; Patino, H.O.; Correa, P.S.; Abdalla, A.L. Encapsulated nitrate and cashew nut shell liquid on blood and rumen constituents, methane emission, and growth performance of lambs. *J. Anim. Sci.* **2014**, *92*, 2214–2224. [CrossRef] [PubMed]
8. Lee, C.; Araujo, R.C.; Koenig, K.M.; Beauchemin, K.A. In situ and in vitro evaluations of a slow release form of nitrate for ruminants: Nitrate release rate, rumen nitrate metabolism and the production of methane, hydrogen, and nitrous oxide. *Anim. Feed Sci. Technol.* **2017**, *231*, 97–106. [CrossRef]
9. Lee, C.; Araujo, R.C.; Koenig, K.M.; Beauchemin, K.A. Effects of encapsulated nitrate on enteric methane production and nitrogen and energy utilization in beef heifers. *J. Anim. Sci.* **2015**, *93*, 2391–2404. [CrossRef] [PubMed]
10. Lee, C.; Araujo, R.C.; Koenig, K.M.; Beauchemin, K.A. Effects of encapsulated nitrate on growth performance, nitrate toxicity, and enteric methane emissions in beef steers: Backgrounding phase. *J. Anim. Sci.* **2017**, *95*, 3700–3711. [CrossRef]
11. Lee, C.; Araujo, R.C.; Koenig, K.M.; Beauchemin, K.A. Effects of encapsulated nitrate on growth performance, carcass characteristics, nitrate residues in tissues, and enteric methane emissions in beef steers: Finishing phase. *J. Anim. Sci.* **2017**, *95*, 3712–3726. [CrossRef]
12. Bruning-Fann, C.S.; Kaneene, J.B. The effects of nitrate, nitrite and N-nitroso compounds on human health—A review. *Vet. Hum. Toxicol.* **1993**, *35*, 521–538. [PubMed]
13. European Commission (EC). Regulation EC No 1831/2003 of the European Parliament and Council of 22 September 2003 on additives for use in animal nutrition. *Off. J. Eur. Commun.* **2003**, *L268*, 29–43.
14. Calsamiglia, S.; Busquet, M.; Cardozo, P.W.; Castillejos, L.; Ferret, A. Invited review: Essential oils as modifiers of rumen microbial fermentation. *J. Dairy Sci.* **2007**, *90*, 2580–2595. [CrossRef]
15. Benchaar, C.; McAllister, T.A.; Chouinard, P.Y. Digestion, ruminal fermentation, ciliate protozoal populations, and milk production from dairy cows fed cinnamaldehyde, quebracho condensed tannin, or Yucca schidigera saponin extracts. *J. Dairy Sci.* **2008**, *91*, 4765–4777. [CrossRef] [PubMed]
16. Patra, A.K. Effects of essential oils on rumen fermentation, microbial ecology and ruminant production. *Asian J. Anim. Vet. Adv.* **2011**, *6*, 416–428. [CrossRef]
17. Cobellis, G.; Trabalza-Marinucci, M.; Yu, Z. Critical evaluation of essential oils as rumen modifiers in ruminant nutrition: A review. *Sci. Total Environ.* **2016**, *545–546*, 556–568. [CrossRef] [PubMed]
18. Oh, J.; Wall, E.H.; Bravo, D.M.; Hristov, A.N. Host-mediated effects of phytonutrients in ruminants: A review. *J. Dairy Sci.* **2017**, *100*, 5974–5983. [CrossRef] [PubMed]
19. Silva, R.B.; Pereira, R.A.N.; Araujo, R.C.; Pereira, M.N. Response of lactating cows to a blend of essential oils and pepper extract. *J. Anim. Sci.* **2015**, *93*, 45 (abstract).
20. Yang, W.Z.; Ametaj, B.N.; Benchaar, C.; Beauchemin, K.A. Dose response to cinnamaldehyde supplementation in growing beef heifers: Ruminal and intestinal digestion. *J. Anim. Sci.* **2010**, *88*, 680–688. [CrossRef]
21. Hori, T. Capsaicin and central control of thermoregulation. *Pharmacol. Ther.* **1984**, *26*, 389–416. [CrossRef]
22. Silva, R.B.; Silva, W.R.; Oliveira, C.D.S.; Lacreta Junior, A.C.C.; Chalfun, L.H.L.; Araujo, R.C.; Pereira, R.A.N.; Pereira, M.N. Supplementation of lactating cows with a blend of essential oils and capsaicin or monensin: Thermoregulation and blood variables. *J. Dairy Sci.* **2017**, *100* (Suppl. 2), 321–322.
23. Soltan, Y.A.; Natel, A.S.; Araujo, R.C.; Morsy, A.S.; Abdalla, A.L. Progressive adaptation of sheep to a microencapsulated blend of essential oils: Ruminal fermentation, methane emission, nutrient digestibility, and microbial protein synthesis. *Anim. Feed Sci. Technol.* **2018**, *237*, 8–18. [CrossRef]
24. Araujo, R.C.; Daley, D.R.; Goodall, S.R.; Jalali, S.; Guimarães Bisneto, O.A.; Budde, A.M.; Wagner, J.J.; Engle, T.E. The effects of a microencapsulated blend of essential oils supplemented alone or in combination with monensin on performance and carcass characteristics of growing and finishing beef steers. *Prof. Anim. Sci.* **2019**, under review.
25. Canadian Council on Animal Care. *CCAC Guidelines on: The Care and Use of Farm Animals in Research, Teaching and Testing*; Canadian Council on Animal Care: Ottawa, ON, Canada, 2009.
26. Manafiazar, G.; Zimmerman, S.; Basarab, J.A. Repeatability and variability of short-term spot measurement of methane and carbon dioxide emissions from beef cattle using GreenFeed emissions monitoring system. *Can. J. Anim. Sci.* **2016**, *97*, 118–126. [CrossRef]

27. Kononoff, P.J.; Heinrichs, A.J.; Buckmaster, D.R. Modification of the Penn State forage and total mixed ration particle separator and the effects of moisture content on its measurements. *J. Dairy Sci.* **2003**, *86*, 1858–1863. [CrossRef]
28. Van Soest, P.J.; Robertson, J.B.; Lewis, B.A. Methods for dietary fiber, neutral detergent fiber, and nonstarch polysaccharides in relation to animal nutrition. *J. Dairy Sci.* **1991**, *74*, 3583–3597. [CrossRef]
29. AOAC. *Official Methods of Analysis*, 18th ed.; AOAC Int.: Gaithersburg, MD, USA, 2005.
30. Lee, C.; Beauchemin, K.A. A review of feeding supplementary nitrate to ruminant animals: Nitrate toxicity, methane emissions, and production performance. *Can. J. Anim. Sci.* **2014**, *94*, 557–570. [CrossRef]
31. Lund, P.; Dahl, R.; Yang, H.J.; Hellwing, A.L.; Cao, B.B.; Weisbjerg, M.R. The acute effect of addition of nitrate on in vitro and in vivo methane emission in dairy cows. *Anim. Prod. Sci.* **2014**, *54*, 1432–1435. [CrossRef]
32. Lee, C.; Araujo, R.C.; Koenig, K.M.; Beauchemin, K.A. Effects of encapsulated nitrate on eating behavior, rumen fermentation, and blood profile of beef heifers fed restrictively or ad libitum. *J. Anim. Sci.* **2015**, *93*, 2405–2418. [CrossRef]
33. Lee, C.; Araujo, R.C.; Koenig, K.M.; Beauchemin, K.A. Effects of feed consumption rate of beef cattle offered a diet supplemented with nitrate ad libitum or restrictively on potential toxicity of nitrate. *J. Anim. Sci.* **2015**, *93*, 4956–4966. [CrossRef]
34. Velazco, J.I.; Cottle, D.J.; Hegarty, R.S. Methane emissions and feeding behaviour of feedlot cattle supplemented with nitrate or urea. *Anim. Prod. Sci.* **2014**, *54*, 1737–1740. [CrossRef]
35. Leng, R.A. The Potential of Feeding Nitrate to Reduce Enteric Methane Production in Ruminants. A Report to Department of Climate Change. Commonwealth Government of Australia, Canberra. 2008. Available online: https://pdfs.semanticscholar.org/2c17/65afdfff583c9d72aec94d3a4facc894e064.pdf (accessed on 20 August 2018).
36. Moate, P.J.; Jacobs, J.L.; Hannah, M.C.; Morris, G.L.; Beauchemin, K.A.; Hess, P.S.; Eckard, R.J.; Liu, Z.; Rochfort, S.; Wales, W.J.; et al. Adaptation responses in milk fat yield and methane emissions of dairy cows when wheat was included in their diet for 16 weeks. *J. Dairy Sci.* **2018**, *101*, 7117–7132. [CrossRef] [PubMed]
37. Ungerfeld, E.M. A theoretical comparison between two ruminal electron sinks. *Front. Microbiol.* **2013**, *4*, 1–15. [CrossRef] [PubMed]
38. Afeefy, H.Y.; Liebman, J.F.; Stein, S.E. Neutral thermochemical data. In *NIST Chemistry WebBook, NIST Standard Reference Database*; Linstrom, P.J., Mallard, W.G., Eds.; National Institute of Standards and Technology: Gaithersburg, MD, USA, 1998; Volume 69.
39. Shi, C.X.; Meng, Q.X.; Hou, X.Z.; Ren, L.P.; Zhou, Z.M. Response of ruminal fermentation, methane production and dry matter digestibility to microbial source and nitrate addition level in an in vitro incubation with rumen microbes obtained from wethers. *J. Anim. Vet. Adv.* **2012**, *11*, 3334–3341. [CrossRef]
40. Nolan, J.V.; Godwin, I.R.; de Raphélis-Soissan, V.; Hegarty, R.S. Managing the rumen to limit the incidence and severity of nitrite poisoning in nitrate-supplemented ruminants. *Anim. Prod. Sci.* **2016**, *56*, 1317–1329. [CrossRef]
41. Beauchemin, K.A.; McGinn, S.M. Methane emissions from beef cattle: Effects of fumaric acid, essential oil, and canola oil. *J. Anim. Sci.* **2006**, *84*, 1489–1496. [CrossRef] [PubMed]
42. Meyer, N.F.; Erickson, G.E.; Klopfenstein, T.J.; Greenquist, M.A.; Luebbe, M.K.; Williams, P.; Engstrom, M.A. Effect of essential oils, tylosin, and monensin on finishing steer performance, carcass characteristics, liver abscesses, ruminal fermentation, and digestibility. *J. Anim. Sci.* **2009**, *87*, 2346–2354. [CrossRef] [PubMed]
43. Rivaroli, D.C.; do Prado, R.M.; Ornaghi, M.G.; Mottin, C.; Ramos, T.R.; Barrado, A.G.; Jorge, A.M.; do Prado, I.N. Essential oils in the diet of crossbred ($\frac{1}{2}$ Angus vs. $\frac{1}{2}$ Nellore) bulls finished in feedlot on animal performance, feed efficiency and carcass characteristics. *J. Agric. Sci.* **2017**, *9*, 205–2015. [CrossRef]
44. Wang, C.J.; Wang, S.P.; Zhou, H. Influences of flavomycin, ropadiar, and saponin on nutrient digestibility, rumen fermentation, and methane emission from sheep. *Anim. Feed Sci. Technol.* **2009**, *148*, 157–166. [CrossRef]
45. Yatoo, M.A.; Chaudhary, L.C.; Agarwal, N.; Chaturvedi, V.B.; Kamra, D.N. Effect of feeding of blend of essential oils on methane production, growth, and nutrient utilization in growing buffaloes. *Asian-Aust. J. Anim. Sci.* **2018**, *31*, 672–676. [CrossRef]
46. Tomkins, N.W.; Denman, S.E.; Pilajun, R.; Wanapat, M.; McSweeney, C.S.; Elliott, R. Manipulating rumen fermentation and methanogenesis using an essential oil and monensin in beef cattle fed a tropical grass hay. *Anim. Feed Sci. Technol.* **2015**, *200*, 25–34. [CrossRef]

1. Introduction

Enteric methane (CH_4) is a final product of ruminal fermentation via methanogenesis, which contributes substantially to atmospheric CH_4 accumulation. As dietary structural carbohydrates (e.g., cellulose and hemicellulose) are degraded by ruminal microorganisms, CH_4 emission represents up to 12% loss of dietary energy ingested in the rumen [1]. Thus, reducing enteric CH_4 emissions will help improve energy utilization efficiency and alleviate environmental pressures for dairy production regimes. According to the national report, CH_4 emissions in 2005 from agriculture sector accounts for 25.5 Tg, of which approximately 57% is from rumen fermentation within ruminant production system in China [2]. As dairy population and milk production increased by 24 and 58 times, respectively, from 1961 to 2010, it is projected that the total CH_4 emissions in 2030 will reach 52.1 Tg [3]. Meanwhile, in response to the domestic and international pressures on sustainable development, the Chinese government has made the commitment to reduce greenhouse gases (GHG), which has been incorporated in to the 2009 Copenhagen Accord [4]. Although Tier-2 methodology from International Panel on Climate Change (IPCC) guidelines is commonly used in many countries for the quantification of CH_4 emission inventories [5], it gives little consideration to the variations of geographic conditions, animal physiology stages, and dietary components [6,7]. In addition, this methodology tries to calculate the CH_4 emission for the whole dairy population using a default value derived from lactating cows. In China, approximately 60% of the dairy population is milking cows, and the remainder is heifers (i.e., 5.68 million of heifers) [8]. Different physiological conditions and varied composition and abundance of ruminal methanogens demonstrated great difference of CH_4 emission of lactating cow and heifer, indicating the importance of quantifying the individual emissions for these animals [3]. However, limited studies have examined the effects of dietary components on CH_4 emissions of heifers under different physiological conditions. Therefore, the objective of the current study is to assess the effect of different dietary concentrate contents on the enteric CH_4 emissions of Holstein heifers at various stage, and develop prediction equations using data collected using sulphur hexafluoride (SF_6) tracer technique.

2. Materials and Methods

2.1. Animals, Experimental Design, and Diets

This study was conducted in 2018 at the Zhongjiayonghong dairy farm located in Fangshan district, (Beijing, China, latitude: N39°39′6″ and longitude: E116°12′21″). Forty-five Chinese Holstein heifers with an initial body weight (BW) of 264.9 ± 25.6 kg (mean ± SD) were used in this study with three measurements taken at age of 9, 12, and 15 months, respectively. In each experiment stage, heifers were balanced by date of birth, age and BW and offered randomly assigned to 1 of 3 treatments (n = 15) in which animals were individually offered diets containing 30, 40, and 50% of concentrate (C30, C40, and C50, respectively). Each experimental period was 32 days in length, including 18 days for adaptation, followed by 8 days for gas measurement and 6 days for nutrient digestibility. Heifers were housed individually with free access to feed and water throughout the whole experiment. All animal care and handling procedures were reviewed and approved by the Animal Ethics Committee of Chinese Academy of Agricultural Sciences (protocol number 019–2018) prior to the start of the experiment.

In period 1, cows received their diet as a total mixed rations (TMR) that composted of corn silage, wildrye, and a typical ration of concentrate on Chinese commercial farms. In period 2 and 3, alfalfa was included in the diet based on the ration of period 1 (Table 1). The TMR were prepared daily using a feed mixer (Belle Engineering Ltd., Derbyshire, UK) and distributed *ad libitum* (5% refusals, on an as-fed basis). All diets in three periods were formulated to meet the recommendation of Ministry of Agriculture of P. R. China. For all of the three periods, cows were fed twice daily between 0600 and 0800 h, and 1600 and 1800 h. Feed refusals were collected and weighted to determine the daily feed intake.

Table 1. Ingredient and chemical composition of diets in the current study.

Item	Period 1 (9 months)			Periods 2 and 3 (12 and 15 months)		
	C30	C40	C50	C30	C40	C50
Ingredient						
Corn silage	42	36	30	42	36	30
Chinese wildrye hay	28	24	20	14	12	10
Alfalfa	-	-	-	14	12	10
Concentrate	30	40	50	30	40	50
Nutrient, DM basis						
Dry matter	93.6	93.5	93.6	93.9	93.7	93.3
Organic matter, %	91.8	91.7	91.2	93.1	92.8	92.2
Gross energy, MJ kg^{-1}	18.0	18.1	18.1	16.8	16.7	16.6
Crude protein, %	15.7	17.8	18.7	14.1	14.5	14.7
Ether extract, %	4.1	4.0	4.0	3.7	3.8	3.8
Ash, %	8.2	8.3	8.8	6.9	7.2	7.8
Neutral detergent fiber, %	37.8	34.2	31.4	36.8	32.6	29.3
Acid detergent fiber, %	15.6	14.0	12.0	19.8	17.5	14.8
Ca, %	0.5	0.5	0.5	0.3	0.5	0.6
P, %	0.2	0.3	0.3	0.2	0.2	0.3

C30 = diet containing 30% of concentrate, C40 = diet containing 40% of concentrate, C50 = diet containing 50% of concentrate; Concentrates were purchased from a commercial company (Beijing Sanyuan Breeding Technology Corporation, Beijing, China), which mainly comprise corn, wheat bran, soybean meal, calcium hydrophosphate, limestone and salt. The nutrient content is: crude protein ≥17%, ether extract ≥2.5%, crude fiber ≤9.0%, Ca = 0.5–1.5%, P = 0.4–1.0%, NaCl = 0.5–2.0%, Lysine ≥0.6%.

2.2. Enteric Methane Emission Measurement

Enteric CH_4 emissions were measured from individual cows using the SF_6 tracer technique with minor modification of Deighton et al. [9]. Generally, empty permeation tubes were filled with 450 mL of 99.999% pure SF_6 by immersing in liquid nitrogen. The release rate was determined by incubating the permeation tubes in an oven at 39 °C and weighting each one twice a week for 4 weeks. The calculated releasing rate of the SF_6 tubes ranged from 3.13 to 3.84 (mean, 3.28 ± 0.175) mg /day in period 1, from 3.10 to 3.70 (3.32 ± 0.266) mg/day in period 2, and from 3.10 to 3.60 (3.20 ± 0.167) mg/day in period 3, respectively. Each cow was randomly administrated with one SF_6 tube using a balling gun three weeks before the commence of the experiment.

A back-mounted harness was used to support the canister (volume = 1.85 L) for continuously sample collection; the canisters were washed by flushing 99.999% pure nitrogen and evacuated to over 98 kPa vacuum. The sampling rate of canister was approximately 0.25 mL/min by crimping a stainless-steel capillary tube within the sampling tubing. Canisters were removed after 24 h and residual vacuum was recorded before addition of nitrogen gas. Background gas samples of SF_6 and CH_4 were also collected daily by using six additional canisters that were either placed on the back of animals or about 2.0 m above ground level of the experimental barn.

Gas samples were analyzed using a gas chromatography system (GC126, Shanghai Precision Instruments Co., Ltd., Shanghai, China) equipped with a flame-ionization detector (FID) and an electron-capture detector (ECD). The ECD operated at 300 °C with a molecular sieve 0.5 nm column and the FID at 150 °C with a Porapack N 80–100 mesh column (Shanghai Precision Instruments Co., Ltd., Shanghai, China) for determination of SF_6 and CH_4, respectively. Ultra-high purity nitrogen gas (99.999%) was used as carrier gas at 40 mL/min flow and analysis was performed after calibration with standard gases for SF_6 and CH_4. The daily CH_4 emission was calculated as follows:

$$CH_4 = SF_6 \times [(CH_{4sample} - CH_{4background})/(SF_{6sample} - SF_{6background})] \times (16/146) \times 1000$$

where CH_4 is the calculated emission (g/d); SF_6 is the measured releasing rate of each SF_6 permeation tube (mg/day); the concentration of CH_4 sample and CH_4 background are expressed in ppm and

30. Zhang, X.M.; Wang, M.; Wang, R.; Ma, Z.Y.; Long, D.L.; Mao, H.X.; Wen, J.N.; Bernard, L.A.; Beauchemin, K.A.; Tan, Z.L. Urea plus nitrate pretreatment of rice and wheat straws enhances degradation and reduced methane production in in vitro ruminal culture. *J. Sci. Food Agric.* **2018**, *98*, 5205–5211. [CrossRef]
31. Lan, W.; Yang, C.L. Ruminal methane production: Associated microorganisms and the potential of applying hydrogen-utilizing bacteria for mitigation. *Sci. Total Environ.* **2019**, *654*, 1270–1283. [CrossRef]
32. Cunha, C.S.; Marcondes, M.I.; Veloso, C.M.; Mantovani, H.C.; Pereira, L.G.R.; Tomich, T.R.; Dill-McFarland, K.A.; Suen, G. Compositional and structural dynamics of the ruminal microbiota in dairy heifers and its relationship to methane production. *J. Sci. Food Agric.* **2019**, *99*, 210–218. [CrossRef]
33. Intergovernmental Panel on Climate Change (IPCC). *Revised 1996 IPCC Guidelines for National Greenhouse Gas Inventories*; Intergovernmental Panel on Climate Change, IPCC/OECD/IEA: Bracknell, UK, 1997.
34. Boadi, D.A.; Wittenberg, K.M. Methane production from dairy and beef heifers fed forages differing in nutrient density using the sulphur hexafluoride (SF_6) tracer gas technique. *Can. J. Anim. Sci.* **2011**, *82*, 201–206. [CrossRef]
35. Van Wyngaard, J.D.V.; Meeske, R.; Erasmus, L.J. Effect of concentrate level on enteric methane emissions, production performance, and rumen fermentation of Jersey cows grazing kikuyu-dominant pasture during summer. *J. Dairy Sci.* **2018**, *101*, 1–13. [CrossRef]
36. Hristov, A.N.; Kebreab, E.; Niu, M.; Oh, J.; Bannink, A.; Bayat, A.R.; Boland, T.B.; Brito, A.F.; Casper, D.P.; Crompton, L.A.; et al. Uncertainties in enteric methane inventories, measurement techniques, and prediction models. *J. Dairy Sci.* **2018**, *101*, 1–20. [CrossRef]
37. Yan, T.; Porter, M.G.; Mayne, C.S. Prediction of methane emission from beef cattle using data measured in indirect open-circuit respiration calorimeters. *Animal* **2009**, *3*, 1455–1462. [CrossRef]
38. Hess, P.S.A.; Eckard, R.J.; Jacobs, J.L.; Hannah, M.C.; Moate, P.J. Comparison of five methods for the estimation of methane production from vented in vitro systems. *J. Sci. Food Agric.* **2019**, *99*, 109–116. [CrossRef]
39. Charmley, E.; Williams, S.R.O.; Moate, P.J.; Hegarty, R.S.; Herd, R.M.; Oddy, V.H.; Reyenga, P.; Staunton, K.M.; Anderson, A.; Hannah, M.C. A universal equation to predict methane production of forage-fed cattle in Australia. *Anim. Prod. Sci.* **2016**, *56*, 169–180. [CrossRef]
40. Knapp, J.R.; Laur, G.L.; Vadas, P.A.; Weiss, W.P.; Tricarico, J.M. Enteric methane in dairy cattle production: Quantifying the opportunities and impact of reducing emissions. *J. Dairy Sci.* **2014**, *97*, 3231–3261. [CrossRef]
41. Mills, J.A.N.; Kebreab, E.; Yates, C.M.; Crompton, L.A.; Cammell, S.B.; Dhanoa, M.S.; Agnew, R.E.; France, J. Alternative approaches to predicting methane emissions from dairy cows. *J. Anim. Sci.* **2003**, *81*, 3141–3150. [CrossRef]
42. Patra, A.K.; Lalhriatpuii, M. Development of statistical models for prediction of enteric methane emission from goats using nutrient composition and intake variables. *Agric. Ecosyst. Environ.* **2016**, *215*, 89–99. [CrossRef]

animals

MDPI

Article

The Effects of System Changes in Grazed Dairy Farmlet Trials on Greenhouse Gas Emissions

Tony van der Weerden [1,*], Pierre Beukes [2], Cecile de Klein [1], Kathryn Hutchinson [3], Lydia Farrell [2,†], Tinke Stormink [1,‡], Alvaro Romera [2], Dawn Dalley [4], Ross Monaghan [1], David Chapman [4], Kevin Macdonald [2] and Robyn Dynes [5]

1 AgResearch, Invermay Agricultural Centre, Mosgiel 9053, New Zealand; cecile.deklein@agresearch.co.nz (C.d.K.); tinke.stormink@outlook.com (T.S.); ross.monaghan@agresearch.co.nz (R.M.)

2 DairyNZ Ltd., Private Bag 3221, Hamilton 3240, New Zealand; pierre.beukes@dairynz.co.nz (P.B.); lydia.farrell64@gmail.com (L.F.); alvaro.romera@dairynz.co.nz (A.R.); kevin.macdonald@xtra.co.nz (K.M.)

3 AgResearch, Grasslands Research Centre, Palmerston North 4410, New Zealand; kathryn.hutchinson@agresearch.co.nz

4 DairyNZ Ltd., Canterbury Agricultural & Science Centre, Lincoln 7608, New Zealand; dawn.dalley@dairynz.co.nz (D.D.); david.chapman@dairynz.co.nz (D.C.)

5 AgResearch, Lincoln Research Centre, Lincoln 7674, New Zealand; robyn.dynes@agresearch.co.nz

* Correspondence: tony.vanderweerden@agresearch.co.nz; Tel.: +64-3-4899012; Fax: +64-3-4893739

† Present Address: School of Agriculture and Environment, College of Sciences, Massey University, Palmerston North 4442, New Zealand.

‡ Present Address: DuPont Industrial Biosciences, 2300 AE Leiden, The Netherlands.

Received: 15 October 2018; Accepted: 4 December 2018; Published: 7 December 2018

Simple Summary: Dairy farm system practices aimed at reducing nitrate leaching can also reduce emissions of the greenhouse gases methane and nitrous oxide. A study comparing 'current' and 'improved' grazed dairy system practices showed that 'improved' systems generally produced lower greenhouse gas emissions while milk production was maintained. The amount of feed eaten per hectare was the key driver of total greenhouse gas emissions per area, with 'improved' systems generally exhibiting lower total enteric methane and less N flowing through the herd.

Abstract: An important challenge facing the New Zealand (NZ) dairy industry is development of production systems that can maintain or increase production and profitability, while reducing impacts on receiving environments including water and air. Using research 'farmlets' in Waikato, Canterbury, and Otago (32–200 animals per herd), we assessed if system changes aimed at reducing nitrate leaching can also reduce total greenhouse gas (GHG) emissions (methane and nitrous oxide) and emissions intensity (kg GHG per unit of product) by comparing current and potential 'improved' dairy systems. Annual average GHG emissions for each system were estimated for three or four years using calculations based on the New Zealand Agricultural Inventory Methodology, but included key farmlet-specific emission factors determined from regional experiments. Total annual GHG footprints ranged between 10,800 kg and 20,600 kg CO_2e/ha, with emissions strongly related to the amount of feed eaten. Methane (CH_4) represented 75% to 84% of the total GHG footprint across all modelled systems, with enteric CH_4 from lactating cows grazing pasture being the major source. Excreta deposition onto paddocks was the largest source of nitrous oxide (N_2O) emissions, representing 7–12% of the total GHG footprint for all systems. When total emissions were represented on an intensity basis, 'improved' systems are predicted to generally result in lower emissions intensity. The 'improved' systems had lower GHG footprints than the 'current' system, except for one of the 'improved' systems in Canterbury, which had a higher stocking rate. The lower feed supplies and associated lower stocking rates of the 'improved' systems were the key drivers of lower total GHG emissions in all three regions. 'Improved' systems designed to reduced N leaching generally also reduced GHG emissions.

Keywords: environmental modelling; pasture systems; nitrous oxide; methane emissions; nitrate leaching

---

1. Introduction

Agriculture is responsible for 47.9% of total greenhouse gas (GHG) emissions in New Zealand (NZ) but contributions to national emissions of methane (CH_4) and nitrous oxide (N_2O) are 86% and 95%, respectively [1]. Dairy farms primarily emit two GHG: (1) CH_4 from enteric fermentation in the cow rumen, and (2) N_2O arising mainly from denitrification of urinary nitrogen (N) in the soil and nitrogen fertiliser application. Methane emissions from dairy cattle have increased 130% from 1990 to 2015 [1]. In 2015, enteric CH_4 was the major contributor (73%) while N_2O from agricultural soils represented 21% of total GHG emissions from the agricultural sector [1]. The main drivers for this change are a doubling of the national dairy cow population since 1990 [2] and an increase in the application of synthetic N fertiliser (>600%) over this same period [1]. In New Zealand, dairy cows graze ryegrass-dominant pastures, of which perennial ryegrass (*Lolium perenne*) is the major species, with supplements (e.g., maize silage, barley grain) typically less than 100 g/kg of feed intake. Cows typically calve at the end of winter (i.e., July–September) and are milked for 8–10 months [3]. Similar pasture-based grazing systems for dairy cows are used in Australia [4,5] and Ireland [6].

An important challenge facing the NZ dairy industry, and globally, is to develop farm systems that can maintain or increase production (to meet increasing population demand) and profitability, while reducing impacts on receiving environments including water and air [7–9]. Various hypotheses have been advanced on changing dairy farm systems to reduce their environmental impact. Lowering stocking rate can result in more feed per cow, resulting in better-fed cows with more production per cow [3]. Fewer cows with better feed conversion efficiency could mean less feed is required for maintenance and more is converted into product. If cow genetic merit can be improved at the same time then these lower-stocked, well-managed systems can produce the same amount of product per hectare as higher-stocked systems [10]. The lower-stocked system will require less feed per area and, therefore, reduce the amount of N consumed and excreted by the herd. Nitrogen leaching will be reduced since the amount of urinary N deposited onto pasture is a major source of nitrate leaching [11]. Also, with lower feed intake from the smaller, more efficient herd it can be expected that the amount of enteric CH_4 emitted will be reduced [3].

A reduction in N fertiliser use will usually reduce N leaching [12,13] and N_2O losses from soils [14]. Less fertiliser will reduce the total amount of pasture grown, and also assist in a small reduction in the total N concentration of the herbage [4,15] and, therefore, reduce the amount of N flowing through the stock, and excreted as urinary and dung N. Reducing N fertiliser can be achieved through optimising its use, by targeting N application to pastures that have the greatest yield potential and to paddocks displaying signs of deficiency (yellowing and poor performance), rather than whole-farm N applications [16]. There is also the possibility of some compensation from less fertiliser through improved clover vigour and soil health resulting in greater natural N fixation [17,18].

Other strategies for reducing environmental impacts of dairy farm systems include improved reproductive performance of the herd, which results in less involuntary culling and lower replacement rates (reviewed in [19]). Replacements produce CH_4 and urinary N without contributing to milk production [20]. Greater use of high energy/low N feed (grain or forage) will reduce total urinary N excreted through lowering N intake [21] and improve the energy intake when pasture growth or pasture quality is low. This strategy also dilutes the effect of excess crude protein supplied by the pasture [22]. Off-paddock facilities can be used to reduce N returns to pasture during periods of low N utilisation and in turn decrease the risk of N leaching in winter and spring [23]. In addition, off-paddock facilities protect wet pasture from treading damage [24,25].

Farmlet systems trials (Pastoral 21–Phase 2 (P21), [26,27]) were run over a five-season period from 2011 to 2016 with the aim of developing industry-accessible, adoptable, system-level solutions for profitably increasing production while reducing N leaching [7]. Four dairy regions were used to provide contrasting challenges to dairy production due to different soil types, climates and local management practices. 'Improved' dairy systems for each region were initially developed to improve water quality outcomes via strategic changes to the current system. Some of these changes were recognised to also deliver reduced emissions of GHG whilst maintaining or increasing milk production and profitability [3]. The key changes included:

1. Using fewer, higher producing, cows
2. Smaller N fertiliser inputs
3. Lower herd replacement rate
4. Greater use of high energy/low N feed
5. Using off-paddock facilities to reduce the time cows spend on pasture (or on forage crops).

These five components were used to design the P21 farmlet systems trials with all or some of them applied to the 'improved' system in each location. Here, we examine whether 'improved' dairy systems designed to reduced N leaching also reduce GHG emissions. The effects of these system changes, or 'stacked mitigation options', are evaluated for total GHG emissions and emissions intensity (kg GHG per kg milksolids) by comparing 'current' and 'improved' dairy systems in these locations. This analysis tests the hypothesis that system changes aimed at reducing nitrate leaching will also reduce total greenhouse gas emissions and emissions intensity (kg GHG per kg milksolids). As part of this study, we determined which of the five key system changes delivered the greatest benefit.

2. Methodology

2.1. Farmlets

Relatively small-scale farms (farmlets ranging from 13 to 39 ha) were used to evaluate the system changes implemented in each of the three regions studied: the Waikato, Canterbury, and Otago (Table 1). These systems' studies ran from 2011 to 2015 in the Waikato, from 2011 to 2014 in Canterbury and from 2012 to 2015 in Otago.

In all systems, replacement stock were removed from the milking platform at birth and reared on support blocks, returning to the milking platform (pasture areas used for feeding milking cows) as rising 2-year old cows (~22 months) before calving. In the modelling analysis, the support was included in the inventory calculations to ensure analysis of a complete system. N fertiliser applications on these blocks were assumed as 100 kg/ha/year, based on expert opinion.

Details of the methodology used for estimating GHG emissions from the farmlets in all three regions can be found in the Supplementary File S1.

Table 1. 'Improved' system changes applied to farmlet system trials in Waikato, Canterbury, and Otago, New Zealand.

Region	Fewer, Higher Producing, Cows	Reduced N Fertiliser Inputs	Reduced Herd Replacement Rate	Greater Use of High Energy/Low N Feed	Off-Paddock Facilities
Waikato	✓	✓	✓	✓	✓
Canterbury	✓	✓		✓	
Otago		✓	✓		✓

2.2. Waikato

Two farmlets (13 ha each) were established at Scott Farm, Hamilton, New Zealand (37°46′ S, 175°22′ E) in June 2011 [7]. One system represented a current Waikato farm system ('current'), while the other employed technologies that might be required in improved farm systems ('improved') to reduce nitrate leaching. The 'improved' system was based on the concept of producing the same amount of milk per ha, but with the highest level of efficiency allowed by available technologies.

The stocking and replacement rates were lower for the 'improved' system (Table 2) and dairy cows with higher genetic merit were used (breeding worth of $170 vs. $90, respectively; Table 2). Reducing the stocking rate (SR) increased the annual feed allowance per cow which, combined with the higher genetic merit of cows in the 'improved' system, led to increased kg MS/cow. All this translated into a reduced need for N input, as less feed was required (i.e., producing the same with less).

An off-paddock facility was used in the 'improved' system where cows were removed from pasture onto a loafing pad for between 8 and 16 h daily from March until June (autumn until early winter). The loafing pad, also called a stand-off pad, was a plastic-lined area with wood chip bedding where cows could lie and where some of the dung and urine could be collected into the effluent system. The goal was to reduce N returns to pasture during a period of low N utilisation, thereby reducing N leaching risk during periods of drainage in winter/spring. All solid excreta deposited to the standoff pad was collected and stored until the following spring. A further goal was to protect wet pastures from treading damage.

The 'improved' farmlet cows were offered up to 3 kg DM/cow/day of low-crude protein grain to improve their energy intake when pasture growth or pasture quality was low. This strategy also had the aim of diluting excess crude protein supplied by the pasture. The 'current' system used bought in pasture and maize silage when DM requirements could not be met from its pasture growth.

Liquid effluent from the milking shed, collecting yard, and loafing pad was spread on the farmlets. Dung solids deposited in the milking shed and collecting yard were mechanically separated from the liquid phase for both systems. Emissions from these solids, and from solids captured by the loafing pad in the 'improved' system, were included in the GHG footprint calculations although these solids were exported from the farmlets to be spread in another location.

2.3. *Canterbury*

The Canterbury systems trial (43°38′ S, 172°28′ E) examined the effect of two 'improved' farm systems with contrasting stocking rates of 3.5 and 5 cows/ha for 'improved(LOW)' and 'improved(HIGH)', respectively (Table 2; [16]). The farmlet size was 8.25 ha milking platform (MP) plus 2 ha wintering crop (WC) and 6.75 ha MP plus 1 ha WC, respectively.

The 'improved(LOW)' system used dairy cows with higher genetic merit than the 'improved(HIGH)' system (breeding worth of 140 vs. 133, respectively).

A combination of 'standard' ryegrass/white clover pasture and 'diverse' pasture (containing chicory, plantain, ryegrass, and clover) was incorporated into the 'improved(LOW)' system, in contrast to the 'improved(HIGH)' system that solely relied on 'standard' ryegrass/white clover pasture. In the 'improved(LOW)' system, non-lactating cows were wintered on forage kale and oats silage, while the 'improved(HIGH)' system cows were wintered on fodder beet and pasture silage. Cow replacement rates were the same for the two systems.

There was insufficient resourcing to include a 'current' system in Canterbury. However, a suitable farm nearby (Lincoln University Dairy Farm, 'LUDF') represented current Canterbury practices from 2011 to 2013, with a stocking rate of 4 cows/ha and non-lactating cows wintered on fodder beet [28]. Therefore, the LUDF was adopted as a representative 'current' system (Table 2), ensuring the methodology used for estimating the GHG footprint is consistent with the P21 farmlets.

Table 2. Key management features of 'current' and 'improved' systems in Canterbury, Waikato, and Otago for assessing differences in, and key drivers of, GHG emissions from dairy systems. Estimated N leaching losses are also included [7,16,27]. HIGH and LOW = high and low stocking rate; OPT = optimised feeding and DCG = duration controlled grazing. See text for further descriptions of each systems.

	Waikato		Canterbury			Otago		
System Features	**Current**	**Improved**	**LUDF (2011–2013)**	**Improved (HIGH)**	**Improved (LOW)**	**Current**	**Improved (OPT)**	**Improved (DCG)**
Milking platform								
Stocking rate (cows/ha)	3.2	2.6	4.0	5.0	3.5	3.0	2.8	2.8
Cow genetic merit ($BW #)	90	170	115	133	140	109	105	104
N fertiliser (kg N/ha/year)	137	52	345	311	158	109	42	73
Replacement rate (%)	22	18	23	23	23	23	18	18
High energy/low N feed	0	0.24 (Grain t DM/cow/year)	N/A	N/A	40% diverse pasture	N/A	N/A	N/A
Stand-off/housing	No	Yes	No	No	No	No	No	Yes
Milksolids production *								
(kg MS/ha/year)	1193	1163	1870	2335	1785	964	931	949
(kg MS/cow/year)	373	447	468	467	510	329	333	334
Wintering								
Winter feed	On platform	On platform	Fodder beet + Pasture silage	Fodder beet + Pasture silage	Kale + Oat silage	Kale	Kale	N/A
N fertiliser (kg N/ha/year)	N/A	N/A	150	200	307	200	200	N/A
Total dairy system N loss								
N leaching (kg N/ha/year)	54	31	54	55	58	18 ^	13 ^	11 ^

LUDF: Lincoln University Dairy Farm; N/A: not applicable; # Breeding worth, $ (May 2011); * Fat + Protein, measured in P21 study; ^ assessed as estimated N leaching [27].

2.4. Otago

In Otago (46°17′ S, 169°43′ E), three farmlet systems were used, consisting of 110 cows each (Table 2). Firstly, a 'current' system (37 ha milking platform) adopted management practices typical of the region. Secondly, a 'improved optimised' ('improved(OPT)') system (39 ha) focused on improved cow feeding without the need for additional spending on costly farm infrastructure, and thirdly, a 'improved duration-controlled grazing' ('improved(DCG)') system (39 ha) that utilised an off-paddock facility (loose-housed deep litter animal shelter) for housing cows periodically during winter, spring, and autumn [27].

There were several key differences between the 'current' and the two 'improved' systems. Both the 'improved(OPT)' and 'improved(DCG)' systems included a lower replacement rate (18%) compared with 23% for the 'current' system, while N fertiliser applications were lower (42–73 kg N/ha/year) on the 'improved' system milking platforms compared with 109 kg N/ha/year on the 'current' milking platform. In the 'improved(OPT)' system, pasture was supplemented with whole crop cereal silage during lactation. Short rotation (Italian) ryegrass pastures were also incorporated to better align pasture growth rates with cow feed demand.

In addition, optimised grazing management of winter brassica crops along with allocation of more feed per cow during winter months were used to improve body condition score (BCS) relative to the 'current' herd. A key management goal of this farmlet was to ensure that cows calved later, and in better condition, onto higher pasture covers to decrease the reliance on N fertiliser and supplements and better match pasture growth with cow demand.

In the 'improved(DCG)' system cows were removed from pasture overnight (12 h) in spring and autumn during the milking season, when critical soil moisture thresholds and grazing times were reached, to protect pastures from damage and, for autumn grazing, reduce urinary nitrogen return to soils prior to winter. Shorter grazing times in spring led to relatively large amounts of pasture requiring conservation as silage on this farmlet; combined with the relatively large amounts of effluent returned to pasture (more details below), N fertiliser inputs were lower compared with the control farmlet and ranged between 63 and 83 kg N/ha/year.

During winter, cows were housed full time in a loose-housed deep litter animal shelter with the aim of improving BCS relative to the 'current' herd through improved utilisation of feed energy. The shelter (767 m^2) initially contained 300 m^3 of woodchip bedding material, with another 150 m^3 added midway through the winter period. Adjacent to the shelter was a feeding apron. Non-lactating cows were housed in the animal shelter fulltime during the winter months (June until mid-August; see Table S2) until calving commenced.

Details of the manure management from the shelter use can be found in the Supplemental File. Briefly, liquid from excreta deposited in the animal shelter was collected in an effluent pond, whereas solid manure was removed from the shelter and stored prior to spreading onto the milking platform paddocks. Manure deposited onto the adjacent feeding apron was scraped and stored behind a weeping wall, with the liquid fraction contained in an effluent pond. Data collected by [29] were used for estimating the greenhouse gas emissions from the manure management.

2.5. Greenhouse Gas Emissions

Annual GHG emissions for each system were estimated for three (Canterbury and Otago) or four (Waikato) years, using calculations based on the New Zealand Agricultural Inventory methodology (NZAI; [1]). In brief, this methodology uses estimates of dry matter intake (DMI), N inputs, and N leaching losses, in combination with CH_4 and N_2O emission factors (EF). In this study a combination of NZAI emission factor values and CH_4 and N_2O emissions factors were used that were measured for key components of the milking platform or the wintering period for each system [30–32]. We used these targeted measurements to provide us with emission factor results for key components in the farmlets that we otherwise would not able to assess as the NZAI emission factors are not sufficiently disaggregated. For example, NZAI uses the same methane emission factor for all feeds and therefore cannot distinguish between different feed types. As some of the key changes in our 'improved' systems were related to different feed types, we used our targeted measurements to get more specific CH_4 (and N_2O) emission factors for these feed types. Similarly, as NZAI uses only one manure management system (anaerobic lagoons), we conducted targeted measurements of key components of the manure management system for the South Otago farmlet that used an off-paddock facility to ensure we could capture any difference in emissions as a result of the off-paddock facility.

The GHG footprint boundary was limited to CH_4 and N_2O emissions, and excluded carbon dioxide (CO_2) emissions from fertiliser manufacturing and use, fuel use, electricity use, and infrastructure construction. This ensured the footprint aligned with the boundaries of the NZAI. The footprint included both on-farm and off-farm sources of CH_4 and N_2O emissions. On-farm sources included enteric CH_4 from the milking platform and wintering paddocks, N_2O from soils receiving N inputs and CH_4 and N_2O emissions derived from manure management. Off-farm sources included N fertiliser use for producing pasture for replacement stock, N-excreta deposited by replacement stock, enteric CH_4 from replacement stock and N fertiliser used for growing crops and supplements.

In addition to farm-scale GHG footprints, key sources of emissions were separated to determine the impact of off-paddock facilities on GHG emissions by combining the sources of emissions that are influenced by the presence or absence of such a facility. These sources included (i) direct and indirect N_2O emissions associated with excreta deposition onto paddocks, because removing cows from paddocks onto off-paddock facilities would directly influence the amount of excreta deposited onto paddocks, (ii) and N_2O and CH_4 emissions associated with manure collected, stored, and subsequently applied to land (i.e., manure management). Both off-paddock facilities included in this study were assessed, the loafing pad in Waikato and animal shelter in Otago.

All CH_4 and N_2O emissions were converted to CO_2-equivalent emissions using the 100-year time horizon global warming potentials of 25 kg CO_2-equivalent per kg CH_4 and 298 kg CO_2-equivalent per kg N_2O [33]. Greenhouse gas footprints for each system were calculated on an area basis (kg CO_2-equivalent per milking platform hectare; kg CO_2e/ha; Table 3) and intensity basis (kg CO_2-equivalent per kg milksolids produced; kg CO_2e/kg MS; Table 4).

Table 3. Total emissions (kg carbon dioxide equivalents per hectare of the milking platform (kg CO_2e/ha MP) calculated for nitrous oxide (N_2O) and methane (CH_4) for improved farm systems tested on small-scale farms in three regions in New Zealand (Waikato, Canterbury, and Otago). HIGH and LOW = high and low stocking rate; OPT = optimised feeding and DCG = duration controlled grazing. See text for further descriptions of each systems. Note: may not sum to total due to rounding.

	Source	Waikato		Canterbury			Otago		
		Current	Improved	LUDF (2011–2013)	Improved (HIGH)	Improved (LOW)	Current	Improved (OPT)	Improved (DCG)
		3.2 Cows/ha	2.6 Cows/ha	4 Cows/ha	5 Cows/ha	3.5 Cows/ha	3 Cows/ha	2.8 Cows/ha	2.8 Cows/ha
N_2O	Urine + dung	1602	1012	2110	2309	1519	1229	1108	760
	Fertiliser	390	150	1108	990	665	448	270	220
	Manure mgmt	99	390	254	31	21	11	9	387
	NH_3 volatilised	320	288	632	442	308	250	202	220
	NO_3 leached	217	162	247	223	241	86	64	44
	Replacement stock	285	189	346	676	460	329	246	249
	Total N_2O	2913	2191	4697	4671	3213	2353	1900	1879
CH_4	Ent Ferm. Pasture	8131	7044	9195	8965	8085	5752	5330	5493
	Ent Ferm. Supplement + Winter crop	1079	1081	2601	4120	2235	1982	2175	1976
	Ent Ferm. Replacement stock	1363	906	1541	2226	1558	1308	976	987
	Manure mgmt	124	183	594	633	490	433	411	1127
	Total CH_4	10,697	9214	13,931	15,944	12,368	9475	8892	9582
	Total CO_2e (kg)	13,610	11,405	18,628	20,615	15,582	11,827	10,792	11,461

Table 4. Emissions intensity (kg carbon dioxide equivalents per milksolids (kg CO_2e/kg MS)) calculated for nitrous oxide (N_2O) and methane (CH_4) for current and improved farm systems tested on small-scale farms in three regions in New Zealand (Waikato, Canterbury, and Otago). HIGH and LOW = high and low stocking rate; OPT = optimised feeding and DCG = duration controlled grazing. See text for further descriptions of each systems. Note: may not sum to total due to rounding.

	Source	Waikato		Canterbury			Otago		
		Current	Improved	LUDF (2011–2013)	Improved (HIGH)	Improved (LOW)	Current	Improved (OPT)	Improved (DCG)
		1193 kg MS/ha MP	**1164 kg MS/ha MP**	**1870 kg MS/ha MP**	**2335 kg MS/ha MP**	**1785 kg MS/ha MP**	**964 kg MS/ha MP**	**931 kg MS/ha MP**	**949 kg MS/ha MP**
N_2O	Urine + dung	1.3	0.9	1.1	1.0	0.9	1.3	1.2	0.8
	Fertiliser	0.3	0.1	0.6	0.4	0.4	0.5	0.3	0.2
	Manure mgmt	0.1	0.3	0.1	<0.1	<0.1	0.0	0.0	0.4
	NH_3 volatilised	0.3	0.2	0.3	0.2	0.2	0.3	0.2	0.2
	NO_3 leached	0.2	0.1	0.1	0.1	0.1	0.1	0.1	<0.1
	Replacement stock	0.2	0.2	0.2	0.3	0.3	0.3	0.3	0.3
	Total N_2O	2.4	1.8	2.5	2.0	1.8	2.4	2.0	2.0
CH_4	Ent Ferm. Pasture	6.8	6.1	4.9	3.8	4.5	6.0	5.7	5.8
	Ent Ferm. Supplement + Winter crop	0.9	0.9	1.4	1.8	1.3	2.1	2.3	2.1
	Ent Ferm. Replacement stock	1.1	0.8	0.8	1.0	0.9	1.4	1.0	1.0
	Manure mgmt	0.1	0.2	0.3	0.3	0.3	0.4	0.4	1.2
	Total CH_4	8.9	8.0	7.4	6.8	6.9	9.8	9.6	10.1
	Total CO_2e (kg)	11.3	9.8	9.8	8.8	8.7	12.3	11.6	12.1

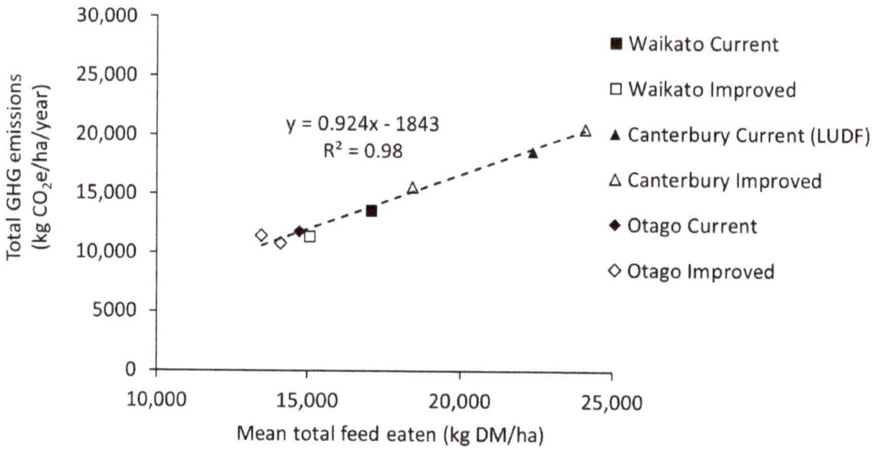

Figure 1. Relationship between mean total feed eaten (kg DM/ha) and total GHG emissions for 'current' and 'improved' systems trialled in three regions of New Zealand. Also included is the Canterbury 'current' system based on an updated analysis of the Lincoln University Dairy Farm (LUDF) system (P. Beukes, pers. comm.).

The Canterbury systems had the largest contrast in feed eaten (18,400 vs. 24,100 kg DM/ha), resulting in the largest difference in GHG emissions per hectare (15,582 vs. 20,615 kg CO_2e/ha; Table 3). Emissions from the LUDF 'current' system [28], were similar to the 'improved(HIGH)' system, at 18,628 kg CO_2e/ha. However, this 'current' system produced 1870 kg MS/ha, which was substantially less than the milk production of the 'improved(HIGH)' system (2335 kg MS/ha; Table 2).

Inclusion of off-paddock facilities in the Waikato and Otago 'improved' systems resulted in a decrease in excreta deposited onto paddocks and an increase in the amount of manure that required active management (see Table S4). To assess the impact of off-paddock facilities on GHG emissions, direct and indirect N_2O emissions associated with excreta deposition onto paddocks and N_2O and CH_4 emissions associated with manure management were collated. Attempting to present these emissions on a per area basis can be difficult to interpret, given these facilities are farm structures; therefore, emissions have been calculated and presented as kg CO_2e/cow/year. Our analysis showed that using an off-paddock facility results in a decrease in emissions per cow from excreta deposited onto paddocks, but this was more than offset by an increase in emissions per cow from manure management, resulting in a net increase in GHG emissions (Figure 2). The degree of the increase in emissions was dependent on the extent of the facility's use. For instance, the loafing pad in Waikato increased manure/excreta-related GHG emissions by 10% while the off-paddock facility in Otago led to a 35% increase in associated emissions per cow (Figure 2).

In order to assess the potential impact of adopting 'site-specific' rather than 'NZ-default' EF values on the calculated GHG emissions, we compared our results to those calculated when adopting the EF values from the NZ inventory methodology (results not shown). Adopting the NZ-default EF values had very limited impact on the relative difference in emissions between 'current' and 'improved' systems. For Waikato, the reduction in emissions remained at 16%. For Canterbury, the increase in GHG emissions from 'current to 'improved(HIGH)' changed from 7% (site-specific EFs) to 8% (NZ-default EFs). The decrease in GHG emissions from 'current' to 'improved(LOW)' changed from 18% to 19%. In Otago, using the NZ-default EF values did not impact on the relative difference between the 'current' and 'improved(OPT)', but it did slightly affect the result for 'improved(DCG)'. When using the site-specific EF values that were based on experimental results, a 3% reduction in total GHG emissions was calculated. However, adopting the inventory approach resulted in a 1%

reduction in GHG emissions between these two systems. This is most likely due to the fact that the NZ inventory methodology only includes the manure management system 'anaerobic lagoons', as this system is applicable to the vast majority of NZ dairy systems. However, given that the manure management system of the 'improved(DCG)' system is very different to the 'anaerobic lagoon' system, we believe the calculations based on the site-specific EF values are more accurate and realistic. Our study, therefore, represents the GHG footprint of dairy systems adopting 'current' and 'improved' management practices, estimated using the NZ inventory approach combined with site specific EF values where appropriate.

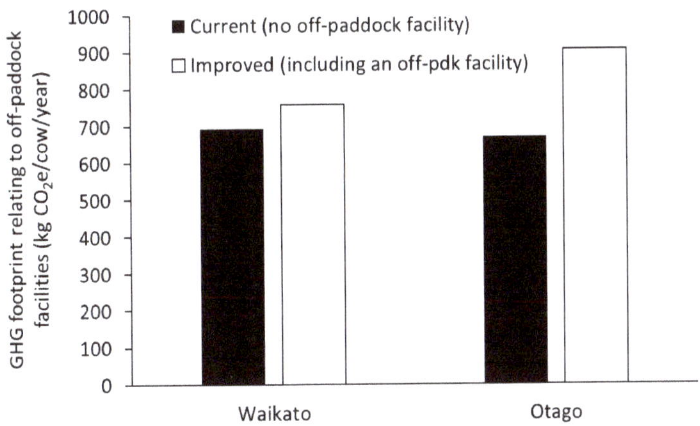

Figure 2. Effect of off-paddock facilities on net greenhouse gas emissions (kg CO_2e/cow/year) associated with N_2O from excreta deposition onto paddocks, and N_2O and CH_4 emissions from manure management.

3.3.2. GHG Emissions Intensity (kg CO_2e/kg MilkSolids)

When total emissions were represented on an intensity basis, emissions ranged from 8.7 kg to 12.3 kg CO_2e/kg MS. The 'improved' systems in all three regions produced lower GHG emission intensities, with reductions of 13%, 11–12%, and 6% being calculated for 'improved' systems in Waikato, Canterbury, and Otago (OPT), when compared with the corresponding 'current' systems (Table 4). In most cases, the lower GHG intensities were largely a result of management practices such as reduced N fertiliser use and lower replacement rates lowering GHG emissions from the 'improved' system, because the difference in MS production between 'improved' and 'current' systems was small (Figure 3). For the 'improved(HIGH)' system in Canterbury, lower GHG intensities were largely a result of higher MS production, rather than reduced inputs. The 'improved(DGC)' system in Otago, based around standing cows off wet paddocks, and wintering cows in an animal shelter, produced a small reduction (2%) in the GHG emissions intensity due to increased emissions from manure management, as shown above. Across all regions, total GHG emissions/ha were strongly related to amount of milk produced/ha (Figure 3). Analysis of the data, determined by comparing a single regression model with a combined model that included system type as a treatment, showed there was no significant difference in the 'current' vs. 'improved' systems ($p > 0.05$). The lack of significance was possibly due to the relatively small dataset, as the results showed lower GHG emissions for the same MS production from the 'improved' systems compared to the 'current' systems.

We compared our GHG emission intensity results with Gerber et al. [34], which presents N_2O and CH_4 emissions intensities from dairy cattle systems in 155 countries. For the comparison, we converted our results from kg MS to kg of fat and protein corrected milk (FPCM; [34]). The NZ dairy cattle systems modelled in our study produced between 4200 and 6700 kg FPCM per year, with CH_4 emissions of have been recent concerns linking consequences of indigenous deforestation with palm oil production, which has placed pressure on restricting the use of PKE on NZ dairy farms [49]. Deforestation results in additional GHG emissions through land use change, among other impacts. While our study has focused solely on biological emissions (i.e., N_2O and CH_4 only) from dairy production, we have estimated the increase in total GHG emissions for the Otago systems when the carbon (C) footprint associated with the use of PKE is included. A recent study suggests a C footprint of 0.506 kg CO_2-equivalents per kg PKE DM used on NZ dairy farms [50]—this value includes emissions due to land use change. Based on this value, the additional C footprint from the use of PKE in the CON, OPT, and DCG systems in Otago averaged over three years was estimated to be 49, 12, and 43 kg CO_2e/ha/year, respectively. Compared to the biological emission-based footprint (Table 3), the inclusion of the LCA-based C footprint associated with PKE use represents an additional 0.4%, 0.1%, and 0.4%, respectively. This additional footprint does not change the emission intensity values shown in Table 4, as the increase was less than <0.1 kg CO_2e/kg MS.

The New Zealand dairy industry aims to identify dairy systems that can maintain or increase production while reducing impacts on receiving environments including water and air [9]. As such, the 'improved' systems designed in this study included a package of measures that incorporated the best available knowledge to reduce impacts on water whilst maintaining or increasing productivity [7,16,27]. Improved systems will be more attractive to farmers if they deliver additional benefits such as reduced impact on water quality, or if a reduction in milk production is not associated with a reduction in profitability. Across the three regions, the implementation of stacked mitigation options aimed at reducing N leaching may have changed the relationship between MS and GHG emissions per hectare. While the analysis across regions showed no significant difference in the regression models describing milk production and GHG emissions for 'current' and 'improved' systems, probably due to the relatively small dataset, it does suggest a reduction in emissions intensity may be possible.

Our modelling suggests, for the systems examined here, total GHG emissions could be reduced by between 4% and 16%. Given the New Zealand dairy industry plan to contribute to meeting this nation's 2030 emissions reduction target of 30% below 2005 levels [51], our modelled reductions are relatively modest. These reductions are based on currently available management options and research into additional agricultural mitigation options is continuing to be an important focus for New Zealand [48]. This includes research into developing low-methane animals, methane vaccines and inhibitors, low GHG feeds, and novel nitrification inhibitors [52–54]. Increasingly 'improved' dairy systems modelled here, and future strategies identified through the current research efforts, will be essential options for farmers to meet social and regulatory requirements in New Zealand. Profitability and cost effectiveness of the 'improved' systems modelled here are yet to be explored.

4. Conclusions

Our analysis indicates that system changes aimed at reducing nitrate leaching can also reduce total greenhouse gas emissions and emissions intensity. The reduced feed supplies and associated lower stocking rates of the 'improved' farmlet systems evaluated here were the key drivers of lower total GHG emissions in all three regions. The main effects of these improved farmlet attributes were smaller total enteric methane emissions and less N flowing through the herd, which lowered N excretion and, therefore, direct and indirect N_2O losses. A system with fewer cows with greater genetic merit contributed to greater milk production per cow per lactation, less N leaching, and lower GHG emission intensities. Smaller but important contributions to lowering emissions were made by dietary changes, e.g., introducing low-protein grain supplements, cereal silages, fodder beet winter feed, herb-containing ryegrass pastures, and by lowering herd replacement rates. Off-paddock facilities contributed to protecting wet soils and reducing N leaching, but resulted in pollution swapping and increased total GHG emissions per cow. Our results suggest that total GHG emissions can be reduced through lower-stocked systems, where individual cow performance is optimised through better feeding of high genetic merit animals to compensate for the lower stocking rates.

Supplementary Materials: The following are available online at http://www.mdpi.com/2076-2615/8/12/234/s1, Supplementary File S1.

Author Contributions: Conceptualization, C.d.K. and R.D.; Methodology, P.B., C.d.K, L.F., T.S. and T.v.d.W.; Validation, P.B., C.d.K., K.H. and T.v.d.W.; Investigation, D.C., D.D., K.M., A.R. and R.M.; Data Curation, P.B., D.D., L.F., T.S. and T.v.d.W.; Writing—Original Draft Preparation, P.B., C.d.K., R.D., K.H. and T.v.d.W.; Writing—Review & Editing, P.B., D.C., D.D., C.d.K., R.D., K.M., A.R., R.M. and T.v.d.W.; Funding Acquisition, C.d.K. and R.D.

Funding: This research was funded by the New Zealand Agricultural Greenhouse Gas Research Centre (http://www.nzagrc.org.nz) (project number 14-IFS8.2) and the Sustainable Land Management and Climate Change Fund (https://www.mpi.govt.nz/funding-and-programs/farming/sustainable-land-management-and-climatechange-research-program/) (contract AGR131402) administered by the Ministry for Primary Industries (Wellington, New Zealand).

Acknowledgments: We thank Dave Clark, Chris Glassey, Chris Roach, Grant Edwards, David Stevens, and Andrew Wall for their involvement in the initial set up and on-going management decision making of these farm system studies. We also thank Alasdair Noble for assistance with statistical analysis. This work could not have been completed without permission to align with work conducted through the Pastoral 21 Programme (C10X1117), jointly funded by the Ministry of Business, Innovation and Employment (MBIE), DairyNZ Inc., Fonterra, Beef + Lamb New Zealand and the Dairy Companies Association of New Zealand.

Conflicts of Interest: The authors declare no conflicts of interest.

References

1. Ministry for the Environment. *New Zealand's Greenhouse Gas Inventory 1990–2015*; Publication Number: ME 1309; Ministry for the Environment: Wellington, New Zealand, 2017; p. 502, ISSN 1179-223X (electronic). Available online: http://www.mfe.govt.nz/node/23304/ (accessed on 8 August 2017).
2. DairyNZ and LIC. New Zealand Dairy Statistics 2015–16. Available online: https://www.dairynz.co.nz/publications/dairy-industry/new-zealand-dairy-statistics-2015-16/ (accessed on 31 July 2017).
3. Beukes, P.C.; Gregorini, P.; Romera, A.J. Estimating greenhouse gas emissions from New Zealand dairy systems using a mechanistic farm model and inventory methodology. *Anim. Feed Sci. Technol.* **2011**, *166–167*, 708–720. [CrossRef]
4. Pembleton, K.G.; Rawnsley, R.P.; Turner, L.R.; Corkrey, R.; Donaghy, D.J. Quantifying the interactions between defoliation interval, defoliation intensity and nitrogen fertiliser application on the nutritive value of rainfed and irrigated perennial ryegrass. *Crop Pasture Sci.* **2017**, *68*, 1100–1111. [CrossRef]
5. Wales, W.J.; Kolver, E.S. Challenges of feeding dairy cows in Australia and New Zealand. *Anim. Prod. Sci.* **2017**, *57*, 1366–1383. [CrossRef]
6. Coffey, E.L.; Delaby, L.; Fleming, C.; Pierce, K.M.; Horan, B. Multi-year evaluation of stocking rate and animal genotype on milk production per hectare within intensive pasture-based production systems. *J. Dairy Sci.* **2018**, *101*, 2448–2462. [CrossRef] [PubMed]
7. Beukes, P.C.; Romera, A.J.; Gregorini, P.; Macdonald, K.A.; Glassey, C.B.; Shepherd, M.A. The performance of an efficient dairy system using a combination of nitrogen leaching mitigation strategies in a variable climate. *Sci. Total Environ.* **2017**, *599–600*, 1791–1801. [CrossRef]
8. O'Brien, D.; Geoghegan, A.; McNamara, K.; Shalloo, L. How can grass-based dairy farmers reduce the carbon footprint of milk? *Anim. Prod. Sci.* **2016**, *56*, 495–500. [CrossRef]
9. Clark, D.A.; Caradus, J.R.; Monaghan, R.M.; Sharp, P.; Thorrold, B.S. Issues and options for future dairy farming in New Zealand. *New Zeal. J. Agric. Res.* **2007**, *50*, 203–221. [CrossRef]
10. Beukes, P.C.; Romera, A.J.; Clark, D.A. Farming within limits—Can systems fundamentals be the key? *N. Z. Soc. Anim. Prod.* **2013**, *74*, 241–244.
11. McDowell, R.W.; Monaghan, R.M.; Dougherty, W.; Gourley, C.J.P.; Vibart, R.; Shepherd, M. Balancing water-quality threats from nutrients and production in Australian and New Zealand dairy farms under low profit margins. *Anim. Prod. Sci.* **2017**, *57*, 1419–1430. [CrossRef]
12. Ledgard, S.F.; Penno, J.W.; Sprosen, M.S. Nitrogen inputs and losses from clover/grass pastures grazed by dairy cows, as affected by nitrogen fertiliser application. *J. Agric. Sci. Camb.* **1999**, *132*, 215–225. [CrossRef]

13. Wachendorf, M.; Büchter, M.; Trott, H.; Taube, F. Performance and environmental effects of forage production on sandy soils. II. Impact of defoliation system and nitrogen input on nitrate leaching losses. *Grass Forage Sci.* **2004**, *59*, 56–68. [CrossRef]
14. Davidson, E.A. The contribution of manure and fertilizer nitrogen to atmospheric nitrous oxide since 1860. *Nat. Geosci.* **2009**, *2*, 659–662. [CrossRef]
15. Shepherd, M.; Lucci, G. A review of the effect of autumn nitrogen fertiliser on pasture nitrogen concentration and an assessment of the potential effects on nitrate leaching risk. *Proc. N. Z. Grassl. Assoc.* **2013**, *75*, 197–202.
16. Chapman, D.; Edwards, G.; Dalley, D.; Cameron, K.; Di, H.; Bryant, R.; Romera, A.; Clement, A.; Malcolm, B.; Curtis, J. Nitrogen leaching, production and profit of irrigated dairy systems using either low or high inputs of fertiliser and feed: The Pastoral 21 experience in Canterbury. In *Science and Policy: Nutrient Management Challenges for the Next Generation*; Currie, L.D., Hedley, M.J., Eds.; Occasional Report No. 30; Fertilizer and Lime Research Centre, Massey University: Palmerston North, New Zealand, 2017.
17. Ledgard, S.F.; Sprosen, M.S.; Penno, J.W.; Rajendram, G.S. Nitrogen fixation by white clover in pastures grazed by dairy cows: Temporal variation and effects of nitrogen fertilization. *Plant Soil* **2001**, *229*, 177–187. [CrossRef]
18. Egan, M.; Galvin, N.; Hennessy, D. Incorporating white clover (*Trifolium repens* L.) into perennial ryegrass (*Lolium perenne* L.) swards receiving varying levels of nitrogen fertilizer: Effects on milk and herbage production. *J. Dairy Sci.* **2018**, *101*, 3412–3427. [CrossRef] [PubMed]
19. Knapp, J.R.; Laur, G.L.; Vadas, P.A.; Weiss, W.P.; Tricarico, J.M. Invited review: Enteric methane in dairy cattle production: Quantifying the opportunities and impact of reducing emissions. *J. Dairy Sci.* **2014**, *97*, 3231–3261. [CrossRef]
20. Waghorn, G.C. Application of greenhouse gas mitigation strategies on New Zealand farms. In Proceedings of the International Conference, Livestock and Global Climate Change 2008, Hammamet, Tunisia, 17–20 May 2008.
21. Kebreab, E.; France, J.; Beever, D.E.; Castillo, A.R. Nitrogen pollution by dairy cows and its mitigation by dietary manipulation. *Nutr. Cycl. Agroecosyst.* **2001**, *60*, 275–285. [CrossRef]
22. Luo, J.; Ledgard, S.F.; de Klein, C.A.M.; Lindsey, S.B.; Kear, M. Effects of dairy farming intensification on nitrous oxide emissions. *Plant Soil* **2008**, *309*, 227–237. [CrossRef]
23. de Klein, C.A.M.; Ledgard, S.F. An analysis of environmental and economic implications of nil and restricted grazing systems designed to reduce nitrate leaching from New Zealand dairy farms. I. Nitrogen losses. *N. Z. J. Agric. Res.* **2001**, *44*, 201–215. [CrossRef]
24. Laurenson, S.; Houlbrooke, D.J.; Beukes, P.C. Assessing the production and economic benefits from preventing cows grazing on wet soils in New Zealand. *J. Sci. Food Agric.* **2016**, *96*, 4584–4593. [CrossRef]
25. van der Weerden, T.J.; Laurenson, S.; Vogeler, I.; Beukes, P.C.; Thomas, S.M.; Rees, R.M.; Topp, C.F.E.; Lanigan, G.; de Klein, C.A.M. Mitigating nitrous oxide and manure-derived methane emissions by removing cows in response to wet soil conditions. *Agric. Syst.* **2017**, *156*, 126–138. [CrossRef]
26. Chapman, D.; Clark, D.; Beukes, P.; Edwards, G.; Dalley, D.; Pinxterhuis, I.; Romera, A.; Shepherd, M.; Cameron, K.; Di, H.; et al. Strategies for increasing dairy systems research approach in two regions of New Zealand. *Proc. Australas. Dairy Sci. Symp.* **2012**, *5*, 417–421.
27. Shepherd, M.; Hedley, M.; MacDonald, K.; Chapman, D.; Monaghan, R.; Dalley, D.; Cosgrove, G.; Houlbrooke, D.; Beukes, P. A summary of key messages arising from the Pastoral 21 Research Programme. In *Science and Policy: Nutrient Management Challenges for the Next Generation*; Currie, L.D., Hedley, M.J., Eds.; Occasional Report No. 30; Fertilizer and Lime Research Centre, Massey University: Palmerston North, New Zealand, 2017.
28. Beukes, P.C. (DairyNZ, Hamilton, New Zealand). Personal communication, 2017.
29. Chrystal, J.; Monaghan, R.; Hedley, M.; Horne, D. Volumes and nutrient concentrations of effluent products generated from a loose-housed wintering barn with woodchip bedding. In *Integrated Nutrient and Water Management for Sustainable Farming*; Currie, L.D., Hedley, M.J., Eds.; Occasional Report No. 29; Fertilizer and Lime Research Centre, Massey University: Palmerston North, New Zealand, 2016.
30. Di, H.J.; Cameron, K.C.; Podolyan, A.; Edwards, G.R.; de Klein, C.A.M.; Dynes, R.; Woods, R. The potential of using alternative pastures, forage crops and gibberellic acid to mitigate nitrous oxide emissions. *J. Soils Sediments* **2016**, *16*, 2252–2262. [CrossRef]

31. Jonker, A.; Scobie, D.; Dynes, R.; Edwards, G.; de Klein, C.; Hague, H.; McAuliffe, R.; Taylor, A.; Knight, T.; Waghorn, G. Feeding fodder beet decreased methane emissions from dry and lactating dairy cows in pastoral systems. *Anim. Prod. Sci.* **2017**, *74*, 145–147.
32. van der Weerden, T.J.; Styles, T.M.; Rutherford, A.J.; de Klein, C.A.M.; Dynes, R. Nitrous oxide emissions from cattle urine deposited onto soil supporting a winter forage kale crop. *N. Z. J. Agric. Res.* **2017**, *60*, 119–130. [CrossRef]
33. Forster, P.; Ramaswamy, V.; Artaxo, P.; Berntsen, T.; Betts, R.; Fahey, D.W.; Haywood, J.; Lean, J.; Lowe, D.C.; Myhre, G.; et al. Changes in atmospheric constituents and in radiative forcing. In *Climate Change 2007: The Physical Science Basis*; Contribution of Working Group I to the Fourth Assessment Report of the Intergovernmental Panel on Climate Change; Cambridge University Press: Cambridge, UK; New York, NY, USA, 2007; p. 996.
34. Gerber, P.; Vellinga, T.; Opio, C.; Steinfeld, H. Productivity gains and greenhouse gas emissions intensity in dairy systems. *Livest. Sci.* **2011**, *139*, 100–108. [CrossRef]
35. Dillon, P.; Roche, J.R.; Shalloo, L.; Horan, B. Optimising financial returns from grazing in temperate pastures. Utilisation of Grazed Grass in Temperate Animal Systems. In Proceedings of the Satellite Workshop, 20th International Grassland Congress, Cork, Ireland, 3–6 July 2005; pp. 131–147.
36. Macdonald, K.A.; Penno, J.W.; Lancaster, J.A.S.; Roche, J.R. Effects of stocking rate on pasture production, milk production and reproduction of dairy cows in pasture-based systems. *J. Dairy Sci.* **2008**, *91*, 2151–2163. [CrossRef] [PubMed]
37. Aguirre-Villegas, H.A.; Passos-Fonseca, T.H.; Reinemann, D.J.; Larson, R. Grazing intensity affects the environmental impact of dairy systems. *J. Dairy Sci.* **2017**, *100*, 6804–6821. [CrossRef]
38. Oenema, J.; Burgers, S.; Verloop, K.; Hooijboer, A.; Boumans, L.; Berge, H.T. Multiscale effects of management, environmental conditions, and land use on nitrate leaching in dairy farms. *J. Environ. Qual.* **2010**, *39*, 2016–2028. [CrossRef]
39. Humphreys, J.; O'Connell, K.; Casey, I.A. Nitrogen flows and balances in four grassland-based systems of dairy production on a clay-loam soil in a moist temperate climate. *Grass Forage Sci.* **2008**, *63*, 467–480. [CrossRef]
40. Monaghan, R.M.; Paton, R.J.; Smith, L.C.; Drewry, J.J.; Littlejohn, R.P. The impacts of nitrogen fertiliser and increased stocking rate on pasture yield, soil physical condition and nutrient losses in drainage from cattle grazed pasture. *N. Z. J. Agric. Res.* **2005**, *48*, 227–240. [CrossRef]
41. Beukes, P.C.; Scarsbrook, M.R.; Gregorini, P.; Romera, A.J.; Clark, D.A.; Catto, W. The relationship between milk production and farm-gate nitrogen surplus for the Waikato region, New Zealand. *J. Environ. Manag.* **2012**, *93*, 44–51. [CrossRef] [PubMed]
42. Dynes, R.A.; Smeaton, D.; Rhodes, A.P.; Fraser, T.J.; Brown, M.A. Modelling farm management scenarios that illustrate opportunities farmers have to reduce greenhouse gas emissions while maintaining profitability. *N. Z. Soc. Anim. Prod.* **2011**, *71*, 167–171.
43. Vibart, R.; Vogeler, I.; Dennis, S.; Kaye-Blake, W.; Monaghan, R.; Burggraaf, V.; Beautrais, J.; Mackay, A. A regional assessment of the cost and effectiveness of mitigation measures for reducing nutrient losses to water and greenhouse gas emissions to air from pastoral farms. *J. Environ. Manag.* **2015**, *156*, 276–289. [CrossRef] [PubMed]
44. de Klein, C.A.M.; Monaghan, R.M.; Ledgard, S.F.; Shepherd, M. A system's perspective on the effectiveness of measures to mitigate the environmental impacts of nitrogen losses from pastoral dairy farming. In *Meeting the Challenges for Pasture-based Dairying: Proceedings of the 4th Australasian Dairy Science Symposium, 31 August–2 September 2010*; Edwards, G.R., Bryant, R.H., Eds.; Lincoln University: Christchurch, New Zealand, 2010; pp. 14–28.
45. Adler, A.A.; Doole, G.J.; Romera, A.J.; Beukes, P.C. Cost-effective mitigation of greenhouse gas emissions from different dairy systems in the Waikato region of New Zealand. *J. Environ. Manag.* **2013**, *131*, 33–43. [CrossRef] [PubMed]
46. Adler, A.A.; Doole, G.J.; Romera, A.J.; Beukes, P.C. Managing greenhouse gas emissions in two major dairy regions of New Zealand: A system-level evaluation. *Agric. Syst.* **2015**, *135*, 1–9. [CrossRef]
47. Basset-Mens, C.; Ledgard, S.; Boyes, M. Eco-efficiency of intensification scenarios for milk production in New Zealand. *Ecol. Econ.* **2009**, *68*, 1615–1625. [CrossRef]

48. Garnsworthy, P.C. The environmental impact of fertility in dairy cows: A modelling approach to predict methane and ammonia emissions. *Anim. Feed Sci. Technol.* **2004**, *112*, 211–223. [CrossRef]
49. Vibart, R.; MacKay, A.; Wall, A.; Vogeler, I.; Beautrais, J.; Dalley, D. A farm-scale framework to assess potential farm- and regional-scale implications of removing palm-kernel expeller as a supplementary feed for dairy cows. *Anim. Prod. Sci.* **2017**, *57*, 1336–1342. [CrossRef]
50. Ledgard, S.; Falconer, S. Total Greenhouse Gas Emissions from Farm Systems with Increasing Use of Supplementary Feeds Across Different Regions of New Zealand. Report Prepared for the Ministry of Primary Industries; SLMACC-AGR30624; Milestone 12; 2015. Available online: https://www.mpi.govt.nz/funding-and-programmes/farming/sustainable-land-management-and-climate-change-research-programme/sustainable-land-management-and-climate-change-slmacc-research-reports/ (accessed on 14 November 2018).
51. United Nations Framework Convention on Climate Change. Intended Nationally Determined Contributions, New Zealand Submission to ADP 2015. Available online: http://www4.unfccc.int/submissions/INDC/Published%20Documents/New%20Zealand/1/New%20Zealand%20INDC%202015.pdf (accessed on 29 October 2018).
52. New Zealand Agricultural Greenhouse Research Centre. Highlights 2017. Available online: https://www.nzagrc.org.nz/knowledge,listing,457,nzagrc-highlights-2017.html (accessed on 2 November 2018).
53. Pastoral Greenhouse Gas Research Consortium. Research 2018. Available online: https://www.pggrc.co.nz/research (accessed on 2 November 2018).
54. Ministry for Primary Industries. Global Partnerships in Livestock Emissions Research, an International Research Fund in Support of the Global Research Alliance on Agricultural Greenhouse Gases 2016. Available online: https://www.mpi.govt.nz/funding-and-programmes/farming/global-partnerships-in-livestock-emissions-research/ (accessed on 2 November 2018).

animals

MDPI

Article

Greenhouse Gas Emissions in Dairy Goat Farming Systems: Abatement Potential and Cost

Alexandra Sintori, Irene Tzouramani and Angelos Liontakis *

Agricultural Economics Research Institute (AGRERI), Hellenic Agricultural Organization (DEMETER), P.C. 11528 Athens, Greece; al_sintori@agreri.gr (A.S.); tzouramani@agreri.gr (I.T.)
* Correspondence: aliontakis@agreri.gr

Received: 31 August 2019; Accepted: 5 November 2019; Published: 10 November 2019

Simple Summary: Agriculture and particularly livestock farming is associated with the production of certain gases that contribute to global warming, commonly referred to as greenhouse gases. These gases are the result of the use of machinery and other inputs such as fertilizers and pesticides or are associated with the digestion process of animals. In this work, we have analyzed data from dairy goat farms in Greece to estimate the amount of greenhouse gases per kilogram of milk produced and identify farming practices that can result in their reduction. We found that greenhouse gases per kilogram of milk are fewer in farms that are characterized by higher milk production per goat. Furthermore, certain practices like the use of homegrown feed instead of purchased feed and the use of compound feedstuffs or oil-rich feedstuffs like cottonseed cake can result in lower greenhouse gases in goat farms. Also, the analysis suggests that the reduction of greenhouse gases can lead to a reduction of farm income, especially in the case of intensive farms. This finding has to be taken into consideration by policy makers and possible measures to compensate for this income loss have to be explored.

Abstract: Dairy goat farming is an important agricultural activity in the Mediterranean region. In Greece the activity offers occupation and income to thousands of families mainly located in mountainous and semi-mountainous areas of the country where it utilizes low productivity pastures and shrub lands. Furthermore, goats are more resilient to climate changes compared to other species, and are often characterized as ideal for keeping in drought areas. However, there is still limited evidence on total greenhouse gases (GHG) emitted from goat farms and their mitigation potential. In this context, this study aims to estimate GHG emissions of goat farms in Greece and explore their abatement options using an economic optimization model. Three case studies are explored i.e., an extensive, a semi-intensive and an intensive goat farm that correspond to the main goat production systems identified in Greece. The analysis aims to assess total GHGs as well as the impact of abatement on the structures, gross margins and labor inputs of the farms under investigation. The issue of the marginal abatement cost is also addressed. The results indicate that the extensive farm causes higher emissions/kg of milk produced (4.08 kg CO_2-eq) compared to the semi-intensive and intensive farms (2.04 kg and 1.82 kg of CO_2-equivelants, respectively). The results also emphasize the higher marginal abatement cost of the intensive farm. In all farm types, abatement is achieved primarily through the reduction of the livestock capital and secondarily by other appropriate farming practices, like substitution of purchased feed with homegrown feed.

Keywords: dairy goat farming; linear programming; GHG emissions; abatement cost; mitigation options; carbon footprint

1. Introduction

Goat farming is an important agricultural activity in Greece since it is mainly located in less favored areas of the country where it utilizes low-productivity pastureland and shrubland. It is estimated that goat farming yields income for 64,049 Greek farms that breed over 3.5 million goats [1]. The activity aims primarily at the production of milk and secondarily at the production of meat. According to Kitsopanidis [2], milk is on average responsible for over 70% of gross revenue of dairy goat farms, with the exception of very hardy, local breeds. It is estimated that 75% of the Greek goat milk is used for the production of cheeses, especially Feta. Furthermore, the activity contributes highly to regional development and helps maintain the population in depressed and marginal areas. Therefore, the preservation of the activity and the income it yields is important not only for farmers but also for policy makers.

The prevailing goat farming system in the country is the extensive one with or without transhumance, in which the nutrition of the livestock is based on grazing. The main characteristic of the extensive breeding farms is the low invested capital and the low-productivity livestock, consisting mainly of native races [3]. More modern and intensive farms that aim to increase their productivity through supplementary feeding, mainly from on-produced cereals and forage, are also present. Specifically, three commercial goat production systems are identified in Greece, namely the traditional extensive farming system, the semi-intensive farming system and the intensive farming system [2]. As mentioned above, in the extensive farming system feed requirements are met mainly through pasturing. In the semi-intensive farming system additional supplementary feed is provided, while in the intensive farming system no pasture is utilized. This heterogeneity among the alternative goat farming systems results in differences in their socioeconomic as well as their environmental sustainability.

One of the main environmental issues associated with livestock farming is the emission of greenhouse gases (GHGs). GHG emissions are particularly high in the case of ruminant livestock farming because of methane production through enteric fermentation [4,5]. The issue of GHG emissions in livestock farms has been addressed in a number of studies that focus mainly on cattle farms [6–9]. On the other hand, studies that focus on the emission of GHGs from sheep and goat farms refer mainly to meat and wool production farming systems that have different technical and economic characteristics from dairy farms (e.g., [10,11]). In the case of small ruminant dairy farming in Greece, limited studies on GHG emissions from sheep farms and their abatement potential are available (see [12]).

This study aims to address the issue of GHG emissions in dairy goat farms, using an economic optimization model, developed to capture and represent the structure and function of Greek goat farms. The use of such model in GHG studies has the advantage that it accounts for all possible sources of GHG emissions in goat farms and therefore reduced emissions from one source at the optimal solution does not result in increased emissions from other sources. Furthermore, since the model is an optimization model, abatement options are explored within the context of gross margin maximization. In other words, abatement practices and options that are proposed by the model are the least-cost options for the farms. Furthermore, this cost is precisely estimated and marginal abatement cost curves are derived. Thus, the analysis and the results it yields can be useful not only for agriculturalists but also for farmers, agricultural advisors and policy makers.

2. Materials and Methods

Optimization models, and specifically linear programming (LP) models are commonly used in agricultural studies (e.g., [13–16]). They yield the optimal amongst all feasible farm plans, taking into account technical and agronomic constraints of the farms. When the matter of GHG emissions in livestock and crop livestock farms is addressed, the complexity of the farm operation, the multiple sources of emissions, and the substitution possibilities between alternative activities require the use of a model that can capture all the interrelationships of these activities. That said, a number of studies use LP models to assess GHGs from various sources and identify cost-effective mitigation strategies (e.g., [9,10,17–21]).

The general expression of a linear programming model is as follows [22]:

$$\text{Max } g(x) = z = c_1x_1 + c_2x_2 + \ldots + c_nx_n \quad (1)$$

Subject to the constraints:

$$a_{11}x_1 + a_{12}x_2 + \ldots + a_{1n}x_n \le b_1$$

$$a_{21}x_1 + a_{22}x_2 + \ldots + a_{2n}x_n \le b_2$$

$$a_{m1}x_1 + a_{m2}x_2 + \ldots + a_{mn}x_n \le b_m$$

$$x_j \ge 0$$

where x_j ($j = 1, 2, \ldots, n$) are the decision variables of the model, they are unknown and determined by the model according to what maximizes gross margin (e.g., number of productive goats, hectares of cereals or forages etc.), c_j are known economic parameters (e.g., gross margin per unit of activity x_j), a_{ij} ($i = 1, 2, \ldots, m$) are known technical parameters (e.g., hours of labor or variable inputs per activity x_j) and b_i are also known parameters that express the availability of inputs (e.g., maximum available labor, capital or land inputs).

The characteristics of the optimization model that was used in this analysis is described in more detail in the following paragraphs. The data used in the analysis is also presented in the same section.

2.1. Model Specification

The model used in this analysis has 241 decision variables and 236 technical and economic constraints and it is presented in Figure 1 which represents the LP matrix. The decision variables of the model (Activities *i* in Figure 1) can be grouped in three main categories. The first one includes all the decision variables that refer to crops, pasture, grassland, shrubland and feeding i.e., to the distribution of produced and purchased feed. The second category refers to labor variables, and the final category to livestock and product variables. In addition to the non-negativity constraint of all decision variables, variables which refer to the livestock capital are restricted to receive only integer numbers and therefore the model is in fact a mixed-integer programming model. The constraints of the model (Constraints *j* in Figure 1) refer mainly to the feeding of the livestock but also to the availability of labor, land and other inputs. In Figure 1, the constraints of the model and their technical parameters (α_{ij}) as well as their right hand side parameters (b_i) are presented in all lines except for the first line that presents the activities (x_j).

The model is built to accurately reflect the structure and function of Greek goat farms and allocates all their available inputs and resources to the alternative economic activities and practices according to what maximizes their total gross margin (objective function). Therefore, the farm plan it suggests is in economic terms the optimal. It should be noted that the model is built according to the dairy sheep model described in detail in Sintori [23].

Constraints i ↓ / Activities j →	Cultivated grains (hect,)	Cultivated forages (hect.)	Grassland (hect.)	Pasture/ shrubland (hect.)	Monthly Consumption of produced grains (kg)*	Monthly Consumption of produced forages (kg)*	Monthly Consumption of grass (kg)*	Monthly Consumption of natural grass/shrubs (kg)*	Monthly Consumption of purchased concentrates (kg)*	Monthly Consumption of purchased forages (kg)*	Monthly Family labor inputs to crops (hours)	Monthly family labor inputs to livestock (hours)	Monthly hired labor inputs to crops (hours)	Monthly hired labor inputs to livestock (hours)	Number of female goats**	Number of male goats	Number of replacement animals**	Milk provided to kids/ suckling month i (kg)**	Milk provided to kids/ suckling month i+1 (kg)**	Milk sold month i (kg)**	Milk sold month i+1 (kg)**	Number of kids sold at each age group**	Crops sold (hect.)	Right hand side
Distribution of produced concentrates (grains)	*aij*				-1																			≥0
Distribution of produced forages		*aij*				-1																		≥0
Production of grass			*aij*				-1																	≥0
Production of shrubs/natural grass				*aij*				-1																≥0
Nutritional requirements					*aij*	*aij*	*aij*	*aij*	*aij*	*aij*					−*aij*	−*aij*	−*aij*	−*aij*	−*aij*	−*aij*	−*aij*	−*aij*		≥0
Concentrates to forage balance					−*aij*	1			−*aij*	1														≥0
Labor requirements crops	−*aij*	−*aij*	−*aij*								1		1										−*aij*	≥0
Labor requirements livestock												1		1	−*aij*	−*aij*	−*aij*			−*aij*	−*aij*	−*aij*		≥0
Family labour availability											1	1												≤*bi*
Kidding															*aij*		-1					-1		≥0
Replacement animals (February/November)															*aij*		-1							≥0
Male to Female ratio															*aij*	-1								≥0
Maximum milk production															*aij*			-1		-1				≥0
Continuous suckling																		1	-1					≥0
Continuous milk production																				1	-1			≥0
Total land	1	1	1	1																			1	≤*bi*
Total irrigated land	1	1	1																				1	≤*bi*
Total scrubland/ pasture				1																				≤*bi*
Total Variable cost	*aij*	*aij*	*aij*						*aij*	*aij*			*aij*	*aij*	*aij*	*aij*	*aij*					*aij*	*aij*	≤*bi*

*Total of 24 variables for each feed, 12 for adult animals and 12 for growing animals, **Two variables for each month representing the two alternative kidding periods

Figure 1. Representation of the original linear programming model.

2.1.1. Crop, Pasture, Grassland and Feeding Variables

Crop activities of the goat farms involve forage and grain production for livestock feeding but also crop production for sale. The main crops cultivated in Greek goat farms are maize for grain and alfalfa for hay. Other crops may also be cultivated like maize for silage or barley for grain. For each crop, pasture and grassland used, one variable is included in the model that expresses the land in hectares allocated to the activity. The economic parameter of the specific variable is negative and expresses the cultivation cost per hectare excluding labor which is represented by a different set of variables. Feeding variables are defined according to month, type of feed and livestock category. In other words, the consumption of each feed (e.g., produced maize, purchased maize, produced alfalfa hay, purchased mixtures) is presented monthly and for two separate categories of livestock i.e., growing and adult animals. This structure allows the model to simulate farm decision making regarding the distribution of homegrown feed and purchase of additional feeds throughout the year. Furthermore, the use of such detailed data on feeding practices of the farms allows the model to be accurate and realistic regarding the predicted type and amount of feed required by the livestock, but also the methane emitted from enteric fermentation. Finally, it should be emphasized that the economic parameters of the variables that refer to purchased feed are negative and represent the price per kilo for each feed, while the economic parameter of each variable that refers to the consumption of homegrown feed is zero, since the cost of production is in fact the economic parameter of the related crop variable and, therefore, it has already been accounted for.

2.1.2. Labor Variables

In order for goat farms to operate one main required input is labor. In Greece, goat farms are in their majority traditional and family owned with very low mechanization degree, since extensive and semi-intensive farms rarely even use milking machines. The amount of labor inputs required in these farm types are also increased because of grazing. Labor variables incorporated in the model represent the amount of labor inputs required each month of the year in hours. The model allocates the available family labor between all crop and livestock activities of the farm. Variables representing the additional hired labor required each month in the above activities are also incorporated in the model. The economic parameters of these variables are negative and represent the wage per hour in livestock and crop activities.

2.1.3. Livestock and Product Variables

Livestock variables incorporated in the model refer to the number of female and male goats, replacement animals and kids that constitute the livestock of the farm. Two variables are used to represent the female goats that are kept in the livestock representing two different kidding periods. One variable refers to the goats that give birth in late November and the other to the goats that give birth in late February. These two kidding periods were chosen to reflect the practices of Greek goat farms, that aim to satisfy the increased demand for goat meat during Christmas and Easter. The economic parameters of these variables are negative and represent the annual variable cost of keeping and breeding one animal for one year, except for labor and feeding cost, since both labor and feeding are represented by the variables already described in previous paragraphs. As far as kids are concerned, they are represented in the model by a set of variables according to their age group in months. Specifically, 12 variables are used in the model to reflect the young kids between the ages of one and six months born during the alternative two kidding periods. The economic parameters of these variables are positive and express the gross margin per animal sold at each specific age. As mentioned above, all livestock variables are allowed to receive only integer numbers.

The final set of variables incorporated in the model represents milk production per month and per kidding period. Milk variables include variables that refer to suckling as well as variables that refer to milk production for sale. The economic parameter of the milk for sale variables is positive and expresses milk price.

2.1.4. Feeding Constraints

The main component of the model reflects the balance of the monthly feed requirements of the livestock. Minimum intake of dry matter, net energy of lactation, nitrogen and fiber matter is ensured through monthly constraints. The feed requirements of the livestock are estimated according to Zervas et al. [24] (see Table 1). For the female productive goats these feed requirements include requirements for preservation, activity and pregnancy. Extra requirements for lactation are estimated per kilogram of produced milk. For male productive goats, the requirements refer to their preservation, activity and reproduction. For the replacement animals, the feed requirements are estimated every month taking into account the live-weight increase. The weight increase is also taken into account in the case of the kids, for which feed requirements are estimated for the period that they remain in the farm.

Table 1. Livestock feed requirements.

Animal Characteristics	Dry Matter (kg/day)	Digestible Nitrogen (g/day)	Net Energy for Lactation (MJ/day)
Productive goats			
Preservation			
Live weight (kilos)			
50	1.6	40	5.2
60	1.8	46	5.8
70	2.0	52	6.6
Pregnancy			
Live weight (kilos)			
50	1.4	105	8.4
60	1.5	120	9.0
70	1.6	140	9.8
Lactation (per kilo of milk)			
Fat content			
3.0%	-	50	2.8
3.5%	-	55	3.0
4.0%	-	60	3.2
4.5%	-	65	3.4
Male goats			
Live weight (kilos)			
80	2.1	63	8.1
100	2.2	75	9.6
Growing animals			
Age (in months)			
0–1		80	3.2
1–2	0.3–0.6	80	3.6
2–3	0.6–0.8	77	4.2
3–4	0.8–1.0	74	4.6
4–5	1.0–1.1	68	4.9
5–6	1.1–1.2	62	5.1
6–7	1.2–1.3	60	5.2

Source: Zervas et al., 2000 [24].

On-produced feed crops, external feed inputs, available grassland and pastureland/shrubland are used for the balance of the feed requirements of the flock. The composition and the nutritional value per kilogram of feedstuff is taken from Kalaisakis [25], Jarrige [26], Zervas et al. [24] and Feedipedia [27]

(Table 2). The nutritional value and the production of grassland and pasture are estimated taking into account Papachristou [28], Zervas et al. [24], F.R.I. [29], Platis and Papanastasis [30], Platis et al. [31]. Additional monthly constraints are incorporated in the model to ensure minimum and realistic intake of concentrate feeds, according to the feeding practices of the farms.

Table 2. Nutritional value of feed.

Type of Feed	Dry Matter (g/kg)	Digestible Nitrogen (g/kg)	Net energy for Lactation (Mj/kg)	Fiber Matter (g/kg)
Maize for grain	0.880	0.073	8.40	0.022
Barley for grain	0.860	0.077	7.60	0.044
Cotton seed	0.922	0.195	7.99	0.211
Alfalfa hay	0.850	0.105	4.10	0.280
Maize silage	0.300	0.018	2.15	0.053
Herbaceous material (pastures)	0.202	0.019	1.13	0.038
Shrubs	0.472	0.021	1.64	0.280
Oat for grazing (grassland)	0.275	0.015	1.45	0.090

2.1.5. Additional Constraints

Another component of the model ensures that monthly labor requirements of all production activities are balanced, mainly with family labor inputs. Additional hired labor can be used, if necessary, in both livestock and crop activities.

Land constraints are also incorporated in the model to ensure that the total area utilized by the various crop activities, grassland and pastureland/shrubland is smaller than the available land of the farm. Moreover, one land constraint refers to total available irrigated land of each farm and another to total available pastureland/shrubland. A final set of constraints reflects the demography of the livestock and the maximum milk and meat production capabilities per goat.

2.1.6. Greenhouse Gas (GHG) Emissions

In order to accurately derive mitigation options for the goat farms, it is important to identify all potential sources of GHGs related to the activity, and include them in the model. The main GHGs, in livestock farms are methane (CH_4) from enteric fermentation and methane and nitrous oxide (N_2O) from manure. In addition, in a crop-livestock farm, nitrous oxide emissions (N_2O) from nitrogen fertilizers should also be accounted for (see for example [10,32]). Carbon dioxide emissions (CO_2) from the use of machinery are an additional source of GHGs. A graphical representation of the model used in the analysis and the emission sources it includes is presented in Figure 2. The emissions sources that the model takes into account are also summarized in Table 3.

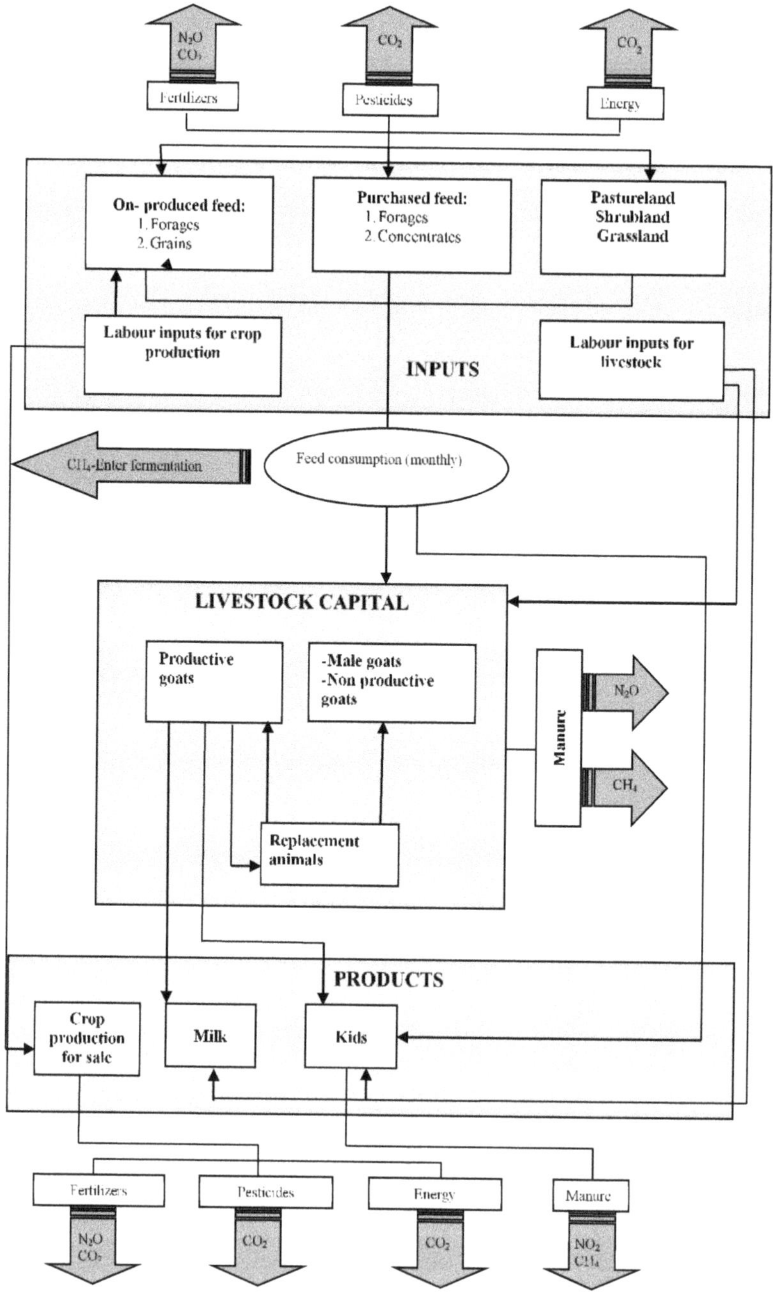

Figure 2. Graphical representation of the mathematical programming model and the emission sources considered in the analysis.

Table 3. Emission sources considered in the analysis.

Emission Sources	Included in the Analysis	Not Included in the Analysis
Livestock emissions		
Enteric CH_4	X	
CH_4 from manure deposited onto pasture	X	
CH_4 from manure management	X	
Direct N_2O emissions from manure deposited onto pasture	X	
Indirect N_2O emissions from manure deposited onto pasture	X	
N_2O emissions from leaching and run-off from manure deposited onto pasture		X
Direct N_2O emissions from manure management	X	
Indirect N_2O emissions from manure management	X	
N_2O emissions from leaching and run-off from manure management		X
Crops		
Direct N_2O emissions from use of fertilizers	X	
Indirect N_2O emissions from use of fertilizers	X	
N_2O emissions from leaching and run-off		X
CO_2 pre-chain emissions associated with the use manufacture and transport of inputs (fertilizers and pesticides)	X	
CO_2 from energy use within the farm	X	
Purchased feed		
Direct N_2O emissions from use of fertilizers	X	
Indirect N_2O emissions from use of fertilizers	X	
N_2O emissions from leaching and run-off		X
CO_2 pre-chain emissions associated with the use manufacture and transport of inputs (fertilizers and pesticides)	X	
CO_2 from energy use required for the cultivation and transport of purchased feed	X	

It should be noted that CH_4 and N_2O have been converted to CO_2-equivalents (CO_2-eq) using the conversion factors proposed by the Intergovernmental Panel on Climate Change (IPCC) [4] i.e., 1 kg of N_2O = 298 kg of CO_2-eq and 1 kg of CH_4 = 25 kg of CO_2-eq. The method used to estimate emissions from various sources in the goat farms is described in more detail in the following paragraphs. Emissions from all sources estimated as CO_2-equivalents are added together to estimate total GHG emissions of the goat farms.

Methane production from enteric fermentation is the most important source of GHGs in small ruminant livestock farms and it is associated with the feeding practices of each farm. Farmers choose to feed their livestock with on-produced feed and purchased feed taking into account their cost and their nutritional value. Linear programming models select the optimal combination of feedstuff and suggest the ration that helps maximize gross margin (least cost ration). For this reason, the ration used in this analysis is not fixed and methane emissions are predicted from intake, taking into account the requirements of the livestock estimated as previously described and the composition of feedstuff, that can be found in Table 2 (see also [33,34]). Specifically, for each of the variables that refer to feed consumption the methane emissions per kilogram have been estimated and included in the model as the technical parameter of this variable in a new constraint regarding GHG emissions. To estimate the percent of gross energy intake lost as methane from enteric fermentation (ECH_4/EB) the following equation was used [33]:

$$ECH_4/EB = 9.84 - 0.0461ADL - 0.0509EE + 0.00366St + 0.00648CP \quad (2)$$

where: ADL = g of lignin/kg of DM, EE = g of ether extract/kg of DM, St = g of starch/kg of DM and CP = g of protein/kg of DM.

Methane emissions from manure are estimated using the Tier 2 methodology proposed by the IPCC [4], which takes into account the management system of manure and the energy consumption of livestock (Equation (3)).

$$EF = (VS \cdot 365) \cdot \left[B_o \cdot 0.67kg/m^3 \cdot \sum_{S,k} \frac{MCF_{S,k}}{100} \cdot MS_{(S,k)} \right] \quad (3)$$

where: EF = annual methane emissions from manure (kg CH_4/head/year), VS = daily volatile solid excreted (kg of dry matter/head/day), B_o = maximum methane producing capacity for manure produced (0.18 m^3 CH_4/kg VS for Greece), $MCF_{(S,k)}$ = methane conversion factors for each manure management system and climate region (1.5% for Greece), $MS_{(S,k)}$ = fraction of manure handled using manure management system S to climate region k (estimated for each farm according to their farming practices) and 365 are the days within the year.

VS is estimated from the gross energy intake (GE) expressed in MJ/head/day, the digestibility of the feed (DE/100) e.g., 65%,the ash content of manure (ASH/100) (8% for goats according to the IPCC) and the conversion factor for dietary GE per kg of dry matter (18.45 MJ/kg), using Equation (4):

$$VS = GE/18.45 \cdot (1 - DE/100) \cdot (1 - ASH/100) \quad (4)$$

The methodology to estimate the energy requirements per livestock category, which is necessary for the implementation of the Tier 2 methodology has already been presented.

Direct N_2O emissions from manure management and pastureland are estimated according to the Tier 1 methodology [4], using the live weight of each livestock category (Equation (5)):

$$N_2O_{D(mm)} = \frac{44}{28} \cdot \sum_{S} Nex \cdot MS_{(S)} \cdot EF_{(S)} \quad (5)$$

where: $N_2O_{D(mm)}$ = direct N_2O emissions from manure management kg/year/head, N_{ex} = annual N excretion (kg of N/head/year), $EF_{(s)}$ = emission factor for direct N_2O emissions from manure management system S (kg N_2O-N/kg N). $EF_{(s)}$ equals 0.02 kg N_2O-N/kg N when manure is managed in solid storage and 0.01 kg N_2O-N/kg N when manure is deposited on pasture [4]. It should be noted that according to the IPCC guidelines N_2O emissions generated by manure deposited on pastures is reported under *Emissions from managed soils*. In this analysis, however, these emissions have been considered, so that comparison between grazing and housed animals can be made.

N_{ex} is estimated taking into account the typical animal mass (TAM) in kg/head and the N excretion rate using the equation (N_{rate} for goats = 1.28 kg of N/1000 kg of animal mass/day):

$$\text{Nex} = N_{rate} \cdot \frac{TAM}{1000} \cdot 365 \quad (6)$$

where 365 are the number of days within the year.

According to the IPCC (2006), for the estimation of indirect N_2O emissions, first the fraction of N that volatilizes as NH_3 and NO_x is estimated according to Equation (7) and then the amount of manure nitrogen that is lost due to volatilization of NH_3 and NOx is estimated using Equation (8):

$$N_{volatilization-MMS} = \sum_{S} Nex \cdot MS_{(S)} \cdot Frac_{GasMS,(s)} \quad (7)$$

$$N_2O_{G(mm)} = (N_{volatilization-MMS} \cdot EF_4) \cdot \frac{44}{28} \quad (8)$$

where: MMS stands for manure management system, $Frac_{GasMS(s)}$ = is the Fraction N that volatilizes as NH_3 and NO_x (0.12) and EF_4 = emissions factor for N_2O from N that volatilizes (0.010 N_2O-N/kg NH_3-N + NO_x –N volatilized).

In our analysis, we have also included direct and indirect N_2O emissions from the use of nitrogen fertilizers. First, the total amount of nitrogen applied in fields has been calculated using the amount and the type of fertilizer (see also [17,34]). Then direct and indirect emissions from the applied N have been estimated according to the Tier 1 methodology and the emission factors proposed by the IPCC [4].

Carbon dioxide linked to energy use is another GHG of crop-livestock farms. The main sources of energy in these farms are fuel (mainly diesel) and electricity (see also [7]). To estimate the emissions from energy use, fuel or electricity requirements for every farm operation and type of machinery are accessed and multiplied by appropriate emission factors [10]. Specifically, as far as electricity is concerned and due to the fact that in Greece lignite is used for the production of electricity, the emission factor considered in the analysis is quite high (0.855 kg of CO_2-eq per KWh). The emission factors per liter of petrol and diesel used in the analysis are 2.23 and 2.66 kg of CO_2-eq, respectively.

Other inputs, like fertilizers and pesticides have also caused GHG emissions when they were manufactured. These emissions have been taken into account as well, using farm-level data to estimate the amount of inputs used and related literature to estimate the emissions caused by the manufacture of these inputs. Carbon dioxide emissions from the manufacture of fertilizers are assumed 0.3 kg of CO_2 eq/kg of N, 0.9 kg of CO_2 eq/kg of P and 0.6 kg of CO_2 eq/kg of K. The energy requirements for the manufacture of herbicides, insecticides and fungicides are 287 MJ/kg, 263 MJ/kg and 195 MJ/kg, respectively [35–39]. Emissions are then calculated by multiplying the total energy requirements with 0.069, which is the amount of CO_2 produced per MJ of energy consumed.

Other pre-chain emissions have also been estimated and included in the analysis, following the work of Olesen et al. [7]. As mentioned above, farmers choose whether to feed their livestock with on-produced or purchased feed. Therefore, N_2O emissions from nitrogen fertilizers and CO_2 emissions from energy requirements have also been estimated per kilogram of purchased feed, according to the methodology that has already been presented in the previous paragraph. However, to estimate the amount of inputs (e.g., fertilizers) required for the production of the purchased feed data from 150 farms producing these feeds and operating in Continental Greece have been used. The data is part of a larger data set obtained during the implementation of the program "Search for Innovative Occupations of Tobacco Producers in the Rural Sector (Measure 9, Reg (EU) 2182/02)" and involve detailed information regarding the practices used to produce feedstuff commonly purchased by goat farms.

The original optimization model presented in Figure 1 was used to obtain the optimal farm plan of the goat farms. GHG emissions from various sources and total GHG emissions were then estimated at this optimal solution and used as the basis of our estimations (0% abatement level). The second step of our methodology is to derive the optimal farm plan across increasing levels of abatement, and assess impact on farm structure and gross margin. Following a number of studies (e.g., [17,23,40]), this was achieved by inserting an additional constraint in the model. Specifically, if a is the level of abatement ($a < 1$) and e_0^* the total emissions at the optimal farm plan, then a new constraint is inserted in the model which restricts total farm emissions below $(1 - a)e_0^*$. The shadow price of this constraint is used to estimate GHG marginal abatement cost for each production system. Additionally, marginal abatement cost curves are derived for each farm type. In order to obtain the marginal abatement cost curves the right-hand side parameter of the emissions constraint was reduced marginally i.e., 1 tone, and the impact on gross margin was estimated. This procedure was performed a number of times to derive the cost curve. It should be emphasized that this kind of sensitivity analysis is usually performed automatically in LP models, but in this analysis some variables are restricted to receive only integer numbers. Therefore, sensitivity analysis could not be performed automatically and marginal costs were obtained manually.

2.2. Materials

To estimate the parameters of the model (*cj*, *aij* and *bi*) data from actual goat farms were used. Studies that implement the LP methodology commonly utilize data from representative or typical farms of the region under study (see for example [14]). LP models are not statistical models and, therefore, data from a large number of farms is not usually required. On the other hand, to increase the predictive ability of such models detailed data was used and the model was validated through the comparison of the predicted values (optimal solution) and the actual values of the representative farms. In this analysis, data from three goat farms were used. The goat farms were selected to represent the common production systems identified in Greece. Table 4, summarizes the main characteristics of these production systems as described in the literature. All of the above characteristics were taken under consideration during the selection of the goat farms, the characteristics of which are also presented in Table 4. All farms are located in Continental Greece, specifically the intensive and the extensive farms, were located in the region of Thessaly (Prefectures of Karditsa and Magnesia, respectively) and the semi-intensive farm in the region of Epirus (Prefecture of Preveza).

Table 4. Main Characteristics of the production systems identified in Greece and of the representative farms used in the analysis.

Characteristics	Farming Systems [2]			Representative Farms		
	Extensive	**Semi-Intensive**	**Intensive**	**Extensive**	**Semi-Intensive**	**Intensive**
Farm size	No significant diversification (extensive usually larger)			350 productive goats	300	300
Breeds	Hardy local breeds,	Improved local breeds	Highly productive breeds foreign breeds or local improved breeds	Local breeds	Improved local breeds	Highly productive local improved breeds
Use of pastures-shrublands	About 80% of feeding requirements, supplementary feeding during winter	50% of the feeding requirements	0% of the feeding requirements	75% of the feeding requirements	30% of the feeding requirements	0% of the feeding requirements
Use of concentrates	About 15% of nutritional requirements	Higher than extensive, lower than intensive	Mainly used to satisfy livestock feeding requirements	20% of the feeding requirements	60% of the feeding requirements	62% of the feeding requirements
Annual milk yield (kg/goat)	Estimated for Makedonitiki breed by Kitsopanides [2] at 134	Estimated for Skopelou breed by Kitsopanides [2] at 292	Estimated for Saanen and Alpine breeds by Kitsopanides [2] at 580–625	115	300	520
Level of mechanization (level of usage of equipments (e.g., for preparation of feed, milking machines etc.)	Low	Moderate (usually no milking machine)	Very high	Low (no milking machine)	Moderate (no milking machine)	Very high
Invested capital/goat	Low-low productivity livestock	Moderate	Very high	Low	Moderate	Very high

Table 4. *Cont.*

Characteristics	Farming Systems [2]			Representative Farms		
	Extensive	Semi-Intensive	Intensive	Extensive	Semi-Intensive	Intensive
Prolificacy index (number of kids per goat per birth)	Estimated for Makedonitiki breed by Kitsopanides [2] at 1.14	Estimated for Skopelou breed by Kitsopanides [2] at 1.37	Estimated for Saanen and Alpine breeds by Kitsopanides [2] at 1.72–1.74	1.2	1.5	1.80
Percent of milk income to total farm income	Estimated for Makedonitiki breed by Kitsopanides [2] at 57%	Estimated for Skopelou breed by Kitsopanides [2] at 74%	Estimated for Saanen and Alpine breeds by Kitsopanides [2] at 80%	60%	75%	86%

As can be seen in Table 4, the livestock of the extensive farm consists of hardy local breeds that are characterized by low productivity. More specifically, the farm breeds 350 reproductive female goats with an annual production of milk of 115 kg/goat. The average live-weight of the goat is 50 kg and the prolificacy index is 1.2.

The farm produces barley and uses only 2 hectares of grassland. Feeding of the livestock is based mainly in pasturing, since the farm uses 100 hectares of summer pasture and 50 hectares of winter pasture (mainly shrub cover). Additional feed is purchased, mainly maize, cotton seed and alfalfa. Additional parameters used in the model regarding the extensive farm can be found in Table 5.

Table 5. Main parameters used in the linear programming (LP) model.

Model Parameter	Extensive Farm	Semi-Intensive Farm	Intensive Farm
Variable costs of cultivated crops (€/hectare)			
Maize for grain	-	2142	1651
Maize for forage	-	-	734
Alfalfa for hay	-	-	1148
Barley for grain	591	-	1071
Crop yield (tones/hectare)			
Maize for grain	-	15	11
Maize for forage	-	-	54
Alfalfa for hay	-	-	15
Barley for grain	3	-	3
Price of purchased feedstuff (€/kg)			
Maize for grain	0.20	0.20	-
Barley for grain	-	0.30	-
Alfalfa hay	0.22	0.22	0.16
Mixture	-	0.40	0.40
Cotton seed	0.25	-	-
Variable cost for livestock (except for feeding and labor) (€/adult goat)	18.96	26,49	39.2
Replacement rate	22%	7%	16%
Average price of meat sold (€/kg)	4.25	2.9	3.43
Average price of milk sold (€/kilo)	0.63	0.60	0.73

The livestock of the semi-intensive farm consists of 300 female productive goats, with an annual milk production of 300 kg/goat and a prolificacy index of 1.5. The farm maintains maize cultivation for grain production and utilizes 50 hectares of pastureland (mainly shrub cover). Additional purchased feed is used, namely maize, barley, alfalfa and ready to buy feed mixes for goats. Table 5, summarizes the main technicoeconomic characteristics of the semi-intensive farm.

Finally, the intensive farm has a livestock of 300 female productive goats with an annual production of milk of about 520 kg/goat, an average live-weight of 70 kg and a prolificacy index of 1.8. For the feeding of the highly productive livestock, maize for grain and forage production, alfalfa and barley are cultivated. Additionally, special feed mixtures and alfalfa is purchased (see also Table 5).

The detailed technical and economic data required from the three farms were obtained in the summer of 2015 and refer to the year 2014.

3. Results

Carbon emissions at the optimal farm plan for the extensive, the semi-intensive and the intensive farms were first estimated and are presented in Tables 6–8, respectively. The constraint on total emissions was then inserted and the emissions at the new optimal farm plans were again obtained for various levels of abatement (α = 10%, 15% and 20%), through parametric optimization. Emissions per source at various levels of abatement are also presented in Tables 6–8. The values of certain variables of the model, that summarize the optimal farm plan at these abatement levels for the extensive, the semi-intensive and the intensive farm, are presented in Tables 9–11, respectively. This way, the best abatement strategy for each farm can be identified. Finally, the marginal abatement cost for each of the farms was estimated and the marginal abatement cost curve is built and presented in Figures 3–5.

3.1. GHG Emissions

As can be seen in Tables 6–8, the results of the analysis emphasize the significance of CH_4 in goat farms. Methane represents 75%, 65% and 52% of total emissions of the extensive, the semi-intensive and the intensive farms, respectively. Methane emissions refer mainly to CH_4 from enteric fermentation, as the CH_4 produced from manure management is negligible. Methane is particularly high in the extensive-farming system, where the feeding of livestock is based on grazing. On the other hand in the case of the intensive farm, methane from enteric fermentation is considered low, because of the high amount of compound feed used in the ration. Nitrous oxide emissions from manure management are also a significant source of GHGs in dairy goat farms, since it accounts for 20%, 25% and 34% of total emissions of the extensive, the semi-intensive and the intensive farming systems.

Table 6. Annual greenhouse gas (GHG) emissions of the extensive farm (in kg of CO_2-eq).

Abatement (α)	0%		10%		15%		20%	
	Total	Per kg of Milk *	Total	Per kg of Milk	Total	Per kg of Milk	Total	Per kg of Milk
Total GHGs	305,576	4.08	275,025	3.95	259,740	3.92	244,461	3.91
CH_4 enteric fermentation	226,471	3.03	203,644	2.92	192,858	2.91	181,971	2.91
CH_4 manure	3207	0.04	2997	0.04	2848	0.04	2690	0.04
N_2O manure	59,771	0.80	55,901	0.80	53,103	0.80	50,158	0.80
N_2O fertilizer	899	0.01	450	0.01	450	0.01	450	0.01
N_2O fertilizer-purchased feed	3250	0.04	2914	0.04	2452	0.04	2050	0.03
CO_2 energy-purchased feed	8435	0.11	7187	0.11	6371	0.10	5420	0.09
CO_2 energy-farm	3543	0.04	1536	0.02	1536	0.02	1536	0.02

* These GHGs refer only to milk production. Meat production related GHGs are not presented separately since milk is the main product of the farms. For the allocation of the GHGs between milk and meat the share in the production value is used.

Table 7. Annual GHG emissions of the semi-intensive farm (in kg of CO_2-eq).

Abatement (α)	0%		10%		15%		20%	
	Total	Per kg of Milk	Total	Per kg of Milk	Total	Per kg of Milk	Total	Per kg of Milk
Total GHGs	284,120	2.04	255,708	2.00	238,268	1.98	227,296	1.96
CH_4 enteric fermentation	182,281	1.31	163,790	1.28	154,348	1.27	146,580	1.27
CH_4 manure	2767	0.02	2547	0.02	2379	0.02	2302	0.02
N_2O manure	71,219	0.51	65,691	0.51	61,153	0.51	59,284	0.51
N_2O fertilizer	1349	0.01	1349	0.01	1349	0.01	1349	0.01
N_2O fertilizer-purchased feed	6178	0.04	4700	0.04	3961	0.03	3400	0.03
CO_2 energy-purchased feed	14,662	0.11	11,967	0.09	9413	0.08	8716	0.08
CO_2 energy-farm	5665	0.04	5665	0.04	5665	0.05	5665	0.05

Table 8. Annual GHG emissions of the intensive farm (in kg of CO_2-eq).

Abatement (α)	0%		10%		15%		20%	
	Total	Per kg of Milk	Total	Per kg of Milk	Total	Per kg of Milk	Total	Per kg of Milk
Total GHGs	332,797	1.82	299,518	1.81	282,878	1.80	266,238	1.79
CH_4 enteric fermentation	169,926	0.93	152,886	0.92	144,684	0.92	136,352	0.92
CH_4 manure	3151	0.02	2846	0.02	2700	0.02	2,549	0.02
N_2O manure	111,501	0.61	100,615	0.61	95,638	0.61	90,150	0.61
N_2O fertilizer	3609	0.02	3100	0.02	2637	0.02	2372	0.02
N_2O fertilizer-purchased feed	7183	0.04	6206	0.04	5587	0.04	5075	0.03
CO_2 energy-purchased feed	23,363	0.13	20,036	0.12	17,882	0.11	16,124	0.11
CO_2 energy-farm	14,065	0.08	13,829	0.08	13,750	0.09	13,616	0.09

Emissions per kg of goat milk are estimated at 4.08, 2.04 and 1.82 kg of CO2-eq for the extensive, the semi-intensive and the intensive dairy goat production system respectively. The carbon footprint of goat milk is particularly high in the case of the extensive-farming system. On the other hand in the semi-intensive and the intensive farming system emissions per kg of milk are low and comparable to the emissions estimated for cow's milk (see also [6,41,42]). The reasons for this variation of carbon footprint among the alternative farming systems are the high productivity of more intensive farms and the significant amount of compounds in the ration used in the semi-intensive and the intensive farms, compared to the low productivity of the extensive farms and the grazing/forage-based nutrition.

As can be seen in Tables 6–8, when emissions are restricted to various levels, the emissions per kg of produced milk were also reduced, in all farm types. In other words, lower levels of total farm emissions correspond not only to lower milk production levels but also to lower carbon footprint of milk. Specifically, emissions from enteric fermentation per kg of milk are reduced across various levels of abatement, as the result of adopting appropriate feeding practices. Carbon dioxide emissions from purchased feed are also reduced, which indicates that either farms purchase fewer feedstuffs or purchase feedstuffs that cause fewer emissions when they are produced. These results depict the optimal abatement plan for goat farms, as will be discussed in more detail in the next paragraph.

3.2. Abatement Cost and Strategies

Tables 9–11 summarize the optimal farm plan of the extensive, semi-intensive and intensive farm. The tables emphasize the fact that in all cases abatement has a negative impact on farm gross margin, particularly in the case of the intensive farm. Specifically, in the intensive farm 10% and 20% reduction in emissions result in 9% and 18% loss in farm gross margin, respectively. The reduction in farm gross margin in the case of the semi-intensive and the extensive farm, when emissions are reduced by 10% and 20%, is about 2% and 5%, respectively.

Table 9. Optimal farm plan of the extensive farm.

Abatement (α)	0%		10%		15%		20%	
	Total	Per Female Goat	Total	Per Female Goat	Total	Per Female Goat	Total	Per Female Goat
Gross margin (€)	27,271	70	26,738	73	26,389	76	25,957	79
Total labour (hours)	5018	13	4672	13	4441	13	4197	13
Female productive goats	390	1	364	1	346	1	327	1
Purchased cottonseed cake (kg)	4312	11	4265	12	3746	11	4530	14
Barley for consumption (hectares)	1.5	0.00	0	0.00	0	0.00	0	0.00
Purchased barley (kg)	0	0.00	0	0.00	0	0.00	0	0.00
Grassland (hectares)	2	0.01	2	0.01	2	0.01	2	0.06
Purchased alfalfa (kg)	9748	25	7477	21	7586	22	5225	16
Purchased maize (kg)	45,849	118	41,271	113	34,082	99	27,740	85
Fresh grass/shrub (kg)	994,584	2550	902,372	2479	871,492	2519	838,790	2565
Winter pasture (hectares)	50	0.13	50	0.14	50	0.15	50	0.15
Summer pasture (hectares)	100	0.26	100	0.28	100	0.29	100	0.31
Crop cultivation for sale (hectares)	0.5		0		0		0	

Table 10. Optimal farm plan of the semi-intensive farm.

Abatement (α)	0%		10%		15%		20%	
	Total	Per Female Goat	Total	Per Female Goat	Total	Per Female Goat	Total	Per female Goat
Gross margin (€)	47,275	136	46,440	146	46,000	151	45,296	157
Total labour (hours)	7140	21	6559	21	6270	21	5,946	21
Female productive goats	348	1	319	1	305	1	289	1
Maize for consumption (hectares)	2	0.006	2	0.006	2	0.007	2	0.007
Pasture (hectares)	50	0.144	50	0.157	50	0.164	50	0.173
Purchased alfalfa (kg)	17,781	51	12,930	41	10,935	36	8,699	30
Purchased maize (kg)	59,692	172	32,812	103	26,419	87	20,570	71
Purchased barley (kg)	0	0	11,334	36	10,010	33	8779	30
Purchased mixture (kg)	33,764	97	27,404	86	24,693	81	22,606	78
Fresh grass/shrub (kg)	294,134	845	293,482	920	294,134	964	291,166	1007
Crop cultivation for sale (hectares)	0		0		0		0	

The impact of abatement in the case of the intensive farm can be explained by the high productivity and specialization of the farm in milk production. Over 85% of the gross production value of the farm comes from milk production, while milk yield/goat and price of milk/kg are very high (520 kg/goat and 0.73 €/kg of milk, respectively). The high productivity of the intensive farm is also emphasized by the high gross margin per goat in Table 11.

Furthermore, the analysis indicates that, in all farm types, the mitigation of GHGs is primarily achieved by the reduction of the herd size, especially when high levels of abatement are imposed. Specifically, in the extensive farm 10% and 20% abatement leads to 7% and 16% reduction in livestock size, respectively. In the case of the semi-intensive farm the reduction in livestock size is 8% and 17%, while in the intensive farm the reduction is even higher, 9% and 19%, respectively. These findings are in accordance with previous studies regarding mitigation of GHGs in livestock farms (see for example [17,23]).

Table 11. Optimal farm plan of the intensive farm.

Abatement (α)	0%		10%		15%		20%	
	Total	Per Female Goat	Total	Per Female Goat	Total	Per Female Goat	Total	Per Female Goat
Gross margin (€)	78.870	259	72,169	261	68,194	261	64,929	263
Total labour (hours)	4071	13	3702	13	3524	14	3345	14
Female productive goats	305	1	276	1	261	1	247	1
Maize for consumption (hectares)	5.5	0.02	4.8	0.02	4.1	0.02	3.7	0.02
Alfalfa for consumption (hectares)	0.8	0.00	1.7	0.01	2.4	0.01	2.8	0.01
Maize silage for consumption (hectares)	0.1	0.00	0	0.00	0.0	0.00	0.0	0.00
Purchased alfalfa (kg)	183,609	602	154,085	558	133,911	513	118,012	478
Purchased mixture (kg)	44,546	146	40,215	146	38,040	146	35,929	145
Barley for consumption (hectares)	0	0.00	0.4	0.00	2.1	0.01	2.6	0.01
Crop cultivation for sale (hectares)	0		0		0		0	

However, adjustments in farming practices may also achieve some level of abatement. Specifically, as previously commented, the results indicate that the reduction of purchased feed and their substitution with on-produced feed is a strategy that can lead to lower emissions in all goat production systems. However, as can be observed in the case of the extensive farm, the use of purchased cottonseed cake is suggested as good practice to reduce emissions, since the amount consumed per goat either remains stable or increases across the various levels of abatement. The explanation for this finding lies in the fact that the inclusion of oil-rich feedstuffs in the ration of ruminants can lead to lower CH_4 emissions from enteric fermentation [43].

Furthermore, in the case of the semi-intensive and the extensive farms, the use of pastureland/shrubland and grassland is also included in the optimal farm plans, when abatement is imposed. In these low productivity farms, the use of pasture and grassland and the switch to on-produced feed reduce the feeding cost and compensate at a great extent the loss in total gross margin caused by abatement.

These results are also confirmed by the marginal abatement cost curve of the extensive, the semi-intensive and the intensive farm type which are presented in Figures 3–5, respectively. As can be seen in the figures the marginal abatement costs of the extensive and the semi-intensive farms are very low compared to the intensive farm. Specifically, the marginal abatement cost of the extensive farm is 11 €/t at the 95% level of the original emissions, 35 €/t at 80% and 76 €/t at 60%. In the case of the semi-intensive farm the marginal abatement cost is about 50 €/t, until 30% of the original emissions are abated and reaches 220 €/t at 40% abatement level. On the other hand, in the case of the intensive farm the marginal abatement cost reaches 250 €/t at only 10% abatement, indicating that intensive farms, already achieve the production of low carbon footprint milk and further abatement comes at a higher cost.

Two scenarios are investigated in this analysis, regarding the potential to restore the gross margin of the goat farms that is reduced as the result of GHG abatement. First, the impact of milk price increase is investigated using parametric optimization. The results indicate that a small price increase of about 5%–6% allows the extensive and the semi-intensive farms to maintain their original gross margin and still abate 20% of their emissions. This price increase may for example come as the result of the labeling of milk as a low-carbon product. In the case of the intensive farm the price increase should be 14% in order for the farm to achieve its original gross margin level.

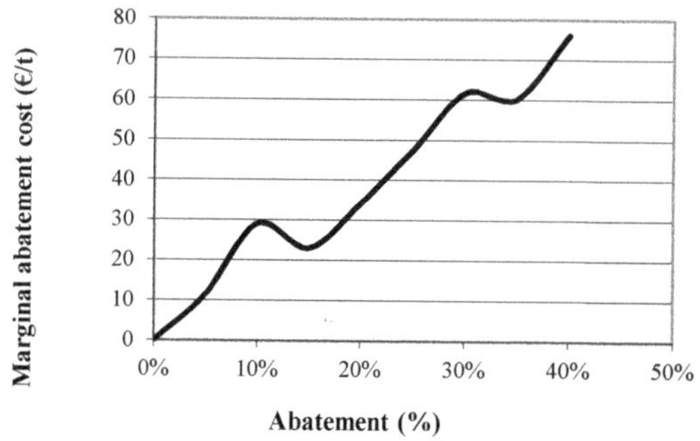

Figure 3. Marginal abatement cost curve of the extensive farm.

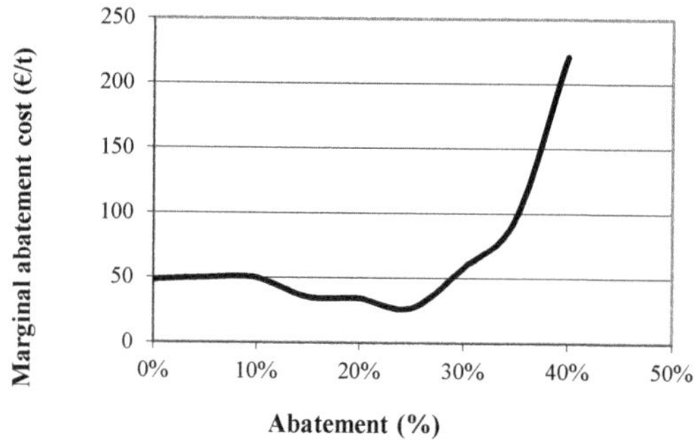

Figure 4. Marginal abatement cost curve of the semi-intensive farm.

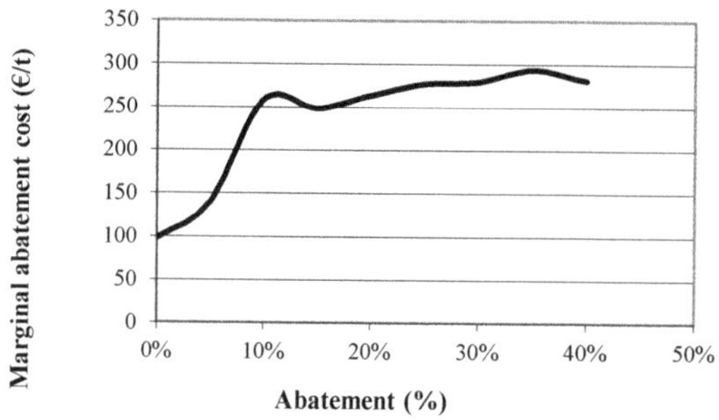

Figure 5. Marginal abatement cost curve of the intensive farm.

Alternatively, as far as policy measures are concerned, the loss in farm gross margin can be restored if farms are offered compensation/subsidy per productive goat. This compensation should be less than 6 €/goat in the cases of the extensive and the semi-intensive farming system but should reach 56 €/goat in the case of the intensive system, at 20% abatement level. It should be noted, however, that the majority of goat farms in Greece are extensive and semi-intensive farming systems, while only a few farms are characterized as intensive. Therefore, the cost of this policy measure may not be prohibitive, though this should be further investigated.

Finally, it should be mentioned that abatement has a significant impact on labor in all three production systems (see Tables 9–11). Specifically, 10% and 20% abatement results in 7%–9% and 16%–19% reduction in required labor inputs of the farms, respectively. This is an important finding, given the fact that the activity is mainly located in less favored areas of the country, where alternative occupations are scarce.

4. Discussion of Results and Conclusions

In this study a mixed-integer programming model was used to estimate GHG emissions in dairy goat farms in Greece and explore their abatement opportunities and cost. The analysis is undertaken in three goat farms that represent the extensive, semi-intensive and intensive production systems and takes into account all potential emission sources within the farm as well as pre-chain emissions.

The results of the analysis indicate that in all production systems, the main source of GHG emissions is enteric fermentation. Emissions per kg of milk are particularly high in the extensive farm, mainly because of its low productivity. The analysis also emphasizes that the intensive-farming system can produce milk with very low carbon footprint, while the carbon footprint of milk produced in semi-intensive farms is also relatively low.

Moreover, the analysis also suggests that imposing high levels of abatement unavoidably leads to the reduction of livestock size and, therefore, milk production. However, lower levels of abatement can be achieved by adjusting farming and especially feeding practices. These mitigation practices include the use of oil-rich feedstuffs, like cottonseed cake, in the ration of livestock and the substitution of purchased feed with on produced feed. These findings are important for farmers who are encouraged to adopt not only economically but also environmentally sound farming practices. The substitution of purchased feed with homegrown feed reduces emissions that are associated with their transportation to the farm, while at the same time reduces the feeding cost of farmers. However, such an adjustment in the feed would entail serious adjustments to the production system of farms, including new investments in land and machinery, as well as a different labor usage. Thus, further investigation is required concerning these implications.

As far as the marginal abatement cost is concerned, it is increasing across various levels of abatement and is significantly higher in the case of the intensive farm. The results reveal that the high productivity of the intensive farm causes a significant loss of gross margin when abatement is imposed. The abatement cost of the extensive farm is smaller, because of its smaller milk yield and, therefore, its smaller gross margin per goat. Abatement also results in a significant reduction of labor required in all farm types, which should also be taken into account when designing environmental policy measures.

Moreover, the results of the analysis indicate that the loss in gross margin caused by abatement may be restored by a small milk price increase in the case of extensive and semi-intensive farms. Further investigation is required to establish whether this price increase is possible from the promotion of milk labeled as a low-carbon footprint product. From the policy makers' point of view, a small compensation offered to farmers per productive goat can also restore the original gross margin of extensive and semi-intensive farms, when abatement is imposed. Finally, it should be emphasized that even though the carbon footprint of milk is higher in extensive farms, other environmental benefits may emerge from these production systems that are beyond the scope of this study but have to be considered when estimating their overall sustainability.

Author Contributions: Conceptualization, A.S.; methodology, A.S., A.L.; software, A.S.; validation, A.S., A.L.; formal analysis, A.S., A.L., I.T.; investigation, A.S., A.L., I.T.; data curation, A.S.; writing—original draft preparation, A.S., A.L.; writing-review and editing, A.S., A.L., I.T.; visualization, A.S., A.L., I.T.; supervision, I.T.; project administration, A.S.; funding acquisition, A.S.

Funding: This study is a part of post-doctoral research, funded under the Project 'Research & Technology Development Innovation Projects'-AgroETAK, MIS 453350, in the framework of the Operational Program 'Human Resources Development'. This study is also co-funded by the European Social Fund through the National Strategic Reference Framework (Research Funding Program 2007–2013) coordinated by the Hellenic Agricultural Organization-DEMETER (Agricultural Economics and Policy Research Institute/Scientific supervisor: Dr Tzouramani).

Conflicts of Interest: The authors declare no conflict of interest. The funders had no role in the design of the study; in the collection, analyses, or interpretation of data; in the writing of the manuscript, or in the decision to publish the results.

References

1. Hellenic Statistical Authority. *Farm Structure Survey*; Press Release: Athens, Greece, 2016. Available online: http://www.statistics.gr/documents/20181/76d51b89-18ed-426b-8227-01790d9c2a3d (accessed on 1 November 2019).
2. Kitsopanides. *Economics of Animal Production*; ZITI Publishing: Thessaloniki, Greece, 2006.
3. Hellenic Ministry of Rural Development and Food (HMRDF). Sheep and Goat Sector Development. Available online: http://www2.minagric.gr/images/stories/docs/ypoyrgeio/dimosieyseis-Arthra/meleti_gia_Nea_KAP/filadia_zoikis/aigoproboatotrofias.pdf (accessed on 1 November 2019).
4. IPCC. *Guidelines for National Greenhouse Gas Inventories*; Institute for Global Environmental Strategies (IGES) for the IPCC: Kanagawa, Japan, 2006.
5. Pitesky, M.E.; Stackhouse, K.R.; Mitloehner, F.M. Chapter 1—Clearing the Air: Livestock's Contribution to Climate Change. In *Advances in Agronomy*; Sparks, D.L., Ed.; Academic Press: Burlington, MA, USA, 2009; Volume 103, pp. 1–40. ISBN 0065-2113.
6. Weiske, A.; Vabitsch, A.; Olesen, J.E.; Schelde, K.; Michel, J.; Friedrich, R.; Kaltschmitt, M. Mitigation of greenhouse gas emissions in European conventional and organic dairy farming. *Agric. Ecosyst. Environ.* **2006**, *112*, 221–232. [CrossRef]
7. Olesen, J.E.; Schelde, K.; Weiske, A.; Weisbjerg, M.R.; Asman, W.A.H.; Djurhuus, J. Modelling greenhouse gas emissions from European conventional and organic dairy farms. *Agric. Ecosyst. Environ.* **2006**, *112*, 207–220. [CrossRef]
8. Veysset, P.; Lherm, M.; Bébin, D. Energy consumption, greenhouse gas emissions and economic performance assessments in French Charolais suckler cattle farms: Model-based analysis and forecasts. *Agric. Syst.* **2010**, *103*, 41–50. [CrossRef]
9. de Oliveira Silva, R.; Barioni, L.G.; Hall, J.A.J.; Moretti, A.C.; Fonseca Veloso, R.; Alexander, P.; Crespolini, M.; Moran, D. Sustainable intensification of Brazilian livestock production through optimized pasture restoration. *Agric. Syst.* **2017**, *153*, 201–211. [CrossRef]
10. Petersen, E.H.; Schilizzi, S.; Bennett, D. *The Impacts of Greenhouse Gas Abatement Policies on the Predominantly Grazing Systems of South-Western Australia*; International and Development Economics Working Paper 02-9; Asia Pacific School of Economics and Government, Australian National University: Canberra, Australia, 2002.
11. Benoit, M.; Laignel, G. Sheep-for-meat farming systems in French semi-upland area. Adapting to new context: Increased concentrates and energy prices, and new agricultural policy. In Proceedings of the 8th European IFSA Symposium: Empowerment of the Rural Actors: A Renewal of Farming Systems Perspectives, Paris, France, 6–10 July 2008; pp. 6–10.
12. Sintori, A.; Tsiboukas, K.; Zervas, G. Evaluating Socio-economic and Environmental Sustainability of the Sheep Farming Activity in Greece: A Whole-Farm Mathematical Programming Approach. In *Methods and Procedures for Building Sustainable Farming Systems*; Costa, M., Alexandra, A., da Silva, E.L.S., Eds.; Springer: Dordrecht, The Netherlands, 2013; pp. 219–235.
13. Alford, A.; Griffith, G.; Cacho, O. *A Northern Tablelands Whole-Farm Linear Program for Economic Evaluation of New Technologies at the Farm-Level*; No. 1160-2016-93030; Economic Research Report No. 13; NSW Agriculture: Armidale, Australia, 2004.
14. Veysset, P.; Bebin, D.; Lherm, M. Adaptation to Agenda 2000 (CAP reform) and optimisation of the farming system of French suckler cattle farms in the Charolais area: A model-based study. *Agric. Syst.* **2005**, *83*, 179–202. [CrossRef]
15. Sofi, N.A.; Ahmed, A.; Ahmad, M.; Bhat, B.A. Decision making in agriculture: A linear programming approach. *Int. J. Mod. Math. Sci.* **2015**, *13*, 160–169.
16. Crosson, P.; O'Kiely, P.; O'Mara, F.P.; Wallace, M. The development of a mathematical model to investigate Irish beef production systems. *Agric. Syst.* **2006**, *89*, 349–370. [CrossRef]
17. De Cara, S.; Jayet, P.A. Emissions of greenhouse gases from agriculture: The heterogeneity of abatement costs in France. *Eur. Rev. Agric. Econ.* **2000**, *27*, 281–303. [CrossRef]
18. Breen, J.P.; Donellan, T. Estimating the Marginal Costs of Greenhouse Gas Emissions Abatement using Irish Farm-Level Data. In Proceedings of the 83rd Annual Conference of Agricultural Economics Society, Dublin, Ireland, 30 March–1 April 2009; Agricultural Economics Society: Dublin, Ireland, 2009.

19. Moraes, L.E.; Wilen, J.E.; Robinson, P.H.; Fadel, J.G. A linear programming model to optimize diets in environmental policy scenarios. *J. Dairy Sci.* **2012**, *95*, 1267–1282. [CrossRef] [PubMed]
20. de Oliveira Silva, R.; Barioni, L.G.; Hall, J.A.J.; Folegatti Matsuura, M.; Zanett Albertini, T.; Fernandes, F.A.; Moran, D. Increasing beef production could lower greenhouse gas emissions in Brazil if decoupled from deforestation. *Nat. Clim. Chang.* **2016**, *6*, 493–497. [CrossRef]
21. Smith, E.G.; Upadhyay, M. Others Greenhouse gas mitigation on diversified farms. In Proceedings of the Western Agricultural Economics Association-Western Economics Association International Joint Annual Meeting, San Francisco, CA, USA, 6–8 July 2005.
22. Kaiser, H.M.; Messer, K.D. *Mathematical Programming for Agricultural, Environmental and Resource Economics*; John Wiley and Sons: New York, NY, USA, 2014; ISBN 9780874216561.
23. Sintori, A. Analysis of the Decision Making Process with Multiple Goals: The Case of Dairy Sheep Farming in Continental Greece. Ph.D. Thesis, Agricutlural University of Athens, Athens, Greece, 2012; p. 365.
24. Zervas, G.; Kalaisaikis, P.; Feggeros, K. *Farm Animal Nutrition*; Stamoulis Publishing: Athens, Greece, 2000.
25. Kalaisakis, P.; Giourdas, M. *Applied Animal Nutrition*; Giourdas M. Publishing: Athens, Greece, 1965.
26. Jarrige, R. *[Alimentation des Ruminants]: Principes de la Nutrition et de L'alimentation des Ruminants: Besoins Alimentaires des Animaux: Valeur Nutritive des Aliments*; INRA Pub.: Versailles, France, 1980; ISBN 2853402886.
27. Feedipedia. An Online Encyclopedia of Animal Feeds. Available online: https://www.feedipedia.org/ (accessed on 1 February 2019).
28. Papachristou, T.G. Utilization of available forage in shrublands from goats: Selection and nutritional value of forage. In Proceedings of the 1st Greek National Conference of the Hellenic Range and Pasture Society, Athens, Greece, 16 June 1996; The Hellenic Range and Pasture Society: Drama, Greece, 1997.
29. FRI (Forest Reearch Institute). *Final Report of the Program "Agrometeorological Models for the Prediction of the Production and Quality of Pastures"*; NAGREF: Thessaloniki, Greece, 2000.
30. Platis, P.D.; Papanastasis, V. Relationship between Shrub Cover and Available Forage in Mediterranean Shrublands. *Agrofor. Syst.* **2003**, *57*, 59–67. [CrossRef]
31. Platis, P.D.; Papanastasis, V.P.; Papachristou, T.G.; Tsiontsis, A.G. Varation of quantity and quality of forage in grasslands of the lowland ecological zone in the Prefecture of Thessaly. In Proceedings of the 4th Greek National Conference of the Hellenic Range and Pasture Society, Volos, Greece, 10–12 November 2004; The Hellenic Range and Pasture Society: Athens, Greece, 2006.
32. Schils, R.L.M.; de Haan, M.H.A.; Hemmer, J.G.A.; van den Pol-van Dasselaar, A.; de Boer, J.A.; Evers, A.G.; Holshof, G.; van Middelkoop, J.C.; Zom, R.L.G. DairyWise, A Whole-Farm Dairy Model. *J. Dairy Sci.* **2007**, *90*, 5334–5346. [CrossRef] [PubMed]
33. Giger-Reverdin, S.; Sauvant, D. Methane Production in Sheep in Relation to Concentrate Feed Composition from Bibliographic Data. In *Sheep and Goat Nutrition: Intake, Digestion, Quality of Products and Rangelands*; Ledin, I., Morand-Fehr, P., Eds.; CIHEAM-IAMZ: Zaragoza, Spain, 2000; pp. 43–46.
34. Petersen, E.; Schilizzi, S.; Bennett, D. The impacts of greenhouse gas abatement policies on the predominantly grazing systems of south-western Australia. *Agric. Syst.* **2003**, *78*, 369–386. [CrossRef]
35. Wells, D. *Total Energy Indicators of Agricultural Sustainability: Dairy Farming Case Study*; Technical Paper; Ministry of Agriculture and Forestry: Wellington, New Zealand, 2001; ISBN 0-478-07968-0.
36. Wood, S.; Cowie, A. *A Review of Greenhouse Gas Emission Factors for Fertiliser Production*. IEA bioenergy task 38. Available online: www.sciencetheearth.com/uploads/2/4/6/5/24658156/2004_wood_a_review_of_greenhouse_gas_emission_factors.pdf (accessed on 1 February 2019).
37. Defra. *Guidelines for Company Reporting on Greenhouse Gas Emissions Annexes Updated July 2005*; Defra: London, UK, 2005.
38. Helsel, Z.R. Energy in pesticide production and use. *Encycl. Pest Manag.* **2006**, *1*, 1–4.
39. Kaltsas, A.M.; Mamolos, A.P.; Tsatsarelis, C.A.; Nanos, G.D.; Kalburtji, K.L. Energy budget in organic and conventional olive groves. *Agric. Ecosyst. Environ.* **2007**, *122*, 243–251. [CrossRef]
40. Smith, E.G.; Upadhyay, B.M. Greenhouse Gas. Mitigation on Diversified Farms. In Proceedings of the Annual Meeting of the Canadian Agricultural Economics Society, San Francisco, CA, USA, July 2005.
41. Lesschen, J.P.; van den Berg, M.; Westhoek, H.J.; Witzke, H.P.; Oenema, O. Greenhouse gas emission profiles of European livestock sectors. *Anim. Feed Sci. Technol.* **2011**, *166–167*, 16–28. [CrossRef]

42. FAO. *Livestock's Long Shadow: Environmental Issues and Options*; FAO: Rome, Italy, 2006.
43. Ludemann, C.; Howden, S.; Eckard, R. What is the best use of oil from cotton (Gossypium spp.) and canola (*Brassica* spp.) for reducing net greenhouse gas emissions-biodiesel, or as a feed for cattle? *Anim. Prod. Sci.* **2016**, *56*, 3442–3450. [CrossRef]

Contents

About the Editors

Hosam M. Saleh

Hosam M. Saleh is a professor of radioactive waste management at the Radioisotope Department, Egyptian Atomic Energy Authority, Egypt. He was awarded MSc and Ph.D. degrees in Physical Chemistry from Cairo University. Saleh has more than 26 years of experience in hazardous waste management with an emphasis on treatment and developing new matrixes for the immobilization of these wastes. He is also interested in studying innovative economic and environmentally friendly techniques for the management of hazardous and radioactive wastes. He authored many peer-reviewed scientific papers and chapters and served as a Book Editor of several books related to international publishers. He has been named among the top 2% of scientists in the world according to the Stanford University report for 2020, 2021 and 2022. He is also a reviewer, an Editor in Chief or Associate Editor for different journals as well as a member of the Technical Program Committee for international conferences. He won the Scientific Encouragement Award from Atomic Energy Authority (2013) and the Encouragement Prize in Advanced Technical Sciences from the Academy of Scientific Research (2014).

Mohammad Mahmoud Dawoud

Dr. Mohammad Mahmoud Dawoud is an environmental researcher at the Radioisotope Department, Egyptian Atomic Energy Authority, Egypt. He graduated from the Agricultural Engineering Department, Faculty of Agriculture, Ainshams University. BSc. Agricultural Engineering (Excellent with Rank of Honors); M.Sc. (Detoxification of Heavy metals From wastewater using Agriculture residues as absorbent material and Nuclear Technique); Ph.D. (Remediation of contaminated water using some physical and radiation techniques). He specializes in environmental chemistry and engineering redox and adsorption interactions, and he has more than ten years of experience in water treatment systems (reverse osmosis, sewage plants and filtration units for poultry farms, livestock and fish farms). He has designed and implemented more than 380 water treatment projects. He has participated in many research projects related to the environmental field and sustainability. Currently, he is considered one of the most important consultants in the field of water treatment in Egypt. He is also an environmental consultant for major international NGOs such as Rotary Egypt and the Alpha foundation for Environmental Technologies, and he has won many awards in this regard.

sustainability

MDPI

Editorial

Radioactivity: Sustainable Materials and Innovative Techniques

Hosam M. Saleh * and Amal I. Hassan

Radioisotope Department, Nuclear Research Center, Egyptian Atomic Energy Authority, Cairo 11787, Egypt
* Correspondence: hosamsaleh70@yahoo.com or hosam.saleh@eaea.org.eg

Radioisotopes, also known as radionuclides, are atoms with unstable nuclei that emit ionizing radiation. Radioisotopes have a wide range of applications in medicine, industry, and scientific research, including in cancer diagnosis and treatment, sterilization of medical equipment, and dating of geological and archaeological materials [1,2]. However, the use of radioisotopes also poses significant risks to human health and the environment due to their potential for radioactive contamination and exposure [3,4].

Radioactive waste management is a critical component of nuclear safety. It involves the safe and sustainable disposal of nuclear waste generated from various sources, including nuclear power plants, medical facilities, research facilities, and industrial processes [5]. With the advancement of technology and scientific research, innovative approaches to radioactive waste management have emerged, ensuring sustainable nuclear safety [6].

Waste management and nuclear safety are critical issues that need to be addressed with utmost seriousness. Proper waste management ensures that the environment and the public are not exposed to harmful materials, while nuclear safety measures are necessary to prevent accidents and incidents that could result in radiation leaks [7–9].

Waste treatment and disposal methods are crucial components of waste management. Several methods are used for waste treatment, including recycling, incineration, and landfilling. Recycling involves converting waste materials into new products or materials, while incineration involves burning waste materials to reduce their volume and weight [10]. Landfilling is the most common method of waste disposal, where waste is buried in a designated area [11].

Nuclear waste, which is generated from nuclear power plants, research facilities, and medical institutions, requires special treatment and disposal methods due to its radioactive nature. The most common method of nuclear waste treatment is reprocessing, which involves extracting useful materials from nuclear waste [12]. The remaining waste is then disposed of in specialized facilities designed for storage or final disposal [13].

Nuclear safety measures are critical to prevent accidents that could lead to radiation leaks. Nuclear power plants are designed with multiple layers of safety systems, including automated shutdown systems and emergency cooling systems. Regular inspections, maintenance, and training of personnel are also necessary to ensure nuclear safety [14].

Overall, waste management and nuclear safety are complex issues that require a comprehensive approach. Governments, industries, and individuals must work together to develop effective waste treatment and disposal methods while also implementing strict safety measures to prevent nuclear accidents [15].

Innovative approaches to radioactive waste management are critical for ensuring sustainable nuclear safety [16]. The use of natural materials, radiation techniques, and waste management methods can ensure the safe and sustainable disposal of radioactive waste generated by various sources, including nuclear power plants, medical facilities, research facilities, and industrial processes [17]. These innovative approaches ensure the protection of the environment and public health, while also promoting sustainable nuclear safety. Several sustainable materials have been used in nuclear applications such as cellulosic waste in treatment [18,19] or cement mixed with bitumen [20,21], cement waste [22,23], glass [24,25], and natural clay [26,27] for radioactive waste stabilization and radiation shielding.

Citation: Saleh, H.M.; Hassan, A.I. Radioactivity: Sustainable Materials and Innovative Techniques. *Sustainability* **2023**, *15*, 5792. https://doi.org/10.3390/su15075792

Received: 21 March 2023
Accepted: 23 March 2023
Published: 27 March 2023

In this Special Issue, Young Jae Jang, Hye Kwon, Seong Hee Park, Yona Choi, Kum Bae Kim, Dong Wook Kim, Suk Ho Bhang, and Sang Hyoun Choi discuss discarding parts of a medical linear accelerator (linac). The study emphasizes the importance of checking the activity of each part to determine the disposal time according to the clearance level for self-disposal. The clearance level for self-disposal for each radionuclide-activated part was applied by three manufacturers to confirm the timing of self-disposal and to predict when workers are not exposed to radiation during disassembly/disposal. In the interest of establishing the framework for unambiguous safety management standards for the disposal of decommissioned linacs, a qualitative and quantitative study of radionuclides was undertaken using high-purity Germanium (HPGe)-based gamma spectroscopy in addition to the Monte (contribution 1).

Natural mixed waste biomass is an innovative approach to the remediation of contaminated sites. The use of natural mixed waste biomass as a radioisotope biosorbent is a cost-effective approach to the remediation of contaminated sites. Natural mixed waste biomass is an excellent radioisotope biosorbent that can remove radioactive contaminants from soil and water. This approach ensures the safe and sustainable disposal of radioactive waste generated by nuclear facilities. Arwa Abdelhamid, Mogeda Badr, Ramadan Mohamed, and Hosam Saleh explain a sustainable treatment system that uses dry biomass from mixed trash to show synergistic benefits such as high efficiency and cost-effectiveness. The removal of stable and radioisotopes of cobalt and cesium from aqueous solutions was examined in this work using biosorption in dried mixed waste of olive waste and water hyacinth as a low-cost and naturally accessible sorbent. The findings indicate that the natural biomass of mixed garbage is a good biosorbent for the isotopes studied. According to the experimental and mathematical results in this study, a mixed dry waste of water hyacinth and olive waste might be proposed as a sustainable low-cost and effective natural adsorbent material for the efficient remediation of radio or stable cobalt and cesium ions from wastewater. In addition, this method is expected to successfully decontaminate toxic metals and radionuclides in an ecologically benign and sustainable way while significantly lowering wastewater treatment costs (contribution 2).

Natural radioactivity is present in many materials, including granite and ceramic tiles, which contain small amounts of naturally occurring radioactive elements such as ^{232}Th and ^{238}U [28]. These elements emit ionizing radiation in the form of alpha, beta, and gamma rays, which can potentially pose health risks to humans if they are exposed to them for prolonged periods. However, research has shown that the use of mainly granite and ceramic tiles in construction and decoration does not present significant risks to human health. To this effect, Essam Sidique, Sedky Hassan, and Mohammad Mahmoud Dawoud evaluated twenty-three well-known brands of Egyptian commercial granites and ceramic tile samples radiologically using an HPGe detector, consisting of 107 samples of typical materials. The concentrations of radioisotopes were found to be greater in most granite samples compared to the ceramic samples. Additionally, the concentration values of terrestrial radionuclides showed considerable variations in granite and ceramic tile samples obtained from various manufacturers. This is important in distinguishing between the brands under consideration. The concentration of ^{40}K was discovered to be the greatest contribution to the overall concentration for all samples, followed by ^{232}Th and ^{238}U (contribution 3).

A study by Mohamed Ehab, Elsayed Salama, Ahmed Ashour, Mohamed Attallah, and Hosam Saleh investigated the radiation shielding capabilities as well as the optical qualities of a prepared SiO_2-ZnO-Na_2CO_3-H_3BO_3-$BaCO_3$ glass composite with varying amounts of barium carbonate (0–30 mol%). Therefore, fabricated glass is an innovative approach to transparent shielding. The use of fabricated glass for applications requiring transparent shielding can reduce the volume of radioactive waste generated by nuclear facilities. This approach ensures the safe and sustainable disposal of radioactive waste while also providing transparent shielding for various applications (contribution 4).

In a study by Prasoon Raj, Nemeer Padiyath, Natalia Semioshkina, Francois Foulon, Ahmed Alkaabi Gabriele Voigt, and Yacine Addad seven date palm plants in Abu Dhabi were tested. Their root zone soils, fruits, and leaves were sampled and analyzed for gamma-emitting naturally occurring radioactive materials (NORM) radionuclide activity. The soil samples exhibit significant levels of ^{40}K, while the levels of ^{238}U and ^{232}Th are substantially lower. The measured soil radioactivity is below the global average (contribution 5).

The goal of Tarek Sayed and El-Sayed Ahmed's research is to see if gamma irradiation, chitosan, and yeast may help promote the sustainable development of marjoram in the context of organic farming. The primary plot has an abiotic elicitor (15 Gy gamma irradiation), two biotic elicitors (500 ppm chitosan, 0.5% yeast, and a non-elicitor (as control), and two organic fertilizers (20 g/m^2 moringa dry leaves, 20 g/m^2 fulvic acid, and 20 g/m^2 (NPK); the latter is a classic agrochemical). They show strong evidence for the potency of biotic elicitors chitosan > yeast > abiotic, gamma irradiation coupled with organic fertilizers moringa > fulvic acid > traditional agrochemical fertilizer NPK, as reliable CO-friendly solutions, significantly improving marjoram biomass, secondary metabolite production, and quality without the use of agrochemical pesticides and/or microbicides. In addition, chitosan > yeast > gamma irradiation in combination with organic fertilizers outperformed their integration with commercial NPK fertilizer (contribution 6).

Elsayed Salama, Dalal Aloraini, Sara El-Khateeb, and Mohamed Moustafa investigated the thermoluminescence properties of natural rhyolite. Dose response has been evaluated across a large dosage range of 0.5–2000 Gy. The minimum detectable dosage and the rate of thermal fading are calculated. After selecting the optimal read-out conditions, glow curve deconvolution was performed. The overlapping peaks were detected using the repeated initial rise (RIR) approach, and the thermoluminescence characteristics of rhyolite were extracted using a glow curve deconvolution procedure. A linear dose response up to 25 Gy was achieved, followed by supra linearity up to 2000 Gy. The TL properties of rhyolite revealed that it exhibits a linear dose response up to 25 Gy, followed by supra linearity up to 2000 Gy, and a comparatively high fading rate of 57% after two weeks with no further thermal fading seen. There is a low detection limit of around 0.5 Gy and a decent reproducibility of about 4% variance in subsequent measurements. These properties may qualify rhyolite as the sustainable natural material employed in this work for a range of radiation dose assessment applications (contribution 7).

Mohammad Khairul Azhar Abdul Razab, Norazlina Mat Nawi, Fara Hana Mohd Hadzuan, Nor Hakimin Abdullah, Maimanah Muhamad, Rosidah Sunaiwi, Fathirah Ibrahim, Farah Amanina Mohd Zin and An'amt Mohamed Noor confirmed that the high sorption capacity and the ability to coagulate with any reactive elements at molecular structures, such as ^{18}F-FDG, makes it a prime option for alternative radionuclides decontamination. The 'wax tissue' nanolayers and vast surface area have been shown to help graphene oxide (GO) wrap and adsorb radionuclides effectively. The adsorption rate was effective at a slow decay rate of ^{18}F-FDG, where more available free electrons are ready for the adsorption interaction with GO functional groups. Graphene oxide nanoparticles are innovative approaches to the safe and sustainable preparation of radiopharmaceuticals. The use of graphene oxide nanoparticles can ensure the safe and sustainable preparation of radiopharmaceuticals, which may contaminate surface areas due to a spill during their preparation or an accident during their transportation from the laboratory to the treatment room. Graphene oxide nanoparticles show a high absorption affinity towards radionuclides, ensuring the safe and sustainable disposal of radioactive waste (Contribution 8).

Hosam Saleh, Ibrahim Bondouk, Elsayed Salama, Hazem Mahmoud, Khalid Omar, and Heba Esawii adopt environmentally friendly techniques to conserve the environment, such as recycling municipal or industrial waste, where the debris of polyvinyl chloride (PVC) pipes and asphaltene is used as cement additives to improve its mechanical properties while stabilizing radioactive waste resulting from the peaceful uses of nuclear materials, or enhancing its efficiency in radiation protection (Contribution 9).

Chelating agents for uranium mine residues: Chelating agents are innovative approaches to eliminating uranium mine residues. The use of chelating agents can remove uranium from mine residues and prevent soil pollution. This approach ensures the safe and

Short Author Biography

Hosam Saleh is a professor of radioactive waste management at the Radioisotope Department, Atomic Energy Authority, Egypt. He was awarded MSc and Ph.D. degrees in Physical Chemistry from Cairo University. Saleh has more than 25 years of experience in hazardous waste management with an emphasis on treatment and developing new matrixes for the immobilization of these wastes. He is also interested in studying innovative economic and environment-friendly techniques for the management of hazardous and radioactive wastes. He authored many peer-reviewed scientific papers and chapters and served as a Book Editor of several books related to international publishers. He has been selected among the top 2% of scientists in the world according to the Stanford University report for 2020, 2021 and 2022. He is also a reviewer, an Editor in Chief or Associate Editor for different journals as well as a member of the Technical Program Committee for international conferences. He gained the Scientific Encouragement Award from Atomic Energy Authority (2013), Encouragement Prize in Advanced Technical Sciences from the Academy of Scientific Research (2014).

sustainability

MDPI

Article

Analysis of Activated Materials of Disposed Medical Linear Accelerators according to Clearance Level for Self-Disposal

Young Jae Jang [1,2,†], Na Hye Kwon [3,4,†], Seong Hee Park [2], Yona Choi [2], Kum Bae Kim [1], Dong Wook Kim [3,4], Suk Ho Bhang [5,*] and Sang Hyoun Choi [1,*]

1 Research Team of Radiological Physics & Engineering, Korea Institute of Radiological & Medical Sciences, Seoul 01812, Republic of Korea
2 Department of Accelerator Science, Korea University, Sejong 30015, Republic of Korea
3 Department of Radiation Oncology, Yonsei Cancer Center, Heavy Ion Therapy Research Institute, Yonsei University College of Medicine, Seoul 03722, Republic of Korea
4 Medical Physics and Biomedical Engineering Lab (MPBEL), Yonsei University College of Medicine, Seoul 03722, Republic of Korea
5 School of Chemical Engineering, Sungkyunkwan University, Suwon 16419, Republic of Korea
* Correspondence: sukhobhang@skku.edu (S.H.B.); shchoi@kirams.re.kr (S.H.C.); Tel.: +82-31-290-7242 (S.H.B.); +82-2-970-1590 (S.H.C.)
† These authors contributed equally to this work.

Abstract: In Korea, when replacing or discarding parts of a medical linear accelerator (linac), self-disposal is required in the consideration of the activity, but there is no standard regulation to manage radioactive waste. The aim of this study is to check the activity of each part to determine the disposal time according to the clearance level for self-disposal. The results of measuring the components of the linac head parts of the disposed Varian, Elekta, and Siemens equipment were reflected in the Monte Carlo simulation to confirm the radionuclide change according to the presence or absence of impurities. To confirm the degree of activation of the linac, the main radionuclides according to the time after the linac shutdown, considering the workloads of 40/80 Gy/day of 10/15 MV linac irradiated with beams for 10 years in the results of the simulation of the linac parts, and the radionuclide concentration was confirmed. As a result of applying the clearance level for self-disposal in the notice of the Korean Nuclear Safety (KINS) to each linac head part, most parts of the 10 MV linac could be dismantled after 1 month, and 15 MV target and primary collimators were stored after a long period of time before being dismantled. Although additional radionuclides were identified according to the presence or absence of impurities, the disposal timing for each part did not change significantly. In this study, the clearance level for self-disposal for each radionuclide was applied to activated parts by three manufacturers to confirm the self-disposal timing and predict the timing at which workers are not exposed to radiation during dismantling/disposal.

Keywords: radioactive waste; medical linear accelerator; clearance level; radionuclide; Monte Carlo simulation

Citation: Jang, Y.J.; Kwon, N.H.; Park, S.H.; Choi, Y.; Kim, K.B.; Kim, D.W.; Bhang, S.H.; Choi, S.H. Analysis of Activated Materials of Disposed Medical Linear Accelerators according to Clearance Level for Self-Disposal. *Sustainability* **2023**, *15*, 4100. https://doi.org/10.3390/su15054100

Academic Editors: Hosam M. Saleh and Mohammad Mahmoud Dawoud

Received: 30 January 2023
Revised: 21 February 2023
Accepted: 21 February 2023
Published: 23 February 2023

1. Introduction

To meet the growing demand for radiotherapy, an increasing number of medical linear accelerators (hereinafter referred to as "linac") are being installed in Korean hospitals every year (n = 185 in 2022) [1]. Alongside new installations, an increasingly higher number of existing linacs has been replaced or disposed of in recent years. The service life of linacs ranges from 10 to 15 years, and they are replaced or disposed of for reasons such as the aging of parts, failure of major parts, and introduction of new treatment equipment [2]. According to the data from the Korean Institute of Nuclear Safety (KINS) in 2018 [3], two or more linacs are replaced or disposed of every year in Republic of Korea. It is recommended that the details of each case of replacement or disposal be reported to the KINS and major parts be subjected to self-disposal or storage for later disposal after evaluating their activity.

The Korean Nuclear Safety and Security Commission (NSSC) Notice No. 2020-6 stipulates the permissible standards for the self-disposal of radioactive waste according to the clearance level of each radionuclide [4]. In the case of medical linacs, however, no accurate information on radionuclides is available and standardized regulations allowing the flexible management of different workloads have yet to be established. Therefore, there is a major need for a management plan for dismantling radioactive parts of medical linear accelerators.

Various radionuclides generated through the activation process in a linac can cause problems to patients and related healthcare workers in general, and workers can be exposed to them during disposal/dismantling in particular. It is, hence, of paramount importance to quantify the radionuclides generated and compute the dose rates [5].

NCRP 79 specifies 10 MeV as the energy threshold for evaluation because only a negligible number of neutrons are generated at energies below 10 MeV, which defines the target, primary collimator, flattening filter, and head parts as the highest risks of activation [6]. In addition, IAEA-TECDOC-1183 suggested the possibility of the surrounding materials being activated by an accelerator using an energy of 10 MeV or greater, and clarified the necessity to assess the radiation hazard when dismantling the head of a medical linac used in radiation oncology [7].

A variety of studies have been conducted to identify and analyze radionuclides to ensure the safety of workers, concluding that radionuclides are mainly generated from the target, primary collimator, flattening filter, jaws, multileaf collimator (MLC), and bending magnet by high-energy photon beams [8,9]. These materials mainly consist of tungsten, gold, copper, and stainless steel. The target to produce photons is made of a high-density metal, such as tungsten or gold, and a primary collimator is made of tungsten, which is used to reduce the amount of scattered radiation. The flattening filter consisting of tungsten or brass is designed to generate a beam with a more uniform intensity across the field, and a jaw (MLC) made of tungsten is used to adjust the size and shape of the treatment beam. The properties of these components are carefully designed to ensure that the radiation beam is accurately directed towards the target area, while minimizing the dose to surrounding healthy tissue.

For the accurate assessment of radionuclides generated by the high-energy photon beams of each part when dismantling a linac, radionuclides emitted from each part should be identified using a gamma spectrometer or a Monte Carlo simulation method based on the data provided by the manufacturer and the component analysis of additional parts using laser-induced breakdown spectroscopy (LIBS) [10].

Adam Konefal et al. and Waller et al. identified and quantified activation products and their estimated activities generated by 18 MV Elekta and 15 MV Varian Truebeam using gamma-ray spectroscopy with HPGe [11,12]. Vichi et al. performed a Monte-Carlo-simulation-based activation study of a 15 MeV linear accelerator, investigating the production of radioactive isotopes and dose rates in various materials [9]. However, these studies did not evaluate the effect of radioactivity caused by impurities in the linear accelerator, nor did they study the change in nuclide and radioactivity over time after shutting down the linear accelerator.

The presence of impurities in components can be a significant source of uncertainty in radiation measurements. Impurities in materials can produce additional gamma rays or other types of radiation that may not be accurately accounted for in the measurement process. There is a limit to studying the evaluation of radioactivity over time after the shutdown of the linear accelerator using HPGe. Even after the linear accelerator is turned off, the materials used in the construction of the accelerator and its components may continue to emit radiation owing to the residual activity of radioactive isotopes. This residual activity can persist for varying lengths of time depending on the specific isotopes involved, making it difficult to assess the radiation exposure levels in the surrounding environment accurately.

In this study, the degrees of activation of the components and impurities of each major part were assessed using LIBS for mapping out their composition in Varian, Elekta, and Siemens linacs, and by performing a Monte Carlo simulation. Based on the assessment results, it is intended to predict the timing of storage and disposal of each part taking into account its clearance level by determining the radionuclides emitted during the accelerator dismantling/disposal and the activation level over time.

2. Materials and Methods

2.1. Component Analysis of the Parts of Decommissioned Linacs by Manufacturer

Components, including the head of three decommissioned medical linacs, were analyzed. Particularly, the linacs investigated were Clinac iX (Varian), Agilty^{TM} (Elekta), and Oncor Expression (Siemens). The study focused on the parts of the linac head that are prone to activation in terms of the beam pathway, which included the target, primary collimator, ion chamber, jaw (MLC), and bending magnet. To confirm the components of each dismantled linac head part, laser-induced breakdown spectroscopy (LIBS) was used. Surface investigation was conducted at least three times for each part using a portable LIBS element analyzer (Z-300 GeoChem Pro, SciAps Inc., Woburn, MA, USA). The results of the measurements were then compared with the specifications provided by each manufacturer, and the levels of impurities were analyzed. Figure 1a–c shows the process of dismantling the linac head parts manufactured by Varian, Elekta, and Siemens, respectively. Figure 1d shows the portable LIBS device measuring the emission from each major component of the dismantled head parts.

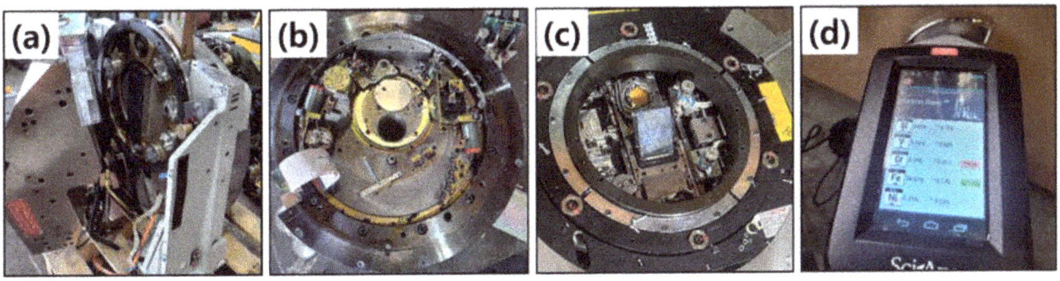

Figure 1. Dismantled linac head parts of (**a**) Clinac iX (Varian); (**b**) Synergy platform (Elekta); (**c**) Oncor Expression (Siemens); and (**d**) LIBS measurement results of the components of their major parts.

2.2. Monte Carlo Simulation of Medical Linacs

Monte Carlo simulation was used to evaluate radionuclides generated by parts of the linear accelerator according to energy and workload. It is a computational method that uses random sampling to obtain numerical solutions to mathematical problems, and it is often used to model the behavior of particles (such as photons, electrons, or neutrons) as they interact with matter. It can simulate the behavior of particles in complex systems to study the effects of different materials, geometries, and other factors.

To perform a Monte Carlo simulation, we obtained the internal structures and specifications of different components of the medical linacs from their respective manufacturers, and devised three-dimensional models using AutoCAD software (Autodesk Inc., San Francisco, CA, USA). Specifically, parts such as target, primary collimator, ion chamber, jaw, and MLC of Varian, Elekta, and Siemens equipment were modeled based on the data provided by the manufacturers, and bending magnets and lead shields were modeled additionally based on the analysis data of the structures of decommissioned linacs. The SuperMC program (Super Monte Carlo simulation program, FDS team, Hefei, China) [13] was used to convert the modeled data into PHITS (Version 3.25) code written in Fortran [14]. The particle and heavy ion transport code system (PHITS) is a widely used general-purpose Monte Carlo simulation code that can handle the transport of all particles over a wide range of energies using various nuclear reaction models and nuclear data libraries. It is used in numerous fields related to particle and heavy ion transport phenomena, including accelerator technology, radiotherapy, and space radiation. Using this code, radionuclides that can be generated for each part of the linear accelerator were simulated according to the energy of 10/15 MV and the workload of 40 and 80 Gy/day.

2.3. Time-Dependent Activity Level

The PHITS code, which was used to analyze radionuclides and their activity levels, was based on the nuclear data in the ENDF-6 format (ENDF-102), and the coding was performed to include all reactions in the 0.001 to 50 MeV energy band for gamma spectroscopic analysis by using source and tally. The source component in the PHITS code was specified as a transverse Gaussian distribution electron beam with energy levels of 10/15 MeV, which collided with a target to generate a photon beam. The nuclear physics module of PHITS included models for simulating nuclear reactions, such as scattering, fission, and fusion, as well as models to produce secondary particles, such as neutrons and gamma rays, in these reactions.

To calculate the neutron flux from the PHITS code, T-DCHAIN and T-Volume Tally were utilized. These include information such as cell number, beam amplitude, number of irradiations, cooling time steps, and the volume of each part to derive the neutron flux. To precisely determine the radionuclide concentration, time-dependent energy and current values were entered in the DCHAIN code, considering the linac's workload. The decay chain analysis code DCHAIN [15] was used to analyze all radionuclides of each component based on the neutron's flux results, whereby nuclear data were extracted from hybrid data libraries, such as JENDL/AD-2017, FENDL/A-3.0, JENDL-4.0, ENDF/B-VIII.0, and JEFF-3.3, to encompass all nuclides [16,17].

To assess the degree of activation of a medical linac, workloads of 80 Gy/day and 40 Gy/day were reflected in the DCHAIN code for equipment irradiated with beams for 10 years, and the time-dependent change in the activation level of radionuclides of each part was monitored by measuring it immediately after the linac shutdown and 6 h, 1 month, 3 months, 1 year, 5 years, and 10 years later. Major radionuclides were selected as those from the components whose activity level accounted for 0.01% or more of the total activity level among those with an activity level of 10^{-6} Bq/g or higher immediately after linac shutdown and a half-life of 5 s or more.

3. Results

3.1. Component Analysis of the Parts of Decommissioned Linacs by Manufacturer

In Table 1, the elements of each of the main parts (target, primary collimator, flattening filter, ion chamber, mirror, jaw, and MLC) that affect major impurities and additional radionuclides, whose element composition ratios account for more than 0.5% of the total element composition, are listed along with the data provided by each manufacturer.

Among the parts with no data provided by the manufacturers and whose components were measured by LIBS, the bending magnet was found to be mainly composed of Fe, with the bending magnet provided by Varian containing large amounts of Cr and Ni, unlike those provided by Elekta and Siemens. In the LIBS measurement results, Varian's flattening filter 15 MV, Elekta's primary collimator, and Siemens' MLC were composed of 100% tungsten, with no impurities detected. As for the target, primary collimator, and jaw, the largest share of impurities was detected in those manufactured by Varian. Only the primary collimator manufactured by Siemens includes an absorber; therefore, it is annotated as Varian and Elekta in Table 1.

Table 1. Elements of the major parts of the medical linacs by manufacturer.

Major Parts	Varian	Elekta	Siemens
Target	Cu W Zn *	W Re Cu	Au Fe Cr Ni Cu *
Bending magnet	Fe * Cr * Ni *	Fe * Si * Ni * Pb * Cu *	Fe * Mo *
Primary collimator	W Fe * Cr * Ni * Mn *	W Ni Fe	W
Absorber	**	**	Al Ag * Cu * Mg *
Flattening filter 15 MV	W	Fe Cr Mn Ni Mo *	Fe Cr Mn Ni Mo *
Flattening filter 10 MV	Cu O Zn *		
Ion chamber	H C O N	Al Si Fe Mn Mg	Al O Mg * Ag * Cu *
Mirror	H C O	H C O	O Si Ca Na
Jaw	W Ni * Cu * Nb * Mo *	W Ni Fe	W
MLC	W Ni Fe Cu	W Ni Fe Cr * Mo *	W

* Components additionally identified in LIBS measurements; ** head part components with no specifications provided by the manufacturers.

3.2. Medical Linac Modeling and Workload-Based Activation Evaluation

The schematic representations in Figure 2 are the results of modeling the main components and shield structures of the head parts of the three manufacturers. Siemens' target, unlike those of Varian and Elekta, is made of different materials, such as graphite, gold, and SST304, and is modeled in such a way that the absorber is placed inside the primary collimator; a beam of 10 MV or greater is applied. While the MLC of Varian is placed after the jaw, that of Elekta is placed before the jaw. For this reason, unlike other manufacturers, Elekta's MLC has a higher activity than the jaw. In addition to the main head part components, radiation shielding mainly composed of lead was additionally modeled to evaluate radionuclides and activity levels.

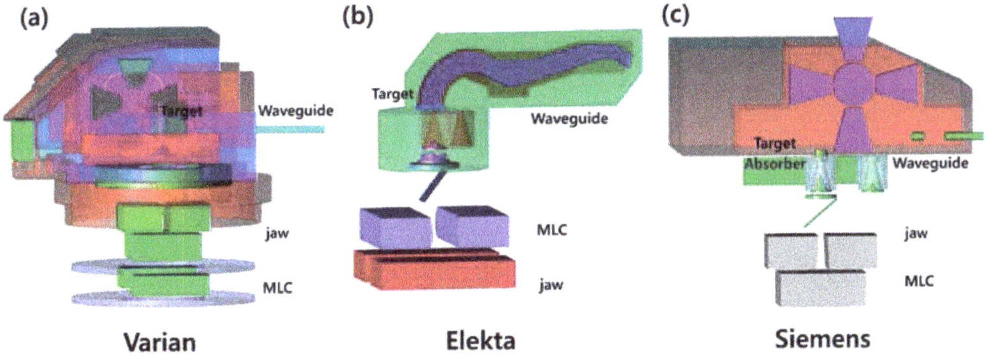

Figure 2. Medical linac modeling results of the three manufacturers: (**a**) Varian; (**b**) Elekta; and (**c**) Siemens.

No workload-dependent differences in the radionuclides were observed among the parts, and an analysis of the activity level revealed that the radiation of beams at a workload of 80 Gy/day was about twice as high as that of 40 Gy/day. In this study, radionuclides and their activity level analysis were analyzed based and compared on the workload of 80 Gy/day, taking a conservative approach. The main radionuclides from each manufacturer's parts are listed in Table 2. Radionuclides in Table 2 are the results of photonuclear reactions, such as (γ, n), (n,γ), (γ, 2n), and (γ, np). Nuclear reactions were confirmed through Monte Carlo simulations, including ^{182}W(γ, n)^{181}W, ^{197}Au(γ, n)^{196}Au, ^{66}Zn(γ, n)^{65}Zn, ^{186}W(n, γ)^{187}W, ^{27}Al(n, γ)^{28}Al, ^{54}Fe(n, p)^{54}Mn, and ^{58}Ni(n, p)^{58}Co. As additional radionuclides were contained in the parts with large proportions of impurities, ^{94m}Nb, ^{58}Co, and ^{99}Mo were detected in the jaw of Varian, ^{99}Mo in the MLC of Elekta, and ^{108}Ag, ^{65}Zn, and ^{58}Co in the absorber of Siemens.

Table 2. List of major radionuclides in the parts of each manufacturer emitted at the workload of 80 Gy/day and activity in each part after shutdown.

Major Parts	Varian			Elekta			Siemens		
	Nuclide ($T_{1/2}$)	Activity [Bq/g]		Nuclide ($T_{1/2}$)	Activity [Bq/g]		Nuclide ($T_{1/2}$)	Activity [Bq/g]	
		15 MV	10 MV		15 MV	10 MV		15 MV	10 MV
Target	^{64}Cu (12.70 h) ^{181}W (121.18 d) ^{185}W (75.12 d) ^{63}Ni (101.15 y)	7.58×10^4	9.71×10^2	^{181}W (121.18 d) ^{64}Cu (12.70 h) ^{65}Zn (243.98 d) ^{58}Co (70.86 d)	1.47×10^7	1.84×10^5	^{196}Au (6.17 d) ^{55}Fe (2.75 y) ^{58}Co (70.86 d) ^{64}Cu * (12.70 h)	1.88×10^7	2.03×10^5
Bending magnet	^{55}Fe * (2.75 y) ^{57}Co * (271.76 d) ^{58}Co * (70.86 d) ^{57}Ni * (1.48 d)	14.47	5.26×10^{-2}	^{55}Fe * (2.75 y) ^{54}Mn * (312.04 d) ^{59}Fe * (44.49 d)	9.65	9.65×10^{-2}	^{55}Fe * (2.75 y) ^{54}Mn * (312.04 d) ^{59}Fe * (44.49 d)	0.52	1.97×10^{-3}
Primary collimator	^{185}W (75.12 d) ^{181}W (121.18 d) ^{179}Ta (1.82 y) ^{55}Fe * (2.75 y)	1.09×10^4	3.52×10^2	^{185}W (75.12 d) ^{181}W (121.18 d)	1.88×10^3	14.46	^{185}W (75.12 d) ^{181}W (121.18 d) ^{179}Ta (1.82 y)	1.58×10^4	6.79×10^2
Absorber		**			**		^{28}Al (134.50 s) ^{108}Ag * (142.90 s) ^{65}Zn * (243.98 d) ^{58}Co * (70.86 d)	7.35×10^2	0.48
Flattening filter	^{181}W (121.18 d) ^{185}W (75.12 d)	19.12		^{58}Co (70.86 d) ^{54}Mn (312.04 d) ^{55}Fe (2.75 y) ^{99}Mo * (2.75 d)	2.63×10^3	2.65×10^{-2}	^{55}Fe (2.75 y) ^{51}Cr (27.70 d) ^{54}Mn (312.04 d) ^{99}Mo * (2.75 d)	6.58×10^3	2.90×10^2
	^{66}Cu (307.20 s) ^{64}Cu (12.70 h)		0.28						
Ion chamber	^{3}H (12.33 y) ^{14}C (5704.59 y)	2.60	2.49×10^{-2}	^{28}Al (134.50 s) ^{56}Mn (2.58 h) ^{65}Zn (243.98 d) ^{63}Ni (101.15 y)	0.62	1.37×10^{-2}	^{28}Al (134.50 s) ^{108}Ag * (142.90 s) ^{66}Cu * (307.20 s) ^{65}Zn * (243.98 d)	17.74	0.17
Mirror	^{3}H (12.33 y) ^{14}C (5704.59 y)	3.00×10^{-7}	4.32×10^{-7}	^{3}H (12.33 y) ^{14}C (5704.59 y)	3.12×10^{-4}	4.12×10^{-5}	^{24}Na (15.00 h) ^{37}Ar (35.03 d) ^{31}Si (2.62 h) ^{45}Ca (162.62 d)	0.12	9.13×10^{-4}
Jaw	^{185}W (75.12 d) ^{94m}Nb * (375.80 s) ^{58}Co * (70.86 d) ^{99}Mo * (2.75 d)	3.51	3.82×10^{-2}	^{181}W (121.18 d) ^{58}Co (70.86 d)	58.29	0.4	^{185}W (75.12 d) ^{181}W (121.18 d) ^{179}Ta (1.82 y) ^{183}W (1.10×10^{17} y)	4.48×10^2	19.43
MLC	^{185}W (75.12 d) ^{181}W (121.18 d) ^{64}Cu (12.70 h) ^{55}Fe (2.75 y)	0.62	3.82×10^{-3}	^{185}W (75.12 d) ^{181}W (121.18 d) ^{93m}Nb (16.14 y) ^{99}Mo * (2.75 d)	87.44	1.14	^{185}W (75.12 d) ^{181}W (121.18 d) ^{179}Ta (1.82 y) ^{183}W (1.10×10^{17} y)	1.59×10^2	6.59

* Components additionally identified in LIBS measurements; ** head part components with no specifications provided by the manufacturers.

3.3. Time-Dependent Activity Level of Each Linac Part

The major radionuclides detected in the target after stopping the beam were ^{185}W, ^{181}W, and ^{64}Cu in Varian, with the highest activity level changing from ^{64}Cu (2.69×10^4 Bq/g) immediately after the linac shutdown to ^{185}W and ^{181}W one day later. This change of the major radionuclide from ^{64}Cu to tungsten isotopes can be explained by the former's half-life of 12.7 h (Figure 3a,b). In the target of Elekta, the major radionuclides changed from ^{181}W, ^{64}Cu, and ^{65}Zn immediately after the linac shutdown to ^{181}W and ^{64}Cu after one day. In particular, since ^{181}W, which is a long half-life radionuclide, maintained a high concentration of 1.83×10^6 Bq/g even after one year, not much lower than the initial concentration of 1.47×10^7 Bq/g (Figure 3c,d).

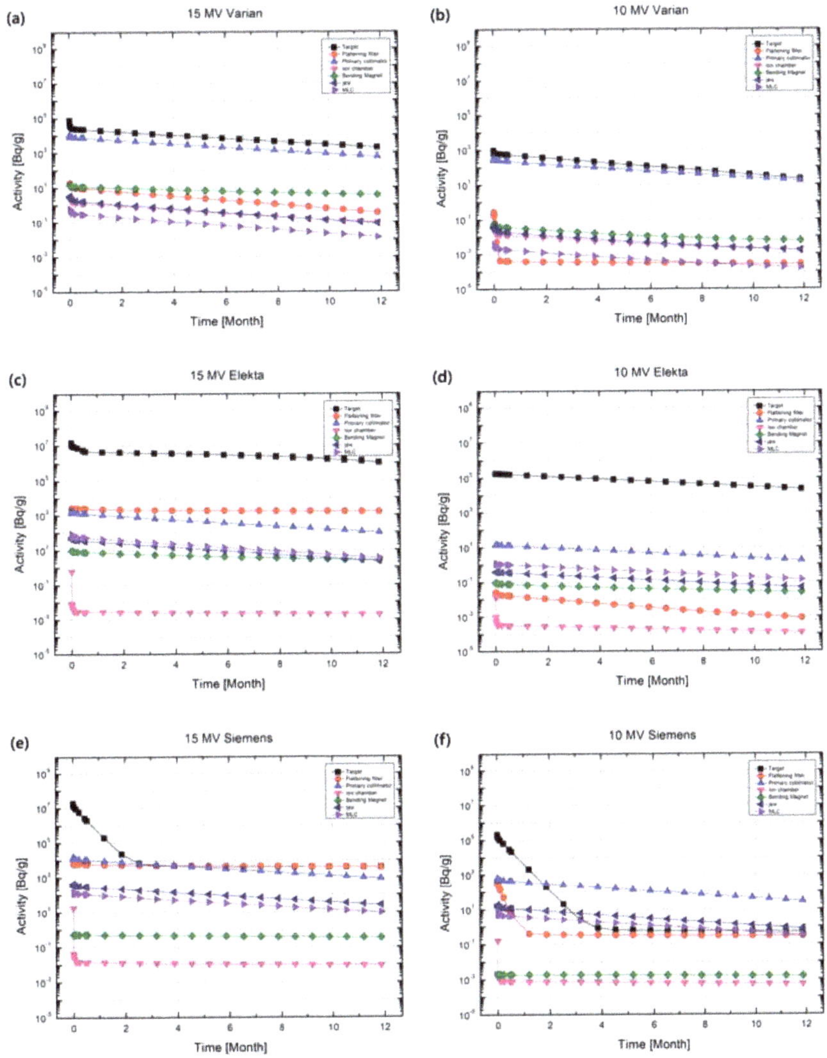

Figure 3. Time-dependent changes in the activity levels of the linac major parts obtained by applying a workload of 80 Gy/day at the energy levels of 15 and 10 MV: (**a**) 15 MV Varian; (**b**) 10 MV Varian; (**c**) 15 MV Elekta; (**d**) 10 MV Elekta; (**e**) 15 MV Siemens; and (**f**) 10 MV Siemens.

Unlike Varian and Elekta, Siemens' target is composed of materials such as graphite, gold and SST304, giving rise to various radionuclide changes for each material. The main radionuclides detected immediately after stopping the beam were ^{3}H and ^{14}C in graphite, ^{196}Au and ^{198}Au in gold, and ^{55}Fe, ^{51}Cr, ^{99}Mo in SST304. In particular, ^{196}Au had a very high initial concentration of 1.23×10^{7} Bq/g, which fell to 1×10^{3} Bq/g two months later because it has a short half-life of 6.183 days (Figure 3e,f). From two months after the linac shutdown onward, ^{55}Fe, ^{51}Cr, and ^{58}Co, which are long half-life radionuclide in SST304, were identified as the major radionuclides. The highest time-dependent activity levels were found in the target up to two months after the linac shutdown in all three manufacturers (Figure 3).

On the other hand, although various radionuclides in impurities were additionally detected (Table 2), the content of the impurities in each part was negligibly low when their activity levels were checked against the composition of the components of the parts provided by the manufacturers. However, the bending magnet, for which no data were provided by the manufacturers, was found to have initial activity levels ranging from 0.1 to 10 Bq/g due to the long half-life radionuclides ^{57}Co and ^{58}Co.

4. Discussion

Leprince assessed ^{187}W, ^{183m}W, ^{185}W, ^{181}W, and ^{178}Ta as major radionuclides in tungsten targets, with an activity level of 4.03×10^{17} Bq, using the MCNPX code, in equipment with a service life of five years [18]. Patil et al. identified radionuclides, such as ^{198}Au, ^{196}Au, ^{181}W, ^{178}W, ^{60}Co, ^{58}Co, ^{57}Co, ^{54}Mn, and ^{51}Cr, in gold and SST304 targets [19–22]. Roig et al. identified ^{196}Au, ^{57}Co, ^{60}Co, ^{54}Mn, and ^{58}Ni in Siemens KDS 18 MV, and specified a separate storage period of at least 30 days for ^{196}Au radionuclide in the target pursuant in accordance with Spanish regulations (IS/05, 2003) until the activity level fell below 10^{6} Bq [23]. While these results are consistent with those of this study, this study differentiates itself from others in that it also investigated the radionuclides additionally obtained by considering impurities.

To present, the allowable disposal time for each part of a dismantled linac, the clearance level of the radionuclides identified in each part should be applied. Regarding the self-disposal standards for radioactive waste, the United States use the technical standards published by the Department of Energy (DOE) for the disposal of radioactive waste from accelerators and accelerator facilities. Accordingly, Varian provides DOE-standards-based recommendations [24,25]. Japan has also established academic society standards for the management of radioactive materials of radiotherapy devices, specifying radioactive waste disposal procedures, parts subject to regulations, and measurement methods [26]. In the Korean notice, pursuant to the regulations on radioactive waste classification and self-disposal standards [4], clearance levels are indicated in Appendix 1 (related to Article 2 (2) and Article 3 (1)), and in the case of mixed radionuclides, the clearance level of each radionuclide should be calculated using Equation (1):

$$\sum_i \frac{C_i}{C_{L,i}} < 1 \quad (1)$$

C_i: activity level (Bq/g) of radionuclide i
$C_{L,i}$: clearance level (Bq/g) of radionuclide i

For the application of the clearance levels in compliance with the methods specified in the notice, the time-dependent concentrations of all radionuclides ($C_{L,i}$) pertaining to each part obtained using the DCHAIN code were added, and the clearance level of a given radionuclide (C_i) was checked, and a part calculated (using Equation (1)) to have a value exceeding its clearance level was analyzed to be stored until it reaches a value below the clearance level. Among the clearance level of the radionuclide specified by the NSSC notice, those on the radionuclides analyzed in this study are listed in Table 3. As recommended in the notice, the concentration of 0.1 Bq/g was applied to radionuclides for which no clearance level is provided.

Table 3. Clearance level of each radionuclide detected in the medical linacs manufactured by Varian, Elekta, and Siemens.

Radionuclides	Clearance Level (Bq/g)
^{24}Na, ^{28}Al, ^{31}Si, ^{37}Ar, ^{45}Ca, ^{54}Mn, ^{57}Ni, ^{65}Zn, ^{66}Cu, ^{108}Ag, ^{179}Ta, ^{183}W, ^{196}Au	0.1
^{14}C, ^{57}Co, ^{58}Co, ^{59}Fe	1
^{56}Mn, ^{93m}Nb, ^{99}Mo, ^{181}W	10
^{3}H, ^{51}Cr, ^{63}Ni, ^{64}Cu	100
^{55}Fe, ^{185}W	1000

Radionuclide ^{64}Cu had the highest initial concentration (2.62×10^4 Bq/g) in Varian's target 15 MV, followed by ^{185}W (1.23×10^4 Bq/g). As mentioned in another study, ^{64}Cu and ^{62}Cu have very short half-lives, and the concentration of ^{185}W had to be considered for their self-disposal [27]. In Elekta's tungsten target 15 MV, ^{181}W had the highest initial concentration (1.47×10^7 Bq/g), followed by ^{62}Cu (2.99×10^3 Bq/g). Likewise, the concentration of ^{181}W was considered for their self-disposal. In Siemens' target, ^{196}Au pertaining to gold material had the highest initial concentration (1.23×10^7 Bq/g), followed by ^{55}Fe (3.17×10^3 Bq/g) among the non-gold materials. As shown in Figure 3e,f, after about 2 months, with major radionuclide changing from Au isotopes with short half-lives to Fe isotopes with long half-lives, the activity level is maintained, and the activity level of ^{55}Fe should be considered for self-disposal.

In Figure 4, Equation (1) and Table 3 were applied to indicate the clearance levels based on the activity levels assessed for the head parts of the equipment of the three manufacturers that operated at 10 and 15 MV energy for 10 years, applying the workload of 80 Gy/day.

Six hours after the linac shutdown, the activity levels of the ion chamber and mirror were found to be lower than the clearance level among all parts of the linacs using 10 and 15 MV energy (Figure 4). Most of the parts of the linacs of the three manufacturers using 10 MV energy were found to have an activity level below the clearance level after one month, but Siemens' 10 MV target reached the clearance level after two months, and Elekta's 10 MV target was found to have an activity level exceeding the clearance level even after one year. In particular, the 15 MV targets of the linacs of the three manufacturers maintained the activity level of 10^3 Bq/g or higher even after one year, exceeding the clearance level. The 15 MV flattening filter of Elekta and Siemens also had to be stored for one year or longer.

As regards the 15 MV primary collimator, which is composed mainly of tungsten, those of Elekta and Varian could be disposed after two and 11 years, respectively, but that of Siemens had an activity level in excess of the clearance level even after 20 years (Figure 4e). This is due to the fact that, unlike Elekta and Varian, Siemens' primary collimator has an absorber attached to it, and its clearance level exceeds 1 due to the radionuclide ^{108m}Ag identified as an impurity. ^{108m}Ag radionuclide has a significantly long half-life of 418 years and a high radionuclide concentration of 6.93 Bq/g immediately after the linac shutdown. Since the notice does not provide its clearance level, 0.1 Bq/g, which is recommended for radionuclides with an unspecified clearance level, was applied, and its activity level still exceeded the reference value of 1 even 20 years after linac shutdown. Since only the radionuclide ^{28}Al was detected in Siemens' absorber without impurities, its tungsten primary collimator could be disposed after 15 years.

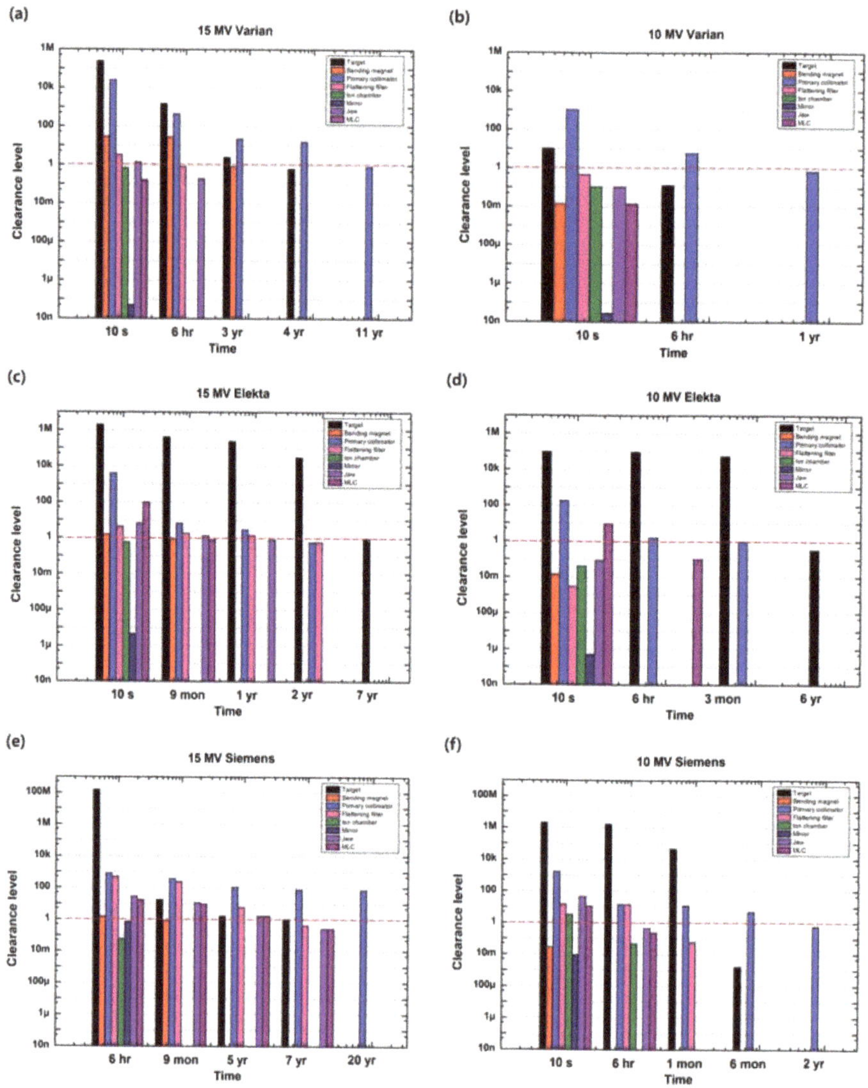

Figure 4. Clearance level by part and by equipment (15 MV/10 MV linacs of three manufacturers), applying a workload of 80 Gy/day (dotted red line: reference clearance level of 1; self-disposal is allowed at the activity level below this line): (**a**) 15 MV Varian; (**b**) 10 MV Varian; (**c**) 15 MV Elekta, (**d**) 10 MV Elekta; (**e**) 15 MV Siemens; (**f**) 10 MV Siemens.

The parts that were analyzed to be stored for six months or longer until self-disposal, were those with radionuclides with mid-to-long half-lives ($\geq 1 \times 10^6$ s) and high initial activity levels, such that their activity levels are still in excess of 10^3 Bq/g even after six months. Typical radionuclides with these attributes are ^{51}Cr, ^{54}Mn, ^{55}Fe, ^{57}Co, ^{58}Co, ^{179}Ta, ^{181}W, and ^{185}W generated from the SST304 and tungsten materials (Table 3).

When a workload of 40 Gy/day is applied, earlier disposal is possible because the activity levels of the parts are lower than when a workload of 80 Gy/day is applied. For the parts that require a long storage time ($\geq$2 years), i.e., target and primary collimator, no great changes occurred in the time for disposal when the workload was halved. However, among those whose clearance level was reached after nine months, Elekta's MLC and bending magnet and Siemens' bending magnet could be disposed after one month when the workload was halved. As mentioned previously, a workload of 80 Gy/day was applied, adopting a conservative approach, and more accurate disposal time can be predicted by reflecting the energy and workload applied in actually decommissioned medical linacs.

5. Conclusions

This study was conducted to provide basic data for preparing regulations on the disposal of linear accelerators (linacs) by performing a component analysis of the linac parts of three manufacturers (Varian, Elekta, and Siemens), and assessing and analyzing the time-dependent activity levels of radionuclides using the Monte Carlo PHITS code and DCHAIN.

Although various radionuclides were additionally detected due to impurities, it was determined that their activity levels were negligible for the total activity level of each part and that the clearance level could be sufficiently examined using the Monte Carlo simulation results based on the data provided by the manufacturers. According to the clearance levels of the radionuclides specified in the pertinent Korean notice, the 15 MV targets of the three manufacturers should be stored at least for 3.5 to 7 years until they reach the clearance levels, whereas the 10 MV targets of Varian, Elekta, and Siemens can be disposed 6 h, 5.5 years, and 6 months after the linac shutdown, respectively. In order to avoid workers' exposure to radiation during dismantling/disposal, it is recommended to work after storing the parts until the clearance level (<1) is reached, as shown in Figure 4.

As demonstrated in this study, if values exceeding the clearance level can be computed through Monte Carlo simulation applying the workload until the dismantling/disposal of a linac, the storage period of each part after which workers can safely work on dismantling/disposal can be predicted. It is planned to perform a qualitative and quantitative analysis of radionuclides using HPGe-based gamma spectroscopy in addition to the Monte Carlo simulation with a view to providing a groundwork for clear safety management regulations for the disposal of decommissioned linacs.

Author Contributions: Conceptualization, Y.J.J. and S.H.C.; methodology, S.H.P. and D.W.K.; software, Y.C.; validation, K.B.K. and S.H.B.; formal analysis, N.H.K. and S.H.B.; investigation, Y.C. and K.B.K.; data curation, Y.J.J. and N.H.K.; writing—original draft, Y.J.J.; writing—review and editing, S.H.C.; supervision, S.H.P. and D.W.K.; funding acquisition, S.H.C. All authors have read and agreed to the published version of the manuscript.

Funding: This research was supported by the National Research Council of Science & Technology (NST) grant by the Korea government (MSIT) (No. CAP22041-000) & Korea Institute of Radiological & Medical Sciences(KIRAMS) grant funded by the Korea government (Ministry of Sciences and ICT) (No. 50572-2023).

Institutional Review Board Statement: Not applicable.

Informed Consent Statement: Not applicable.

Data Availability Statement: Not applicable.

Conflicts of Interest: The authors declare no conflict of interest.

References

1. Kwon, N.H.; Jang, Y.J.; Kim, D.W.; Shin, D.O.; Kim, K.B.; Kim, J.S.; Choi, S.H. Trend Analysis on Korean and International Management for Activated Material Waste from Medical Linear Accelerator. *Prog. Med. Phys.* **2020**, *31*, 194–204. [CrossRef]
2. Kwon, N.H.; Shin, D.O.; Kim, J.S.; Yoo, J.; Park, M.S.; Kim, K.B.; Kim, D.W.; Choi, S.H. Current status of disposal and measurement analysis of radioactive components in linear accelerators in Korea. *Nucl. Eng. Technol.* **2022**, *54*, 507–513. [CrossRef]
3. Korea Institute of Nuclear Safety. *Improving Radiation Safety Management in the Medical Sector [Workshop]*; Korea Institute of Nuclear Safety: Daejeon, Republic of Korea, 2018.
4. Radioactive Waste Safety Division, Nuclear Safety and Security Commission. *Nuclear Safety and Security Commission No-Tification No. 2020-6, Regulations on Radioactive Waste Classification and Clearance Standards*; (Implementation and Revision in 2020.05.26); Radioactive Waste Safety Division, Nuclear Safety and Security Commission: Seoul, Republic of Korea, 2020.

5. MOTIE. *Radioactive Waste Control Act No. 15082*; (Implementation and Revision in 2017.11.28); Ministry of Trade Industry and Energy (MOTIE): Seoul, Republic of Korea, 2017.
6. Olsen, K.J. Neutron Contamination from Medical Electron Accelerators (NCRP Report No. 79). *Med. Phys.* **1986**, *13*, 968–969. [CrossRef]
7. International Atomic Energy Agency. *Management of Radioactive Waste from the Use of Radionuclides in Medicine*; IAEA-TECDOC-1183; IAEA: Vienna, Austria, 2000.
8. Naseri, A.; Mesbahi, A. A review on photoneutrons characteristics in radiation therapy with high-energy photon beams. *Rep. Pract. Oncol. Radiother.* **2010**, *15*, 138–144. [CrossRef] [PubMed]
9. Vichi, S.; Dean, D.; Ricci, S.; Zagni, F.; Berardi, P.; Mostacci, D. Activation study of a 15MeV LINAC via Monte Carlo simulations. *Radiat. Phys. Chem.* **2020**, *172*, 108758. [CrossRef]
10. Jang, Y.-J.; Kwon, N.H.; Park, S.H.; Choi, Y.; Yu, H.; Kim, K.B.; Kim, D.W.; Choi, S.H. Activation evaluation of Siemens linear accelerator using Monte Carlo simulation. *J. Korean Phys. Soc.* **2022**, *81*, 1107–1114. [CrossRef]
11. Konefał, A.; Bieniasiewicz, M.; Wendykier, J.; Adamczyk, S.; Wrońska, A. Additional radiation sources in a treatment and control room of medical linear accelerators. *Radiat. Phys. Chem.* **2021**, *185*, 109513. [CrossRef]
12. Waller, E.; Ram, R.; Steadman, I. Preliminary Determination of Activation Products for a Varian Truebeam Linear Accelerator. *Health Phys.* **2017**, *113*, 227–233. [CrossRef] [PubMed]
13. Wu, Y.; Song, J.; Zheng, H.; Sun, G.; Hao, L.; Long, P.; Hu, L. CAD-based Monte Carlo program for integrated simulation of nuclear system SuperMC. *Ann. Nucl. Energy* **2015**, *82*, 161–168. [CrossRef]
14. Sato, T.; Iwamoto, Y.; Hashimoto, S.; Ogawa, T.; Furuta, T.; Abe, S.-I.; Kai, T.; Tsai, P.-E.; Matsuda, N.; Iwase, H.; et al. Features of Particle and Heavy Ion Transport code System (PHITS) version 3.02. *J. Nucl. Sci. Technol.* **2018**, *55*, 684–690. [CrossRef]
15. Ratliff, H.N.; Matsuda, N.; Abe, S.-I.; Miura, T.; Furuta, T.; Iwamoto, Y.; Sato, T. Modernization of the DCHAIN-PHITS activation code with new features and updated data libraries. *Nucl. Instrum. Methods Phys. Res. Sect. B Beam Interact. Mater. Atoms* **2020**, *484*, 29–41. [CrossRef]
16. Zheng, L.; Huang, S.; Wang, K. Criticality benchmarking of ENDF/B-VIII.0 and JEFF-3.3 neutron data libraries with RMC code. *Nucl. Eng. Technol.* **2020**, *52*, 1917–1925. [CrossRef]
17. Shibata, K.; Iwamoto, O.; Nakagawa, T.; Iwamoto, N.; Ichihara, A.; Kunieda, S.; Chiba, S.; Furutaka, K.; Otuka, N.; Ohsawa, T.; et al. JENDL-4.0: A New Library for Nuclear Science and Engineering. *J. Nucl. Sci. Technol.* **2012**, *48*, 1–30. [CrossRef]
18. Leprince, A.; David, J.-C.; Ene, D.; Leray, S. Reliability and use of INCL4.6-Abla07 spallation model in the frame of European Spallation Source target design. *Prog. Nucl. Sci. Technol.* **2014**, *4*, 491–495. [CrossRef]
19. Patil, B.; Chavan, S.; Pethe, S.; Krishnan, R.; Bhoraskar, V.; Dhole, S. Estimation of neutron production from accelerator head assembly of 15MV medical LINAC using FLUKA simulations. *Nucl. Instrum. Methods Phys. Res. Sect. B Beam Interact. Mater. Atoms* **2011**, *269*, 3261–3265. [CrossRef]
20. Saeed, M.K.; Poppe, B.; Fischer, H. Direct air activation measurements at a 15-MV medical linear accelerator. *Radiat. Prot. Dosim.* **2015**, *163*, 233–237. [CrossRef] [PubMed]
21. Šagátová, A.; Fülöp, M.; Pavlovič, M.; Sedlačková, K.; Nečas, V. Electron-beam accelerator with conversion to X-rays: Optimal radiation type according to application. *Radiat. Phys. Chem.* **2020**, *172*, 108789. [CrossRef]
22. Długosz-Lisiecka, M.; Biegała, M.; Jakubowska, T. Activation of medical accelerator components and radioactive waste classification based on low beam energy model Clinac 2300. *Radiat. Phys. Chem.* **2023**, *205*, 110730. [CrossRef]
23. Roig, M.; Panettieri, V.; Ginjaume, M.; Sánchez-Reyes, A. Photonuclear isotope characterization of a Siemens KDS 18 MV linac head. *Phys. Med. Biol.* **2004**, *49*, N243–N246. [CrossRef]
24. U.S. Department of Energy. *Clearance and Release of Personal Property from Accelerator Facilities*; DOE-STD-6004-2016; U.S. Department of Energy: Washington, DC, USA, 2016.
25. U.S. Department of Energy. *Radiation Protection of the Public and the Environment*; DOE O 458.1; U.S. Department of Energy: Washington, DC, USA, 2013.
26. Japanese Society of Radiation Oncology. *Society Standard for the Management of Radioactive Substances in Radiotherapy Equipment*; Japanese Society of Radiation Oncology: Tokyo, Japan, 2014.
27. Dellepiane, G.; Casolaro, P.; Mateu, I.; Scampoli, P.; Braccini, S. Alternative routes for 64Cu production using an 18 MeV medical cyclotron in view of theranostic applications. *Appl. Radiat. Isot.* **2023**, *191*, 110518. [CrossRef] [PubMed]

sustainability

MDPI

Article

Using Agricultural Mixed Waste as a Sustainable Technique for Removing Stable Isotopes and Radioisotopes from the Aquatic Environment

Arwa A. Abdelhamid [1], Mogeda H. Badr [1], Ramadan A. Mohamed [2,3] and Hosam M. Saleh [4,*]

[1] Department of Chemistry, Faculty of Science, Al-Azhar University, Nasr City 11884, Cairo, Egypt
[2] Central Laboratory for Elemental and Isotopic Analysis-Nuclear Research Center, Egyptian Atomic Energy Authority, Nasr City 11787, Cairo, Egypt
[3] Ion Sources and Accelerators Department, Nuclear Research Center, Egyptian Atomic Energy Authority, Nasr City 11787, Cairo, Egypt
[4] Radioisotope Department, Nuclear Research Center, Egyptian Atomic Energy Authority, Nasr City 11787, Cairo, Egypt
* Correspondence: hosam.saleh@eaea.org.eg or hosamsaleh70@yahoo.com; Tel.: +20-10051-91018; Fax: +20-23749-3042

Abstract: In the current study, dried ground mixed waste of olive waste and water hyacinth was subjected to the treatment of wastewater contaminated with stable or radioactive cobalt and cesium. Contact time, temperature, amount of adsorbent, metal ion concentration and pH-value were evaluated as the most potent factors that affect the adsorption process. Concentrations of Co and Cs ions before and after treatment are measured using ICP-OES technique. Moreover, kinetic and equilibrium isotherm parameters were investigated by explaining the equilibrium data by induction of two isotherms, "Langmuir" and "Freundlich". Experimental results indicated that more than 85% of ^{60}Co and ^{134}Cs were efficiently removed from spiked wastewater after one hour of contact time by using 0.2 g of dried mixed waste. The optimal time to remove Cs^+ and Co^{2+} was 50 and 90 min, respectively. The kinetic study showed that the adsorption of Cs^+ and Co^{2+} were better suited to the second order. The most favorable pH value was at a range of 6–7 for Cs^+ and 5–6 for Co^{2+}. R^2 values were higher at the Freundlich isotherm, indicating that the adsorption process was taking place according to Freundlich isotherm at all temperatures for Co^{2+} and Cs^+ ions.

Keywords: olive waste; water hyacinth; cobalt; cesium; adsorption; kinetics; biomass; wastewater; ICP-OES spectroscopy

Citation: Abdelhamid, A.A.; Badr, M.H.; Mohamed, R.A.; Saleh, H.M. Using Agricultural Mixed Waste as a Sustainable Technique for Removing Stable Isotopes and Radioisotopes from the Aquatic Environment. *Sustainability* **2023**, *15*, 1600. https://doi.org/10.3390/su15021600

Academic Editor: Farooq Sher

Received: 7 November 2022
Revised: 6 January 2023
Accepted: 10 January 2023
Published: 13 January 2023

1. Introduction

Water covers the vast majority of the Earth's surface (71%), including only a small portion (3%) of freshwater that plays a vital role in the global economy. This natural essential resource became insufficient in numerous areas. The water scarcity was a major economic and social concern, especially in the agriculture sector, which consumes about 70% of the freshwater [1]. In many attempts in developing countries, 90% of the wastewater remains untreated in freshwater bodies, rendering it unsafe for human consumption, resulting in scarcity or toxic impact [2]. Clean soil and pure water are the most in-demand resources all over the world nowadays [3,4]; also in demand are more efficient selective methods for analysis [5,6].

Pollution is an important issue around the world, especially the contamination with chemical elements, including stable isotopes or radioisotopes. Chemical pollution is the most significant problem because of the easy dispersal of harmful metals, for example, it can be transferred into edibles; however, water contamination is considered a serious ecological problem around the world as well [7].

Physicochemical approaches, such as precipitation, redox-reaction, ion exchanging [8,9] and electrochemically [10], have been reported. Ultrafiltration [11], photocatalysis [12], reverse osmosis [13] and electro flotation [14] are some of the other important techniques. The disadvantages of these techniques are their cost and the production of secondary waste that requires extra treatment [15].

Developing sustainable, simple, efficient, and cost-effective techniques for removing different elements from wastewater is recent worldwide challenge. Because it is a straightforward means of removing metal ions from wastewater, phytoremediation [16,17], using aquatic plants, and adsorption process [18–20], using agricultural waste, become viable choices. For example, the cost of excavation and disposal of contaminated soil based on the Federal Remediation Technologies Roundtable (FRTR) data was in the range of 270 to 460 $/ton, while the cost of phytoextraction, including plant selection and position, irrigation, soil amendment, field monitoring, harvesting, and residue management, could be in the range 10–35 $/ton [21].

Although phytoremediation using aquatic plants achieved great effectiveness in the treatment process, the findings of the experiments revealed that dried plants are more effective than fresh plants, and consequently, adsorption techniques have recently been investigated in the management of contaminated wastewater spiked with arsenate, nitrate, phosphate, cadmium, and other heavy metals [7].

Increased cobalt concentrations injure humans, causing lung irritations, asthma, pneumonia, weight loss, paralysis, diarrhea, vomiting, thyroid hormones, liver damage, and nausea, due to the resemblance of cesium-134 to sodium; the ingestion of cesium causes its deposition in soft tissues throughout the body, posing internal hazards [22].

Physical procedures, such as adsorption by natural or manufactured materials, have been used to remove cesium and cobalt. As one of the traditional adsorbents, ion exchangers have limitations because of their competition with other monovalent cations found in liquid radioactive waste, such as sodium and potassium, which can inhibit the adsorption of cesium or other competing divalent cations capable of occupying cobalt adsorption sites [23]. Because the metal adsorption depends on cation exchange sites, it requires the development of new and more active materials with safe and sustainable features for the economical removal of stable or radio pollutants from wastewater.

Water hyacinth is an aquatic plant weed that grows quickly and floats freely in water. It has a reputation of having a very high metal tolerance. The adsorption of Pb^{2+} from liquid effluents was conducted using the dried biomass of Eichhornia crassipes. The performance of the adsorption mechanism was determined to be depended on pH and the adsorbent's uneven surface may aid in the adsorption of Pb^{2+} on the surface of the adsorbent [24].

The current study focuses on the utilization and effectiveness of biomass originated from dried ground mixed waste of water hyacinth and olive waste to remove cesium and cobalt from wastewater due to their high availability and low operating cost instead of traditional techniques, which are difficult to control and require primarily expensive requirements. Cobalt compounds should be considered genotoxic carcinogens with a practical threshold, and chronic inhalation of cobalt compounds can induce respiratory tumors [25], while the toxicity of cesium can induce gastrointestinal distress, hypotension, syncope, numbness or tingling of the lips [26]. Moreover, radiotoxicity was added on the chemical toxicity when in contact with radio cobalt or radio cesium.

In order to identify the ideal physicochemical conditions for the most effective sorption for treatment applications, the adsorption process was studied in terms of the impact of pH-value, adsorbent dose, metal ion, and interaction time. Additionally, the adsorption process followed the "Langmuir" and "Freundlich" adsorption models, and the kinetic studies were developed to describe the isothermal trend and constants of the adsorption process, with the pH-value being the most important factor. An experiment, using the nominated adsorbent to eliminate radiocobalt and radiocesium, was performed. These treatment processes generated secondary radioactive waste that had to be stabilized and remediated for safe disposal, as reported in the previous literature, by using cement [27,28], cement polymer composite [29,30], and cement mixed with natural clay [31].

2. Experimental Approach

2.1. Materials

2.1.1. Adsorptive Materials

Green plants of water hyacinth have been harvested from Egyptian waterways, dehydrated in a 70 °C oven for 2 days and ground to produce a fine powder using an electric mixer. Other types of unimportant materials, such as solid waste derived from the olive oil industry, were prepared by washing thoroughly with methanol, rinsing with pure water, and introducing them to drying at 70 °C. The dry matter was crushed and ground. The two dried ground plant wastes were mixed with a ratio of 50 to 50%. As shown in Figure 1, the chart of spectra confirmed the presence of major functional groups in ground dry single waste and mixed plant waste that were identified and characterized using FTIR spectroscopy with a Shimadzu FTIR-8201 PC. FTIR analysis was performed to identify the functional groups in the dried single waste and mixed waste using the KBr disc method. This investigation showed a strong broad absorption band of O-H stretching vibrations in the region of 3272–3177 cm^{-1} and another peak at 3445 cm^{-1}. Other peaks appeared at 1739 cm^{-1} and 1322 cm^{-1}, which corresponded to the C=O stretching vibrations and the phenolic hydroxyl groups, respectively. At 1547 cm^{-1}, a stretching vibration regarding the C–N of the peptidic bond of proteins was present; a peak at around 1500 cm^{-1} indicated C=C bending. The range 1130–1000 cm^{-1} is a vibration of C–O–C and O–H of polysaccharides. The intensity of the adsorption bands confirmed no significant difference between olive waste alone and that mixed with water hyacinth; the functional groups were responsible for the adsorption of many elements.

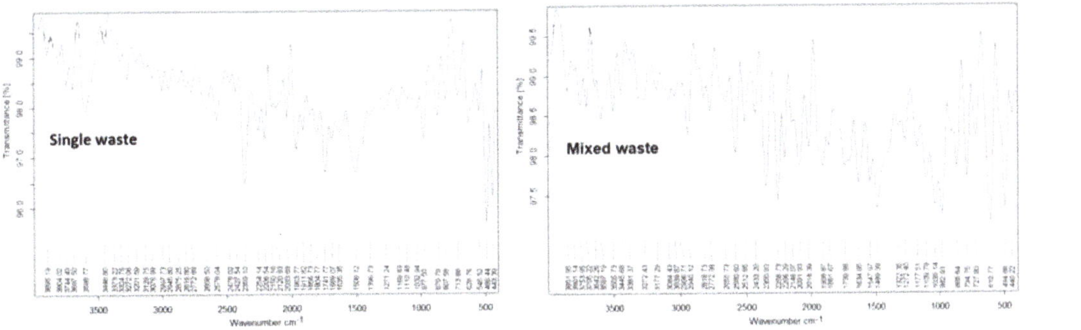

Figure 1. FTIR spectra of single waste and mixed waste.

2.1.2. Adsorbate Elements

Radioisotopes of ^{134}Cs and ^{60}Co were produced at the Egyptian Atomic Energy Authority, Egypt. The activity content of two radioisotopes was diluted to various activities in the range of (1400–4850 Bq). High purity and analytical-grade salts of stable cesium and cobalt were purchased from Sigma-Aldrich. To achieve a suitable value of concentration for the subsequent experimentations, dilution was used to adjust the concentration of stable elements in the range of (5–100 mg/L). Additionally, gradual addition of NaOH and 0.1 N of HCl were performed to adjust the acidity or alkalinity of the spiked solutions for various pH values.

2.2. Experimentation

2.2.1. Adsorption Studies

All experiments have been performed in a bathwater with a constant shaking speed to investigate the influence of contact time, pH-value, adsorbent weight, concentration of Cs^+ or Co^{2+} and temperature.

Adsorption of Single Waste and Mixed Waste

A primary experiment was performed to evaluate adsorption efficiency using different weights of single waste (olive waste) and mixed waste (olive waste) to uptake Cs or Co elements for 2 h at the constant temperature of 25 °C, pH value of 7 and shaking speed of 100 rpm. q_e was performed according to the recent literature [32] and Equation (1).

$$q_e = \left(\frac{C_0 - C_e}{m}\right) \times V \quad (1)$$

q_e is the adsorption capacity (mg/g), C_0 is the initial concentration of ions (mg/L), C_e is the equilibrium concentration of ions (mg/L), m is the mass of biosorbent used in the process (g) and V is the volume of the solution (L).

Contact Time

The effect of time on the adsorption of (50 mg/L) Cs^+ or Co^{2+} on (0.1 g) mixed waste was evaluated at the constant temperature of 25 °C while maintaining the pH value at 7 and the shaking speed of 100 rpm at variable time during two hours (0, 1, 5, 10, 15, 20 and 30, 45, 60, 90 and 120 min). At each contact time, the concentration of non-adsorbed ions was determined by taking 1 mL of clear solution for subsequent analysis using atomic absorption spectroscopy.

pH Effect

Under varying pH conditions, stable elements were adsorbed at the constant temperature and a shaking rate of 25 °C and 100 rpm, respectively. The pH of non-radioactive solutions was maintained by the addition of 0.1 N of HCl and NaOH to adjust the pH. The removal percentage of both ions (concentration, 50 mg/L) at various pH (3 to 8) by using 0.1 g of dry ground mixed waste was determined at 60 min.

Adsorbent Dose

Six extended weights of dry mixed waste (0.025, 0.05, 0.1, 0.25, 0.5, and 1.0 g) were poured to six bottles including 100 mL of Cs^+ or Co^{2+} at 25 °C, shaken at 100 rpm, with pH-value (7 for Cs^+ and 6 for Co^{2+}) for 60 min. Then, 1 mL of each bottle was collected and analyzed to determine the remaining concentration of each element after completing the process.

2.2.2. Inductively Coupled Plasma Optical Emissions Spectroscopy (ICP-OES)

ICP-OES, inductively coupled plasma optical emission spectroscopy (Prodigy High Dispersion ICP, Leeman in the United States) was used to evaluate the initial concentrations of Co and Cs, as well as the concentration of the remaining metal in the solution. The liquid was converted into an aerosol inside the instrument via a process known as nebulization. The sample aerosol was then transported to the plasma and desolvated, vaporized, atomized, excited, and/or ionized by the plasma. The excited atoms and ions emitted their characteristic radiation, which was collected and sorted by wavelength by a device [33].

The limit of detection (LOD) was primarily determined by instrumental sensitivity, spectral interferences, memory effects, digestion vessel cleanliness, and analytical reagent blank level, and it is possible to define the lowest concentrations that can be reliably detected and quantified.

The LODs for each element were calculated as the concentration equivalent of three times the standard deviation of the ion counts obtained from duplicate runs of reagent blank solutions (3% *v*/*v* HNO_3 including the internal standard spikes) [34].

ICP-OES has the following features:

Available in radial, axial and dual view configurations;
Wavelength range from 165 to 1100 nm;
Resolution < 0.008 nm;
High precision elemental ratio measurements (0.01–0.02% RSD).

2.2.3. Adsorption Isotherms and Data Analysis

At various temperatures and constant pH and shaking speed (pH = 7, 100 rpm), adsorption isotherms were extrapolated for Cs and Co ions at the two contact times of 45 and 90 min, respectively. At 15, 25, 35, and 45 °C, 0.1 g of dried mixed waste was added to 100 mL of Cs or Co ions of various initial concentrations (C_0), (5, 10, 20, 50, 75, and 100 mg/L), respectively, until equilibrium time was reached. A total of 1 mL of the remaining clear solution was analyzed by an atomic absorption apparatus to determine the non-adsorbed ions under various conditions.

Batch data from various contact times with variable initial concentrations of metal ions were analyzed using kinetic studies (pseudo-first order, pseudo-second order, and intra-particular diffusion).

Furthermore, different models of isothermal absorption, such as "Langmuir" and "Freundlich", were applied to investigate the adsorption potential of different ions.

2.2.4. Adsorption of Radioisotopes (^{60}Co or ^{134}Cs)

Absorption processes at constant room temperature were followed by using 50 mL of ^{60}Co or ^{134}Cs solutions of various initial activity in the range of (1400 to 4350 Bq for ^{134}Cs, and 3800 to 11000 Bq for ^{60}Co). In this experiment, extended doses of adsorbent (0.5, 1.0, 1.5, 2.0, 2.5 g) were mixed with ^{60}Co and ^{134}Cs for contact time (120 min.), constant shaking speed (100 rpm), and constant pH-value (7). The removal of radionuclides was assessed by periodically recording the radioactivity content of the remaining solution using a multichannel analyzer of a NaI detector, PCAP, USA.

2.2.5. Statistical Analysis

All results are expressed as means $\pm$ SEM and regression models with the SPSS software (Statistical Package for the Social Sciences), version 24 (IBM SPSS Statistics for Windows, Armonk, NY, USA: IBM Corp.). Comparisons made between more than two groups were performed using multifactorial analysis of variance (ANOVA) followed by the Least Significant Difference (LSD) test. Differences were considered statistically significant when the alpha probability was $\leq$0.05.

3. Results and Discussions

3.1. *Single Waste and Mixed Waste*

The potential of agricultural waste has been conducted in recent decades according to its ability to accumulate some elements from contaminated water or soil [35]. At least the same potential, such as terrestrial plants, can eliminate heavy metal contaminants that are commonly found in the environment, such as Cd, Cr, Cu, Hg, Pb, Ni, Zn and Fe [36].

Some of these elements are essential for plants but the others are toxic or still have unknown biological effects. A primary experiment was performed to evaluate the efficiency of using single waste, such as dry olive waste, and mixed waste of the same waste with dry water hyacinth to uptake Cs or Co elements after 2 h of contact. Figure 2 shows the superiority of mixed waste in the accumulation of constant concentration of Cs and Co by different amounts of adsorbents. According to the observed results, it is confirmed that mixed waste has a tendency to accumulate more than single waste, especially in the case of Co. From the statistical analysis, it is clear that F for treatment = 19.336 at $p = 0.00$ (<0.05) and this confirms the statistically significant differences between the averages in the case of treatment. Consequently, in this study, the mixed waste of olive waste and water hyacinth was selected to evaluate its efficiency to eliminate cobalt and cesium elements from artificial contaminated water. This result is consistent with the recently published literature conducted to evaluate the use of mixed biomass (*Aspergillus campestris* and agro waste) [37].

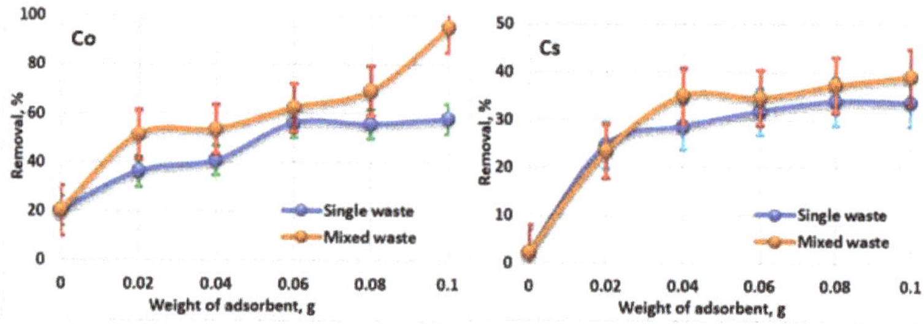

Figure 2. Adsorption of both ions separately on dry single waste and mixed waste.

3.2. Contact Time

Contact time acts as a vital key for determining the equilibrium point at the highest adsorption value. The optimal contact time is the time spent to achieve the highest removal percent with respect to different adsorption concentrations by evaluating the biosorption kinetics [38]. In this experiment, the statistical analysis confirmed that F for time = 70.681 when $p = 0.000$ (<0.05) and this confirms the statistically significant differences between the averages in the case of the time factor.

At a constant pH (7) and constant temperature (25 °C), the adsorption capacities of both ions (Cs^+ or Co^{2+}) onto dry biowaste was evaluated. As shown in Figure 3, as the contact time increases, the rate of adsorption first increases and then becomes constant due to the aggregation of adsorbed ions that diffuse deeper into the adsorbent structure at higher energy sites with time. This aggregation negates the influence of contact time as the pores get filled up and start offering resistance to the diffusion of aggregated ions in the adsorbents [39]. However, the elimination efficiency was increased and maximum biosorption capacity was achieved in 50 and 90 min, respectively.

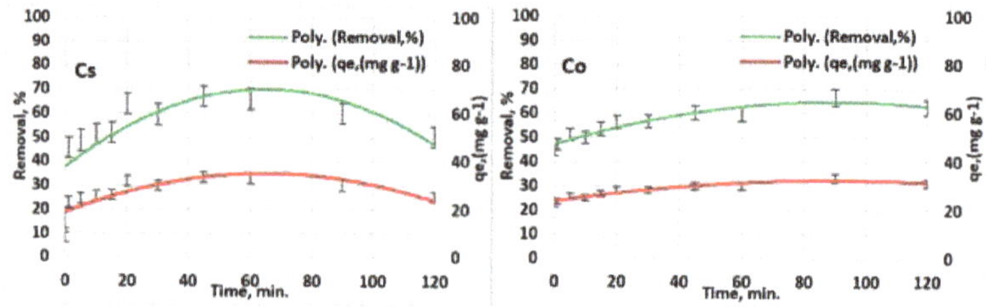

Figure 3. Removal (%) and adsorption capacity(q_e) of both ions separately using dry mixed waste during contact time.

The data are means ± SE ($n = 5$). ANOVA followed by LSD's multiple comparison post hoc test; the differences were considered significant at $p \leq 0.05$. A statistically significant difference F (contact time) was 70.681 at $p = 0.000$ (<0.05).

The calculated data of adsorption/time were employed to both kinetic models; pseudo-first order as shown in Equation (2) and pseudo-second order as shown in Equation (3).

$$\text{Log}\,(q_e - q_t) = \log q_e - (k_1/2.303)\,t \tag{2}$$

$$\frac{t}{q_t} = \frac{1}{k_2\,{q_e}^2} + \frac{1}{q_e} \tag{3}$$

where k_1 (min^{-1}) is the pseudo-first order rate constant of adsorption, q_e and q_t (mg/g) refer to amount of ions adsorbed at equilibrium and at time t, respectively. K_2 (g mg^{-1} min^{-1}) is the pseudo-second order rate constant of adsorption.

The suitable model was investigated by computing the correction coefficient value of straight line (R^2). As indicated in Figure 4, the adsorption of Cs^+ ion processes were in pseudo second order greater than in pseudo-first order because the high regression coefficient (R^2) of the pseudo-second order model was higher than that in the pseudo-first order model. Similarly, the adsorption of Co^{2+} ion processes fit the second-order model more than the first-order model according to the higher value of regression coefficient (R^2) in the case of the pseudo-second order model, as shown in Figure 5 [40]. The adsorption kinetics of Cs^+ and Co^{2+} suited the pseudo-second order model, indicating that the key step in the adsorption mechanism was the ion exchange by electrostatic interactions.

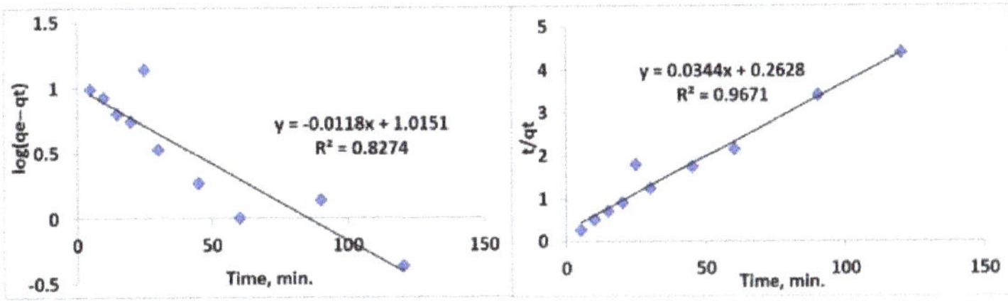

Figure 4. Adsorption kinetics modeled using pseudo-first and pseudo-second orders for Cs^+.

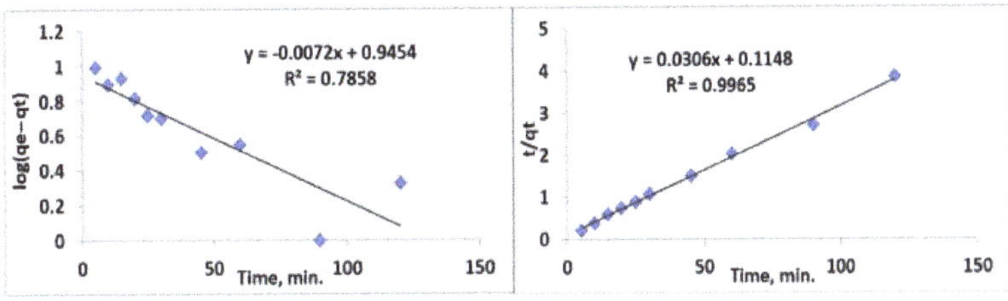

Figure 5. Adsorption kinetics modeled using pseudo-first and pseudo-second orders for Co^{2+}.

3.3. pH Effect

pH is one of the most important factors that largely affects the adsorption processes due to the effect of H^+ ions on the biosorption of Cs^+ and Co^{2+} ions by using dry plant mixed waste [41]. The potential of changing the pH from 3 to 8 was evaluated and presented in Figure 6, significantly pH could change the relative distribution of ions in the solution, as well as the surface properties of the adsorbent. The adsorption of the mixed waste was clearly affected by the varying solution pH values, which considerably increased by enhancing the solution's pH from 2 to 5, and basically remained unchanged in the case of the Co solution. In the case of the Cs solution, at pH 7, the removal capacity reached maximum value; however, afterwards rapidly decreased at a solution pH of more than 7.0. This behavior can be explained by the fact that at lower pH values, excess H^+ ions compete with Cs^+ and Co^{2+} for free adsorption sites, reducing their adsorption. With low pH, the solubility of some heavy elements in water increases, which causes an increase in their spread; pH plays a critical role in the transfer of heavy metals between the liquid and solid phases [42]. Another factor that could contribute to enhancing metal ion adsorption is the increasing pH, which encourages metal ion precipitation from the solution in the form of hydroxides [43]. From the statistical analysis it is clear that the F value (for pH) was 7.274 when $p = 0.016$ (<0.05) and this confirms the statistically significant differences between the averages in the case of this factor.

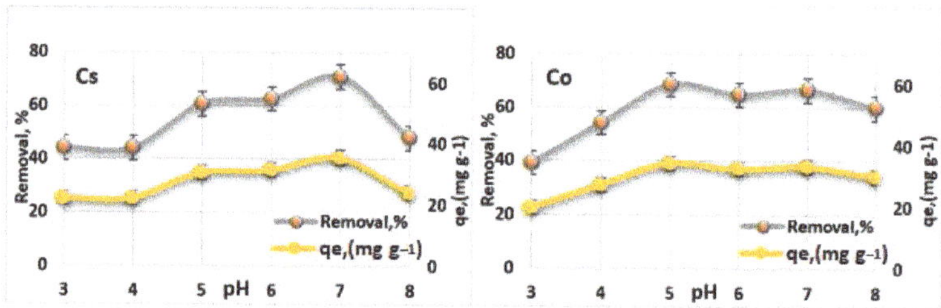

Figure 6. Removal (%) and adsorption capacity(q_e) of both ions separately in dry mixed waste at different pH-values.

The data are means ± SE (n =5). ANOVA followed by LSD's multiple comparison post hoc test; the differences were considered significant at $p \leq 0.05$. A statistically significant difference F (pH) was 7.274 at p= 0.016 (<0.05).

3.4. Dosage Weight of Dried Mixed Waste

Knowledge of the effect of the adsorbent dose on adsorption is critical for optimal adsorbent use. Figure 7 shows that batch studies with varying quantities of mixed waste were carried out to determine the best adsorbent dose. The adsorption capacity (q_e) was increased by increasing the adsorbent dose and then proceeded steadily at a high level in the case of Co, while in case of Cs it increased with the increasing the dose, then dropped suddenly, and then settled at a low level. The rise in removal with the dose is associated with the increased number of active sites, it is readily understood that the number of available adsorption sites and the surface area increase by increasing the adsorbent dose [44].

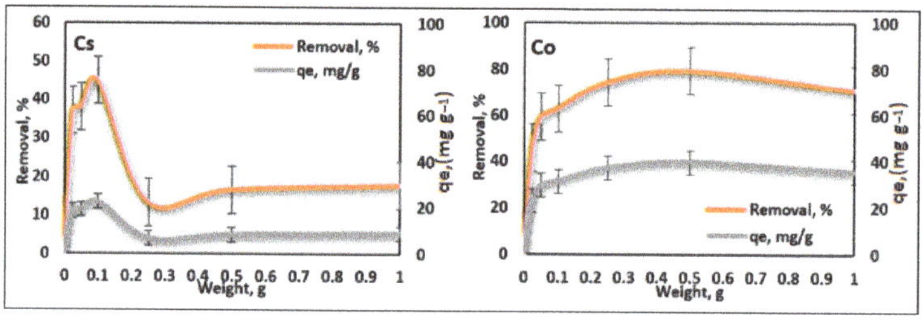

Figure 7. Removal (%) and adsorption capacity(q_e) of both ions separately at different loading weights of mixed waste.

The steady removal thereafter is an indication of the adsorption equilibrium, and the decline in removal at a high dose could have resulted from the aggregation of adsorbent particles [45]. At low adsorbent dosages, however, the adsorbed amount (q_e) reduced dramatically in the case of Cs. The optimum adsorbent doses of Cs^+ and Co^{2+} were 0.1 g for Cs^+ and 0.5 g for Co^{2+} ions. Statistically, F (for weight) was 0.041 at p = 0.998 (>0.05) non-significant.

The data are means ± SE (n = 5). A significant difference F (loading weight) was 0.041 at p = 0.998 (>0.05) non-significant.

3.5. Comparison of the Study Findings with Other Similar Published Work

The economic and ecological features of the adsorption technique, various adsorbents have been studied in the literature. Some of these adsorbents have been collected in Table 1, where they are compared with the mixed waste used in the current study to evaluate the adsorption capacity.

Table 1. Adsorption capacity of mixed waste relative to other adsorbents for both ions separately.

Material	Adsorption Capacity (q_e) of Co	Literature
Kaolinite	0.92	[46]
Soil	1.50	[47]
Marine bacterium	4.38	[48]
Nedalco sludge	11.71	[49]
Eerbeek sludge	12.34	[49]
Coir pith	12.82	[50]
Brown seaweed	20.63	[51]
Myriophyllum spicatum	43.40	[18]
Mixed waste	37.45	Current study
	Adsorption Capacity (q_e) of Cs	
Ceiling tiles	0.21	[52]
Coal and chitosan	3.00	[53]
Bure mudrock	13.30	[54]
Modified akadama clay	16.10	[55]
Kaolinite clay	17.10	[56]
Coir pith	32.00	[57]
Bentonite	40.00	[55]
Myriophyllum spicatum	58.00	[18]
Mixed waste	48.30	Current study

3.6. Adsorption Isotherm at Different Teperatures

At various temperatures, the adsorption behavior attained a maximum uptake value of 35 mg/g at 25 °C and 45 mg/g at 45 °C for both elements Cs^+ or Co^{2+}, respectively, as shown in Figure 8. Due to the ratio of the interaction energies of the cations with water molecules between the layers and with the charges on the surface of the biomass layer, ions with a small radius, (Co), have lower competitive advantages over those with a large radius, (Cs) [58]. However, the change in the cation diffusion coefficient with increasing temperature is slightly indicated in the case of Cs.

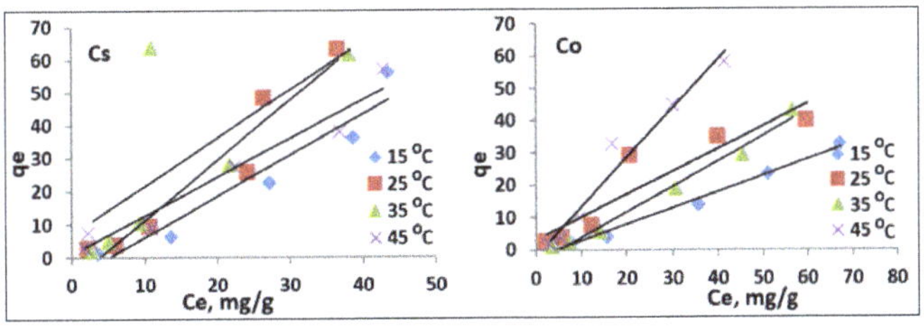

Figure 8. Adsorption capacity (q_e) of both ions separately on dry mixed waste at various temperatures.

The Langmuir adsorption isotherm implies the presence of a homogenous monolayer at all sorbent surface sites with no adsorbed molecules interacting with nearby adsorption sites. The Langmuir model was applied by Equation (4).

$$\frac{1}{q_e} = \frac{1}{k_L\, q_{max}} \frac{1}{C_e} + \frac{1}{q_{max}} \tag{4}$$

where q_e (mg/g) is the uptake at equilibrium concentration, C_e (mg/L), and q_{max} (mg/g) is the maximum number of ions required to create a monolayer. The linearized Langmuir adsorption isotherm was used to examine the equilibrium data, as shown in Figures 9 and 10. Table 2 shows the Langmuir constants, K_L, and monolayer sorption capacity, q_{max}, which were computed using the slope and intercept of the curve between $1/q_e$ and $1/C_e$. The q_{max} values obtained from "Langmuir" plots do not agree with the experimental result [59].

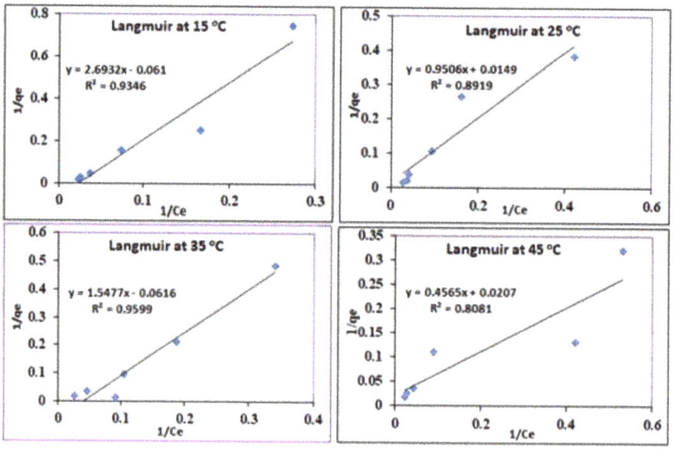

Figure 9. Adsorption of Cs^+ ions on mixed waste at different temperatures (Langmuir isotherm).

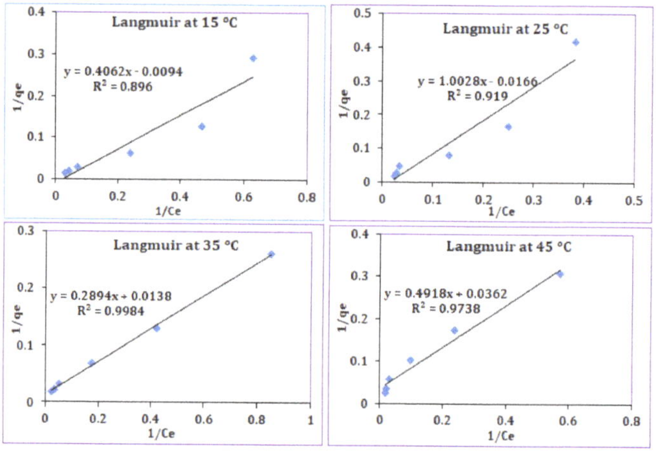

Figure 10. Adsorption of Co^{2+} ions on mixed waste at different temperatures (Langmuir isotherm).

Table 2. Value of Langmuir parameters for adsorption of both ions separately on mixed waste.

	Cs^+				Co^{2+}			
Temp.	q_e	q_{max}	k_L	R^2	q_e	q_{max}	k_L	R^2
15 °C	43	16.3	0.022	0.934	67	21.55	0.021	0.896
25 °C	36	67.0	0.015	0.891	59	37.45	0.029	0.919
35 °C	38	16.2	0.039	0.959	56	21.23	0.013	0.998
45 °C	42	48.3	0.045	0.808	41	14.80	0.034	0.973

An empirical equation used to define heterogeneous schemes is the Freundlich isotherm. The "Freundlich" equation is represented as

$$\ln q_e = \ln K_f + \frac{1}{n} \ln C_e \tag{5}$$

The Freundlich equation's linear formula is given in Equation (5), where K_f and n are Freundlich constants, with K_f (mg/g (L/mg) 1/n denoting the adsorption capability and n denoting how beneficial the adsorption progression is. The adsorption favorability is determined by the extent of the exponent, 1/n. $n > 1$, which indicates that the adsorption conditions are good. The intercept and slope of the curve illustrated in Figures 11 and 12 are used to estimate K_f and n, which are provided in Table 3.

The explanation behind the isotherm's compatibility with the Freundlich isotherm model rather than the Langmuir isotherm model is that R^2 was higher at Freundlich, which indicates that the adsorption process were taking place according to the Freundlich isotherm at all temperatures for Co^{2+} and Cs^+ ions. In this study, the Freundlich model is more significantly correlated with high R^2 than the Langmuir model, assuming that the stronger binding sites are occupied first and that binding strength decreases with the increasing degree of site occupation [60].

Table 3. Value of Freundlich parameters for adsorption of both ions separately on mixed waste.

Temp.	Cs^+			Co^{2+}		
	n	K_f	R^2	n	K_f	R^2
15 °C	0.719	0.239	0.972	0.926	0.298	0.957
25 °C	0.935	0.951	0.935	1.048	0.953	0.926
35 °C	0.750	0.653	0.766	0.734	0.166	0.995
45 °C	1.29	2.3	0.895	0.703	0.372	0.970

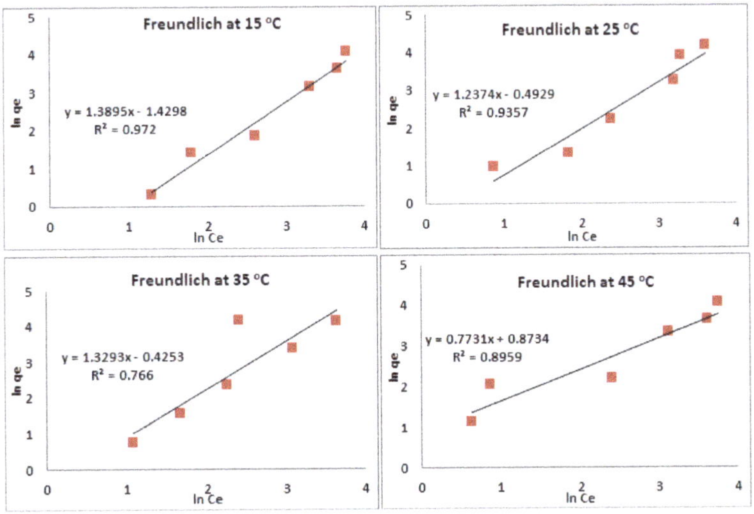

Figure 11. Adsorption of Cs^+ ions on mixed waste at different temperatures (Freundlich isotherm).

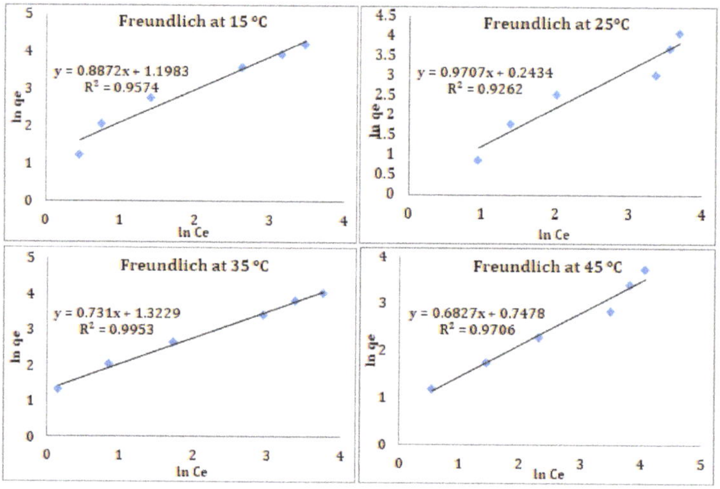

Figure 12. Adsorption of Co^{2+} ions on mixed waste at different temperatures (Freundlich isotherm).

To acquire the thermodynamic parameters of the adsorption reaction, k_f values were processed at different temperatures according to van't Hoff Equation (6):

$$\ln k_f = \frac{-\Delta H^0}{RT} + \frac{\Delta S^0}{R} \quad (6)$$

where ΔH^0 (KJ·mol^{-1}) and ΔS^0 (KJ·mol·k^{-1}) are enthalpy and entropy changes, respectively, R is gas constant. Plotting k_f against 1/T, as shown in Figure 13, gives a straight line with a slope and intercept $-\Delta H^0/R$ and $\Delta S^0/R$, respectively. The value of ΔH^0 and ΔS^0 were calculated and listed in Table 4.

The positive value of ΔH^0 corresponds to the endothermic adsorption process of Cs^+ and the negative value of ΔH^0 corresponds to the exothermic adsorption process Co^{2+} ions. The Gibbs free energy of adsorption was estimated from the following relation Equation (7) Table 4.

$$\Delta G^0 = \Delta H^0 - T\Delta S^0 \quad (7)$$

Table 4 indicated that little change in $T\Delta S^0$ has been conducted at all temperatures for Cs^+ ion and $T\Delta S^0 < \Delta H^0$. This finding implied that enthalpy, rather than entropic change, is the driving force of adsorption. These findings are consistent with prior research indicating that the adsorption process is non-spontaneous due to positive ΔG^0 values obtained [40]. Result for Co^{2+} show that $T\Delta S^0 > \Delta H^0$ suggest an entropic rather than an enthalpic change in the adsorption process. The positive value of ΔG^0 obtained indicated that the adsorption process of Cs^+ ion is nonspontaneous.

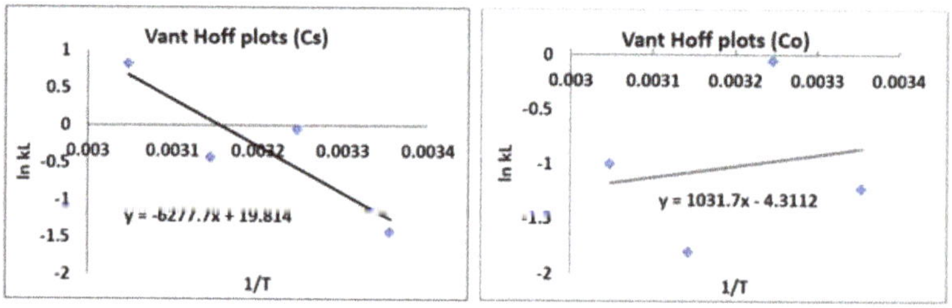

Figure 13. Van't Hoff plots related to adsorption of both ions separately on mixed waste.

Table 4. Thermodynamic parameters of adsorption of both ions on mixed waste.

Element	ΔS	ΔH	T(K)	TΔS	ΔG
Co^{2+}	−0.035	−8.5	288	−10.08	1.58
			298	−10.43	1.39
			308	−10.78	2.28
			318	−11.13	2.63
Cs^{+}	0.16	52.1	288	46.08	6.02
			298	47.68	4.42
			308	49.28	2.82
			318	50.88	1.22

3.7. Adsorption of Radioisotopes (^{60}Co and ^{134}Cs) on Dry Mixed Waste

Figure 14 depicts the adsorption of radioactive cesium and cobalt; the removal efficiency increases with time until it reaches a constant value at equilibrium. The removal percentage of ^{60}Co after 120 min with different weights of dried mixed waste (0.05, 0.1, 0.2, and 0.25 g) was 55, 70, 80, and 80%, respectively. While the removal percentage of ^{134}Cs after 120 min with different weights of dried mixed waste (0.05, 0.1, 0.2, and 0.25 g) was between 78.5 and 84.6%. Because there are no chemical differences between stable and radioactive elements, the same adsorption pattern was produced as in a previously reported study of the stable and radioisotopes [18]. With the difference in the radii of cesium and cobalt elements, the ability to impact diffusion mobility and to improve the adsorption process is different, and consequently, the diffusive mobility of the large radius of ^{134}Cs from the solution to the surface of the biosorbent in aqueous media is lower than the diffusive mobility of the small radius of ^{60}Co [61].

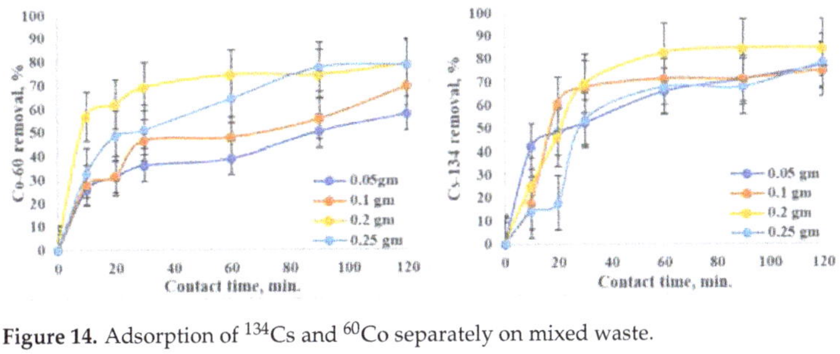

Figure 14. Adsorption of ^{134}Cs and ^{60}Co separately on mixed waste.

4. Conclusions

The novelty in this study lies in a sustainable treatment technology because it employs dry biomass from mixed waste to demonstrate its synergistic benefits, such as high efficiency and cost effectiveness.

In this study, the removal of stable and radioisotopes of cobalt and cesium from aqueous solutions by biosorption in dried mixed waste of olive waste and water hyacinth as a low-cost and natural available sorbent was investigated. The results show that the natural biomass of mixed waste is an excellent biosorbent for the examined isotopes. The pH experiments showed that the significant biosorption takes place in the acidic range in the case of Co and in the neutral medium in case of Cs. Contact times of 50 and 90 min were the optimal times for the maximum biosorption efficiency of Cs^{+} or Co^{2+}. Generally, the increase in mass biosorbent leads to the increase in biosorption due to an increase in the number of biosorption sites. Experimental data were better described by the pseudo-second order model. The adsorption isotherm could be well-fitted to the Freundlich isotherm model. The biosorption capacity decreases with an increase in solution temperature.

Based on the experimental and mathematical data obtained, mixed dry waste of water hyacinth and olive waste could be nominated as a sustainable low cost and efficient natural adsorbent material suggested for the efficient remediation of radio or stable cobalt and cesium ions from wastewater. However, this technology is predicted to achieve the successful decontamination of hazardous metals and radionuclides in an environmentally friendly and sustainable manner with the extensively decreasing the cost of wastewater treatment.

Author Contributions: Conceptualization, H.M.S.; Methodology, A.A.A., R.A.M. and H.M.S.; Writing—original draft, H.M.S.; Writing—review & editing, H.M.S.; Supervision, M.H.B. and H.M.S. All authors have read and agreed to the published version of the manuscript.

Funding: This research received no external funding.

Institutional Review Board Statement: Not applicable.

Informed Consent Statement: Not applicable.

Data Availability Statement: Not applicable.

Conflicts of Interest: The authors declare no conflict of interest.

References

1. Baroni, L.; Cenci, L.; Tettamanti, M.; Berati, M. Evaluating the Environmental Impact of Various Dietary Patterns Combined with Different Food Production Systems. *Eur. J. Clin. Nutr.* **2007**, *61*, 279–286. [CrossRef]
2. Corcoran, E. *Sick Water?: The Central Role of Wastewater Management in Sustainable Development: A Rapid Response Assessment*; UNEP/Earthprint: Geneva, Switzerland, 2010; ISBN 8277010753.
3. Saleh, H.M.; Mahmoud, H.H.; Abdou, M.I.; Eskander, S.B. Health Risk Assessment Based on Metal Analysis of Soil and Crops in Al-Dakhla Oasis. *Arab. J. Geosci.* **2021**, *14*, 260. [CrossRef]
4. Saleh, H.M.; Eskander, S.B.; Mahmoud, H.H.; Abdou, M.I. Groundwater Quality and Health Assessments Based on Heavy Metals and Trace Elements Content in Dakhla Oasis, New Valley Governorate, Egypt. *Water Sci.* **2022**, *36*, 1–12. [CrossRef]
5. Aglan, R.F.; Saleh, H.M.; Mohamed, G.G. Potentiometric Determination of Mercury (II) Ion in Various Real Samples Using Novel Modified Screen-Printed Electrode. *Appl. Water Sci.* **2018**, *8*, 141. [CrossRef]
6. Aglan, R.F.; Mahmoud, H.H.; Rashad, A.M.; Saleh, H.M. Novel Coated Wire Potentiometric Sensor for Selective Determination of Mn (II) Ions in Various Authentic Samples. *J. Iran. Chem. Soc.* **2021**, *18*, 1567–1579. [CrossRef]
7. Al Naggar, Y.; Khalil, M.S.; Ghorab, M.A. Environmental Pollution by Heavy Metals in the Aquatic Ecosystems of Egypt. *Open Acc. J. Toxicol.* **2018**, *3*, 555603.
8. Dąbrowski, A.; Hubicki, Z.; Podkościelny, P.; Robens, E. Selective Removal of the Heavy Metal Ions from Waters and Industrial Wastewaters by Ion-Exchange Method. *Chemosphere* **2004**, *56*, 91–106. [CrossRef]
9. Shaidan, N.H.; Eldemerdash, U.; Awad, S. Removal of Ni (II) Ions from Aqueous Solutions Using Fixed-Bed Ion Exchange Column Technique. *J. Taiwan Inst. Chem. Eng.* **2012**, *43*, 40–45. [CrossRef]
10. Dermentzis, K.; Christoforidis, A.; Valsamidou, E. Removal of Nickel, Copper, Zinc and Chromium from Synthetic and Industrial Wastewater by Electrocoagulation. *Int. J. Environ. Sci.* **2011**, *1*, 697–710.
11. Molinari, R.; Poerio, T.; Argurio, P. Selective Separation of Copper (II) and Nickel (II) from Aqueous Media Using the Complexation–Ultrafiltration Process. *Chemosphere* **2008**, *70*, 341–348. [CrossRef]
12. Shirzad Siboni, M.; Samadi, M.T.; Yang, J.K.; Lee, S.M. Photocatalytic Reduction of Cr (VI) and Ni (II) in Aqueous Solution by Synthesized Nanoparticle ZnO under Ultraviolet Light Irradiation: A Kinetic Study. *Environ. Technol.* **2011**, *32*, 1573–1579. [CrossRef] [PubMed]
13. Mohsen-Nia, M.; Montazeri, P.; Modarress, H. Removal of Cu2+ and Ni2+ from Wastewater with a Chelating Agent and Reverse Osmosis Processes. *Desalination* **2007**, *217*, 276–281. [CrossRef]
14. Belkacem, M.; Khodir, M.; Abdelkrim, S. Treatment Characteristics of Textile Wastewater and Removal of Heavy Metals Using the Electroflotation Technique. *Desalination* **2008**, *228*, 245–254. [CrossRef]
15. Saeed, A.; Iqbal, M.; Akhtar, M.W. Removal and Recovery of Lead (II) from Single and Multimetal (Cd, Cu, Ni, Zn) Solutions by Crop Milling Waste (Black Gram Husk). *J. Hazard. Mater.* **2005**, *117*, 65–73. [CrossRef] [PubMed]
16. Saleh, H.M.; Aglan, R.F.; Mahmoud, H.H. Ludwigia Stolonifera for Remediation of Toxic Metals from Simulated Wastewater. *Chem. Ecol.* **2019**, *35*, 164–178. [CrossRef]
17. Saleh, H.M.; Mahmoud, H.H.; Aglan, R.F.; Bayoumi, T.A. Biological Treatment of Wastewater Contaminated with Cu(II), Fe(II) and Mn(II) Using Ludwigia Stolonifera Aquatic Plant. *Environ. Eng. Manag. J.* **2019**, *18*, 1327–1336. [CrossRef]

18. Saleh, H.M.; Moussa, H.R.; El-Saied, F.A.; Dawoud, M.; Nouh, E.S.A.; Abdel Wahed, R.S. Adsorption of Cesium and Cobalt onto Dried Myriophyllum Spicatum L. from Radio-Contaminated Water: Experimental and Theoretical Study. *Prog. Nucl. Energy* **2020**, *125*, 103393. [CrossRef]
19. Dawoud, M.M.A.; Hegazi, M.M.; Saleh, H.M.; El Helew, W.K. Removal of Stable and Radio Isotopes from Wastewater by Using Modified Microcrystalline Cellulose Based on Taguchi L16. *Int. J. Environ. Sci. Technol.* **2022**, 1–12. [CrossRef]
20. Dawoud, M.M.A.; Hegazy, M.M.; Helew, W.K.; Saleh, H.M. Overview of Environmental Pollution and Clean Management of Heavy Metals and Radionuclides by Using Microcrystalline Cellulose. *J. Nucl. Energy Sci. Power Gener. Technol.* **2021**, *3*, 2.
21. Chen, W.; Li, H. Cost-Effectiveness Analysis for Soil Heavy Metal Contamination Treatments. *Water Air Soil Pollut.* **2018**, *229*, 126. [CrossRef]
22. Ghaly, M.; Metwally, S.S.; El-Sherief, E.A.; Saad, E.A.; Abdel Rahman, R.O. Utilization of Synthetic Nano-Cryptomelane for Enhanced Scavenging of Cesium and Cobalt Ions from Single and Binary Solutions. *J. Radioanal. Nucl. Chem.* **2022**, *331*, 1821–1838. [CrossRef]
23. Avery, S.V.; Codd, A.; Gadd, M. Caesium Accumulation and Interactions with Other Monovalent Cations in the Cyanobacterium Synechocystis PCC 6803. *Microbiology* **1991**, *137*, 405–413. [CrossRef]
24. Singh, A.; Kumar, S. Application of Aquatic Plants Dead Biomass in Remediation of Heavy Metals Pollution by Adsorption: A Review. *Indian J. Sci. Technol.* **2022**, *15*, 729–735. [CrossRef]
25. Lison, D.; Van Den Brûle, S.; Van Maele-Fabry, G. Cobalt and Its Compounds: Update on Genotoxic and Carcinogenic Activities. *Crit. Rev. Toxicol.* **2018**, *48*, 522–539. [CrossRef] [PubMed]
26. Melnikov, P.; Zanoni, L.Z. Clinical Effects of Cesium Intake. *Biol. Trace Elem. Res.* **2010**, *135*, 1–9. [CrossRef]
27. Saleh, H.M.; Moussa, H.R.; El-Saied, F.A.; Dawoud, M.; Bayoumi, T.A.; Abdel Wahed, R.S. Mechanical and Physicochemical Evaluation of Solidified Dried Submerged Plants Subjected to Extreme Climatic Conditions to Achieve an Optimum Waste Containment. *Prog. Nucl. Energy* **2020**, *122*, 103285. [CrossRef]
28. Saleh, H.M.; Aglan, R.F.; Mahmoud, H.H. Qualification of Corroborated Real Phytoremediated Radioactive Wastes under Leaching and Other Weathering Parameters. *Prog. Nucl. Energy* **2020**, *119*, 103178. [CrossRef]
29. Saleh, H.M.; Eskander, S.B. Toxicity and Remediation of Radioactive Pollutants. *J. Nucl. Energy Sci. Power Gener. Technol.* **2020**, *9*, 197. [CrossRef]
30. Saleh, H.; Salman, A.; Faheim, A.; El-Sayed, A. Polymer and Polymer Waste Composites in Nuclear and Industrial Applications. *J. Nucl. Energy Sci. Power Gener. Technol.* **2020**, *9*, 1000199.
31. Saleh, H.M. Some Applications of Clays in Radioactive Waste Management. In *Clays and Clay Minerals: Geological Origin, Mechanical Properties and Industrial Applications*; Wesley, L.R., Ed.; Nova Science Pub. Inc.: Hauppauge, NY, USA, 2014; pp. 403–415; ISBN 9781631177804.
32. Aijaz, M.O.; Yang, S.B.; Karim, M.R.; Naseer, I.A.; Alahmari, A.D.; Mubaddel, F.A.; Assaifan, A.K. Preparation and Characterization of Electrospun Poly (Lactic Acid)/Poly (Ethylene Glycol)–b–Poly (Propylene Glycol)–b–Poly (Ethylene Glycol)/Silicon Dioxide Nanofibrous Adsorbents for Selective Copper (II) Ions Removal from Wastewater. *Membranes* **2023**, *13*, 54. [CrossRef]
33. Boss, C.B.; Fredeen, K.J. *Concepts, Instrumentation and Techniques in Inductively Coupled Plasma Optical Emission Spectroscopy*; Perkin Elmer: Waltham, MA, USA, 1999.
34. Liu, Y.; Diwu, C.; Zhao, Y.; Liu, X.; Yuan, H.; Wang, J. Determination of Trace and Rare-Earth Elements in Chinese Soil and Clay Reference Materials by ICP-MS. *Chin. J. Geochem.* **2014**, *33*, 95–102. [CrossRef]
35. Bhatnagar, A.; Sillanpää, M.; Witek-Krowiak, A. Agricultural Waste Peels as Versatile Biomass for Water Purification–A Review. *Chem. Eng. J.* **2015**, *270*, 244–271. [CrossRef]
36. He, Z.L.; Yang, X.E.; Stoffella, P.J. Trace Elements in Agroecosystems and Impacts on the Environment. *J. Trace Elem. Med. Biol.* **2005**, *19*, 125–140. [CrossRef] [PubMed]
37. Saravanan, A.; Sundararaman, T.R.; Jeevanantham, S.; Karishma, S.; Kumar, P.S.; Yaashikaa, P.R. Effective Adsorption of Cu (II) Ions on Sustainable Adsorbent Derived from Mixed Biomass (*Aspergillus campestris* and Agro Waste): Optimization, Isotherm and Kinetics Study. *Groundw. Sustain. Dev.* **2020**, *11*, 100460. [CrossRef]
38. Fubara, A.G.; Uche, C.C.; Nwoko, C.O.; Tony-Njoku, R.F.; Ojiaku, A.A.; Edo, F.A. Assessment of the Effectiveness of Water Hyacinth (E. Crassipes) in the Biosorption of Heavy Metals from Aluminium Extruding Company Effluents. *J. Appl. Sci. Environ. Manag.* **2022**, *26*, 37–46. [CrossRef]
39. Sharma, P.; Kaur, H. Sugarcane Bagasse for the Removal of Erythrosin B and Methylene Blue from Aqueous Waste. *Appl. Water Sci.* **2011**, *1*, 135–145. [CrossRef]
40. Ghanavati, B.; Bozorgian, A.; Ghanavati, J. Removal of Copper (II) Ions from the Effluent by Carbon Nanotubes Modified with Tetrahydrofuran. *Chem. Rev. Lett.* **2022**, *5*, 68–75.
41. Xu, Z.; Xing, Y.; Ren, A.; Ma, D.; Li, Y.; Hu, S. Study on Adsorption Properties of Water Hyacinth-Derived Biochar for Uranium (VI). *J. Radioanal. Nucl. Chem.* **2020**, *324*, 1317–1327. [CrossRef]
42. Hanaf, R.A.-K. Effect of Salinity and PH on the Absorption of Cadmium in Lemna Minor. *Int. J. Aquat. Biol.* **2021**, *9*, 383–387.
43. Adeolu, A.T.; Okareh, O.T.; Dada, A.O. Adsorption of Chromium Ion from Industrial Effluent Using Activated Carbon Derived from Plantain (Musa Paradisiaca) Wastes. *Am. J. Env. Prot.* **2016**, *4*, 7–20.

44. Sartape, A.S.; Mandhare, A.M.; Jadhav, V.V.; Raut, P.D.; Anuse, M.A.; Kolekar, S.S. Removal of Malachite Green Dye from Aqueous Solution with Adsorption Technique Using Limonia Acidissima (Wood Apple) Shell as Low Cost Adsorbent. *Arab. J. Chem.* **2017**, *10*, S3229–S3238. [CrossRef]
45. Azam, M.G.; Kabir, M.H.; Shaikh, M.A.A.; Ahmed, S.; Mahmud, M.; Yasmin, S. A Rapid and Efficient Adsorptive Removal of Lead from Water Using Graphene Oxide Prepared from Waste Dry Cell Battery. *J. Water Process Eng.* **2022**, *46*, 102597. [CrossRef]
46. Yavuz, Ö.; Altunkaynak, Y.; Güzel, F. Removal of Copper, Nickel, Cobalt and Manganese from Aqueous Solution by Kaolinite. *Water Res.* **2003**, *37*, 948–952. [CrossRef] [PubMed]
47. Rawat, J.P.; Iraqi, S.M.U.; Singh, R.P. Sorption Equilibria of Cobalt (II) on Two Types of Indian Soils—The Natural Ion Exchangers. *Colloids Surfaces A Physicochem. Eng. Asp.* **1996**, *117*, 183–188. [CrossRef]
48. Iyer, A.; Mody, K.; Jha, B. Biosorption of Heavy Metals by a Marine Bacterium. *Mar. Pollut. Bull.* **2005**, *50*, 340–343. [CrossRef]
49. van Hullebusch, E.D.; Peerbolte, A.; Zandvoort, M.H.; Lens, P.N.L. Sorption of Cobalt and Nickel on Anaerobic Granular Sludges: Isotherms and Sequential Extraction. *Chemosphere* **2005**, *58*, 493–505. [CrossRef]
50. Parab, H.; Joshi, S.; Shenoy, N.; Lali, A.; Sarma, U.S.; Sudersanan, M. Determination of Kinetic and Equilibrium Parameters of the Batch Adsorption of Co (II), Cr (III) and Ni (II) onto Coir Pith. *Process Biochem.* **2006**, *41*, 609–615. [CrossRef]
51. Vijayaraghavan, K.; Jegan, J.; Palanivelu, K.; Velan, M. Biosorption of Cobalt (II) and Nickel (II) by Seaweeds: Batch and Column Studies. *Sep. Purif. Technol.* **2005**, *44*, 53–59. [CrossRef]
52. Miah, M.Y.; Volchek, K.; Kuang, W.; Tezel, F.H. Kinetic and Equilibrium Studies of Cesium Adsorption on Ceiling Tiles from Aqueous Solutions. *J. Hazard. Mater.* **2010**, *183*, 712–717. [CrossRef]
53. Wang, T.-H.; Li, M.-H.; Yeh, W.-C.; Wei, Y.-Y.; Teng, S.-P. Removal of Cesium Ions from Aqueous Solution by Adsorption onto Local Taiwan Laterite. *J. Hazard. Mater.* **2008**, *160*, 638–642. [CrossRef]
54. Melkior, T.; Yahiaoui, S.; Motellier, S.; Thoby, D.; Tevissen, E. Cesium Sorption and Diffusion in Bure Mudrock Samples. *Appl. Clay Sci.* **2005**, *29*, 172–186. [CrossRef]
55. Ding, D.; Lei, Z.; Yang, Y.; Zhang, Z. Efficiency of Transition Metal Modified Akadama Clay on Cesium Removal from Aqueous Solutions. *Chem. Eng. J.* **2014**, *236*, 17–28. [CrossRef]
56. Rani, R.D.; Sasidhar, P. Sorption of Cesium on Clay Colloids: Kinetic and Thermodynamic Studies. *Aquat. Geochem.* **2012**, *18*, 281–296. [CrossRef]
57. Parab, H.; Sudersanan, M. Engineering a Lignocellulosic Biosorbent–Coir Pith for Removal of Cesium from Aqueous Solutions: Equilibrium and Kinetic Studies. *Water Res.* **2010**, *44*, 854–860. [CrossRef] [PubMed]
58. Wang, C.; Myshkin, V.F.; Khan, V.A.; Poberezhnikov, A.D.; Baraban, A.P. Effect of Temperature on the Diffusion and Sorption of Cations in Clay Vermiculite. *ACS Omega* **2022**, *7*, 11596–11605. [CrossRef] [PubMed]
59. Elemile, O.O.; Akpor, B.O.; Ibitogbe, E.M.; Afolabi, Y.T.; Ajani, D.O. Adsorption Isotherm and Kinetics for the Removal of Nitrate from Wastewater Using Chicken Feather Fiber. *Cogent Eng.* **2022**, *9*, 2043227. [CrossRef]
60. Anah, L.; Astrini, N. Isotherm Adsorption Studies of Ni (II) Ion Removal from Aqueous Solutions by Modified Carboxymethyl Cellulose Hydrogel. In *Proceedings of the IOP Conference Series: Earth and Environmental Science*; IOP Publishing: Bristol, England, 2018; Volume 160, p. 12017.
61. Dakroury, G.A.R.; Abo-Zahra, S.F.; Hassan, H.S.; Ali, H.E.A. Improvement of the Sorption Behavior of Aluminum Silicate Composite toward 134Cs and 60Co Radionuclides by Non-Living Biomass of Chlorella Vulgaris. *Environ. Sci. Pollut. Res.* **2020**, *27*, 21109–21125. [CrossRef]

sustainability MDPI

Article

Natural Radioactivity Measurements and Radiological Hazards Evaluation for Some Egyptian Granites and Ceramic Tiles

Essam Sidique [1,*], Sedky H. A. Hassan [2,3] and Mohammad Mahmoud Dawoud [4,*]

1 Physics Department, Faculty of Science, New Valley University, El-Kharga 72511, Egypt
2 Biology Department, College of Science, Sultan Qaboos University, Muscat 123, Oman
3 Botany & Microbiology Department, Faculty of Science, New Valley University, El-Kharga 72511, Egypt
4 Radioisotopes Department, Nuclear Research Center, Egyptian Atomic Energy Authority, Giza 12311, Egypt
* Correspondence: esamsadik@yahoo.com (E.S.); mohammed.h.aea@gmail.com (M.M.D.)

Abstract: All over the world, people widely use granites and ceramic tiles in their residential establishments. Information concerning the radiological properties of such materials reveals how to ensure the sustainability of their safe use in terms of these properties. In the present work, the distribution of the terrestrial radioisotopes U-238 (Ra-226), Th-232, and K-40 for 23 different brands of Egyptian commercial granites and ceramic tiles samples (widely used domestically and exported) was determined using gamma radiation spectroscopy. This process pinpoints the possible radiological health risks related to gamma ray exposure and radon gas resulting from the use of these materials indoors. The concentration values of the aforementioned radioisotopes in the examined samples were compared to the corresponding global average values (GAVs) of the UNSCEAR and to those available in other countries. The overall average concentrations for U-238, Th-232, and K-40 in the total samples were observed to be 46.17 $\pm$ 2.81 (less than its GAV), 51.65 $\pm$ 2.35 (slightly above its GAV), and 701.62 $\pm$ 40.60 Bq/kg (1.4 times greater than the GAV), respectively. The related radiological parameters and indices were calculated and compared to the prescribed limits set by commissions and organizations concerned with radiation protection (the WHO, ICRP, UNSCEAR, and EC) to ensure the safe use of the investigated granites and ceramic tiles. The assessed indices and parameters fall within the recommended values and safety limits. In conclusion, there is no risk from using the granites and ceramic tiles under investigation in residential facilities.

Keywords: natural radioactivity; gamma rays; radon gas; radiation exposure; building materials; granites; ceramic

Citation: Sidique, E.; Hassan, S.H.A.; Dawoud, M.M. Natural Radioactivity Measurements and Radiological Hazards Evaluation for Some Egyptian Granites and Ceramic Tiles. *Sustainability* **2022**, *14*, 14611. https://doi.org/10.3390/su142114611

Academic Editor: Changhyun Roh

Received: 24 September 2022
Accepted: 30 October 2022
Published: 7 November 2022

1. Introduction

For humans, exposure to natural ionizing radiation is unavoidable. Exposure arises primarily from both terrestrial and cosmogenic radioisotopes. Terrestrial radioisotopes (K-40, U-238, and Th-232) exist naturally in all different environmental media, including air, water, food, soil, rock, building materials, etc. [1–3]. The existence of the abovementioned radioisotopes in building materials is responsible for delivering about 85% of the radiation dosage to the world's population [3,4]. Accordingly, providing information on terrestrial radioisotope concentrations and distributions in building materials is essential and it is required to monitor contamination originating from their radioactivity in residential environments.

The natural radioactivity level in building materials is confined in the terrestrial radionuclide concentrations in the geological materials from which they are derived [4,5], i.e., it varies according to the geological origin and geochemical characteristics of the constituent materials. Additionally, the radiation dose received is controlled by several factors, including residences places ventilation, and kinds [1].

In fact, gamma rays and radon gas (Rn-222 and Rn-220) are the most significant products of the terrestrial radioisotopes' radioactivity in building materials, in light of the radiological hazards to the population [6]. Gamma rays are responsible for the external exposure of populations and their dose comes mainly from Tl-208 and Ac-228 of the Th-232 disintegration chain, from Pb-214 and Bi-214 derived from R-226 of the U-238 disintegration chain, and from K-40. On the other hand, radon gas (specifically Rn-222) which originates as a result of radium disintegration (Ra-226 of the U-238 chain), is responsible for the internal exposure [7]. According to the UNSCEAR [6], the average yearly external exposure indoors due to gamma rays was assessed as 0.41 mSv, while the internal exposure from Rn-222 inhalation was about 1.15 mSv. It is worth mentioning that several epidemiological investigations conducted in many countries have demonstrated substantial evidence linking raised levels of radon exposure in houses to an increased risk of lung cancer [8–10].

Construction materials that typically come from the soil and rocks of the earth may be categorized into three groups: structural materials, covering materials, and additive raw materials. Structural materials such as cement, concrete, mortar, and clay bricks, etc., are primarily used for building structures. The covering materials (granite, ceramic, marble, etc.) are employed for ornamentation and insulation purposes, whereas fly ash, bauxite, phosphogypsum, etc., are the additive raw materials used as optional components for modifying certain properties of building materials [1]. In light of global recommendations, determining the natural radioactivity levels of construction materials is crucial for assessing the radiological risks owing to radiation exposures as well as for developing national standards and guidelines for these materials. Recently, as a result of rising social concern, there has been a high worldwide interest in studying the natural radioactivity of construction materials as well as investigating their impact on the public (e.g., [4,5,11–20]). To the best of the authors' knowledge, important international studies on natural radioactivity measurements and radiological hazards evaluation for granites and ceramic tiles were published in the Refs. [1,2,9,18,19,21–39], just to name a few. Nevertheless, although granites and ceramic tiles of various brands are widely used in residences for interior and exterior ornamentation and decoration purposes in Egypt, no detailed studies have been conducted to determine the activity levels of terrestrial radioisotopes in these materials.

The present work investigates in depth the natural radioactivity of twenty-three well-known Egyptian brands of commercial granites and ceramic tiles samples widely used in Egypt and abroad, as well as the potential health risks associated with their use indoors. Moreover, its overall goal is to complete a radiometric study on some sample Egyptian commercial granites and ceramic tiles which have not been previously covered. Hopefully, the findings of the present work and the accompanying assessments will establish baseline data for monitoring radioactive pollution in residential environments and will provide adequate public protection recommendations.

2. Experimental Arrangements

2.1. Sample Preparation

A total of 107 tile samples (42 granite samples plus 65 ceramic samples) from 23 different brands were purchased from Egyptian building materials markets and suppliers. The commercial ceramics and granites studied are among the best widely used decorative building materials brands in Egypt. Before transporting the samples to the lab, they were properly catalogued, labeled, and named according to their popular names known in both the global and domestic markets (Table 1). More detailed information, particularly for granite, is available at (www.stonecontact.com, accessed on 19 June 2022). The samples have been given identification numbers in brackets, which are (1 to 42) for granites, and (43 to 107) for ceramic tiles. Then, each sample was individually ground to a powder, to avoid contamination between samples, and sieved through a sieve (200 μm mesh). All the samples were oven-dried for 5 h at 105 °C to remove the moisture content. These prepared samples were subsequently weighed (between 600 g and 850 g) and sealed in plastic cylindrical beakers (48 mm radius, 82 mm height, and 0.5 mm thickness) for more than 4 weeks to guarantee access to the secular equilibrium between parent radioisotopes and daughters in the natural disintegration series (^{232}Th and ^{238}U).

Table 1. Granites and ceramic tiles of various brands used in Egypt.

Tiling Material	Brand Name	Brand ID	Sample Size	Sample Origin
Granite	Bianco Halayeb	GBiHa	3	Abu Ghusun, Red Sea, Egypt
	Brown Hurgada	GBrHu	3	Hurghada, Egypt
	Imperial Red	GIR	3	Aswan, South of Egypt
	Karnak Grey	GKG	3	Aswan, South of Egypt
	Negro Aswan	GNA	4	Aswan, South of Egypt
	Rosa Aswan Dark	GRAD	3	Aswan, South of Egypt
	Rosa El Hody Light	GRHL	3	Aswan, South of Egypt
	Rosa Abu Simble	GRAS	4	Wadi Halfa, Aswan, Egypt
	Rosa Sardo Sinai	GRSS	3	Sinai, Egypt
	Red Aswan	GRA	3	Aswan, South of Egypt
	Red Nefertary	GRN	4	Aswan, South of Egypt
	Red Forsan	GRF	3	Wadi Forsan, northeastern Egypt
	Yellow Ghazal	GYG	3	Sinai, Egypt
Ceramic	Alfa	CAL	7	6th of October City (2), Giza, Egypt
	Art	CAR	7	6th of October City (2), Giza, Egypt
	Cleopatra	CCL	5	10th of Ramadan City (1), Cairo, Egypt
	Gemma (Al-Jawhara)	CGE	7	El Sadat City Desert, Menoufia, Egypt
	Gloria	CGL	6	Nasr City, Cairo, Egypt
	Labotie	CLA	6	10th of Ramadan City, Sharqia, Egypt
	Pharaohs (Alfaraeina)	CPH	7	Al Azbakeya, Cairo, Egypt
	Prima	CPR	7	5th Industrial zone, Menoufia, Egypt
	Royal	CRO	6	Al Obour, Al Qalyubia, Egypt
	Venezia	CVE	7	6th of October City (2), Giza, Egypt

2.2. Gamma Spectrometric Analysis

To measure the activity concentrations of the radioisotopes (γ-emitters) in the samples, a low-background γ-rays spectroscopy system consisting of a semiconductor HPGe (Hyper-Pure Germanium) detector (Model GR4020, Canberra, Meriden, CT, USA) with a 40% relative efficiency and energy resolution (FWHM) of 2 keV at the 1332 keV γ-line (Co-60) was used. In addition, the system contains a suitable lead shield (Model 747E, Canberra, USA) surrounding the detector to prevent more than 98% of the external background radiation from reaching the detector during the analysis. The gamma spectrums were acquired and analyzed utilizing the Genie-2000 software (Version 3.3, Canberra, USA) [40] coupled with a multichannel analyzer (Model DAS-1000, Canberra, USA). For calibrating the detector's energy and efficiency, the LabSOCS (Laboratory Sourceless Calibration Software) designed using the features of geometry composer and gamma analysis within the Genie-2000 software, was used. The Genie-2000 software also contains the detector's characterization files created based on the system manufacturer's fundamental calibration tests (Canberra). To authenticate the efficiency data provided by LabSOCS, measurements were completed in our laboratory with a set of gamma calibration sources (Co-60, Cs-137, Ba-133, Mn-54, Zn-65, and Na-22), which revealed a significant agreement (90%) between empirical and mathematical peak efficiency.

Each prepared sample was put on the detector for a time period no less than 12 h in order to obtain an accurate counting statistic for gamma lines (photo-peaks) of importance. Additionally, the background level in the laboratory was measured using an empty beaker in similar conditions. The K-40 radionuclide was determined directly by its own gamma line intensity (1460.8 keV). As for U-238 (Ra-226), it was specified through its progeny Bi-214 (1764.5, 1120.3, and 609.3 keV) and Pb-214 (351.9 and 295.2 keV). On the other hand, Th-232 was identified via its daughters Ac-228 (968.9, 911.2, and 338.3 keV), Tl-208 (2614.5 and 583.2 keV), and Pb-212 (238.6 keV). The activity concentration (AC) and the uncertainty in activity concentration (U_{AC}) of the previously mentioned radionuclides in each sample were calculated from their corresponding gamma line intensities taking into account the sample mass, counting time, gamma decay transition probabilities, and detector efficiencies [41–43].

Activity Concentrations Estimation

The activity concentration (AC) in the granites and ceramic tiles samples under investigation was estimated as follows [41]:

$$AC[Bq/kg] = \frac{N_{c,E}}{P_{\gamma,E} \cdot \varepsilon_E \cdot M_s} \quad (1)$$

where $N_{c,E}$ is the net count rate resulting from subtracting the count rate of the peak at energy E in the sample spectrum minus that of the background spectrum at the same energy E, $P_{\gamma,E}$ denotes the probability of emitting gamma radiation with energy E for the radioisotope of interest, ε_E is the detector absolute efficiency at energy E, and M_s refers to the sample mass. Moreover, using the equation below, the uncertainty in activity concentration, U_{AC}, was calculated based on uncertainties in $N_{c,E}$, $P_{\gamma,E}$, ε_E, and M_s [43]:

$$U_{AC} = AC\sqrt{\left[\frac{U_{N_{c,E}}}{N_{c,E}}\right]^2 + \left[\frac{U_{P_{\gamma,E}}}{P_{\gamma,E}}\right]^2 + \left[\frac{U_{\varepsilon_E}}{\varepsilon_E}\right]^2 + \left[\frac{U_{M_s}}{M_s}\right]^2} \quad (2)$$

3. Estimation of Radiological Risks

For understanding the effect of radiological hazards on human health when using the investigated granites and ceramic tiles as tiling materials in buildings, several radiation hazard indices were estimated. The radium equivalent activity (Ra_{eq}), gamma index (I_γ), indoor absorbed gamma dose rate (D_{in}), yearly effective gamma dose rate (E_{in}), and excess lifetime cancer risk (ELCR) were used to investigate gamma radiation risks, while the alpha index (I_α), radon exhalation rate (RX), radon concentration (C_{Rn}), and yearly effective dose due to radon (E_{Rn}) were evaluated in order to investigate the potential radon risks.

3.1. Radium Equivalent Activity (Ra_{eq})

The Ra_{eq} is one of the most effective radiological indices for determining gamma radiation hazards due to the radioisotopes K-40, U-238 and Th-232 content in tiling materials, considering the non-uniform distribution of these radioisotopes in matter. As per Beretka et al. [44], Ra_{eq} is described by the following equation:

$$Ra_{eq}\left[Bq\ kg^{-1}\right] = \left(\frac{AC_U}{370} + \frac{AC_{Th}}{259} + \frac{AC_K}{4810}\right) \times 370 \quad (3)$$

where AC_U, AC_{Th}, and AC_K are the specific activity concentrations of the radioisotopes U-238 (Ra-226), Th-232, and K-40, respectively. In fact, the Ra_{eq} reflects the weighted total of the abovementioned three radioisotopes' concentrations within materials under the premise that gamma dose rates from 4810 Bq/kg of K-40, 259 Bq/kg of Th-232, and 370 Bq/kg of U-238 (Ra-226) are almost equal.

From a radiation protection perspective, the ceramic and granite tiles studied herein are safe provided that their Ra_{eq} levels are not above 370 Bq/kg [44] (permissible limit) corresponding to a yearly effective dosage of 1.5 mSv [41,45].

3.2. Gamma Index (I_γ)

The I_γ is taken into consideration as a monitoring tool specifying whether construction materials are safe to use or not. Considering that the external exposure due to gamma radiation from the tiling (superficial or covering) materials has a limit of 1 mSv/year, the I_γ is adopted by the European Commission [46] to be estimated via the following equation:

$$I_\gamma = \frac{AC_U}{300\ Bq\ kg^{-1}} + \frac{AC_{Th}}{200\ Bq\ kg^{-1}} + \frac{AC_K}{3000\ Bq\ kg^{-1}} \quad (4)$$

According to the European Commission [46], for covering materials such as the ceramic and granite tiles under investigation, if they have a $I_\gamma \leq 2$, this leads to an increase in the annual gamma dose rate with an amount ≤ 0.3 mSv/y resulting from these materials. In other words, these materials fall within the exemption level for building materials from all limitations about their radioactivity. Furthermore, if materials achieve criteria $2 < I_\gamma \leq 6$, they will contribute to the annual gamma dose rate with an amount ≤ 1 mSv/y and fall within the recommended action level. Eventually, materials with $I_\gamma > 6$ are not suitable for safe use in buildings [46].

3.3. Indoor Absorbed Gamma Dose Rate (D_{in}) and Yearly Effective Dose (E_{in})

Estimation of the indoor absorbed gamma dose rate (D_{in}) and its associated yearly effective dose (E_{in}) are significant mechanisms for determining the external exposure caused by terrestrial radioisotopes (Th-232, U-238, and K-40). According to the European Commission [46], Equations (5) and (6) can be used to estimate the D_{in} and E_{in} in the air within rooms as a result of using the investigated granites and ceramic tiles as superficial construction materials:

$$D_{in}\left[\mathrm{nGy\ h^{-1}}\right] = (12AC_{Ra} + 14AC_{Th} + 0.96AC_{K}) \times 10^{-2} \quad (5)$$

$$E_{in}\left[\mathrm{mSvy^{-1}}\right] = D_{in}\left[\mathrm{nGy\ h^{-1}}\right] \times F_1 \times F_2 \times F_3 \times 10^{-6} \quad (6)$$

where F_1 (=0.7 Sv/Gy), F_2 (=0.8), and F_3 (=8766 h) represent the conversion factor from the absorbed dose to the effective dose in the air, indoor residency factor, and hours of the year, respectively.

3.4. Excess Lifetime Cancer Risk (ELCR)

The ELCR is an important quantity through which the incidence of cancer for an individual exposed to a low gamma radiation dose over their lifetime (66-years) can be figured out. Depending on the E_{in} incurred by individuals from the studied ceramic and granite tiles (superficial building materials) when used in buildings, the ELCR can be estimated as follows [4,41]:

$$\mathrm{ELCR} = E_{in}[\mathrm{Sv/y}] \times C_1 \times C_2 \quad (7)$$

where C_1(=66 y [47]) and C_2 (=0.05 $\mathrm{Sv^{-1}}$ for the general population) stand for life expectancy on average and fatal cancer risk, respectively [42,48].

3.5. Alpha Index (I_α)

The I_α given by the equation below [41] is used for estimating the risk of exposure to internal alpha radiation owing to the inhalation of radon. The estimation of I_α is fundamentally dependent on the U-238 (Ra-226) activity concentration (AC_{Ra}) in construction materials, considering that materials with a concentration of ^{226}Ra < 200 Bq $\mathrm{kg^{-1}}$ cannot emit an indoor radon concentration >200 Bq $\mathrm{m^{-3}}$, i.e., these materials come within the range of the recommended action level of indoor radon exposure for buildings, as previously agreed upon by the European Commission [46] and ICRP [49].

$$I_\alpha = \frac{AC_{Ra}}{200\mathrm{Bq\ kg^{-1}}} \leq 1 \quad (8)$$

3.6. Radon Concentration (C_{Rn}) and Yearly Effective Dose Resulting Therefrom (E_{Rn})

In this study, the parallelepiped room model (4 × 5 × 2.8 m) is taken into account to assess the concentration of radon and the associated absorbed dose rate delivered to residents, assuming the floor is made of the investigated ceramic or granite tiles. Thus, Equation (9) is applied, according to the European Commission [46], to determine the increase in indoor radon concentration (C_{Rn}) brought on by the radon exhalation rate (RX) from the investigated ceramic or granite tiles used in buildings:

$$C_{Rn}\left[Bq\ m^{-3}\right] = \frac{RX \cdot S}{(\lambda + \lambda_o) \cdot V} \quad (9)$$

where RX [Bq m^{-2} h^{-1}], λ (=0.0076 h^{-1}), λo [h^{-1}], S [m^2], and V [m^3] represent the radon exhalation rate per unit area, Ra-226 decay constant, ventilation rate, tiled floor surface, and volume of the room, respectively. Based on the determined Ra-226 concentration (AC_{Ra}), Equation (10) can be applied to estimate the RX for the investigated tiles with thickness d (=3 cm and 1 cm for granites and ceramic tiles, respectively), density ρ (=2600 kg/m^3), and emanation coefficient η (=0.45), as reported in Refs. [20,41,50]:

$$RX = AC_{Ra} \cdot \lambda \cdot \rho \cdot \eta \cdot d \cdot 0.5 \quad (10)$$

It is worth noting that the room ventilation rate (λ_o) was chosen with 0.5 h^{-1} for normal ventilation and 0.1 h^{-1} for poor ventilation. [9,41]. Furthermore, the (S/V) ratio of the room surface tiled with the investigated ceramic and granite tiles was chosen to be 2 m^{-1} [41].

As per the UNSCEAR [6] report, the yearly effective dose (E_{Rn}) that dwellers obtain from the indoor radon concentration (C_{Rn}) was estimated by the following formula:

$$E_{Rn}\left[mSv\ y^{-1}\right] = C_{Rn}\left[Bq\ m^{-3}\right] \times C_1 \times 8766\left[h\ y^{-1}\right] \times C_2 \times C_3 \times 10^{-6} \quad (11)$$

where C_1 (=9 nSv per Bq m^{-3} h), C_2 (=0.8), and C_3 (=0.4) stand for the factors of dose conversion, dwellers' indoor residency, and equilibrium equivalent radon concentration indoors, respectively [6,51].

4. Results and Discussion

4.1. Radioisotope (^{238}U, ^{232}Th, and ^{40}K) Concentrations

Table 2 displays the ranges and averages, as well as standard errors, for the estimated concentrations values of U-238 (Ra-226), Th-232, and K-40 in the granites and ceramic tiles collected. Furthermore and by way of illustration, Figure 1 depicts the variations of the abovementioned radioisotopes concentrations in the samples under examination. Evidently, Figure 1 and Table 2 show that the concentrations of the considered radioisotopes fluctuate from 8.40 $\pm$ 2.09 in GBiH (sample No. 3) to 196.01 $\pm$ 37.84 Bq kg^{-1} in GRSS (sample No. 28), 11.62 $\pm$ 1.28 in GBiH (sample No. 3) to 140.32 $\pm$ 27.41 Bq kg^{-1} in GRSS (sample No. 28), and 141.01 $\pm$ 12.83 in CAR (sample No. 52) to 1785.78 $\pm$ 125.00 Bq kg^{-1} in GRN (sample No. 35) for ^{238}U, ^{232}Th, and ^{40}K, respectively. Conceivably, the observed variations in radioisotopes' concentrations could be attributed to the samples' various origins and compositions. Evidently, most of the ceramic samples (samples No. 43 to 107) have lower concentrations of the three radioisotopes (U-238, Th-232, and K-40) than those of the granite samples (samples No. 1 to 42). This materializes the granite's naturally high level of terrestrial radioisotopes [3].

Table 2. Mean concentrations of ^{238}U (^{226}Ra), ^{232}Th, and ^{40}K (mean value ± standard error) in the investigated samples of the considered granite and ceramic brands, compared to their global average values (AGVs) in building materials.

Tiling Material	Brand ID	Sample Size	Activity Concentration [Bq kg^{-1}]					
			Ra-226		Th-232		K-40	
			Range	Mean ± SE	Range	Mean ± SE	Range	Mean ± SE
Granite	GBiHa	3	8.40–25.37	16.95 ± 4.90	11.62–86.75	37.31 ± 24.73	207.21–1260.66	598.67 ± 332.83
	GBrHu	3	75.12–101.20	88.90 ± 7.56	77.66–88.53	82.05 ± 3.31	1631.50–1760.39	1695.38 ± 37.21
	GIR	3	27.20–47.05	35.93 ± 5.85	44.37–52.01	46.99 ± 2.51	1124.02–1500.22	1302.47 ± 109.03
	GKG	3	48.81–69.49	56.98 ± 6.35	33.32–62.66	45.38 ± 8.86	516.76–678.58	612.54 ± 49.02
	GNA	4	21.65–48.85	33.49 ± 5.88	39.69–99.06	62.28 ± 13.02	827.73–1391.41	1044.82 ± 125.30
	GRAD	3	40.01–55.74	46.55 ± 4.73	57.57–86.32	73.44 ± 8.43	1045.65–1068.33	1060.42 ± 7.39
	GRHL	3	35.98–67.08	50.06 ± 9.10	41.40–92.11	65.34 ± 14.71	771.84–1065.73	925.86 ± 85.13
	GRAS	4	21.13–67.08	48.99 ± 10.19	18.67–86.71	51.89 ± 14.15	350.61–1073.63	817.88 ± 166.70
	GRSS	3	157.24–196.01	174.55 ± 11.38	122.33–140.32	131.07 ± 5.20	1139.85–1411.17	1311.92 ± 86.37
	GRA	3	15.31–28.13	22.46 ± 3.78	58.88–86.32	69.42 ± 8.54	1128.03–1380.24	1267.88 ± 74.09
	GRN	4	87.24–126.01	100.37 ± 8.82	42.37–116.83	82.22 ± 19.52	1346.94–1785.78	1604.53 ± 94.24
	GRF	3	35.03–64.76	48.78 ± 8.65	45.05–78.71	63.70 ± 9.89	962.71–1198.79	1067.69 ± 69.39
	GYG	3	24.08–41.92	33.48 ± 5.17	17.74–79.20	44.09 ± 18.28	842.65–1291.41	1039.86 ± 132.37
Ceramic	CAL	7	19.65–54.90	36.77 ± 4.18	22.25–49.10	36.59 ± 3.25	268.80–580.00	447.99 ± 37.90
	CAR	7	27.67–72.44	47.97 ± 5.83	39.71–85.67	58.16 ± 6.75	141.01–967.61	553.36 ± 97.53
	CCL	5	33.67–50.24	41.70 ± 3.32	31.00–62.56	41.93 ± 5.51	314.01–508.89	385.20 ± 36.68
	CGE	7	33.25–58.88	46.13 ± 4.10	34.57–61.48	48.98 ± 3.59	332.64–667.11	515.68 ± 43.22
	CGL	6	26.41–42.62	32.37 ± 2.60	31.18–51.31	38.88 ± 2.94	318.72–480.82	404.54 ± 28.08
	CLA	6	28.05–41.58	33.26 ± 2.22	34.78–48.75	41.87 ± 2.34	407.88–719.45	535.22 ± 51.80
	CPH	7	20.50–44.17	35.47 ± 3.06	27.04–49.82	38.96 ± 3.45	230.25–565.65	391.33 ± 45.91
	CPR	7	23.11–46.87	34.64 ± 3.19	26.84–60.17	44.59 ± 4.38	354.45–606.42	506.75 ± 31.46
	CRO	6	31.46–53.77	40.73 ± 3.64	32.94–54.47	41.58 ± 3.02	222.48–467.20	366.72 ± 35.20
	CVE	7	26.40–39.61	33.09 ± 1.92	26.11–39.63	32.96 ± 1.94	206.35–317.18	269.36 ± 13.15
	Total	107	8.40–196.01	46.17 ± 2.81	11.62–140.32	51.65 ± 2.35	141.01–1785.78	701.62 ± 40.60
GAV (Global Average Value) UNSCEAR [52])			—	50	—	50	—	500

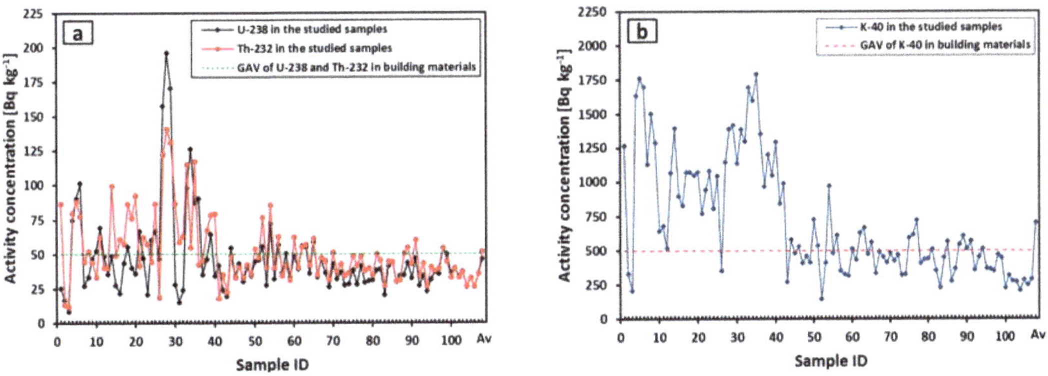

Figure 1. U-238, Th-232 and K-40 concentrations in the investigated granites and ceramic tiles samples.

Regarding the mean values of the radioisotope concentrations inserted in Table 2 and plotted in Figure 2, it is clear that CVE samples contain the lowest mean concentrations of both K-40 and Th-238, with levels of 269.36 ± 13.15 and 32.96 ± 1.94 Bq/kg, respectively, while GBiHa samples have the lowest mean concentration of U-238 with a level of 16.95 ± 4.90 Bq/kg. In contrast, GRSS samples appear to have the highest mean concentrations of both U-238 and Th-232, with levels of 174.55 ± 11.38 and 131.07 ± 5.20 Bq/kg, respectively, whereas, GBrHu samples have the highest mean concentration of K-40, with a level of 1695.38 ± 37.21 Bq/kg.

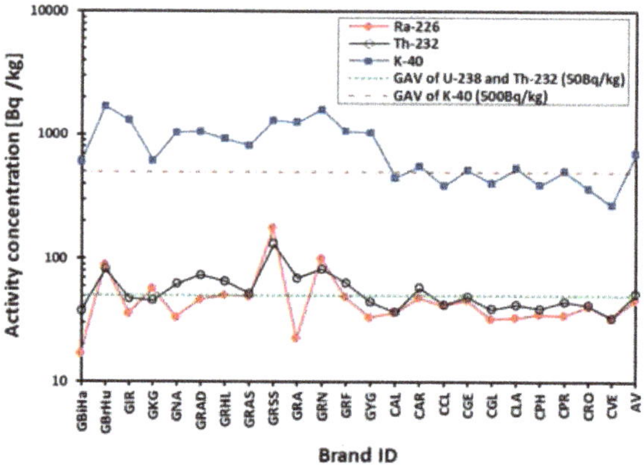

Figure 2. Comparison of the average values of U-238 (Ra-226), Th-232 and K-40 in the investigated ceramic and granites tiles to their corresponding GAVs in the building materials.

Figure 2 and Table 2 also display the mean values of the aforementioned radioisotope concentrations in the materials under investigation against their corresponding global average values (GAVs) in building materials as given by the UNSCEAR [52]. Notably, the GAVs for U-238, Th-232, and K-40 in building materials are 50, 50, and 500 Bq/kg, respectively, according to the UNSCEAR [52]. Apparently, the mean concentration of U-238 in all granites and ceramic tiles samples herein, except for GBrHu, GKG, GRHL, GRSS, and GRN samples, is lower than its GAV of 50 Bq/kg in building materials [52], (Figure 2 and Table 2). Similarly, the average concentration of Th-232 in the investigated samples, except for the GBrHu, GNA, GRAD, GRHL, GRAS, GRSS, GRA, GRN, GRF, and CAR samples, is lower than its GAV in building materials of 50 Bq/kg [52]. However, the average concentration of K-40 for all of the granite and ceramic brands, except for the CAL, CCL, CGL, CPH, CRO, and CVE samples, is greater than the GAV (500 Bq/kg) of building materials [52], (Figure 2 and Table 2).

The overall average concentrations for U-238, Th-232, and K-40 of the total samples were observed to be 46.17 $\pm$ 2.81 (less than its GAV), 51.65 $\pm$ 2.35 (slightly above its GAV), and 701.62 $\pm$ 40.60 Bq/kg (1.4 times greater than the GAV), respectively, as demonstrated in Table 2 and Figure 2. Moreover, in all of the investigated brands' samples (Figure 2), the K-40 concentration is the greatest among the concentrations of the three studied radionuclides, as the granites contain about 33% potash feldspar minerals [21]. Moreover, it was found that the K-40 concentration is the largest contributor to the total concentration for all samples (Figure 3). Both U-238 and Th-232 contribute roughly the same percentage (6%), to the overall concentration of samples, while K-40 contributes a larger percentage (88%), as shown in Figure 3.

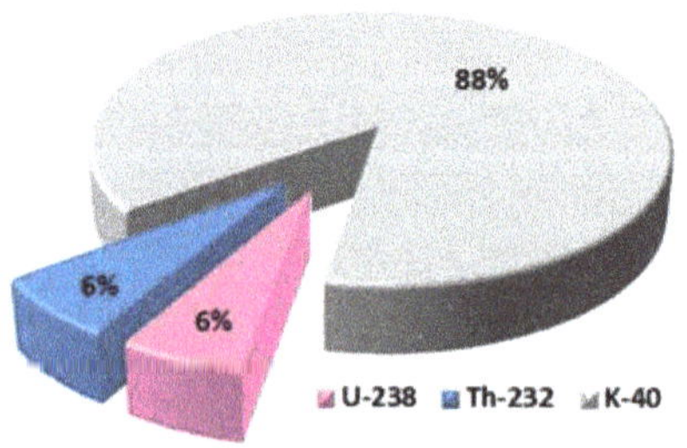

Figure 3. Contributions of the three radioisotopes, ^{238}U, ^{232}Th, and ^{40}K, to the studied samples' overall concentration.

A comparison between the radioactivity levels in the studied samples and those in other previous relevant studies on granites and ceramics used in Egypt and other countries is illustrated in Table 3. From Table 3, one can deduce that many of the data from the relevant literature are comparable to our findings. This can be confirmed using the dendrogram (Figure 4) derived from the hierarchical cluster analysis (HCA) based on the three variables (U-238, Th-232, and K-40 concentrations). In HCA, the cluster method between-groups linkage is used coupled with the square Euclidean distance. Accordingly, the countries in which the studies were accomplished have been divided into homogeneous groups based on the similarity in concentrations of the three radionuclides (U-238, Th-232, and K-40) (Figure 4). Figure 4a shows that the Egyptian granite tiles studied herein come close to the granites used in Bangladesh [22] and Nigeria [2] in terms of the content of the three aforementioned radionuclides, as they were grouped into a homogeneous group. However, they are far from those used in Pakistan [25], as illustrated in Figure 4a. Likewise, Figure 4b exhibits that the ceramic tiles examined herein are matched to the ceramics used in Italy [23] and Egypt [30] but are very different to those used in Poland [37], Serbia [9], and Nigeria [2] in terms of radionuclide content.

Table 3. The radioisotopes concentrations (present study) in comparison to other similar previous international research.

Tiling Materials	Countries' Names	Concentrations [Bq/kg]			References
		U-238	Th-232	K-40	
Granite	Turkey	45.40	82.30	931.60	[1]
	Saudi Arabia	54.50	43.40	677.70	[21]
	Nigeria	74	100	1098	[2]
	China	355.9	317.9	1636.5	[19]
	Bangladesh	49.51	75.50	1122.15	[22]
	Italy	81.33	129	1065	[23]
	Iran	38	47	917	[18]
	USA	31	61	1210	[24]
	Serbia	200	77	1280	[9]
	Pakistan	659	598	1218	[25]
	Jordan	41.52	58.42	897	[26]
	India	82	112	1908	[27]
	Brazil	31	73	1648	[28]
	Spain	101	48	1293	[29]
	Egypt	65	60	885	[30]
	Egypt	137	82	1082	[31]
	Egypt	15.25	15.35	399.39	[38]
	Egypt	58.46	65.76	1107.55	Current study
Ceramic	Turkey	43.5	37.9	310.9	[1]
	Saudi Arabia	47.18	80.70	590.2	[21]
	Nigeria	85	77	877	[2]
	China	172.35	135.5	351.4	[19]
	Algeria	55	41	410	[32]
	Italy	52	42.5	450	[23]
	Iran	32	27	292	[18]
	India	17.52	38.93	298.59	[33]
	Malaysia	92	68	673	[34]
	Serbia	67	61	828	[9]
	Yemen	131.88	83.55	400.7	[35]
	Jordan	33.86	28.82	411	[26]
	Japan	82.7	63.9	527	[36]
	Poland	50	50	963	[37]
	Egypt	52	33	450	[30]
	Egypt	38.23	42.54	439.33	Current study

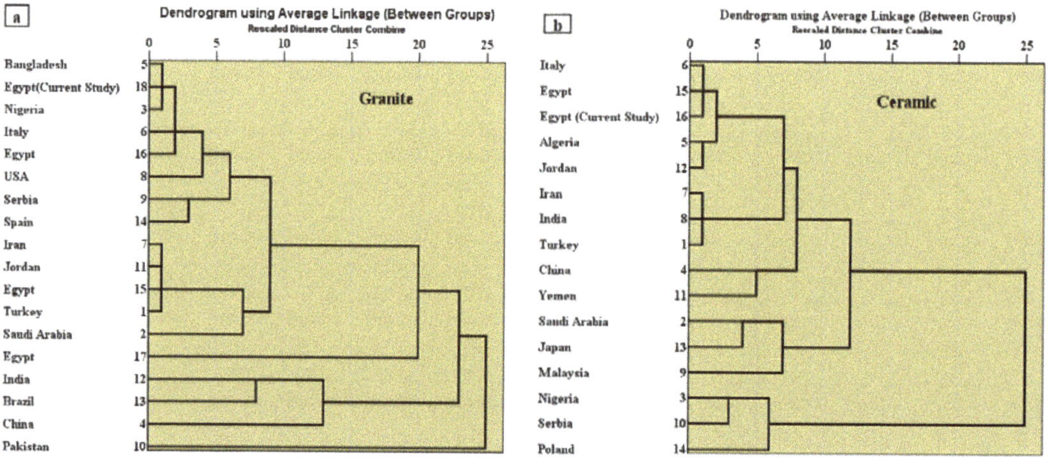

Figure 4. The dendrogram for comparing the Egyptian granites and ceramic tiles investigated herein with those used in Egypt and other countries in terms of the natural radioactivity.

Evidently, in many of the investigated granite and ceramic samples, the radioisotopes' displayed concentrations were more elevated than the GAVs (Figure 2). For instance, in granites of "Rosa Sardo Sinai" (GRSS) samples, concentrations of uranium, thorium, and potassium exceed their corresponding GAVs by about three times, confirming the previous study on the granite of the Sinai area by Fares [39]. Therefore, it was crucial to assess the likely radiological risks to peoples' health owing to the use of these materials in buildings.

4.2. Gamma Radiation Impact Estimation

Table 4 displays some of the evaluated radiological variables for the granite and ceramic samples under consideration. Through the estimated R_{eq}, I_{γ}, D_{in}, E_{in}, and ELCR (Table 4), the gamma ray hazards posed by the materials under investigation when used as tiling in buildings can be judged. It was found that all Ra_{eq} values in the samples of ceramic tiles fell within ranges below the threshold standard of 370 Bq/kg [44]. Similarly, the Ra_{eq} values for all the samples of the investigated granite tiles were in ranges lower than the criterion limitation of 370 Bq/kg, with the exception of three samples (No. 27 to 29) of GRSS and two samples (No. 33 and 35) from GRN samples. Furthermore, the mean Ra_{eq} values varied between 100.96 Bq/kg in the CVE samples and 463 Bq/kg in the GRSS samples, with an overall mean of 174.06 Bq/kg. Accordingly, the granites and ceramic tiles of the different brands herein don't constitute any considerable radiological risks for individuals when used as tiling materials, except for the GRSS samples which may be a cause for concern due to the mean value of Ra_{eq} going beyond 370 Bq/kg (recommended value) (Figure 5a).

Table 4. Ranges and mean values of the radiological parameters showing the gamma impact of the studied granite and ceramic brands.

Tiling Material	Brand ID	Sample Size	Parameters Showing Gamma Impact				
			Ra_{eq} [Bq/kg]	I_γ	D_{in} [nGy/h]	E_{in} [mSv/y]	ELCR/10^3
Granite	GBiHa	3	(40.97–246.48) 116.39	(0.16–0.94) 0.44	(4.62–27.29) 13.00	(0.02–0.13) 0.06	(0.07–0.44) 0.21
	GBrHu	3	(315.09–352.52) 336.77	(1.19–1.33) 1.27	(35.87–40.14) 38.43	(0.18–0.20) 0.19	(0.58–0.65) 0.62
	GIR	3	(177.53–223.44) 203.42	(0.69–0.87) 0.79	(20.30–25.71) 23.39	(0.10–0.13) 0.11	(0.33–0.42) 0.38
	GKG	3	(146.02–211.34) 169.04	(0.54–0.77) 0.62	(16.44–23.63) 19.07	(0.08–0.12) 0.09	(0.27–0.38) 0.31
	GNA	4	(167.45–297.64) 202.99	(0.64–1.12) 0.77	(18.86–33.09) 22.77	(0.09–0.16) 0.11	(0.31–0.54) 0.37
	GRAD	3	(208.47–261.36) 233.22	(0.79–0.97) 0.88	(23.58–29.02) 26.05	(0.12–0.14) 0.13	(0.38–0.47) 0.42
	GRHL	3	(185.71–249.75) 214.78	(0.69–0.94) 0.80	(21.25–27.44) 24.04	(0.10–0.12) 0.12	(0.34–0.44) 0.39
	GRAS	4	(100.68–271.20) 186.17	(0.37–1.00) 0.70	(11.62–30.18) 20.99	(0.06–0.15) 0.10	(0.19–0.49) 0.34
	GRSS	3	(419.94–503.29) 463.00	(1.52–1.82) 1.67	(46.94–56.46) 51.89	(0.23–0.28) 0.25	(0.76–0.91) 0.84
	GRA	3	(205.78–238.43) 219.35	(0.81–0.90) 0.84	(23.33–26.29) 24.59	(0.11–0.13) 0.12	(0.38–0.43) 0.40
	GRN	4	(254.79–391.81) 341.48	(0.96–1.47) 1.28	(29.72–44.00) 38.96	(0.15–0.22) 0.19	(0.48–0.71) 0.63
	GRF	3	(173.58–257.52) 222.08	(0.66–0.96) 0.84	(19.75–28.79) 25.02	(0.10–0.14) 0.12	(0.32–0.47) 0.41
	GYG	3	(132.17–247.14) 176.59	(0.51–0.94) 0.68	(15.60–27.62) 20.17	(0.08–0.14) 0.10	(0.25–0.45) 0.33
Ceramic	CAL	7	(72.17–169.77) 123.58	(0.27–0.62) 0.45	(8.05–19.03) 13.83	(0.04–0.09) 0.07	(0.13–0.31) 0.22
	CAR	7	(117.76–269.45) 173.75	(0.43–0.99) 0.64	(12.99–29.98) 19.21	(0.06–0.15) 0.09	(0.21–0.49) 0.31
	CCL	5	(102.18–177.47) 131.31	(0.37–0.65) 0.48	(11.39–19.50) 14.57	(0.06–0.1) 0.07	(0.18–0.32) 0.24
	CGE	7	(108.30–190.11) 155.88	(0.39–0.69) 0.57	(12.02–21.07) 17.34	(0.06–0.1) 0.09	(0.19–0.34) 0.28
	CGL	6	(101.24–153.02) 119.11	(0.37–0.56) 0.44	(11.18–16.91) 13.21	(0.05–0.08) 0.06	(0.18–0.27) 0.21
	CLA	6	(114.95–166.65) 134.34	(0.42–0.62) 0.50	(12.76–18.72) 14.99	(0.06–0.09) 0.07	(0.21–0.30) 0.24
	CPH	7	(76.90–151.06) 121.31	(0.28–0.56) 0.44	(8.46–16.93) 13.47	(0.04–0.08) 0.07	(0.14–0.27) 0.22
	CPR	7	(100.94–176.82) 137.43	(0.38–0.65) 0.51	(11.41–19.52) 15.27	(0.06–0.10) 0.07	(0.18–0.32) 0.25
	CRO	6	(100.61–167.64) 128.42	(0.36–0.61) 0.47	(11.11–18.56) 14.23	(0.05–0.09) 0.07	(0.18–0.30) 0.23
	CVE	7	(80.21–120.70) 100.96	(0.29–0.44) 0.36	(8.86–13.35) 11.17	(0.04–0.07) 0.05	(0.14–0.22) 0.18
	Total	107	(40.97–503.29) 174.06	(0.16–1.82) 0.65	(4.62–56.46) 19.51	(0.02–0.28) 0.10	(0.07–0.91) 0.32
Allowable Value or GAV			370 [a]	2 [b]	84 [a] or 70 [b]	0.41 [a] or 1 [b]	1.16 [c]

[a] GAV given by the UNSCEAR [3]. [b] GAV reported by the European Commission [46]. [c] GAV indicated by Sidique et al. [41] and Qureshi et al. [53].

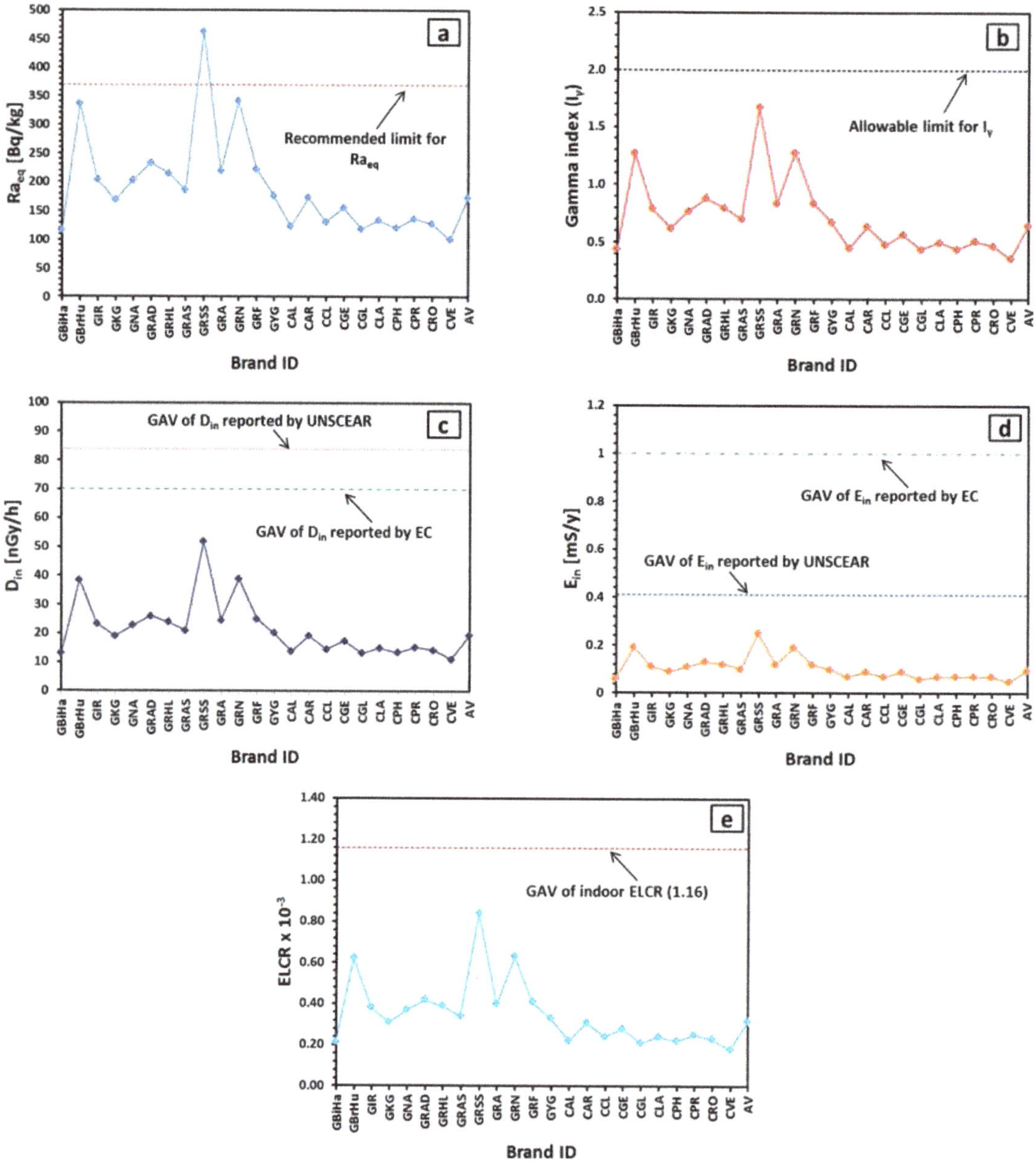

Figure 5. The radiological parameters showing the gamma impact from the ceramic and granite tiles of the brands under investigation.

According to the calculated values of I_γ (Table 4), none of the values in the samples under investigation go beyond the exemption limit of 2. This implies that the yearly effective gamma dose resulting from the investigated materials herein when used as covering or superficial building materials indoors is less than 0.3 mSv/y (exemption limit indicated by the European Commission [46]). Thus, the granites and ceramic tiles of the brands under investigation are suitable for use without any restrictions. Furthermore, and by way of illustration, Figure 5b displays the estimated mean values of I_γ for the granites and ceramic tiles where they do not go beyond their allowable limit, indicating that all of the investigated materials do not raise any cause for concern when used in buildings.

Considering the estimated D_{in} and E_{in} values for the samples under investigation (Table 4), neither the granites nor the ceramic samples of the various brands examined herein had D_{in} and E_{in} values going beyond their corresponding worldwide average values of 84 nGy/h and 0.41 mSv/y [3] and of 70 nGy/h and 1 mSv/y [46]. Notably, the mean values of D_{in} and E_{in} for the studied granites and ceramic tiles compared with the corresponding global average values (GAVs) are illustrated in Figs. 5c and 5d, where they are below the adopted limits, reflecting their safe use.

The indoor ELCR values based on E_{in}, as well as their mean values in the granites and ceramics under investigation, are shown in Table 4 and plotted in Figure 5e. Evidently, all the ELCR values are within the range of 0.07 to 0.91, with an overall mean of 0.32. Namely, the mean ELCR values in all the brands of tiling materials are less than their corresponding GAV of 1.16 indicated by Sidique et al. [41] and Qureshi et al. [53], as shown in Figure 5e. Thus, in buildings where the materials under investigation are used, over a 66-year lifespan, their residents are at a very insignificant risk of developing cancer resulting from exposure to gamma rays emitted by these materials.

4.3. Radon Impact Assessment

The potential radon (Rn-222) risks posed by the studied materials when used as tiles in buildings can be investigated through the estimated I_α, RX, C_{Rn}, and E_{Rn}. For the samples of each brand, Table 5 displays ranges and averages of I_α, RX, C_{Rn}, and E_{Rn} values, while Figure 6 compares averages of these parameters with their corresponding allowable limits. Table 5 reveals that the I_α values oscillate between 0.04 and 0.98, with an overall mean value of 0.23. Furthermore, no mean value of I_α for any of the brands' samples under investigation goes beyond the unity, as shown in Figure 6a. Thus, arguably, the materials under consideration come under the range of the indoor radon safe exposure action level for buildings, as indicated by the European Commission [46] and ICRP [49].

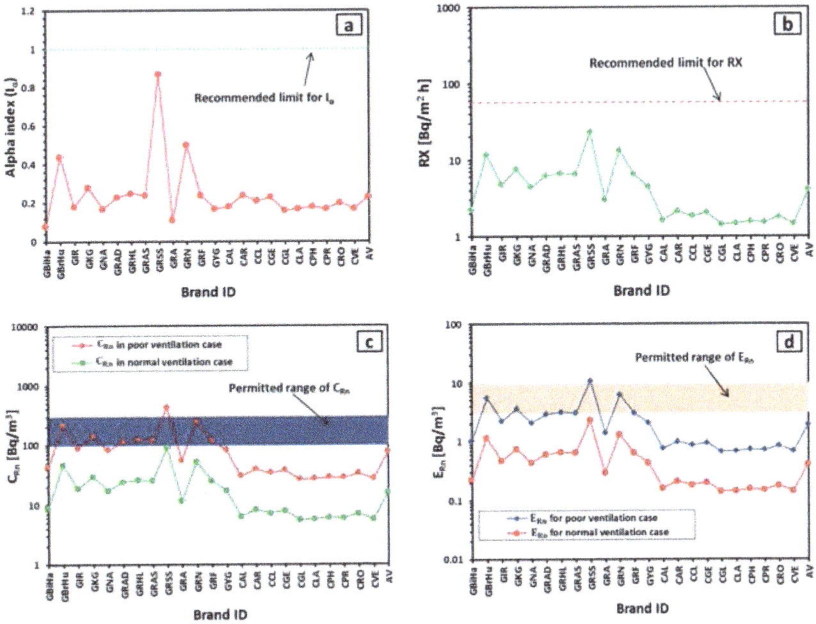

Figure 6. The radiological parameters showing the Rn impact from the studied ceramic and granite tiles.

Table 5. Ranges and mean values of the radiological parameters showing the Rn impact of the studied granite and ceramic brands.

Tiling Material	Brand ID	Sample Size	I_α	RX [Bq/m^2 h]	Poor Ventilation Case		Normal Ventilation Case	
					C_{Rn} [Bq/m^3]	E_{Rn} [mSv/y]	C_{Rn} [Bq/m^3]	E_{Rn} [mSv/y]
Granite	GBiHa	3	(0.04–0.13) 0.08	(1.12–3.38) 2.26	(20.81–62.88) 42.01	(0.53–1.59) 1.06	(4.41–13.33) 8.91	(0.11–0.34) 0.22
	GBrHu	3	(0.38–0.51) 0.44	(10.02–13.50) 11.86	(186.24–250.89) 220.39	(4.70–6.33) 5.56	(39.48–53.18) 46.72	(1.00–1.34) 1.18
	GIR	3	(0.14–0.24) 0.18	(3.63–6.28) 4.79	(67.43–116.65) 89.09	(1.70–2.94) 2.25	(14.29–24.73) 18.88	(0.36–0.62) 0.48
	GKG	3	(0.24–0.35) 0.28	(6.51–9.27) 7.60	(121.01–172.28) 141.27	(3.05–4.35) 3.57	(25.65–36.52) 29.95	(0.65–0.92) 0.76
	GNA	4	(0.11–0.24) 0.17	(2.89–6.52) 4.47	(53.67–121.11) 83.02	(1.36–3.06) 2.10	(11.38–25.67) 17.60	(0.29–0.65) 0.44
	GRAD	3	(0.20–0.28) 0.23	(5.34–7.43) 6.21	(99.19–138.19) 115.39	(2.50–3.49) 2.91	(21.03–29.29) 24.46	(0.53–0.74) 0.62
	GRHL	3	(0.18–0.34) 0.25	(4.80–8.95) 6.68	(89.20–166.29) 124.10	(2.25–4.20) 3.13	(18.91–35.25) 26.31	(0.48–0.89) 0.66
	GRAS	4	(0.11–0.34) 0.24	(2.82–8.95) 6.53	(52.38–166.29) 121.45	(1.32–4.20) 3.07	(11.10–35.25) 25.75	(0.28–0.89) 0.65
	GRSS	3	(0.79–0.98) 0.87	(20.97–26.14) 23.28	(389.83–485.94) 432.75	(9.84–12.27) 10.93	(82.63–103.01) 91.73	(2.09–2.60) 2.32
	GRA	3	(0.08–0.14) 0.11	(2.04–3.75) 3.00	(37.94–69.74) 55.67	(0.96–1.76) 1.41	(8.04–14.78) 11.80	(0.20–0.37) 0.30
	GRN	4	(0.44–0.63) 0.50	(11.64–16.81) 13.39	(216.28–312.40) 248.83	(5.46–7.89) 6.28	(45.85–66.22) 52.75	(1.16–1.67) 1.34
	GRF	3	(0.18–0.32) 0.24	(4.67–8.64) 6.51	(86.85–160.55) 120.93	(2.19–4.05) 3.05	(18.41–34.03) 25.63	(0.46–0.86) 0.65
	GYG	3	(0.12–0.21) 0.17	(3.21–5.59) 4.47	(59.69–103.91) 83.00	(1.51–2.62) 2.10	(12.65–22.03) 17.59	(0.32–0.56) 0.44
Ceramic	CAL	7	(0.10–0.27) 0.18	(0.87–2.44) 1.63	(16.24–45.37) 30.38	(0.41–1.15) 0.77	(3.44–9.62) 6.44	(0.09–0.24) 0.16
	CAR	7	(0.14–0.36) 0.24	(1.23–3.22) 2.13	(22.87–59.86) 39.64	(0.58–1.51) 1.00	(4.85–12.69) 8.40	(0.12–0.32) 0.21
	CCL	5	(0.17–0.25) 0.21	(1.50–2.23) 1.85	(27.82–41.52) 34.46	(0.70–1.05) 0.87	(5.90–8.80) 7.30	(0.15–0.22) 0.18
	CGE	7	(0.17–0.29) 0.23	(1.48–2.62) 2.05	(27.48–48.66) 38.12	(0.69–1.23) 0.96	(5.82–10.31) 8.08	(0.15–0.26) 0.20
	CGL	6	(0.13–0.21) 0.16	(1.17–1.89) 1.44	(21.83–35.22) 26.75	(0.55–0.89) 0.68	(4.63–7.47) 5.67	(0.12–0.19) 0.14
	CLA	6	(0.14–0.21) 0.17	(1.25–1.85) 1.48	(23.18–34.36) 27.48	(0.59–0.87) 0.69	(4.91–7.28) 5.83	(0.12–0.18) 0.15
	CPH	7	(0.10–0.22) 0.18	(0.91–1.96) 1.58	(16.94–36.50) 29.31	(0.43–0.92) 0.74	(3.59–7.74) 6.21	(0.09–0.20) 0.16
	CPR	7	(0.12–0.23) 0.17	(1.03–2.08) 1.54	(19.10–38.73) 28.63	(0.48–0.98) 0.72	(4.05–8.21) 6.07	(0.10–0.21) 0.15
	CRO	6	(0.16–0.27) 0.20	(1.4–2.39) 1.81	(26.00–44.44) 33.66	(0.66–1.12) 0.85	(5.51–9.42) 7.13	(0.14–0.24) 0.18
	CVE	7	(0.13–0.20) 0.17	(1.17–1.76) 1.47	(21.82–32.73) 27.35	(0.55–0.83) 0.69	(4.62–6.94) 5.80	(0.12–0.18) 0.15
	Total	107	(0.04–0.98) 0.23	(0.87–26.14) 4.09	(16.24–485.94) 76.08	(0.41–12.27) 1.92	(3.44–103.01) 16.13	(0.09–2.60) 0.41
Allowable Value or GAV			1 [a]	57.6 [b]	(100–300) [c]	(3–10) [a]	(100–300) [c]	(3–10) [a]

[a] GAV reported by the ICRP [49]. [b] GAV declared by the UNSCEAR [3]. [c] GAV recommended by the WHO (World Health Organization) [0].

Regarding the mean values of the radon exhalation rate (RX) for the samples under investigation (Table 5 and Figure 6b), they stretch between 1.44 Bq/m^2 h in the CGL samples and 23.28 Bq/m^2 h in the GRSS samples, with an overall mean value of 4.09 Bq/m^2 h.

Moreover, no mean value of RX for samples of any brand of ceramic or granite exceeds the global average value (GAV) of 57.6 Bq/m^2 h (0.016 Bq/m^2 s) as declared by the UNSCEAR [3], as shown in Figure 6b. It is therefore expected that the materials under investigation do not constitute any health risks to dwellers.

Based on the RX values, the indoor radon concentration (C_{Rn}) was assessed from the indirect mathematical model of a typical residence room tiled by the studied ceramics and granites in both poor and normal ventilation cases as per Equation (9). Accordingly, for the room model with normal ventilation, the mean values of C_{Rn} of granites and ceramics from the different brands under investigation fluctuate between 5.67 Bq/m^3 in the CGL samples to 91.73 Bq/m^3 in the GRSS samples, with an overall mean value of 16.13 Bq/m^3 (Table 5 and Figure 6c). Consequently, for the normal ventilation case, none of the mean values of C_{Rn} in the studied brands of granites and ceramic tiles, as shown in Figure 6c, go beyond the acceptable range (100–300 Bq/m^3) declared by the WHO (World Health Organization) [8]. On the other hand, for the room model with poor ventilation, the mean values of the C_{Rn} of granites and ceramics from the brands under examination oscillate between 26.75 Bq/m^3 in the CGL samples and 432.75 Bq/m^3 in the GRSS samples, with an overall mean value of 76.08 Bq/m^3 (Table 5 and Figure 6c). Thus, for the poor ventilation case, none of the mean values of C_{Rn} in the studied brands of ceramics and granites, as shown in Figure 6c, exceed the acceptable range (100–300 Bq/m^3) declared by the WHO [8], except for GRSS which as a result is not recommended for poorly ventilated buildings.

Regarding the yearly effective dose rate (E_{Rn}) due to radon concentration (Table 5 and Figure 6d), the mean values span a range from 0.14 mSv/y in the CGL samples to 2.32 mSv/y in the GRSS samples, with an overall mean value of 0.41 mSv/y in the case of normal ventilation. Moreover, all of these values fluctuated below the acceptable range of 3–10 mSv/y documented by ICRP [49], (Figure 6d). On the other hand, for the poor ventilation case, the mean values of E_{Rn} for the granites and ceramics from the different brands under investigation stretch between 0.68 mSv/y in the CGL samples to 10.93 mSv/y in the GRSS samples, with an overall mean value of 1.92 mSv/y. Furthermore, as shown in Figure 6d for poor ventilation, the granite from the brand GRSS is the only one that has a mean value of C_{Rn} going beyond the recommended range of 3–10 mSv/y [49]. Therefore, it is not recommended for poorly ventilated rooms.

The hierarchical cluster analysis (HCA) coupled with the Pearson correlation method was performed to effectively prove the relationship among all considered radiological variables. The dendrogram obtained from the HCA shows the relationship between the radionuclide concentrations and the relevant radiological parameters (Figure 7). Depending on the similarities in existence, all considered variables are gathered into two principal clusters. Cluster I comprised U-238 and Th-232 concentrations as well as all radiological parameters with a highly similar correlation (Figure 7). This reflects that the slight radioactivity level arising in both of the examined granites or ceramic tiles is ascribable to U-238 and Th-232 concentrations. On the other hand, cluster II is only comprised K-40, reflecting the weak relationship of K-40 with the radiological parameters, i.e., despite the high concentration of K-40 in the examined granites and ceramic tiles, K-40 contributes very little to the radioactivity level.

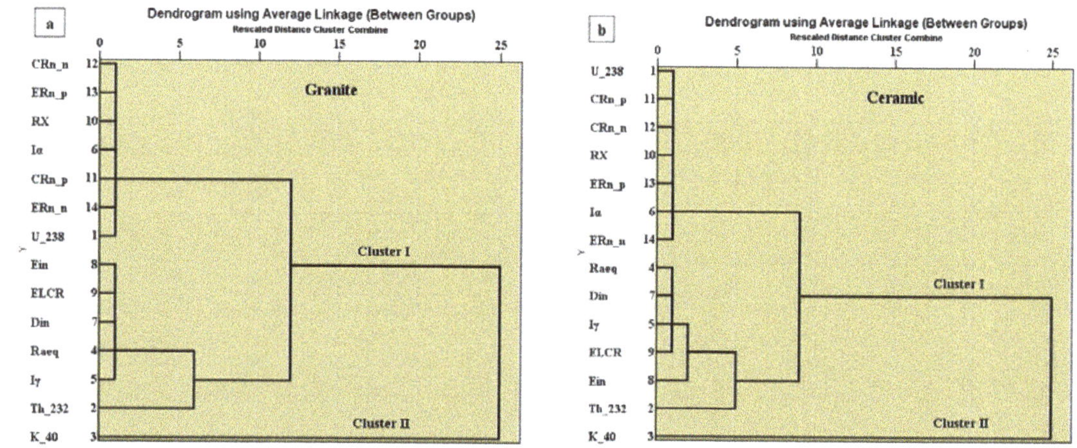

Figure 7. The dendrogram for exhibiting the relationship among the studied radiological variables of the granites and ceramic tiles under consideration.

5. Conclusions

A radiological evaluation for 23 well-known brands of Egyptian commercial granites and ceramic tile samples was performed for hazard and dose estimations. A total of 107 samples representative of the materials under investigation were analyzed using a HPGe detector. The terrestrial radionuclides (U-238 (Ra-226), Th-232, and K-40) concentrations in the investigated samples together with their radiological indices (Ra_{eq}, I_{γ}, D_{in}, E_{in}, ELCR, I_{α}, C_{Rn}, E_{Rn}) were determined. It was found that the concentrations of the aforementioned radioisotopes were higher in most of the granite samples than in the ceramic samples. Furthermore, the concentration values of the terrestrial radionuclides indicated significant differences in the granite and ceramic tile samples collected from the different brands. This is significant in differentiating between the considered brands. K-40 concentration was found to be the biggest contributor to the total concentration for all samples, followed by Th-232 and U-238. Generally, the terrestrial radioisotope concentrations in the materials under investigation are comparable to many of those from the relevant literature and come within the worldwide range. Although the average concentration values for K-40 and Th-232 were higher than their GAVs, the obtained results for the majority of the radiological parameters showed that the studied granites and ceramic tiles are safe to use indoors except "Rosa Sardo Sinai" granite (GRSS). The GRSS samples go beyond the recommended values in terms of their high radium equivalent (Ra_{eq}) mean values, and indoor radon concentration (C_{Rn}), as well as their associated yearly effective dose rate (E_{Rn}) in poorly ventilated buildings, which may be a cause for concern. Therefore, it is not recommended for poorly ventilated buildings. In line with the HCA conducted herein, it reflects the weak relationship of K-40 with all different radiological parameters without exception. In other words, the insignificant risk levels originating from the use of the concerned granites and ceramic tiles are principally due to Th-232 and U-238, with only a weak contribution of K-40.

Our data herein are important for two reasons: firstly, they may raise awareness among the general population of the natural radioactivity of the materials under investigation, and secondly, they are required for developing the standards, rules, and management of tiling materials used in Egypt and in any other country to which such materials are exported.

Author Contributions: All the authors contributed to the conception of the study, study design, methodology, data analysis, validation, writing of the original draft, revision of the manuscript, and final publication. All authors have read and agreed to the published version of the manuscript.

Funding: This research received no external funding.

Institutional Review Board Statement: Not applicable.

Informed Consent Statement: Not applicable.

Data Availability Statement: Not applicable.

Acknowledgments: The authors sincerely thank Mervat A. Elhaddad, Department of Geology, Faculty of Science, Assiut University, Egypt, for her support during the editing and revision of this research article.

Conflicts of Interest: The authors declare that they have no competing interest.

References

1. Turhan, Ş.; Kurnaz, A.; Karataşlı, M. Evaluation of natural radioactivity levels and potential radiological hazards of common building materials utilized in Mediterranean region, Turkey. *Environ. Sci. Pollut. Res.* **2022**, *29*, 10575–10584. [CrossRef] [PubMed]
2. Aladeniyi, K.; Arogunjo, A.M.; Pereira, A.J.S.C.; Khandaker, M.U.; Bradley, D.A.; Sulieman, A. Evaluation of radiometric standards of major building materials used in dwellings of South-Western Nigeria. *Radiat. Phys. Chem.* **2021**, *178*, 109021. [CrossRef]
3. UNSCEAR. *Sources and Effects of Ionizing Radiation: UNSCEAR 2000 Report to the General Assembly, with Scientific Annexes*; United Nations: New York, NY, USA, 2000; Volume 1.
4. Abdullahi, S.; Ismail, A.F.; Samat, S. Determination of indoor doses and excess lifetime cancer risks caused by building materials containing natural radionuclides in Malaysia. *Nucl. Eng. Technol.* **2019**, *51*, 325–336. [CrossRef]
5. Al-Sulaiti, H.; Alkhomashi, N.; Al-Dahan, N.; Al-Dosari, M.; Bradley, D.A.; Bukhari, S.; Matthews, M.; Regan, P.H.; Santawamaitre, T. Determination of the natural radioactivity in Qatarian building materials using high-resolution gamma-ray spectrometry. *Nucl. Instrum. Methods Phys. Res. Sect. A* **2011**, *652*, 915–919. [CrossRef]
6. UNSCEAR. *Sources and Effects of Ionizing Radiation, United Nations Scientific Committee on the Effects of Atomic Radiation (UNSCEAR) 2008 Report*; United Nations: New York, NY, USA, 2010; Volume I. [CrossRef]
7. Aykamış, A.Ş.; Turhan, Ş.; Aysun Ugur, F.; Baykan, U.N.; Kılıç, A.M. Natural radioactivity, radon exhalation rates and indoor radon concentration of some granite samples used as construction material in Turkey. *Radiat. Prot. Dosim.* **2013**, *157*, 105–111. [CrossRef] [PubMed]
8. WHO; Zeeb, H. *WHO Handbook on Indoor Radon: A Public Health Perspective*; World Health Organization: Geneva, Switzerland, 2009.
9. Kuzmanović, P.; Todorović, N.; Filipović Petrović, L.; Mrđa, D.; Forkapić, S.; Nikolov, J.; Knežević, J. Radioactivity of building materials in Serbia and assessment of radiological hazard of gamma radiation and radon exhalation. *J. Radioanal. Nucl. Chem.* **2020**, *324*, 1077–1087. [CrossRef]
10. Axelsson, G.; Andersson, E.M.; Barregard, L. Lung cancer risk from radon exposure in dwellings in Sweden: How many cases can be prevented if radon levels are lowered? *Cancer Causes Control* **2015**, *26*, 541–547. [CrossRef]
11. Higgy, R.H.; El-Tahawy, M.S.; Abdel-Fattah, A.T.; Al-Akabawy, U.A. Radionuclide content of building materials and associated gamma dose rates in Egyptian dwellings. *J. Environ. Radioact.* **2000**, *50*, 253–261. [CrossRef]
12. Rizzo, S.; Brai, M.; Basile, S.; Bellia, S.; Hauser, S. Gamma activity and geochemical features of building materials: Estimation of gamma dose rate and indoor radon levels in Sicily. *Appl. Radiat. Isot.* **2001**, *55*, 259–265. [CrossRef]
13. Moghazy, N.M.; El-Tohamy, A.M.; Fawzy, M.M.; Awad, H.A.; Zakaly, H.M.H.; Issa, S.A.M.; Ene, A. Natural Radioactivity, Radiological Hazard and Petrographical Studies on Aswan Granites Used as Building Materials in Egypt. *Appl. Sci.* **2021**, *11*, 6471. [CrossRef]
14. Kumar, A.; Kumar, M.; Singh, B.; Singh, S. Natural activities of 238U, 232Th and 40K in some Indian building materials. *Radiat. Meas.* **2003**, *36*, 465–469. [CrossRef]
15. Roy, S.; Alam, M.S.; Begum, M.; Alam, B. Radioactivity in building materials used in and around Dhaka city. *Radiat. Prot. Dosim.* **2005**, *114*, 527–532. [CrossRef] [PubMed]
16. Al-Zahrani, J.H. Estimation of natural radioactivity in local and imported polished granite used as building materials in Saudi Arabia. *J. Radiat. Res. Appl. Sci.* **2017**, *10*, 241–245. [CrossRef]
17. Çetinkaya, H.; Manisa, K.; Işık, U. Radioactivity content of building materials used in kutahya province, turkey. *Radiat. Prot. Dosim.* **2022**, *198*, 167–174. [CrossRef] [PubMed]
18. Imani, M.; Adelikhah, M.; Shahrokhi, A.; Azimpour, G.; Yadollahi, A.; Kocsis, E.; Toth-Bodrogi, E.; Kovács, T. Natural radioactivity and radiological risks of common building materials used in Semnan Province dwellings, Iran. *Environ. Sci. Pollut. Res.* **2021**, *28*, 41492–41503. [CrossRef] [PubMed]
19. Tuo, F.; Peng, X.; Zhou, Q.; Zhang, J. Assessment of natural radioactivity levels and radiological hazards in building materials. *Radiat. Prot. Dosim.* **2020**, *188*, 316–321. [CrossRef]

20. Kuzmanović, P.; Todorović, N.; Nikolov, J.; Hansman, J.; Vraničar, A.; Knežević, J.; Miljević, B. Assessment of radiation risk and radon exhalation rate for granite used in the construction industry. *J. Radioanal. Nucl. Chem.* **2019**, *321*, 565–577. [CrossRef]
21. Al-Hamarneh, I.F. Radiological hazards for marble, granite and ceramic tiles used in buildings in Riyadh, Saudi Arabia. *Environ. Earth Sci.* **2017**, *76*, 516. [CrossRef]
22. Asaduzzaman, K.; Khandaker, M.; Amin, Y.; Bradley, D. Natural radioactivity levels and radiological assessment of decorative building materials in Bangladesh. *Indoor Built Environ.* **2016**, *25*, 541–550. [CrossRef]
23. Righi, S.; Bruzzi, L. Natural radioactivity and radon exhalation in building materials used in Italian dwellings. *J. Environ. Radioact.* **2006**, *88*, 158–170. [CrossRef]
24. Kitto, M.E.; Haines, D.K.; Menia, T.A. Assessment of gamma-ray emissions from natural and manmade decorative stones. *J. Radioanal. Nucl. Chem.* **2009**, *282*, 409. [CrossRef]
25. Asghar, M.; Tufail, M.; Sabiha, J.; Abid, A.; Waqas, M. Radiological implications of granite of northern Pakistan. *J. Radiol. Prot.* **2008**, *28*, 387–399. [CrossRef]
26. Sharaf, J.M.; Hamideen, M.S. Measurement of natural radioactivity in Jordanian building materials and their contribution to the public indoor gamma dose rate. *Appl. Radiat. Isot.* **2013**, *80*, 61–66. [CrossRef] [PubMed]
27. Sonkawade, R.G.; Kant, K.; Muralithar, S.; Kumar, R.; Ramola, R.C. Natural radioactivity in common building construction and radiation shielding materials. *Atmos. Environ.* **2008**, *42*, 2254–2259. [CrossRef]
28. Anjos, R.M.; Veiga, R.; Soares, T.; Santos, A.M.A.; Aguiar, J.G.; Frascá, M.H.B.O.; Brage, J.A.P.; Uzêda, D.; Mangia, L.; Facure, A.; et al. Natural radionuclide distribution in Brazilian commercial granites. *Radiat. Meas.* **2005**, *39*, 245–253. [CrossRef]
29. Guillén, J.; Tejado, J.J.; Baeza, A.; Corbacho, J.A.; Muñoz, J.G. Assessment of radiological hazard of commercial granites from Extremadura (Spain). *J. Environ. Radioact.* **2014**, *132*, 81–88. [CrossRef]
30. Medhat, M.E. Assessment of radiation hazards due to natural radioactivity in some building materials used in Egyptian dwellings. *Radiat. Prot. Dosim.* **2009**, *133*, 177–185. [CrossRef]
31. Amin, R.M. Gamma radiation measurements of naturally occurring radioactive samples from commercial Egyptian granites. *Environ. Earth Sci.* **2012**, *67*, 771–775. [CrossRef]
32. Amrani, D.; Tahtat, M. Natural radioactivity in Algerian building materials. *Appl. Radiat. Isot.* **2001**, *54*, 687–689. [CrossRef]
33. Rajamannan, B.; Viruthagiri, G.; Suresh Jawahar, K. Natural radionuclides in ceramic building materials available in Cuddalore district, Tamil Nadu, India. *Radiat. Prot. Dosim.* **2013**, *156*, 531–534. [CrossRef]
34. Abdullahi, S.; Ismail, A.F.; Mohd Fadzil, S.; Samat, S. Assessment of the long-term possible radiological risk from the use of ceramic tiles in Malaysia. *Nucl. Sci. Tech.* **2019**, *30*, 19. [CrossRef]
35. Amin, S.A.; Naji, M. Natural radioactivity in different commercial ceramic samples used in Yemeni buildings. *Radiat. Phys. Chem.* **2013**, *86*, 37–41. [CrossRef]
36. Iwaoka, K.; Hosoda, M.; Suwankot, N.; Omori, Y.; Ishikawa, T.; Yonehara, H.; Tokonami, S. Natural radioactivity and radon exhalation rates in man-made tiles used as building materials in Japan. *Radiat. Prot. Dosim.* **2015**, *167*, 135–138. [CrossRef] [PubMed]
37. Zalewski, M.; Tomczak, M.; Kapala, J. Radioactivity of building materials available in northeastern Poland. *Pol. J. Environ. Stud.* **2001**, *10*, 183–188.
38. Shohda, A.M.; Draz, W.M.; Ali, F.A.; Yassien, M.A. Natural radioactivity levels and evaluation of radiological hazards in some Egyptian ornamental stones. *J. Radiat. Res. Appl. Sci.* **2018**, *11*, 323–327. [CrossRef]
39. Fares, S. Natural Radioactivity Emitted from Granite and Marble Samples Collected from Sinai Area Egypt and Excess Lifetime Cancer Risk. *Int. J. Sci. Eng. Res.* **2016**, *7*, 432–444.
40. Canberra. *The Genie-2000 Operations Manual, Including the S501 Gamma Analysis Option*; Canberra Industries: Canberra, Australia, 2013.
41. Sidique, E.; Elhaddad, M.A.; Abdelwahab, S.F.; El Hadek, H.H. Health Hazards Assessment and Geochemistry of ElSibai-Abu ElTiyur Granites, Central Eastern Desert, Egypt. *Appl. Sci.* **2021**, *11*, 12002. [CrossRef]
42. El-Gamal, H.; Sidique, E.; El-Haddad, M.; Farid, M.E.-A. Assessment of the natural radioactivity and radiological hazards in granites of Mueilha area (South Eastern Desert, Egypt). *Environ. Earth Sci.* **2018**, *77*, 691. [CrossRef]
43. El-Gamal, H.; Sidique, E.; El-Haddad, M. Spatial Distributions and Risk Assessment of the Natural Radionuclides in the Granitic Rocks from the Eastern Desert, Egypt. *Minerals* **2019**, *9*, 386. [CrossRef]
44. Beretka, J.; Matthew, P.J. Natural radioactivity of Australian building materials, industrial wastes and by-products. *Health Phys.* **1985**, *48*, 87–95. [CrossRef]
45. NEA-OECD. *Exposure to Radiation from Natural Radioactivity in Building Materials*; Report by NEA Group of Experts of the Nuclear Energy Agency; OECD: Paris, France, 1979.
46. European Commission. Radiological protection principles concerning the natural radioactivity of building materials. *Radiat. Prot.* **1999**, *112*, 5–16.
47. Khan, A.R.; Rafique, M.; Jabbar, A.; Rahman, S.U.; Shahzad, M.I.; Khan, M.E.; Yasin, M. Radionuclide concentrations in sand samples from riverbanks of Muzaffarabad, Azad Kashmir. *Nucl. Sci. Tech.* **2018**, *29*, 93. [CrossRef]
48. ICRP. *ICRP Publication 60: 1990 Recommendations of the International Commission on Radiological Protection*; SAGE Publications: Southend Oaks, CA, USA, 1990.
49. ICRP. *ICRP Publication 65: Protection Against Radon-222 at Home and at Work*; SAGE Publications: Southend Oaks, CA, USA, 1994.

50. Papadopoulos, A.; Christofides, G.; Koroneos, A.; Papadopoulou, L.; Papastefanou, C.; Stoulos, S. Natural radioactivity and radiation index of the major plutonic bodies in Greece. *J. Environ. Radioact.* **2013**, *124*, 227–238. [CrossRef] [PubMed]
51. Salaheldin, G.; Elhaddad, M.; Sidique, E. Radon concentration and exhalation rate for granitic rocks, Central Eastern Desert, Egypt. *Arab. J. Geosci.* **2022**, *15*, 1221. [CrossRef]
52. UNSCEAR. *Sources and Effects of Ionizing Radiation: UNSCEAR 1993 Report to the General Assembly, with Scientific Annexes*; United Nations: New York, NY, USA, 1993.
53. Qureshi, A.A.; Tariq, S.; Din, K.U.; Manzoor, S.; Calligaris, C.; Waheed, A. Evaluation of excessive lifetime cancer risk due to natural radioactivity in the rivers sediments of Northern Pakistan. *J. Radiat. Res. Appl. Sci.* **2014**, *7*, 438–447. [CrossRef]

sustainability

Article

Optical Properties and Gamma Radiation Shielding Capability of Transparent Barium Borosilicate Glass Composite

Mohamed Ehab [1,*], Elsayed Salama [2,*], Ahmed Ashour [1], Mohamed Attallah [3] and Hosam M. Saleh [4]

1 Engineering Mathematics and Physics Department, Faculty of Engineering and Technology, Future University in Egypt (FUE), New Cairo 11845, Egypt
2 Basic Science Department, Faculty of Engineering, The British University in Egypt (BUE), El Sherouk City 11837, Egypt
3 Basic Science Department, Higher Technological Institute, 10th of Ramadan City 44629, Egypt
4 Radioisotope Department, Nuclear Research Center, Egyptian Atomic Energy Authority, Cairo 11787, Egypt
* Correspondence: mehab@fue.edu.eg (M.E.); elsayed.salama@bue.edu.eg (E.S.); Tel.: +20-1223870849 (E.S.); Fax: +20-2-24665630 (E.S.)

Abstract: In this study, both radiation shielding capability and optical properties of prepared SiO_2-ZnO-Na_2CO_3-H_3BO_3-$BaCO_3$ glass composite with different concentrations of barium carbonate (0–30 mol%) have been studied. Gamma attenuation properties, such as the mass attenuation coefficient (MAC), mean free path (MFP), and exposure build-up factor (EBF), are experimentally and theoretically investigated. The detected XRD patterns for the prepared glass composites confirm their amorphous nature. It is evident from the obtained data that all tested parameters, such as mass density, molar volume, refractive index, dielectric constant, refraction loss (%), and molar refraction, have been increased as $BaCO_3$ mol% increased. At the same time, the results of the optical bandgap show a gradual decrease with increasing barium concentration. It was also found that the mass attenuation coefficients increased with $BaCO_3$ concentration from 0.078 at zero mol% $BaCO_3$ to 0.083 cm^2/g at 30 mol%. Moreover, the half-value layer (HVL) and the exposure build-up factor (EBF) up to 40 mfp penetration depth were investigated in addition to the effective atomic number (Z_{eff}) and the corresponding equivalent atomic number (Z_{eq}) at the energy range of 0.015–15 MeV. The produced glass composite might be considered for many shielding applications based on the obtained results that require a transparent shielding material.

Keywords: mass attenuation coefficient; effective atomic number; build-up factor; borosilicate glass; radiation shielding

Citation: Ehab, M.; Salama, E.; Ashour, A.; Attallah, M.; Saleh, H.M. Optical Properties and Gamma Radiation Shielding Capability of Transparent Barium Borosilicate Glass Composite. *Sustainability* **2022**, *14*, 13298. https://doi.org/10.3390/su142013298

Academic Editor: Changhyun Roh

Received: 7 July 2022
Accepted: 9 October 2022
Published: 16 October 2022

1. Introduction

Shielding materials have a significant role in radiation protection during the wide medical use of radioactive isotopes and X-ray machines and in many industrial applications, such as petroleum and gas extraction [1]. The shielding of ionizing radiation has significantly changed over the last 60 years. As a result of these ongoing developments in anti-ionizing radiation technology, the significance of composite materials for radiation shielding has been acknowledged. In shielding applications, composite materials are desirable because secondary radiation must be considered in radiation shielding design. Therefore, a functional shield's composition must be such that it can efficiently absorb both primary and secondary radiation rays. In addition to their ability to absorb all primary and likely secondary radiation, other properties might restrict the use of particular materials for radiation shielding, such as space, cost, mechanical strength, chemical stability, and thermal stability.

There is a continuing need for new materials to be employed as shielding materials under testing nuclear radiation exposure circumstances [2–5]. The most often utilized protective material so far is concrete [5–10], which is cement mixed with various additives, such as cellulosic waste [11,12], bitumen [13,14], glass [15], polymers [16–18], nanomaterials [19,20], and cement wastes [21,22]. However, the trouble in accomplishing homogeneity, the presence of water and the need for transparency of the shield have persuaded scientists to use glass rather than concrete [23–27]. Borate glasses are used for their exceptionally high transparency, low melting point, and thermal stability. The role of glass additives is different depending on the kind of the enhancement property. Adding sodium carbonate (Na_2CO_3) to the borate glass improves its glass nature characterizations by changing the coordination of the boron coordination group [19], while zinc oxide (ZnO) is added to borate glass to boost its thermal stability, reduce crystallization, and enhance the glass matrix's chemical resistance [28–32]. Heavy elements, such as lead, barium, bismuth, and tungsten are used to increase the borate glass density, which in turn greatly enhancing their radiation-shielding abilities [5,27,33,34]. Borated glasses doped with lead and other specific elements to enhance their densities and radiation attenuation characteristics are successfully produced [32,35–37]. Borosilicate glasses have further advantages, such as chemical durability, better heat stability, extremely low thermal expansion coefficients, and high capacities for substantial visible light transmission [38–41].

Due to its high atomic number and good attenuation coefficient, barium-based glasses have a very promising gamma radiation attenuation coefficient. Several borosilicate glass systems were tested as gamma-ray shielding materials doped with different metal oxides, such as PbO, BaO, Bi_2O_3, BaO, TiO_2, and SrO at different concentrations, showed better shielding efficiencies compared with those previously used in the industry [26,27,42–46]. Adding iron (III) oxide to sodium-barium-vanadate glass has an impact on its physical, optical, mechanical, and radiation absorption features where significant effect on the interaction parameters of thermal neutrons and gamma radiation absorption are observed [47]. The prepared titanium borosilicate glass modified with various ratios of barium oxide showed that adding barium increased the attenuation parameters and enhanced the durability of the prepared sample [48]. The density of sodium zinc borate glasses doped with dysprosium and barium oxide was observed to increase from 2.30 to 4.02 g/cm^3, and the glass's hygroscopic property considerably decreased with the addition of barium [49]. The addition of BaO to zinc barium tellurite glasses enhances its polarization by increasing the bond length, hence the glass network expansion resulting in volume.

The current work aims to prepare and investigate highly transparent, lead-free barium borosilicate glass composite to be used for gamma shielding applications.

2. Materials and Methods

The glass composite of 10 Na_2CO_3 − 20 SiO_2 − 10 ZnO − (60 − x) H_3BO_3 − x $BaCO_3$ (where x = 0, 5, 10, 20, and 30 mol%) was fabricated using the conventional fast melt-quenching technique. High purity grade powders of Na_2CO_3, SiO_2, ZnO, H_3BO_3, and $BaCO_3$ were utilized as a starting material, as shown in Table 1. The compositions were mashed in an agate mortar and melted at 1100 °C for one hour in a porcelain crucible and twirled a lot until a homogenous bubble-free liquid was formed. The melts were poured into preheated stainless-steel molds and annealed at ∼400 °C for 4 h to reduce the cracking and thermal stresses of the samples and then left to cool to room temperature. The photos of the obtained samples before polishing are shown in Figure 1. The samples were then manually polished to obtain maximum flatness.

Table 1. Samples compositions (mol%).

Sample	Na_2CO_3	SiO_2	ZnO	H_3BO_3	$BaCO_3$
S0	10	20	10	60	0
S1	10	20	10	55	5
S2	10	20	10	50	10
S3	10	20	10	40	20
S4	10	20	10	30	30

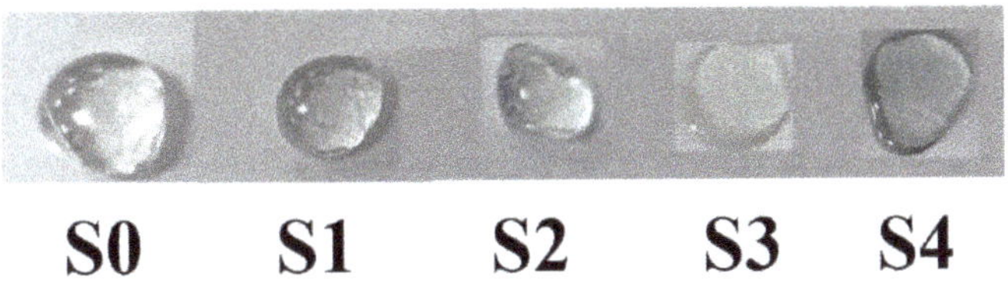

Figure 1. Photos of the prepared samples.

X-ray diffraction (XRD) was accomplished for the prepared glass powders by utilizing a Philips X'pert Pro X-ray powder diffractometer (Malvern Panalytical, Almelo, The Netherlands) at room temperature. The X-ray diffraction patterns were analyzed in 2θ scan from 10° to 90° with CuKα as a target and Ni as a filter (λ = 1.5418 Å) at 40 KV and 30 mA with a speed of scanning reaching to 0.3 s.

Fourier transformation of the infrared absorption spectra (FTIR) of the produced samples were measured in the spectral region 400–4000 cm^{-1} using a JASCOFT-IR6200 spectrometer with the KBr pellet method.

The densities of the prepared glass specimens were measured at room temperature by a simple Archimedes technique that utilizes xylene as a submerged liquid according to the following formula [50]:

$$\rho = \frac{W_a}{W_a - W_b}\rho_b \quad (1)$$

where W_a is the sample's weight in air, W_b represents its weight in xylene, and ρ_b is xylene's density ($\rho_b = 0.863\ g/cm^3$). Using the results of the mass density, the molar volume of the glasses can be calculated according to the following formula [51]:

$$V_m = M_W/\rho \quad (2)$$

where M_W is the molecular weight, and ρ is the glass sample density. Subsequently, the refractive index of the prepared samples can be calculated by the following relation [39,51]:

$$n = \left(\frac{\rho + 10.4}{8.6}\right) \quad (3)$$

Other features depending on the refractive index can be acquired, such as the dielectric constant, which can be calculated according to the following formula [52]:

$$\varepsilon = n^2 \quad (4)$$

Additionally, the reflection loss (R) has been calculated by using the following Fresnel's formula [53]:

$$R = \left(\frac{n-1}{n+1}\right)^2 \quad (5)$$

The ratio of molar volume to molar refractivity (R_M), which is acquired and calculated via the following equation [52], is another structural correlation that can be used to forecast glass propensity that would be metallic or insulating.

$$R_M = V_m\left(\frac{n^2-1}{n^2+2}\right) \quad (6)$$

A recording double beam UV-VIS spectrophotometer (type JASCO Crop., V-770, Japan) encompassing the wavelength range from 200 to 1100 nm was used to evaluate the optical absorption spectra of the polished samples. The absorption coefficient and optical bandgap of the samples were determined according to the optical absorption data. The optical absorption coefficient α (ν) was calculated utilizing the following equation [52]:

$$\alpha(\nu) = 2.303A/d \quad (7)$$

where A is the absorbance, and d is the thickness of the glass sample. Then, the optical bandgap (E_{opt}) is determined through the well-known relation [54]:

$$\alpha(\nu) = B\,\frac{\left(h\nu - E_{opt}\right)^{n}}{h\nu} \quad (8)$$

where hν is the incident photon energy, and B is a constant relating to the band tailing's extent.

For direct allowed, indirect allowed, direct forbidden, and indirect forbidden transitions, respectively, the index n has the values 1/2, 2, 3/2, and 3. Since there is no transition symmetry in the case of indirect transitions, the electron's wave vector might change during the optical transition, and phonons will either take or give up the momentum shift. [55]. In the above-mentioned case, $(\alpha h\nu)^{0.5}$ renders a linear relation with the photon energy. Extrapolating of the linear part of the overhead relation shows the optical bandgap E_{opt} where $(\alpha h\nu)^{0.5} = 0$ in case of indirect transition.

A NaI (Tl) scintillation detector (Teledyne Isotopes "2 × 2" NaI (Tl) Scintillation Detector, AL, USA) with an energy resolution of 8% at 662 keV was used to test the gamma-ray shielding properties of the set glass samples. The generated glass samples were measured at four distinct gamma energies under the correct geometrical constraints: 0.662 MeV from a Cs-137-point source, 0.239 MeV, 0.911 MeV from a 232 Th point source, and 1.332 MeV from a Co-60-point source. All these sources are provided by a spectrum techniques company. The investigated samples were polished and formed to have cylindrical shape of about 2 cm diameter and 1 cm thickness. During the measurements of the gamma attenuation coefficients, the sample was in contact with the point source, and the distance between the source and the detector was fixed at about 10 cm.

3. Theoretical Background

Modified Lambert-Beer law was utilized for the calculation of the linear attenuation coefficients as follows [23]:

$$I = I_0 \times B \times e^{-\mu x} \quad (9)$$

where I_0 and I are the initial and transmitted photon intensities, respectively, μ is a linear attenuation coefficient (cm^{-1}), and B (E, x) is the build-up factor depending on the thickness x (cm) of the used material and the energy E of the incident photon. The mass attenuation coefficient (μ_m) can be determined utilizing the measured linear attenuation coefficient and the mass density (ρ) values by the following relationship [36]:

$$\mu_m = \frac{\mu}{\rho} \quad (10)$$

The following formula can be used to determine μ_m for a compound or mixture [54]:

$$\mu_m = \sum\nolimits_i w_i (\mu_m)_i \quad (11)$$

where $(\mu_m)_i$ is the mass attenuation coefficient of the examined mixture's ith element and w_i stands for its weight percentage. The half-value layer (HVL) of the prepared glasses can be calculated by the following formula [53,56]:

$$HVL = \frac{0.693}{\mu} \quad (12)$$

where μ is the material's linear attenuation coefficient, which obviously relies on the material's type, mass density, and beam energy.

The National Institute of Standard and Technology (NIST) created a photon cross-sections database called XCOM that contains the attenuation coefficients of all elements in the periodic table at various energies in order to calculate the values of the mass attenuation coefficients for the glass samples over a broad range of energies from 0.015 to 15 MeV [57]. The following equation was used to compute the mean free path (MFP) values using the linear attenuation coefficient [58]:

$$\text{MFP} = \frac{1}{\mu} \tag{13}$$

The effective atomic number of a material (Z_{eff}) is defined as the ratio of an object's electronic cross-section (σ_a) to its effective atomic cross-section (σ_e). For the produced glass samples, the following relationship may be used to estimate the values of Z_{eff} based on the obtained data of μ_m [33]:

$$Z_{eff} = \frac{\sigma_a}{\sigma_e} = \frac{\sum_i f_i A_i (\mu_m)_i}{\sum_i f_i \frac{A_i}{Z_i} (\mu_m)_i} \tag{14}$$

where A_i is the atomic weight, Z_i is the atomic number, $(\mu_m)_i$ is the mass attenuation coefficient for the ith element, and f_i represents ith element fractional abundance concerning the number of atoms. To calculate the build-up factor, we must first obtain the Compton partial attenuation coefficient ($(\mu_m)_{comp}$) and total attenuation coefficient ($(\mu_m)_{total}$) values for the constituent elements and compounds of the examined glass samples in the energy range of 0.015–15.0 MeV. The values of the equivalent atomic number (Z_{eq}) for the produced glass samples may then be computed by comparing the ratio $(\mu_m)_{comp}/(\mu_m)_{total}$ at a certain energy with comparable ratios of elements at the same energy. The interpolation of the equivalent atomic number was determined using the following logarithmic interpolation algorithm [59] where the ratio $(\mu_m)_{comp}/(\mu_m)_{total}$ lies between two subsequent ratios of elements:

$$Z_{eq} = \frac{Z_1(\log R_2 - \log R) + Z_2(\log R - \log R_1)}{\log R_2 - \log R_1} \tag{15}$$

where the atomic numbers of the pure elements corresponding to the ratios R_1 and R_2 are Z_1 and Z_2, respectively, and R is the ratio for studied glass samples at certain energy [60]. Using the general progressive (G-P) interpolation in the energy range of 0.015–15 MeV up to 40 mfp, the exposure build-up factors EBF were calculated for the prepared above-mentioned glass samples utilizing the following equations as mentioned in Harima et al. (1993) [6,61,62]:

$$B(E, X) = 1 + \frac{b-1}{K-1}(K^X - 1) \text{ for } K \neq 1 \tag{16}$$

$$B(E, X) = 1 + (b-1)X \text{ for } K = 1 \tag{17}$$

$$K(E, X) = cX^a + d\frac{\tanh\left(\frac{X}{X_K} - 2\right) - \tanh(-2)}{1 - \tanh(-2)} \tag{18}$$

where E is the photon energy, X is the separation between the detector and the source as a function of MFP, B is the EBF value at 1 MFP, K (E, X) is the dosage multiplicative factor, and b, c, a, X_K and d are the calculated G-P fitting parameters that rely on the attenuating medium and source energy. The prepared glasses' b, c, a, X_K and d G-P fitting parameters can be interpolated logarithmically using the following equation-like method for the 0.015–15 MeV gamma-ray energy range up to 40 mfp [63,64].

$$P = \frac{P_1\left(\text{loglog } Z_2 - \text{loglog } Z_{eq}\right) + P_2\left(\text{loglog } Z_{eq} - \text{loglog } Z_1\right)}{\text{loglog } Z_2 - \text{loglog } Z_1} \tag{19}$$

P_1 and P_2 are the values of the G-P fitting parameters that correspond to the Z_1 and Z_2 atomic numbers at the specified energy, respectively. The American Nuclear Society's study criteria for G-P fit for the elements were used [65].

4. Results and Discussion

4.1. XRD Analysis and FTIR

The XRD patterns for the prepared glass samples were obtained and are shown in Figure 2. The absence of sharp peaks in the XRD results demonstrates that the prepared specimens have an amorphous nature. The two humps seen at 2θ° equal 25° and 45° for each sample and serve as a strong piece of evidence for the constructive interferences at variance of two and the aggregation of atoms in the glass matrix in two separate ways. A typical peak for borosilicate matrices was previously seen in several publications [45,66,67].

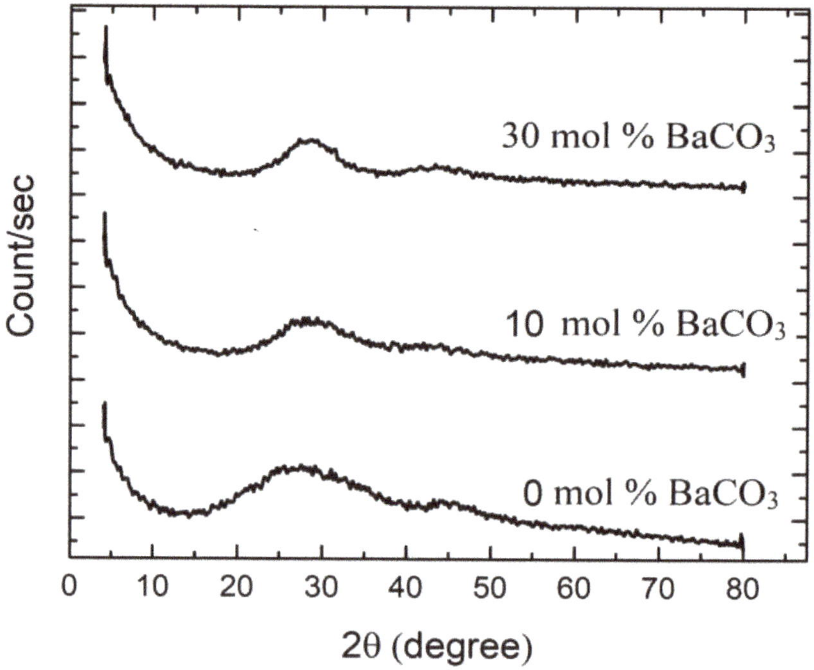

Figure 2. Patterns of X-ray diffraction for ZnO borosilicate glasses doped with $BaCO_3$.

The FTIR transmission spectra of ZnO borosilicate glasses doped with different concentrations of $BaCO_3$ are shown in Figure 3. Table 2 displays the results of the FTIR absorption bands and the associated vibrational modes.

Four distinct bands can be found in the observed data. The band located between 800 and 1200 cm^{-1} represented the BO_4 structural units. Two more bands were visible in the range of 600 and 800 cm^{-1} and 1200 to 1600 cm^{-1} and were returned to BO_3 structural units. Finally, the band of metal ion vibrations was observed at 400 to 600 cm^{-1}. The stretching relaxation modes of B–O bonds of trigonal BO_3 band centered at 1364 cm^{-1} is observed with a small shoulder edge around 1260 cm^{-1} [68]. While the strong broad band from 1176–755 cm^{-1} centered at 955 cm^{-1} are attributed to asymmetric stretching of B–O bonds of tetrahedral BO_4 units. The higher intensity observed may be due to the formation of Si–O–Si and B–O–Si bonds, which contribute vibrational modes at the BO_4 band [70–72]. A moderate band centered around 700 cm^{-1} may be due to bending vibrations of B–O–B of linkages in a borate network [73,74]. The band centered at 440 cm^{-1} and the shoulder noticed at 500 cm^{-1} may be attributed to vibrational modes of all metal cations Ba^{+2} and Zn^{+2} [75–77]. It is noticed that the increase in the $BaCO_3$ mol% in the composite shifts the bands to a lower wavenumber, which denotes a reduction in the BO_3 group and formation of the BO_4 in the glass structure. As a result, an increase in non-bridging oxygens (NBO) and a decrease in the degree of localization of electrons are produced.

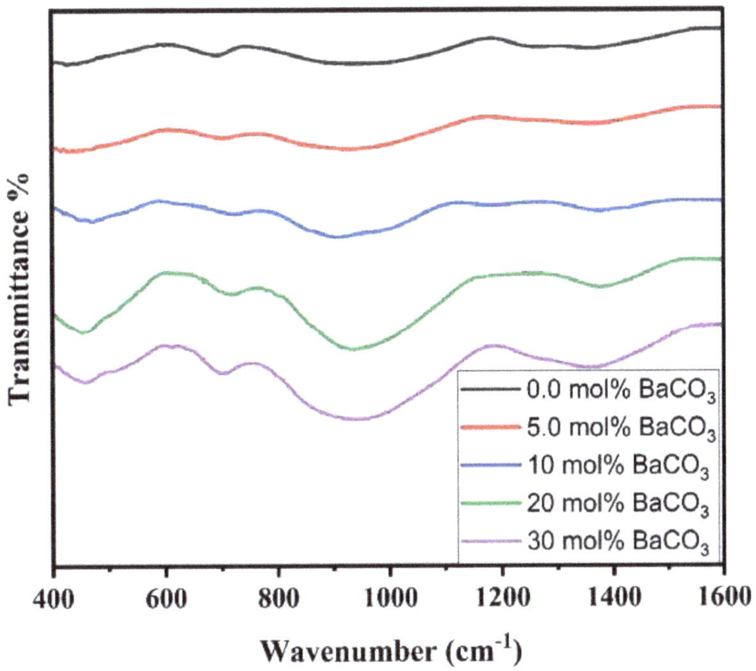

Figure 3. FTIR for ZnO borosilicate glasses doped with $BaCO_3$.

Table 2. The assigned infrared bands to the produced glass samples' spectra.

Peak Position (cm^{-1})	Assignment	Reference Range
1364	Stretching relaxation of B–O bonds of trigonal BO_3 units	1170–1600 [68,69]
950	Stretching vibrations of B–O–Si linkages	950–1050 [70,71]
926	Stretching vibrations of B–O bonds of tetrahedral BO_4 units.	800–1200 [43,71,72]
705	B–O–B vibrations of linkages in a borate network	~700 [73,74]
451	Vibrations of the metal cations Ba^{+2} and Zn^{+2}	400–600 [75,77]

4.2. Density and Molar Volume

Figure 4 displays the mass densities and molar volumes of the produced glasses. Both the density and the molar volume show an equivalent trend increase with increasing the $BaCO_3$ mol% in the composites. By increasing the $BaCO_3$ mol% in the composite, the structure becomes more compact. The larger molecular weight of BaCO3 relative to the other elements may have contributed to the rise in density. On the contrary, the increases in molar volume may be related to the creation of non-bridging oxygen ions (NBOs), which tend to increase the randomizer in the network and convert triangular (BO_3) structure units into tetrahedral (BO_4) structure units [46,78].

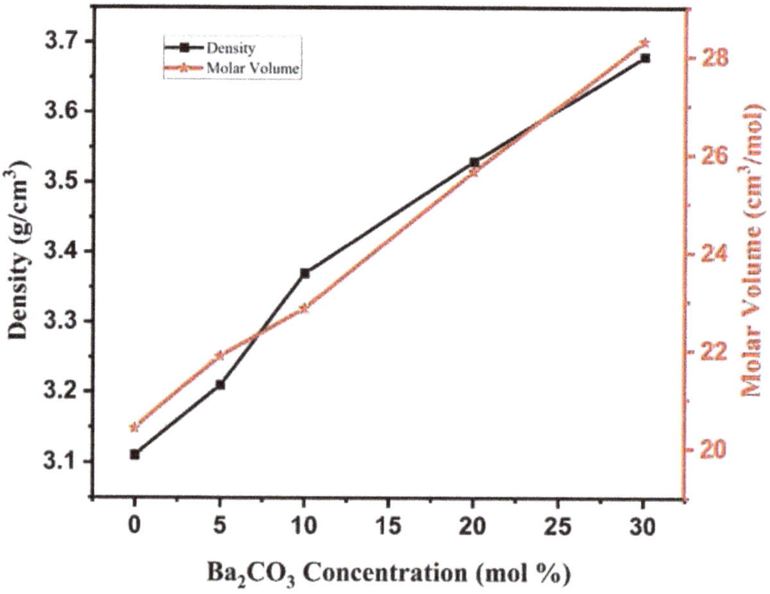

Figure 4. Density and molar volume for ZnO borosilicate glasses doped with $BaCO_3$.

4.3. Optical Absorption Spectra

A powerful technique used to express the optical transitions and electronic band configuration of the amorphous materials is the absorption edge in the region of UV-Vis. Therefore, the optical absorption spectra for set glass samples are shown in Figure 5. The consistency of each sample has been kept as small as possible to evade the inherent absorbance resulted from the long optical path length.

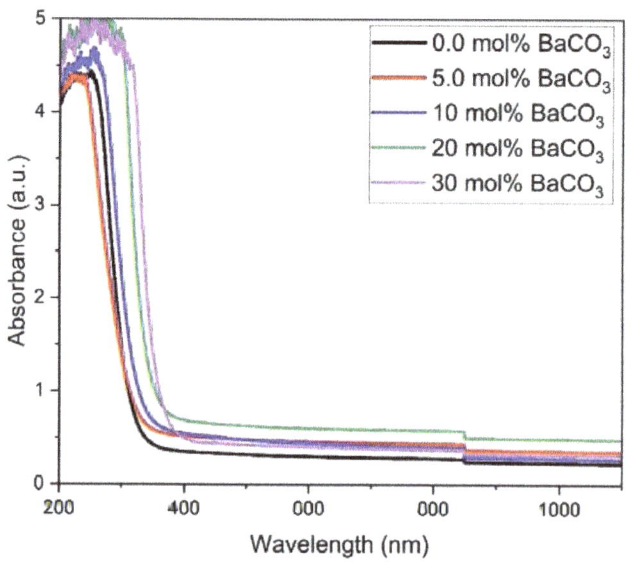

Figure 5. Absorbance versus wavelength for ZnO borosilicate glasses as a function of $BaCO_3$ mol%.

The linear relation between $(\alpha h\nu)^{0.5}$ and the photon energy is shown in Figure 6. Extrapolating of the linear part of the overhead relation shows the optical bandgap E_{opt} energies of 3.55, 3.42, 3.29, 3.21, and 3.13 eV. The values of optical bandgap performance show a gradual decrease with increasing barium concentration. The observed decrease in bandgap with improved $BaCO_3$ mole% concentration could be attributable to potential flaws in the glass network as non-bridging oxygen (NBO) ranges rise. It also indicates the formation of new localized states formed between the valence and conductive bands. Finally, because of the usage of $BaCO_3$ rather than boron oxide, the glass matrix is densified, which is well compatible with the resulting density and changes in the optical bandgap.

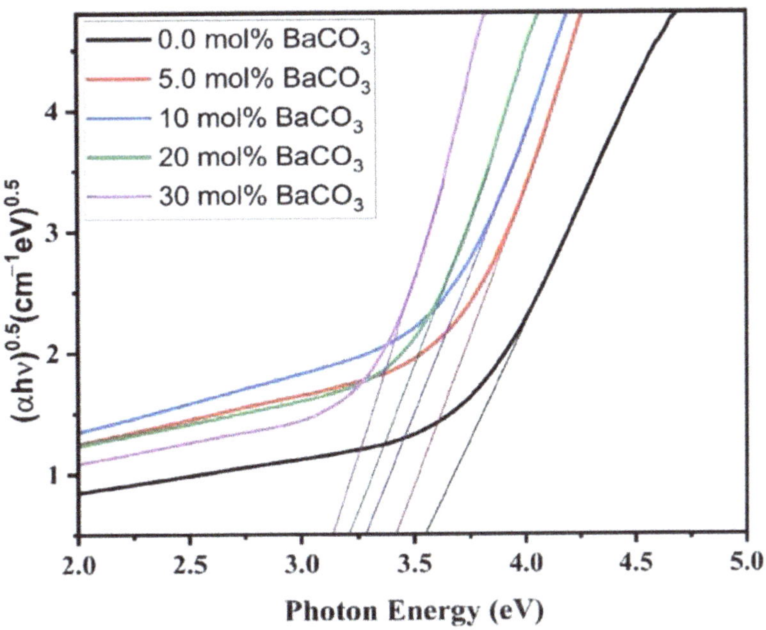

Figure 6. Optical bandgap for ZnO borosilicate glasses as a function of $BaCO_3$ mol%.

The received data of the refractive index and its related parameters are summarized in Table 3. It is evident from the obtained data that all these parameters have the same trend. With increasing $BaCO_3$ mol%, the parameters grew linearly. All estimated parameters confirm the role of barium oxide in the glass network. It has been observed that the refractive index increases as the $BaCO_3$ mol% increases inside the structure grow. The compactness that rises in the glass samples can be linked to this boom in the refractive index.

4.4. Mass Attenuation Coefficient

Table 4 compares the values of the mass attenuation coefficient (μ_m) that were derived theoretically and experimentally. It has been established that there is a close correlation between experimental and theoretical values. Figure 7 displays the fluctuation of m for the produced glass samples with photon energies ranging from 0.015 to 15 MeV. The obtained values of μ_m significantly boost the growth in $BaCO_3$ concentration at the same photon energy while mimicking the chemical composition and photon energy. Based on the interaction of gamma radiation with the examined material, it is possible to explain the inverse relationship between m and the rise in energy for all samples. The photoelectric effect is the most common interaction at low photon energies (E), with an interaction probability proportional to $E^{-3.5}$. While Compton scattering is the dominant interaction at intermediate energies, its probability of interaction is proportional to E^{-1}. Pair production is most prevalent at very high photon energies over 1.022 MeV, where the chance of contact is proportional to E^2. The little variation in the mass density of the prepared glass samples with an increase in BaCO3 concentration from 0 to 30 mol% is what causes the Compton mass attenuation fraction to remain constant. As a result, the mass attenuation coefficient of the produced glass samples has been significantly increased in low-energy areas where x-ray shielding applications are advantageous. The K-absorption edge of barium is what causes the observed peak at around 0.04 MeV (0.037 MeV).

Table 3. Physical parameters of the prepared glass system.

Physical Parameter	$BaCO_3$ mol%				
	0	5	10	20	30
Density (g/cm^3)	3.11	3.21	3.37	3.53	3.68
Molar volume (cm^3 mol^{-1})	33.69	33.77	33.91	33.96	34.52
Refractive index	1.57	1.58	1.59	1.61	1.63
Dielectric constant	2.46	2.5	2.54	2.62	2.68
Refraction loss (%)	0.049	0.050	0.052	0.055	0.058
Molar refraction (cm^3)	11.07	11.29	11.51	11.91	12.39
The optical bandgap (eV)	3.55	3.42	3.29	3.21	3.13

Table 4. The mass attenuation coefficients (cm^2/g) of the created glass samples, both theoretically and experimentally.

$BaCO_3$ (mol%)	0.662 MeV			1.173 MeV			1.332 MeV		
	Exp.	Theo.	% Diff,	Exp.	Theo.	% Diff	Exp.	Theo.	% Diff
0	0.078 ± 0.006	0.078	0.0	0.056 ± 0.006	0.059	5.4	0.055 ± 0.003	0.056	1.8
5	0.078 ± 0.006	0.078	0.0	0.059 ± 0.004	0.059	0.0	0.054 ± 0.002	0.055	1.9
10	0.078 ± 0.006	0.078	0.0	0.049 ± 0.004	0.058	17	0.056 ± 0.003	0.054	3.6
20	0.079 ± 0.006	0.078	1.3	0.054 ± 0.004	0.057	5.6	0.054 ± 0.002	0.053	1.9
30	0.083 ± 0.005	0.078	6.0	0.051 ± 0.004	0.056	7.8	0.054 ± 0.002	0.053	1.9

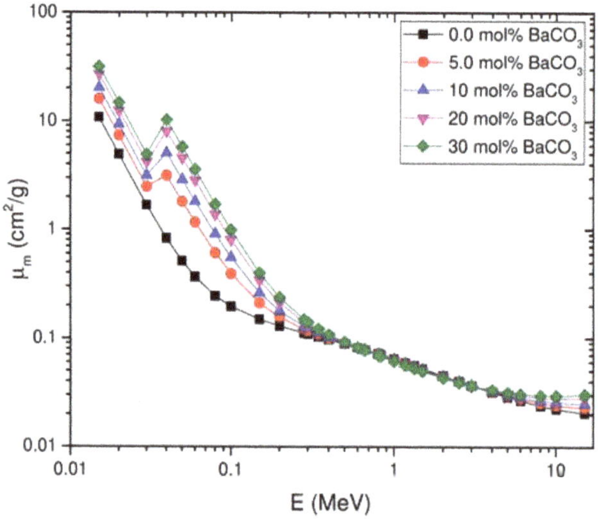

Figure 7. Mass attenuation coefficients of 10 Na_2CO_3 − 20 SiO_2 − 10 ZnO−(60 − x) H_3BO_3 − x $BaCO_3$ glass system in the energy ranges from 0.015–15 MeV and x = 0, 5, 10, 20, 30.

4.5. Half Value Layer (VL) and Effective Atomic Number (Z_{eff})

Figure 8 depicts the fluctuation in the produced glass composite's effective atomic number (Zeff) at energies between 0.015 and 15 MeV and at various concentrations of $BaCO_3$. The observed increase in the Z_{eff} with the increase in the barium concentration can be attributed to the higher atomic number for barium compared with boron (barium is added on the expense of boron) while the change in Z_{eff} of the prepared glass composite in the investigated energy range 0.015–15 MeV can be explained based on the probability of gamma radiation interaction at each energy photon. At low energy range, photoelectric reaction dominates, with the probability proportional to Z^4. As the incident photon energy increases, photo electric interaction probability will decrease, and therefore, Z_{eff} will also decrease. At the intermediate energy range, Compton interaction dominates, with the probability proportional to Z. As the incident photon energy increases, Compton interaction probability will decrease (Compton interaction probability proportional with E^{-1}), and therefore, Z_{eff} will also decrease. At the higher energy range, more than 1.022 MeV pair production interaction dominates, with the probability proportional to Z^2. As the incident photon energy increases, pair production interaction probability will increase, and therefore, Z_{eff} will also increase [79]. At about 0.04 MeV, the ultimate Z_{eff} value was detected in all the prepared glass samples. As discussed in the attenuation curve, maximum absorption occurred at the K-absorption edge of barium at about 0.037 MeV.

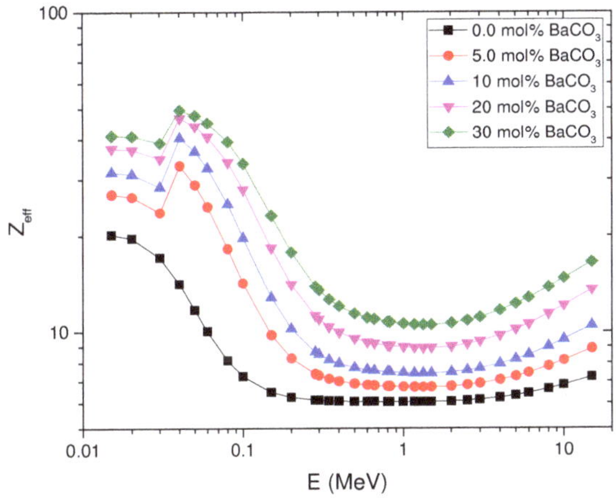

Figure 8. Z_{eff} results of 10 Na_2CO_3 − 20 SiO_2 − 10 ZnO − (60 − x) H_3BO_3 − x $BaCO_3$ glass system.

The results of the calculated values of the half value layer (HVL) at the same energy range 0.015–15 MeV are shown in Figure 9. The discussion of these results is the same as mentioned in the case of the mass attenuation coefficients.

4.6. The Exposure Build-Up Factor (EBF)

As shown in Figure 10, the exposure build-up factor (EBF) values for the prepared glass samples (S0–S4) were calculated using the geometrical progression (G-P) method with depth penetration of up to 40 mfp and photon energies of up to 15 MeV. The picture also demonstrates that, according to the photoelectric effect interaction mechanism, the EBF values of the produced glass samples are negligible at low photon energies. Additionally, within the intermediate energy range when numerous scatterings about the Compton interactions has happened, the samples' EBF significances rise with the energy of the photons. The calculated EBF values for the pair formation process increase at high photon energies. Additionally, strong peaks can be seen at 0.04 MeV in Figure 8 due to the K-absorption edge of barium (0.037 MeV).

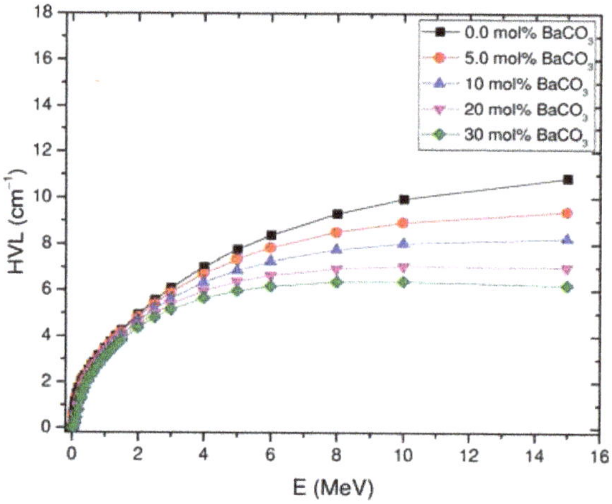

Figure 9. HVL results of 10 Na_2CO_3 − 20 SiO_2 − 10 ZnO − (60 − x) H_3BO_3 − x $BaCO_3$ glass system.

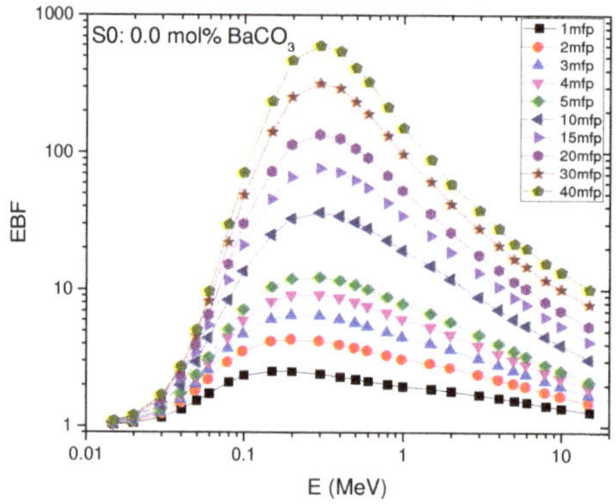

(**a**)

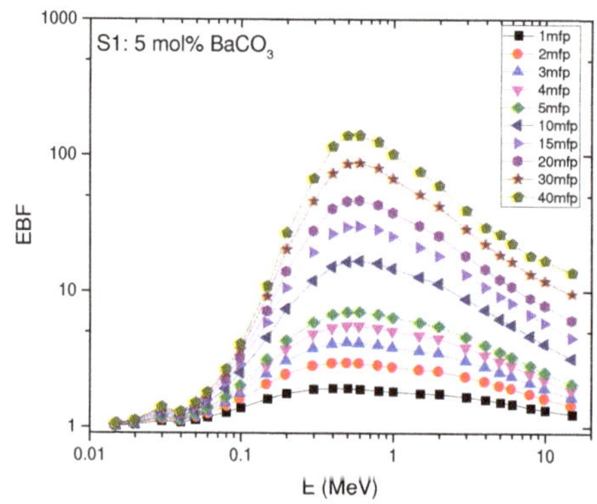

(**b**)

Figure 10. *Cont.*

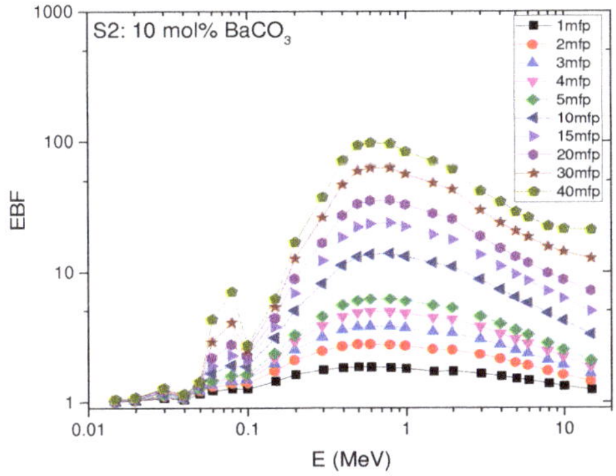

(**c**)

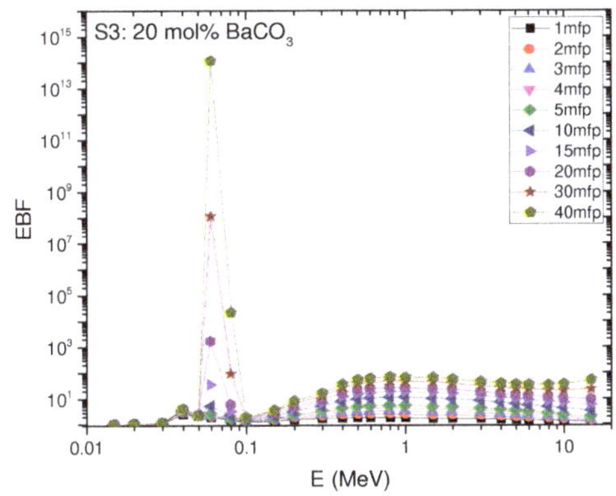

(**d**)

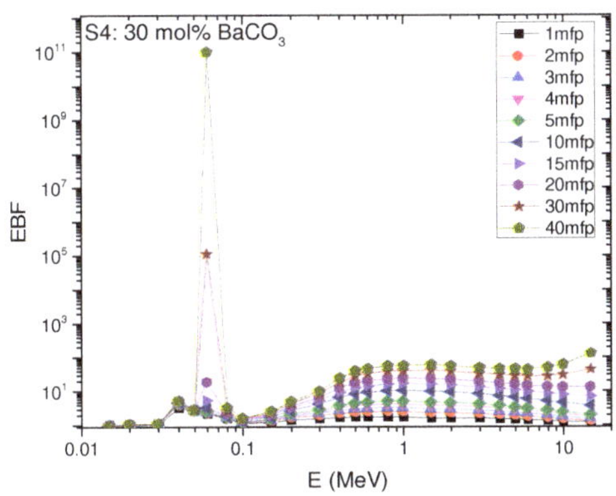

(**e**)

Figure 10. EBF of the produced borosilicate glass composite at 0.015 to 15 MeV up to 40 mfp photon energies with (**a**) 0 mol% $BaCO_3$, (**b**) 5 mol% $BaCO_3$, (**c**) 10 mol% $BaCO_3$, (**d**) 20 mol% $BaCO_3$, and (**e**) 30 mol% $BaCO_3$.

5. Conclusions

The glass composites 10 Na_2CO_3 − 20 SiO_2 − 10 ZnO − (60 − x) H_3BO_3 − x $BaCO_3$ in the current study have been made using the traditional melting procedure, where x = 0, 5, 10, 20, 30. The parameters for structural, optical, and gamma attenuation are established. With the addition of $BaCO_3$, the molar volume and mass density measurements revealed an improvement in compactness. Different vibrational bonding modes, including B–O, B–O–B, and B–O–Si, were seen in the produced glasses according to the FTIR data. The effective atomic number (Z_{eff}), mass attenuation coefficients (MAC), and exposure build-up factors (EBF) of the previously described prepared glass samples were computed at various photon energies between 0.015 and 15 MeV. The findings attained may recommend the manufactured glasses for applications requiring transparent shielding.

Author Contributions: Formal analysis, E.S.; Methodology, M.E., E.S., A.A. and M.A.; Supervision, E.S. and H.M.S.; Writing—original draft, M.E. and M.A.; Writing—review & editing, E.S. All authors have read and agreed to the published version of the manuscript.

Funding: This research received no external funding.

Institutional Review Board Statement: Not applicable.

Informed Consent Statement: Not applicable.

Data Availability Statement: Not applicable.

Conflicts of Interest: The authors declare no conflict of interest.

References

1. Majeed, K.F.; Salama, E.; Elfiki, S.A.; Al-Bakhat, Y.M.Z. Natural radioactivity assessment around the petroleum-producing areas of The-Qar province, Iraq. *Environ. Earth Sci.* **2021**, *80*, 64. [CrossRef]
2. Farah, K.; Mejri, A.; Hosni, F.; Ben Ouada, H.; Fuochi, P.G.; Lavalle, M.; Kovács, A. Characterization of a silicate glass as a high dose dosimeter. *Nucl. Instrum. Methods Phys. Res. Sect. A Accel. Spectrometers Detect. Assoc. Equip.* **2010**, *614*, 137–144. [CrossRef]
3. DWK Life Sciences Glassware Technical Information. *Glas. Types Prop.* **2022**, 12–19.
4. Jubier, N.J. Estimation of Radiation Shielding Properties for Composites Material Based Unsaturated Polyester Filled with Granite and Iron Particles. *J. Multidiscip. Eng. Sci. Stud.* **2017**, *3*, 1309–1316.
5. Khanna, A.; Bhatti, S.S.; Singh, K.J.; Thind, K.S. Gamma-ray attenuation coefficients in some heavy metal oxide borate glasses at 662 keV. *Nucl. Instrum. Methods Phys. Res. Sect. B Beam Interact. Mater. Atoms* **1996**, *114*, 217–220. [CrossRef]
6. Oto, B.; Gür, A.; Kavaz, E.; Çakır, T.; Yaltay, N. Determination of gamma and fast neutron shielding parameters of magnetite concretes. *Prog. Nucl. Energy* **2016**, *92*, 71–80. [CrossRef]
7. Mesbahi, A.; Ghiasi, H. Shielding properties of the ordinary concrete loaded with micro-and nano-particles against neutron and gamma radiations. *Appl. Radiat. Isot.* **2018**, *136*, 27–31. [CrossRef]
8. Shimizu, A.; Onda, T.; Sakamoto, Y. Calculation of gamma-ray buildup factors up to depths of 100 mfp by the method of invariant embedding, (III). *J. Nucl. Sci. Technol.* **2004**, *41*, 413–424. [CrossRef]
9. Singh, K.; Singh, H.; Sharma, V.; Nathuram, R.; Khanna, A.; Kumar, R.; Singh Bhatti, S.; Singh Sahota, H. Gamma-ray attenuation coefficients in bismuth borate glasses. *Nucl. Instrum. Methods Phys. Res. Sect. B Beam Interact. Mater. Atoms* **2002**, *194*, 1–6. [CrossRef]
10. Durani, L. Update to ANSI/ANS-6.4.3-1991 for Low-Z and Compound Materials and Review of Particle Transport Theory. Master's Thesis, University of Nevada, Las Vegas, NV, USA, 2009.
11. Dawoud, M.M.A.; Hegazi, M.M.; Saleh, H.M.; El Helew, W.K. Removal of stable and radio isotopes from wastewater by using modified microcrystalline cellulose based on Taguchi L16. *Int. J. Environ. Sci. Technol.* **2022**, 1–12. [CrossRef]
12. Dawoud, M.M.A.; Hegazy, M.M.; Helew, W.K.; Saleh, H.M. Overview of Environmental Pollution and Clean Management of Heavy Metals and Radionuclides by using Microcrystalline Cellulose. *J. Nucl. Energy Sci. Power Gener. Technol.* **2021**, *3*, 2.
13. Saleh, H.M.; Bondouk, I.I.; Salama, E.; Esawii, H.A. Consistency and shielding efficiency of cement-bitumen composite for use as gamma-radiation shielding material. *Prog. Nucl. Energy* **2021**, *137*, 103764. [CrossRef]
14. Reda, S.M.; Saleh, H.M. Calculation of the gamma radiation shielding efficiency of cement-bitumen portable container using MCNPX code. *Prog. Nucl. Energy* **2021**, *142*, 104012. [CrossRef]
15. Eid, M.S.; Bondouk, I.I.; Saleh, H.M.; Omar, K.M.; Sayyed, M.I.; El-Khatib, A.M.; Elsafi, M. Implementation of waste silicate glass into composition of ordinary cement for radiation shielding applications. *Nucl. Eng. Technol.* **2021**, *54*, 1456–1463. [CrossRef]
16. Eskander, S.B.; Saleh, H.M.; Tawfik, M.E.; Bayoumi, T.A. Towards potential applications of cement-polymer composites based on recycled polystyrene foam wastes on construction fields: Impact of exposure to water ecologies. *Case Stud. Constr. Mater.* **2021**, *15*, e00664. [CrossRef]

17. Saleh, H.; Salman, A.; Faheim, A.; El-Sayed, A. Polymer and polymer waste composites in nuclear and industrial applications. *J. Nucl. Energy Sci. Power Gener. Technol.* **2020**, *9*, 1000199.
18. Saleh, H.M.; Eskander, S.B. Impact of water flooding on hard cement-recycled polystyrene composite immobilizing radioactive sulfate waste simulate. *Constr. Build. Mater.* **2019**, *222*, 522–530. [CrossRef]
19. Saleh, H.M.; El-Saied, F.A.; Salaheldin, T.A.; Hezo, A.A. Influence of severe climatic variability on the structural, mechanical and chemical stability of cement kiln dust-slag-nanosilica composite used for radwaste solidification. *Constr. Build. Mater.* **2019**, *218*, 556–567. [CrossRef]
20. Saleh, H.M.; El-Sheikh, S.M.; Elshereafy, E.E.; Essa, A.K. Performance of cement-slag-titanate nanofibers composite immobilized radioactive waste solution through frost and flooding events. *Constr. Build. Mater.* **2019**, *223*, 221–232. [CrossRef]
21. Saleh, H.M.; Salman, A.A.; Faheim, A.A.; El-Sayed, A.M. Sustainable composite of improved lightweight concrete from cement kiln dust with grated poly (styrene). *J. Clean. Prod.* **2020**, *277*, 123491. [CrossRef]
22. Saleh, H.M.; Salman, A.A.; Faheim, A.A.; El-Sayed, A.M. Influence of aggressive environmental impacts on clean, lightweight bricks made from cement kiln dust and grated polystyrene. *Case Stud. Constr. Mater.* **2021**, *15*, e00759. [CrossRef]
23. Singh, V.P.; Badiger, N.M.; Kaewkhao, J. Radiation shielding competence of silicate and borate heavy metal oxide glasses: Comparative study. *J. Non-Cryst. Solids* **2014**, *404*, 167–173. [CrossRef]
24. Sayyed, M.I.; Tekin, H.O.; Kılıcoglu, O.; Agar, O.; Zaid, M.H.M. Shielding features of concrete types containing sepiolite mineral: Comprehensive study on experimental, XCOM and MCNPX results. *Results Phys.* **2018**, *11*, 40–45. [CrossRef]
25. Hanumantharayappa, C. Study of gamma, X-ray and neutron shielding parameters of some alloys. *IJPAP* **2018**, *56*, 631–634.
26. Abouhaswa, A.S.; Tekin, H.O.; Ahmed, E.M.; Kilicoglu, O.; Rammah, Y.S. Synthesis, physical, linear optical and nuclear radiation shielding characteristics of B_2O_3–BaO–PbO–SrO_2 glasses. *J. Mater. Sci. Mater. Electron.* **2021**, *32*, 18163–18177. [CrossRef]
27. Al-Hadeethi, Y.; Sayyed, M.I. Analysis of borosilicate glasses doped with heavy metal oxides for gamma radiation shielding application using Geant4 simulation code. *Ceram. Int.* **2019**, *45*, 24858–24864. [CrossRef]
28. Zaid, M.H.M.; Matori, K.A.; Abdul Aziz, S.H.; Zakaria, A.; Ghazali, M.S.M. Effect of ZnO on the physical properties and optical band gap of soda lime silicate glass. *Int. J. Mol. Sci.* **2012**, *13*, 7550–7558. [CrossRef]
29. Dong, M.G.; Sayyed, M.I.; Lakshminarayana, G.; Çelikbilek Ersundu, M.; Ersundu, A.E.; Nayar, P.; Mahdi, M.A. Investigation of gamma radiation shielding properties of lithium zinc bismuth borate glasses using XCOM program and MCNP5 code. *J. Non-Cryst. Solids* **2017**, *468*, 12–16. [CrossRef]
30. Tekin, H.O.; Altunsoy, E.E.; Kavaz, E.; Sayyed, M.I.; Agar, O.; Kamislioglu, M. Photon and neutron shielding performance of boron phosphate glasses for diagnostic radiology facilities. *Results Phys.* **2019**, *12*, 1457–1464. [CrossRef]
31. Sayyed, M.I.; Rammah, Y.S.; Abouhaswa, A.S.; Tekin, H.O.; Elbashir, B.O. ZnO-B_2O_3-PbO glasses: Synthesis and radiation shielding characterization. *Phys. B Condens. Matter* **2018**, *548*, 20–26. [CrossRef]
32. Hussein, K.I.; Alqahtani, M.S.; Alzahrani, K.J.; Alqahtani, F.F.; Zahran, H.Y.; Alshehri, A.M.; Yahia, I.S.; Reben, M.; Yousef, E.S. The Effect of ZnO, MgO, TiO2, and Na2O Modifiers on the Physical, Optical, and Radiation Shielding Properties of a TeTaNb Glass System. *Materials* **2022**, *15*, 1844. [CrossRef]
33. Tijani, S.A.; Kamal, S.M.; Al-Hadeethi, Y.; Arib, M.; Hussein, M.A.; Wageh, S.; Dim, L.A. Radiation shielding properties of transparent erbium zinc tellurite glass system determined at medical diagnostic energies. *J. Alloys Compd.* **2018**, *741*, 293–299. [CrossRef]
34. Shams, T.; Eftekhar, M.; Shirani, A. Investigation of gamma radiation attenuation in heavy concrete shields containing hematite and barite aggregates in multi-layered and mixed forms. *Constr. Build. Mater.* **2018**, *182*, 35–42. [CrossRef]
35. Kindrat, I.I.; Padlyak, B.V.; Drzewiecki, A. Intrinsic luminescence of un-doped borate glasses. *J. Lumin.* **2017**, *187*, 546–554. [CrossRef]
36. Singh, N.; Singh, K.J.; Singh, K.; Singh, H. Comparative study of lead borate and bismuth lead borate glass systems as gamma-radiation shielding materials. *Nucl. Instrum. Methods Phys. Res. Sect. B Beam Interact. Mater. Atoms* **2004**, *225*, 305–309. [CrossRef]
37. AlBuriahi, M.S.; Hegazy, H.H.; Alresheedi, F.; Olarinoye, I.O.; Algarni, H.; Tekin, H.O.; Saudi, H.A. Effect of CdO addition on photon, electron, and neutron attenuation properties of boro-tellurite glasses. *Ceram. Int.* **2021**, *47*, 5951–5958. [CrossRef]
38. Kaur, R.; Singh, S.; Pandey, O.P. Structural variation in gamma ray irradiated PbO-Na_2O-B_2O_3-SiO_2 glasses. *Solid State Commun.* **2014**, *188*, 40–44. [CrossRef]
39. Marzouk, S.Y.; Seoudi, R.; Said, D.A.; Mabrouk, M.S. Linear and non-linear optics and FTIR characteristics of borosilicate glasses doped with gadolinium ions. *Opt. Mater.* **2013**, *35*, 2077–2084. [CrossRef]
40. Bootjomchai, C.; Laopaiboon, J.; Yenchai, C.; Laopaiboon, R. Gamma-ray shielding and structural properties of barium-bismuth-borosilicate glasses. *Radiat. Phys. Chem.* **2012**, *81*, 785–790. [CrossRef]
41. Salama, E.; Soliman, H.A.; Youssef, G.M.; Hamad, S. Thermoluminescence Properties of Borosilicate Glass Doped with ZnO Thermoluminescence properties of borosilicate glass doped with ZnO. *J. Lumin.* **2018**, *186*, 164–169. [CrossRef]
42. Alajerami, Y.S.M.; Hashim, S.; Ghoshal, S.K.; Bradley, D.A.; Mhareb, M.; Saleh, M.A. Copper doped borate dosimeters revisited. *J. Lumin.* **2014**, *155*, 141–148. [CrossRef]
43. Mhareb, M.H.A.; Alqahtani, M.; Alshahri, F.; Alajerami, Y.S.M.; Saleh, N.; Alonizan, N.; Sayyed, M.I.; Ashiq, M.G.B.; Ghrib, T.; Al-Dhafar, S.I.; et al. The impact of barium oxide on physical, structural, optical, and shielding features of sodium zinc borate glass. *J. Non-Cryst. Solids* **2020**, *541*, 120090. [CrossRef]

44. Rammah, Y.S.; Tekin, H.O.; Sriwunkum, C.; Olarinoye, I.; Alalawi, A.; Al-Buriahi, M.S.; Nutaro, T.; Tonguc, B.T. Investigations on borate glasses within SBC-Bx system for gamma-ray shielding applications. *Nucl. Eng. Technol.* **2021**, *53*, 282–293. [CrossRef]
45. Aly, P.; El-Kheshen, A.A.; Abou-Gabal, H.; Agamy, S. Structural investigation and measurement of the shielding effect of borosilicate glass containing PbO, SrO, and BaO against gamma irradiation. *J. Phys. Chem. Solids* **2020**, *145*, 109521. [CrossRef]
46. Sayed El-Ahll, L.; Salama, E.; Saudi, H.A.; Alazab, H.A.; Ghany, H.A.A. The Effect of Barium on the Nuclear Radiation Shielding Capabilities of Nickel-Reinforced Borosilicate Glasses. *Silicon* **2022**, *14*, 8909–8917. [CrossRef]
47. Kavas, T.; Alsufyani, S.J.; Alrowaili, Z.A.; Tamam, N.; Kurtulus, R.; Olarinoye, I.O.; Al-Buriahi, M.S. Influence of iron (III) oxide on the optical, mechanical, physical, and radiation shielding properties of sodium-barium-vanadate glass system. *Optik* **2022**, *257*. [CrossRef]
48. Mhareb, M.H.A.; Alqahtani, M.; Alajerami, Y.S.M.; Alshahri, F.; Sayyed, M.I.; Mahmoud, K.A.; Saleh, N.; Alonizan, N.; Al-Buriahi, M.S.; Kaky, K.M. Ionizing radiation shielding features for titanium borosilicate glass modified with different concentrations of barium oxide. *Mater. Chem. Phys.* **2021**, *272*, 125047. [CrossRef]
49. Aboalatta, A.; Asad, J.; Humaid, M.; Musleh, H.; Shaat, S.K.K.; Ramadan, K.; Sayyed, M.I.; Alajerami, Y.; Aldahoudi, N. Experimental investigation of zinc sodium borate glass systems containing barium oxide for gamma radiation shielding applications. *Nucl. Eng. Technol.* **2021**, *53*, 3058–3067. [CrossRef]
50. Abdel-Baki, M.; Salem, A.M.; Abdel-Wahab, F.A.; El-Diasty, F. Bond character, optical properties and ionic conductivity of $Li_2O/B_2O_3/SiO_2/Al_2O_3$ glass: Effect of structural substitution of Li_2O for LiCl. *J. Non-Cryst. Solids* **2008**, *354*, 4527–4533. [CrossRef]
51. Singh, D.; Singh, K.; Singh, G.; Manupriya; Mohan, S.; Arora, M.; Sharma, G. Optical and structural properties of ZnO-PbO-B_2O_3 and ZnO-PbO-B_2O_3-SiO_2 glasses. *J. Phys. Condens. Matter* **2008**, *20*, 075228. [CrossRef]
52. Rammah, Y.S.; Sayyed, M.I.; Ali, A.A.; Tekin, H.O.; El-Mallawany, R. Optical properties and gamma-shielding features of bismuth borate glasses. *Appl. Phys. A Mater. Sci. Process.* **2018**, *124*, 832. [CrossRef]
53. Rammah, Y.S.; Sayyed, M.I.; Abohaswa, A.S.; Tekin, H.O. FTIR, electronic polarizability and shielding parameters of B_2O_3 glasses doped with SnO_2. *Appl. Phys. A Mater. Sci. Process.* **2018**, *124*, 650. [CrossRef]
54. Bashter, I.I. Calculation of radiation attenuation coefficients for shielding concretes. *Ann. Nucl. Energy* **1997**, *24*, 1389–1401. [CrossRef]
55. Thakur, S.; Thakur, V.; Kaur, A.; Singh, L. Structural, optical and thermal properties of nickel doped bismuth borate glasses. *J. Non-Cryst. Solids* **2019**, *512*, 60–71. [CrossRef]
56. Gaafar, I.; El-Shershaby, A.; Zeidan, I.; El-Ahll, L.S. Natural radioactivity and radiation hazard assessment of phosphate mining, Quseir-Safaga area, Central Eastern Desert, Egypt. *NRIAG J. Astron. Geophys.* **2016**, *5*, 160–172. [CrossRef]
57. Umar, S.A.; Halimah, M.K.; Chan, K.T.; Amirah, A.A.; Azlan, M.N.; Grema, L.U.; Hamza, A.M.; Ibrahim, G.G. Optical and structural properties of rice husk silicate incorporated borotellurite glasses doped with erbium oxide nanoparticles. *J. Mater. Sci. Mater. Electron.* **2019**, *30*, 18606–18616. [CrossRef]
58. Sayyed, M.I.; Qashou, S.I.; Khattari, Z.Y. Radiation shielding competence of newly developed TeO_2-WO_3 glasses. *J. Alloys Compd.* **2017**, *696*, 632–638. [CrossRef]
59. Sathiyaraj, P.; Samuel, E.J.J.; Valeriano, C.C.S.; Kurudirek, M. Effective atomic number and buildup factor calculations for metal nano particle doped polymer gel. *Vacuum* **2017**, *143*, 138–149. [CrossRef]
60. Kavaz, E.; Yorgun, N.Y. Gamma ray buildup factors of lithium borate glasses doped with minerals. *J. Alloys Compd.* **2018**, *752*, 61–67. [CrossRef]
61. El-Kameesy, S.U.; Youssef, G.M.; El-Zaiat, S.Y.; Saudi, H.A.; Abd El-Kawy, F.S. Gamma Rays Attenuation Properties and the Associated Optical and Mechanical Behavior of Development (70-x) B_2O_3-$10Al_2O_3$-$10Na_2O$-10ZnO-x PbO Glasses. *Silicon* **2018**, *10*, 1881–1886. [CrossRef]
62. Harima, Y. An historical review and current status of buildup factor calculations and applications. *Radiat. Phys. Chem.* **1993**, *41*, 631–672. [CrossRef]
63. Kaplan, M.F. *Concrete Radiation Shielding: Nuclear Physics, Concrete Properties, Design and Construction*; Longman Scientific & Technical: New York, NY, USA, 1989; ISBN 0470213388.
64. Singh, V.P.; Badiger, N.M. Gamma ray and neutron shielding properties of some alloy materials. *Ann. Nucl. Energy* **2014**, *64*, 301–310. [CrossRef]
65. *ANSI/ANS-6.4.3*; Gamma-Ray Attenuation Coefficients and Buildup Factors for Engineering Materials. American Nuclear Society: La Grange Park, IL, USA, 1991.
66. Gomaa, H.M.; Yahia, I.S.; Zahren, H.Y.; Saudi, H.A.; El-Dosokey, A.H. Effect of replacement of SiO_2 with $BaTiO_3$ on the cadmium calcium-borate glass: Aiming to obtain an active glass for optical and shielding applications. *Radiat. Phys. Chem.* **2022**, *193*, 109955. [CrossRef]
67. Huang, W.J.; Wen, Z.X.; Li, L.J.; Ashraf, G.A.; Chen, L.P.; Lei, L.; Guo, H.; Li, X.M. Photoluminescence and X-ray excited scintillating properties of Tb3+-doped borosilicate aluminate glass scintillators. *Ceram. Int.* **2022**, *48*, 17178–17184. [CrossRef]
68. Shajan, D.; Murugasen, P.; Sagadevan, S. Analysis on the structural, spectroscopic, and dielectric properties of borate glass. *Dig. J. Nanomater. Biostruct.* **2016**, *11*, 177–183.
69. Mustafa, I.S.; Kamari, H.M.; Wan Yusoff, W.M.D.; Aziz, S.A.; Rahman, A.A. Structural and optical properties of lead-boro-tellurite glasses induced by Gamma-ray. *Int. J. Mol. Sci.* **2013**, *14*, 3201–3214. [CrossRef]

70. Yadav, A.K.; Gautam, C.R. Structural and optical studies of Fe2O3 doped barium strontium titanate borosilicate glasses. *Indian J. Pure Appl. Phys.* **2015**, *53*, 42–48.
71. Rani, S.; Sanghi, S.; Agarwal, A.; Seth, V.P. Study of optical band gap and FTIR spectroscopy of $Li_2O{\cdot}Bi_2O_3{\cdot}P_2O_5$ glasses. *Spectrochim. Acta—Part A Mol. Biomol. Spectrosc.* **2009**, *74*, 673–677. [CrossRef]
72. Yadav, A.K.; Gautam, C.R. Synthesis, structural and optical studies of barium strontium titanate borosilicate glasses doped with ferric oxide. *Spectrosc. Lett.* **2015**, *48*, 514–520. [CrossRef]
73. Gautam, C.; Yadav, A.K.; Mishra, V.K.; Vikram, K. Synthesis, IR and Raman Spectroscopic Studies of (Ba,Sr)TiO_3 Borosilicate Glasses with Addition of La_2O_3. *Open J. Inorg. Non-Metallic Mater.* **2012**, *2*, 47–54. [CrossRef]
74. Gautam, C.R.; Kumar, D.; Parkash, O. IR study of Pb-Sr titanate borosilicate glasses. *Bull. Mater. Sci.* **2010**, *33*, 145–148. [CrossRef]
75. Mandal, A.K.; Agrawal, D.; Sen, R. Preparation of homogeneous barium borosilicate glass using microwave energy. *J. Non-Cryst. Solids* **2013**, *371–372*, 41–46. [CrossRef]
76. Study, S.; Consumption, L.E.; Part, D.U.; Utilization, D.; Yoshizawa, N.; Harimoto, K.; Ichihara, M.; Miki, Y.; Takase, K.; Inoue, T. Synthesis, Characterization and Bioactive Study of Borosilicate Sol-Gel Glass Khairy. *Nat. Sci.* **2015**, *13*, 475–476.
77. Rada, S.; Dehelean, A.; Culea, E. FTIR and UV-VIS spectroscopy investigations on the structure of the europium-lead-tellurate glasses. *J. Non-Cryst. Solids* **2011**, *357*, 3070–3073. [CrossRef]
78. Saudi, H.A.; Abd-Allah, W.M.; Shaaban, K.S. Investigation of gamma and neutron shielding parameters for borosilicate glasses doped europium oxide for the immobilization of radioactive waste. *J. Mater. Sci. Mater. Electron.* **2020**, *31*, 6963–6976. [CrossRef]
79. Alzahrani, J.S.; Alrowaili, Z.A.; Olarinoye, I.O.; Alothman, M.A.; Al-Baradi, A.M.; Kebaili, I.; Al-Buriahi, M.S. Nuclear shielding properties and buildup factors of Cr-based ferroalloys. *Prog. Nucl. Energy* **2021**, *141*, 103956. [CrossRef]

sustainability

Article

Transfer of Natural Radionuclides from Soil to Abu Dhabi Date Palms

Prasoon Raj [1,*], Nemeer Padiyath [1], Natalia Semioshkina [2], Francois Foulon [1], Ahmed K. Alkaabi [1], Gabriele Voigt [2] and Yacine Addad [1]

[1] Emirates Nuclear Technology Center (ENTC), Department of Nuclear Engineering, Khalifa University (KU), Abu Dhabi P.O. Box 127788, United Arab Emirates
[2] Cognitive Radioecology, r.e.m. GbR, Liebigstr. 3, 80538 Munich, Germany
* Correspondence: prasoon.raj@ku.ac.ae or prasoon.raj@outlook.com

Abstract: Nuclear power and modern agriculture are two crucial sectors for sustainable development in the United Arab Emirates (UAE). As these industries mature rapidly in the country, their long-term inter-compatibility needs monitoring with local data on transfer of radionuclides from arid sandy soils to farm products. Date palms, main crop from the Arabian Peninsula, remain largely unstudied for radioecological impact assessments. This paper reports the first measurement of soil to UAE date palms concentration ratios for natural radionuclides. Representative samples of soils, fruits, and leaves from seven palms in Abu Dhabi have been studied using gamma-spectrometry. Average activity concentrations in the soils are around 278.9 Bq kg^{-1} for ^{40}K, 15.5 Bq kg^{-1} for ^{238}U, and 8.3 Bq kg^{-1} for ^{232}Th. The latter two decay chains, in the plant samples, are close to detection limits, signifying their lower levels in the UAE flora and the need for upgrading analytical techniques. The geometric means of soil to fruit concentration ratios are 1.12 for ^{40}K, but negligibly low for the others—approximately 0.08 for ^{238}U and 0.17 for ^{232}Th chains. The respective ratios for the leaves are approximately 0.13, 0.36, and 0.77. Personal radiation doses due to soils and dates are very low, posing no danger to the public.

Keywords: radioecology; nuclear power; soil radioactivity; plant sciences; agricultural sustainability; energy studies; arid land; impact assessment; NORM; potassium; middle east

Citation: Raj, P.; Padiyath, N.; Semioshkina, N.; Foulon, F.; Alkaabi, A.K.; Voigt, G.; Addad, Y. Transfer of Natural Radionuclides from Soil to Abu Dhabi Date Palms. *Sustainability* **2022**, *14*, 11327. https://doi.org/10.3390/su141811327

Academic Editors: Hosam M. Saleh and Mohammad Mahmoud Dawoud

Received: 18 July 2022
Accepted: 3 September 2022
Published: 9 September 2022

Publisher's Note: MDPI stays neutral with regard to jurisdictional claims in published maps and institutional affiliations.

1. Introduction

1.1. Radioecology vis-à-vis Nuclear Power in the Arabian Peninsula

In wake of the increasing energy demands in the Arab countries, nuclear power programs are being accelerated in the region, towards a more sustainable future energy mix [1]. Legacy and future nuclear activities are known to put risks of trace radioactivity releases in the environments [2]. The so-released radionuclides may undergo a complex, long-term transport to human food, via wind, rain, waterbodies, soil, plants, animals, etc., as studied under a collective science called radioecology. It is critical to conduct thorough radiological impact assessments (RIA) [3] and to develop nuclear emergency-preparedness systems, for combatting adverse bioecological impacts of such a transfer chain. To that end, at Emirates Nuclear Technology Center (ENTC), we are undertaking baseline radioecological studies in the United Arab Emirates (UAE). A particularly novel challenge is present in the UAE's modern agriculture sector, which is simultaneously progressing to address the food self-sufficiency concerns of the Arab nations [4].

In agricultural farms, radionuclides deposited on the soil are absorbed by plants through their roots [2]. This transfer process is termed as root uptake and is characterized by the radionuclide activity concentrations (AC, units: Bq kg^{-1}) in plant parts and rhizosphere soils, as well as their relative value, i.e., the *plant:soil* concentration ratio (CR). The AC and CR parameters are important inputs for computational prediction of the radioisotope phytoextraction efficiencies following a radioactivity release scenario. Such a calculation assists in predicting cumulative doses to the humans and biota, and furthermore, in optimizing the immediate and long-term mitigation solutions. It also supports the future land decontamination and waste management efforts using phytoremediation. In this regard, the soil–plant bio-transfer kinetics of native plants is crucial to investigate further, but remains severely understudied for arid lands [5,6].

Arid region radioecology (ARRE) is a developing subject to gather more radionuclide transfer factor (TF) data, such as CR, specific to crops, climates, and practices prevalent in arid nations, which are substantially different from temperate regions [7]. It attends to the less explored features [6], such as the physiochemistry of sandy soils, bio-saline growth media, perennial crops such as date palms, heavy irrigation, large usage of chemicals, etc. ARRE not only supports the Arab nations, but also the drylands globally, covering around half of the earth's territory. Moreover, it equips us for better decision-making under changing climatic conditions, which may push many more regions towards aridity in the years to come. The ARRE knowledge will help establish the long-term inter-compatibility of nuclear power and modern agriculture, advocate safety of the nuclear operations and waste management plans to the experts and the public, and finally, support the wider adoption of nuclear electricity for sustainable development.

1.2. Soil and Date Palm Studies in the UAE

Having made substantial progress in nuclear energy through the Barakah Nuclear Power Plant (BNPP) [8], as well as arid agriculture, the UAE serves as a great platform for radioecological activities. Presently, the focus sources of environmental radionuclides in the UAE are naturally occurring radioactive materials (NORM) from fossil fuel industries, and artificial radionuclides (ARN) from the global fallout (mainly ^{137}Cs) [9]. For emergency-preparedness, academic consideration of reactor emissions is also warranted in future activities, especially radionuclides such as ^{134,137}Cs, ^{89,90}Sr, ^{58,60}Co, ^{131}I, etc.

In the UAE's agriculture, date palms (*Phoenix dactylifera* L.) make the main crop, topmost exported entity, and most traditional food. Native date palms are systematically propagated for sustainable land and biodiversity managements. Palm cultivars occupy more than 390 km^2 of land, amounting to an annual produce of approximately 340 kton, valued at USD 500–800 million [10]. With about 40 million trees of over 100 gene-varieties, the UAE's dates constitute 12% of the world's total supply [11]. Besides the fruits, its leaves, pits, wood, and other parts are also used for consumer products [12]. So, we have identified the root uptake CR and bioaccumulation of natural radionuclides from soil to the date palms as an important baseline data to measure in the UAE [6].

Past studies on the UAE soils report approximately mean activity concentrations of the key NORMs; of ^{238}U as 17.8 Bq kg^{-1}, of ^{232}Th as 5.2 Bq kg^{-1}, and of ^{40}K as 182 Bq kg^{-1} [9]. Limited measurements of the date fruits from the UAE's commercial producers have concluded low activities of NORMs, with mean values ~0.66 Bq kg^{-1} for ^{226}Ra (^{238}U series), ~0.15 Bq kg^{-1} for ^{228}Ra (^{232}Th series), and 277 Bq kg^{-1} for ^{40}K [9,13,14]. However, *plant:soil* CR estimates of any kind are practically absent from the UAE's literature. For the date palms, CR values are scarce even from the worldwide literature, except for a few cases [15,16].

1.3. Objectives of the Research

This paper reports a pilot field experiment to estimate baseline *plant:soil* CR of the UAE date palms. The research aims to develop and assess an internationally traceable procedure for CR experiments in arid areas, to obtain the first comprehensive estimates of CRs of natural gamma-emitting radionuclides, to document the farm practices unique to the arid geographies such as the UAE, and, to improve understanding of the NORM migration kinetics in date palms. Seven different palm samples with their surrounding soils have been collected and gamma-assayed in Abu Dhabi (the capital city of the UAE). As the first of its kind, this paper details the consistent and reproducible protocols for sample processing, specific logistical challenges, as well as a detailed analysis of the measured data on soil physiochemistry and CRs. It discusses some essential correlations between soil characteristics and radionuclide speciation and provides conservative risk factors due to background radioactivity in soils and date fruits. Finally, a future program to create a database of radioecological parameters for the UAE agricultural crops is outlined, to complete gaps of the existing global ARRE databases [7].

2. Materials and Methods

The pilot study on date palm CR measurements was conducted in the *Sas-al-Nakhl* locality (SAN in short, the term literally translated from Arabic stands for 'village of the date palms') of Abu Dhabi (UAE) (Figure 1A). Date palms are perennial, woody trees, living up to 60 years, fruiting from as early as 4 years of age, and are particularly tolerant towards very harsh summer temperatures as well as high soil salinities [17]. Date fruits are rich in nutrients, and often, particularly in urban settings, are allowed to ripen on the tree. Trees can grow to 10–20 m in height. They have compound leaves, 2–4 m long, comprised of a thick midrib, and green-coloured pinnae, also known as leaflets. Depending on the location, the roots extend in soil to a depth of 50–60 cm. Date palms, prime among the few native trees, hold an indispensable position in the Arabic food and traditions, and are important for soil, land, and biodiversity conservation. For their safe cultivation and propagation alongside the radioisotope remnants in environments, detailed research is needed in the UAE on date palm phytoremediation capabilities.

Figure 1. (**A**) Sampling location with tree spots (numbered I–VII) in KU SAN Campus, and the nearby Abu Dhabi landmarks in inset; (**B**) a sampled date palm tree with soil sampling spots (yellow circles); (**C**) canopy of a sampled tree with fruit bunches and compound leaves; (**D**) method of soil sampling using a manual Auger; (**E**) photographs of the representative portions of date fruit samples. Several species varieties could be seen in the photographs in (**E**).

2.1. Sampling

The sampling location is in the SAN campus of Khalifa University (KU), south-east of the Abu Dhabi city (Figure 1A), which has around 100 date palms. Soils are primarily

Torripsaments [18] transported from agricultural areas of the UAE, treated monthly with bio compost and macronutrient fertilizers, and watered daily with drip irrigation. These are common practices in the commercial farms as well. Due to saline infertile soil cover in the city, artificial assembly of desert soil layers is done in most of the date palm farms. This makes the KU locations representative for the baseline data measurements.

Date palms typically follow a long growth period, from pollination in February–March to final harvest in July–August. Because of the COVID-19 related restrictions in 2020, the harvests were delayed. In September 2020, 14 palm trees were sampled. Fruits were plucked in their tamr stage (fully ripe), but some kimri and khalal stages (mature but unripe) could also be collected. From 2 kg to 4 kg fresh bunches were carefully detached, from two directions of the tree. Two large leaves with midribs were sampled from two directions. For the barks, around 20 cm long pieces were sliced from surface, but deeper layers could not be sampled to avoid damage to the palms. For each tree, soil was collected at 4–6 randomized spots under the canopy (Figure 1B), to an average depth of 22 cm, using a 10 cm diameter soil Auger (Figure 1D), and mixed to make one composite sample per tree.

2.2. *Sample Processing*

Date fruits were detached from the pedicels (see Figure 1C,E). Likewise, green leaflets were separated from the thick midribs of each compound leaf. A sample processing methodology was developed following the internationally acclaimed technique for radionuclide measurements in biological samples [3], and a careful adaptation of similar studies conducted elsewhere on fruits and vegetables [19,20]. In the end, for each of the 14 palms under study, 1 sample each of fruit (pulp with skin), pit/seed, pedicel, leaflet, leaf midrib, upper bark, and soil were obtained. In this phase of the experiment, fruit, leaflet, and soil samples from only seven of the palm trees were completed and reported. Numbered I to VII, in Figure 1, samples with fully ripe fruits were selected. In this paper, only ripe edible pulp (with skin) of the dates are referred to as 'fruit', and the green leaflets as 'leaf'. Quantities of fruit pits, midribs and barks are relatively more difficult to analyse in the current setting of detector-sample geometry and are postponed for the next phase of this study.

Fruits were gently washed 2–3 times under tap water to remove dust and dirt. Then they were washed with distilled water and left on paper towels to air-dry for one night. The following day, they were cut lengthwise into multiple 1–2 cm wide pieces. Date-pits were separated and stored, while the fruit pieces were spread on steel plates and allowed to dry to a constant weight, in a forced-air convection oven at 100–105 °C. For the seven fruit samples, the net weight reduction due to heating, which amounts to the water content of the fruits, varied between 24% and 34%, and averaged around 26%. Dried fruit pieces were ground using a heavy-duty centrifugal mill. For some samples, the fruits were too sticky because of their high sugar-content. Those samples were briefly embrittled by dipping in liquid nitrogen bath, then immediately ground, and re-dried for the removal of any additional moisture absorbed during air-condensation [21]. Leaves were cleaned multiple times, cut, oven-dried, ground, and packed in the same manner as the fruits. The average moisture content in the leaves was found to be around 38%. Finally, the dried ground powders of fruits and leaves were sieved to <2 mm grain size and packed in 1 L polyethylene bottles. Homogenous and tight pack geometry was ensured. The average net weight of fruits packed in a bottle was about 704 g, while that of leaves was about 382 g. The bottles were properly capped. A two-layer air-tight sealing was done for the caps using aluminium-backed tapes topped with two rounds of polyvinyl tapes.

Soil samples were at first sieved to remove larger chunks of rocks, debris, roots, etc. Then they were thoroughly homogenized by mixing in large plastic drums. Around 1 kg units of wet soil were packed in air-tight zip lock bags for physiochemical analyses. The rest were oven dried like the plant samples, sieved to <2 mm, and packed in 1 L tin cans, each typically packing 1.2 kg to 1.5 kg of dried soils. After securing the caps, a thin ~5 mm layer of molten wax was applied to seal the containers. All the sealed sample bottles and cans were stored for around one month (time > ten half-lives of $^{222}Rn/^{224}Ra$) to ensure radioactivity secular equilibrium between ^{226}Ra and its progenies [22].

2.3. Analytical Techniques

Wet (homogenized, unprocessed) soil samples were studied for their physiochemical properties (Table 1) such as pH, organic matter content (OM), cation exchange capacity (CEC), electrical conductivity (EC), moisture content, and USDA texture class [23], using standardized soil analysis techniques [24,25].

Table 1. Minimum (Min.), maximum (Max.) and AM (Avg.) of the physicochemical characteristics of the soil samples. Parameters include pH, electrical conductivity (EC, unit: dS m^{-1}), organic matter content (OM, unit: weight %), percent moisture content (Mois.), cation exchange capacity (CEC, unit: meq (100 g)$^{-1}$), and soil texture, i.e., the percentage contents of clay, silt, and sand.

	pH	EC	OM	Mois.	CEC	Clay	Silt	Sand
	Aggregate of Sample Numbers I to V and VII (excluding Sample number VI)							
Min.	7.4	1.2	1.8	2.0	4.8	1.6	1.2	91.3
Max.	7.6	2.8	2.8	16.0	8.2	3.9	4.9	96.5
Avg.	7.5	1.9	2.2	9.6	6.8	2.7	2.6	94.4
	Aggregate of all seven samples							
Sample# VI	7.5	52.3	2.0	6.9	5.8	4.8	7.8	87.3
Avg. (I to VII)	7.5	9.1	2.1	9.2	6.6	3.0	3.3	93.4

The processed samples of soils, fruits and leaves were assayed for their gamma spectra using two cylindrical high-purity germanium (HPGe) detectors, located at the Emirates Nuclear Energy Corporation (ENEC) [8] Environmental Radiochemistry Laboratory (ERL). The detectors' relative efficiencies are 40% and 60%, and the energy resolution is 1.9 keV at 1332 keV. They are calibrated for energy and efficiency using a calibration standard containing 105 kBq mixed gamma radionuclides (EZAG 8503 mixture), in the same geometries as the sample containers. Counting times on HPGe detectors were limited to 4 h for soil samples in tin cans and 16 h for plant samples in plastic bottles. The measurement characteristics are optimized for higher processing speeds as per the program requirements of the ERL. Quality assurance for natural radionuclide quantification in environmental matrices is achieved by frequent comparison with certified reference materials, successful participation in standardized proficiency tests (PT) conducted by International Atomic Energy Agency (IAEA) and other top bodies.

NORM measurement protocols as per Ref. [22] were followed. A wide range of radionuclides with progenies from each series were set to be identified, including but not limited to $^{235/238}U$, $^{228/232/234}Th$, $^{224/226/228}Ra$, $^{210/212/214}Pb$, $^{212/214}Bi$, ^{228}Ac, ^{208}Tl, and ^{40}K. A priori minimum detectable activities (MDA), in units of Bq kg^{-1}, for the five key radionuclides are: [^{40}K: 2.41, ^{214}Bi: 1.53, ^{214}Pb: 1.65, ^{228}Ac: 2.72, and ^{208}Tl: 2.23] in 1 L plastic bottles used for plant samples, while [^{40}K: 11.7, ^{214}Bi: 2.61, ^{214}Pb: 2.85, ^{228}Ac: 4.65, and ^{208}Tl: 3.70] in 1 L tin cans used for soil samples. The spectra were further analysed using GENIE 2000 [26] to obtain the nuclide's activity concentrations, along with the necessary corrections as needed, such as interference and coincidence summing, background and Compton corrections, etc.

2.4. Data Treatment

To establish a benchmark for multivariate analyses necessary for the TF experiments in the UAE, the data on soil characteristics were employed in non-parametric Kendall's τ_B Rank Correlation tests [27,28]. The results were studied in terms of coefficients of correlation (r) and statistical significance (p), where $r > 0.7$ and $p < 0.05$ have been taken as a measure of strong correlation. This partially helps in understanding radionuclide speciation and fate in the UAE soils, as further nuclear activities and waste remediation efforts are implemented.

Based on theoretical considerations, biogeological distributions such as the CRs often follow lognormal statistics. Therefore, central tendencies are reported alternatively as arithmetic means (AM) with 1σ standard deviations (SD), as well as geometric means (GM) with geometric standard deviations (GSD) [2,7].

3. Results and Discussion

3.1. Physiochemistry and Radioactivity of Soils

Table 1 reports the range (minimum and maximum) and average values (excluding sample number VI) of the main characteristics for the six soil samples (I–V and VII). Preliminary studies have concluded that sample number VI has higher salinity due to disposal of wastewater and construction debris in the sampling position. So, in Table 1, characteristics of soil sample VI are reported separately, along with the respective averages for all seven samples. A major difference can be noted in the average values of electrical conductivities. Overall, the sampled soils are slightly alkaline, but near neutral, with an average pH of about 7.5. Their organic matter content is between 1.8% and 2.8%, and moisture is between 2% and 16% of the weight. They have been classified under the USDA texture class: sandy (with an average sand content of 93.4%).

Gamma-spectrometry of the soil samples (each one assayed for about 4 h) yielded clear measurements of all the NORM chains. Box-whisker plots in Figure 2 depict variations of the respective radionuclides over the seven soil samples. For ^{238}U chain, ^{214}Pb and ^{214}Bi have been consistently recorded and their activity concentrations agree with each other very well. Likewise, for the ^{232}Th chain, progeny radionuclides measured accurately are ^{228}Ac and ^{208}Th, again with excellent inter-agreement. The weighted AM $\pm$ SD values for activity concentrations of the soils in units Bq kg^{-1} are: 15.5 $\pm$ 2.8 for ^{238}U, 8.3 $\pm$ 0.8 for ^{232}Th, and 278.9 $\pm$ 57.7 for ^{40}K.

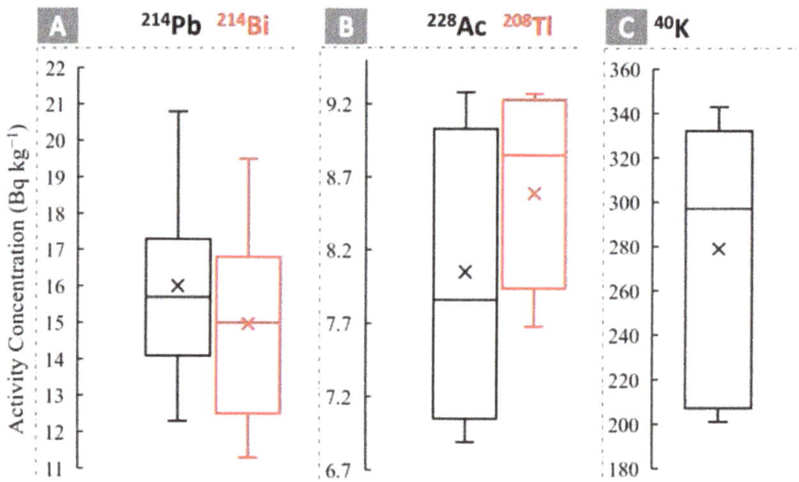

Figure 2. Box-whisker plots showing variabilities of activity concentrations (in unit: Bq kg^{-1}) in the seven soil samples for the main measured radionuclides under NORM decay chains of (**A**) ^{238}U, (**B**) ^{232}Th, and (**C**) ^{40}K. X shows the AM of the data.

Radionuclide speciation in soil solids vs. in aqueous solutions in a plant growing soil water system, are known to be associated with the soil's physiochemical properties [29]. However, such relationships have rarely been established for soils primarily composed of desert sands, as we encountered in this study. A cross-correlation test between various reported parameters have provided some interesting observations, albeit with a very limited number of samples in a small population. There is a clear relationship between concentrations of ^{40}K and ^{214}Bi. The former is also impacted by moisture content in the soil. Moisture, furthermore, has a strong correlation with the organic matter/carbon retention in the soils. Finally, electrical conductivity is found to be positively associated with silt content of the soils, also observable for sample VI in Table 1.

Figure 3 compares the ranges and mean values NORM activity concentrations between various soil studies in the UAE, nearby countries, and world average. Soil NORM activities as measured in this study agrees well with previous measurements conducted in the UAE, with samples specifically from agricultural lands [30], from coastlines [31], as well as those averaging over all territories within the UAE [9].

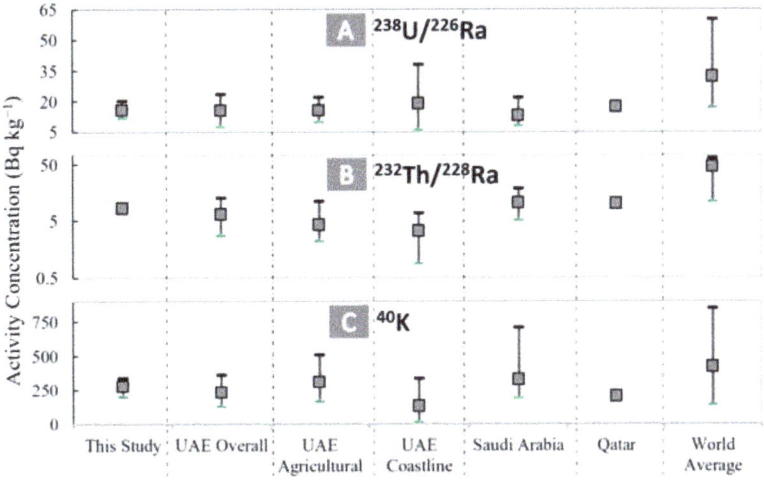

Figure 3. Range (shown with error bars, black dash showing maximum and green the minimum) and AMs (square markers) of soil activity concentrations for (**A**) ^{238}U-series, (**B**) ^{232}Th-series, and (**C**) ^{40}K, as measured in this study, and compared with other measurements in the region. **Sources**: UAE overall [9], UAE agricultural [30], UAE coastline [31], Saudi Arabia [16], Qatar [32], and World [33].

The eastern and western coastlines on the UAE feature some differences, due to differences in geologies, soil makeup [10,18], more rains on the east, and gradient of salinity between the two seas (Arabian Gulf on west, and Arabian Sea on east). The averaging across all coastlines in Ref. [31] has led to relatively lower potassium and thorium contents than urban soils. Abu Dhabi soil radioactivity is very similar to those in Saudi Arabia and Qatar, within the uncertainty margins.

Compared to the world average [33], the activity concentrations of the NORMs in the UAE soils are consistently lower, and well within the recommended limits. Low retention behaviour of sandy soils can be a reason behind the low U and Th values. A comparatively higher content of potassium can be attributed to large-scale application of potassium-containing fertilizers [2,34] in farms, gardens, etc. However, the mean ^{40}K in soils in the UAE and nearby countries such as Qatar, are lower than the global average. Because caesium and potassium uptakes are known to compete [2], in scenarios of large $^{137/134}Cs$ depositions in the local soils, it is probable that low K-levels in soils can promote the uptake of radiocaesium by plants [35].

3.2. NORM Concentrations in Date Palms

Unlike the soils, the fruit and the leaf samples from date palms have considerably lower concentrations of NORM radionuclides, except for ^{40}K. Gamma-spectrometry has quantified the activity concentrations for ^{238}U and ^{232}Th in only 3–5 samples each of fruits and leaves. This is based on the weighted means of the same progenies as in Figure 2 and as shown in Table 2. AM of ^{238}U and ^{232}Th activity concentrations in both leaves and fruits vary between 1.1 and 3.2 Bq kg^{-1}, with the average concentrations in leaves greater than those in fruits. In general, ^{232}Th measures slightly higher than ^{238}U for the plant samples. One of the progenies of ^{238}U chain, ^{210}Pb, has been consistently found in 6/7 of the leaf samples, with AM $\pm$ SD as 30.3 $\pm$ 5.7 Bq kg^{-1}, an order of magnitude higher than the weighted mean of the decay chain's radioactivity. This can be due to the superficial contamination of leaves with lead contained in the aerial pollutants, but it requires further experimental verification.

Table 2. Measured ranges and averages (in parentheses) of activity concentrations of NORMs radionuclides in leaf and fruit samples from this study. BDL stands for below detection limit. A comparison (bottom part) is shown with similar studies on date fruits in the UAE and Saudi Arabia.

Ref.	Country	$^{238}U/^{226}Ra$	^{232}Th	^{40}K
Date palm leaf				
This study [a]	UAE	BDL–2.85 (2.05)	BDL–4.04 (3.18)	75.2–393 (234.2)
Date palm fruit				
This study [a]	UAE	BDL–1.47 (1.06)	BDL–1.79 (1.43)	277–370 (308)
[9] [b]	UAE	BDL	BDL	173–302 (219)
[14]	UAE	BDL–0.8 (0.66)	BDL–0.23 (0.15)	191–362 (277)
[13]	UAE	BDL	0.31–0.62	244.8–302.3
[16] [c]	Saudi Arabia	1.3–7 (5.6)	1.4–6.3 (2.8)	89.5–252 (181)
[36]	Saudi Arabia	1.3–1.4 (1.3)	0.5–0.7 (0.6)	137–198 (163)
[15]	Saudi Arabia	1.1–1.5 (1.37)	1.15–1.3 (1.22)	123–131 (127)

[a] weighted averages of respective progenies are reported. [b] values reported for fresh dates; [c] values reported for date-pits.

The measured concentrations of ^{40}K are much higher (average of leaves and fruits around 271 Bq kg^{-1}) than the other two NORMs. This is due to the large appetite of desert plantations towards macronutrients such as potassium. It is also enhanced by the frequent fertigation using NPK supplements in the soils and waters to sustain the growth and improve yields [10,34]. It is worthwhile to reiterate the competition between the Cs and K uptakes in plants [2]. Larger extraction of potassium, even at ^{40}K soil concentrations lower than the world average, could mean much higher risks of $^{137/134}Cs$ transfers to date palms and other trees in the UAE and nearby countries [35]. Furthermore, phosphates from the fertilizers, depending on their original production process, are known to add considerable amounts of uranium to the soil. This must impact uptake concentrations, which requires a detailed study in future projects.

A pragmatic approach has been taken in estimating the low levels of U and Th concentrations in plant samples. These radionuclides have often suffered with large relative uncertainties in the primary gamma-ray peak area (going to 200%). As they are extremely important data to evaluate, all detected cases providing relative uncertainty up to 50–60%, and within gamma line MDA limits have been accepted, to generate at least their rough estimates. In the future, either much higher efficiency gamma-detection setups should be employed, or alternative techniques with lower limits of detection should be developed. However, it is important to point out that the natural contents of NORMs in the UAE's dates have been verified to be extremely low in other works as well (see Table 2). The literature has merely a few instances of the UAE's date fruits gamma-assays, and the U- and Th-series are often below detection limit (BDL) and difficult to quantify. No measurements on leaves are available for comparison. However, for the fruits, averages of previously measured values are consistently lower than the ones reported in this study, although within the statistical variance margins.

3.3. Concentration Ratio Estimations

Soil to plant NORM CRs for date palm fruits and leaves have been estimated by taking the ratios of the activity concentrations (Bq kg^{-1}) in the plant part to that in the corresponding soil. After calculating the ratios for each sample, GM and GSD values are calculated to provide central tendencies for each NORM. A complete demonstration of the process is shown for ^{40}K in Figure 4, with sample-wise variation of Bq kg^{-1} in soils, fruits, and leaves, along with CR values for *fruit:soil* and *leaf:soil*, as well as the *fruit:leaf* translocation factor.

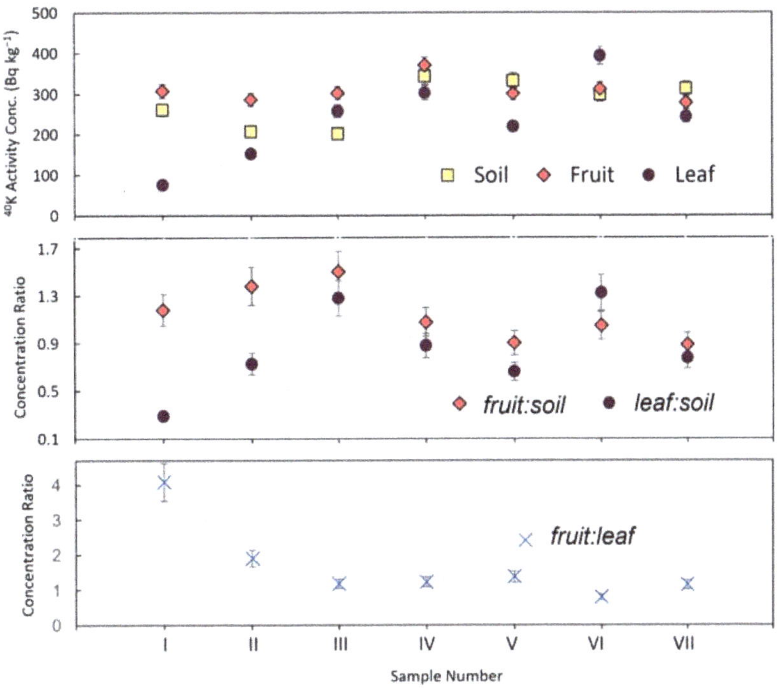

Figure 4. Sample wise distribution of ^{40}K measurements: (**top**) activity concentrations in soil, fruit, and leaf, (**middle**) CR values *fruit:soil* and *leaf:soil*; and (**bottom**) CR *fruit:leaf*. The connecting lines are for guidance only.

For the fruits, ^{40}K CR has a GM of 1.12 ± 1.2, and for leaves it is 0.77 ± 1.6 (see Table 3). This indicates a preferential bioconcentration of ^{40}K in the fruits. Dates and nuts are typically potassium-rich fruits [19], which is a possible explanation for CR > 1. A comparison has been done for the CR values from this study with the much lower ones available in the literature in Table 3. Ref. [16] cites a recorded CR~0.22, while Ref. [15] gives CR~0.55 for ^{40}K. A main driver for potassium uptake in the urban Abu Dhabi date palms is the larger application of NPK fertilizers (with potassium in concentrations of 22 kton [11]—much higher than the global average). This observation highlights the need for a detailed study of the impact of fertilizers on the CR values.

Estimation of CRs for the U- and Th-chains has been difficult due to the low activity concentrations of the corresponding radionuclides in the plant samples. For all the available cases, the measured CRs with 1σ uncertainties are plotted in Figure 5. Regarding both of these decay chains, five samples could provide estimates for *leaf:soil* CR, three for *fruit:soil* CR, and two for *fruit:leaf* CR. The mean values of all CRs are reported in Table 3, and are compared with the two literature cases of date palm CR measurements [15,16]. The *fruit:soil* CR for ^{238}U (GM = 0.08 ± 1.6) is only 20% deviant from the literature value on Saudi dates. The same comparison is more pronounced for CR of ^{232}Th, which deviates by more than 40%. Uncertainties in the case of Th chain are much higher than the U chain (see Figure 5). This is mainly because the inherent thorium content in UAE soils (see Figure 2) is lower by an order of magnitude than uranium, adversely impacting the transfer factor measurements. For some samples, the large uncertainty margins render the ^{232}Th CR < 0, which is a physically impossible extreme. This brings us to the caveat that the estimated CRs for Th- and U-decay chains in date palms are indicative only.

Table 3. Overall values expressed using frequency of observations (n), GM and GSD, for the soil-to-plant CRs for leaf and fruit, with fruit:leaf CR. Literature CR value are also shown for comparison.

Ref.	CR type	^{238}U/^{226}Ra			^{232}Th/^{228}Ra			^{40}K		
		n	GM	GSD	n	GM	GSD	n	GM	GSD
This Study	*leaf:soil*	5	0.13	1.4	5	0.36	1.3	7	0.77	1.6
	fruit:soil	3	0.08	1.6	3	0.17	1.3	7	1.12	1.2
	fruit:leaf	2	0.39 [a]	0.3 [b]	2	0.39 [a]	0.1 [b]	7	1.46	1.6
[15]	*fruit:soil*	5	0.10	1.4	5	0.07	1.2	5	0.22	1.1
[16]	*pits:soil*	9	0.33	2.1	9	0.22	1.8	9	0.51	2.0

[a,b] arithmetic mean and arithmetic standard deviation are reported, as $n \leq 2$.

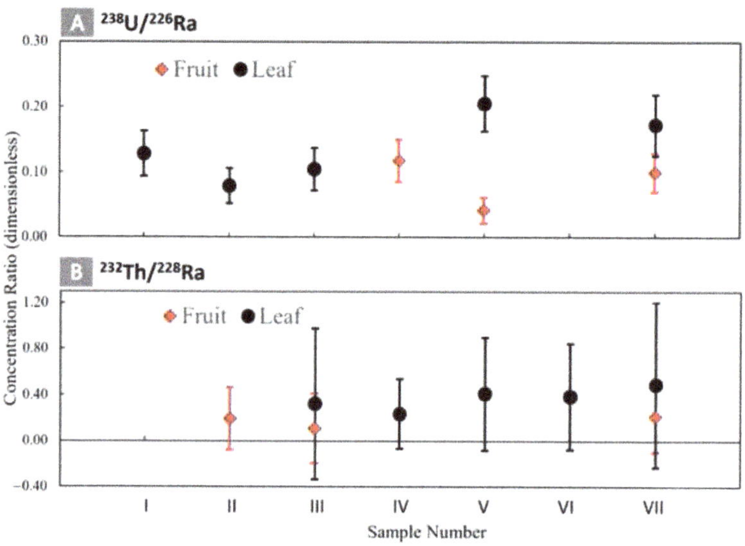

Figure 5. Sample wise variation of *fruit:soil* CR and *leaf:soil* CR values for (**top**) ^{238}U and (**bottom**) ^{232}Th. Unlike ^{40}K in Figure 4, these NORMs were observed in a few samples only. For some cases, very high 1σ uncertainties yield CR < 0 as (meaningless) limits.

3.4. Aggregated Risk Factors

Obtained average activity concentrations of the soils and plant samples have been used for estimating the conservative ranges of radiation exposure 1 m above the earth surface using the standard methods proposed in Refs. [33,36,37]. For soil and date palm (leaf and fruit), the radium equivalent (Ra_{eq}), internal hazard index (H_{in}), absorbed dose (D_{abs}), and effective dose equivalent (D_{eff}) are reported in Table 4.

For the fruits, additional calculations are done for per capita annual radioactivity intake based on the measured indicative activity concentrations. This is further used to evaluate the associated committed effective doses. For the risk factors due to fruits, FAOSAT data on the total annual production of dates in the UAE and total population have been employed [11], adding up to give the maximum committed effective dose from consumption of the studied dates, providing conservative risk estimates. All the aggregated dose factors and hazard indices are well within the UAE national [9] and global limits [33] and pose no risks to the public, as specified in the table. This demonstrates that the phytoextraction of natural radionuclides in the Abu Dhabi date palms are below the safety limits, promising its large-scale propagation in agriculture and landscaping.

Table 4. Summary of aggregated doses and risk factors from exposure to NORM baseline radioactivity in soil, leaf, and fruit of date palms, with (bottom part) ingested doses from date fruits, presented with respective global control limits for comparison.

	Average Activity Concentration (Bq kg^{-1})					**D_{abs}**	**D_{eff}**
	$^{226}Ra/^{238}U$	**$^{228}Ra/^{232}Th$**	**^{40}K**	**Ra_{eq}**	**H_{in}**	**(nGy h^{-1})**	**(µSv y^{-1})**
Soil	15.5	8.3	278.9	48.9	0.17	24.0	29.4
Leaf	2.0	3.2	234.2	24.6	0.07	12.7	15.6
Fruit	1.1	1.4	307.6	26.8	0.08	14.2	17.4
		Global Limit		[370]	[1]	[55]	[1000]
	Per Capita Intake Radioactivity (Bq y^{-1})			**Committed D_{eff} (µSv y^{-1})**			
	$^{226}Ra/^{238}U$	**$^{228}Ra/^{232}Th$**	**^{40}K**	**$^{226}Ra/^{238}U$**	**$^{228}Ra/^{232}Th$**	**^{40}K**	**Total**
Fruit	35.1	47.5	10,220.9	9.8	32.8	63.4	106.0
		Global Limit		[120]	[120]	[170]	[290]

3.5. Perspectives and Possible Expansions

This experiment has strengthened the avenues for future CR measurements in the UAE, but it has also highlighted the necessary developments in our laboratory capabilities. A major shortcoming for measurements of low levels of NORMs in UAE plants is the high detector counting uncertainties. Improved gamma-spectrometry setups, and alternative radioanalytical techniques such as liquid scintillation counting, gross alpha/beta spectrometry, and sequential extraction based isotopic analyses, need to be employed to fill the gaps and improve the confidence in measurements. Using counting times much longer than the one employed in this study (16 h), will also improve the uncertainties and detection limits greatly. Our sample processing steps require longer instrument times and can work with larger quantities (few kg) of samples only. This reduces the speed of our studies, increases manual work, and leads to higher instrumental expenses. Experimental setups adaptable to various quantities and geometries of samples are under development, to reduce the time and expenditure for all steps, from sampling to analysis.

In this pilot study, a limited geographical distribution and statistically smaller number of samples could provide only indicative values of activity concentrations and CRs. In the following stages, many more locations will be sampled, with better focus on the representative, large-scale, commercial farms. An important aspect is to include local agricultural practices in hyper-arid countries, such as the use of modern greenhouses/net houses. Dedicated sampling efforts are necessary to account for the factors capable of systematically impacting the average CR estimates such as the species of fruits sampled, age of trees, seasonal dependency, irrigation, and fertigation practices in the fields, etc. The impact of fertilizers is a highly significant research topic for root uptake studies in UAE farms. More fieldwork is demanded to ascertain the detailed biochemical kinetics of transfer/translocation of radionuclides in date palms, with full-body studies on multiple tree samples. Further sophistication of experiments is expected for the CR estimation for other major crops of the UAE, viz.: vegetables (cucumber, tomato, capsicum, etc.), animal fodder (maize, Rhodes grass, Alfalfa mixture, etc.) and desert plants (Rimth, Ghaf, etc.). Studies addressing the above issues have been planned under the ENTC's coordinated project with IAEA [38].

4. Conclusions

This paper describes the development and results of a field experiment towards the first estimation of natural radionuclides' soil to plant concentration ratios (CR) in the date palms in the UAE. Date palms are one of the topmost crops, exported entities, and traditional foods of Arab nations. Native date palms have been adopted for enhanced cultivation and propagation with modern agriculture and are utilized in sustainable land and biodiversity managements in arid lands. Upcoming nuclear power development in the region raises the risks of radioactivity release to the environments. In this regard, date palms have not been studied thoroughly, making it a critical subject for obtaining the associated radioecological transfer factors and phytoextraction safety limits.

As part of a pilot study in Abu Dhabi (UAE), seven date palm trees were sampled. Samples of their rootzone soils, fruits and leaves were processed and assayed for gamma-emitting NORM radionuclide activities. The soil samples have high concentrations of ^{40}K, but values for ^{238}U and ^{232}Th are much lower. The measured soil radioactivity is below the global averages. The uptake of ^{40}K is considerably high in the dates, leading to GM of CR ~1.12. Only rough estimates of ^{238}U (~0.08) and ^{232}Th (~0.17) could be obtained due to their very low inherent contents in the local dates. The experiment also recorded the uptake to leaf, also a consumer product, as 0.77 for ^{40}K, 0.13 for ^{238}U and 0.36 for ^{232}Th. The values obtained here compare very well with the limited literature measurements but makes a novel and complementary input to the global arid region radioecology database. The developed protocol will be replicated for more accurate estimations of date palm CRs, which can be further utilized in national exercises of risk prediction, radioactive waste management, and nuclear emergency-preparedness. Preliminary calculations suggest negligible personal dose factors due to the baseline radioactivity in Abu Dhabi soils and date palms. More detailed experiments and computational impact assessments are planned based on this experience, expanding to new geographies in the country, as well as new measurement techniques, and other important crops.

Author Contributions: Conceptualization, formal analysis, investigation, software, P.R.; methodology, P.R., N.P., N.S. and G.V.; resources, P.R., N.P., Y.A. and A.K.A.; writing—original draft preparation, P.R.; writing—review and editing, N.P., N.S., G.V. and Y.A.; visualization, P.R.; supervision, F.F. and Y.A.; project administration, F.F.; funding acquisition, F.F. and Y.A. All authors have read and agreed to the published version of the manuscript.

Funding: This research was sponsored by Khalifa University and received no external funding.

Institutional Review Board Statement: Not applicable.

Data Availability Statement: All relevant data are included in the article.

Acknowledgments: The authors would like to thank the Research and Development team of Emirates Nuclear Energy Corporation (ENEC) for providing access to the radioanalytical instruments in the Environmental Radiochemistry Laboratory, located outside Al Ruwais. Their scientists Michael Millinor and Andrew Maddison are dearly acknowledged for their involvement in the gamma spectrometry and for their several discussions and comments. Likewise, the authors extend their gratitude to the Soil Analysis Laboratory of the International Centre for Biosaline Agriculture (ICBA) Dubai for their support in physiochemical analysis of the soil samples. Sincere thanks are due to Greendreams Landscaping for participation in the field surveys and sampling campaigns.

Conflicts of Interest: The authors declare no conflict of interest.

References

1. Bollino, C.A.; Asdrubali, F.; Polinori, P.; Bigerna, S.; Micheli, S.; Guattari, C.; Rotili, A. A Note on Medium- and Long-Term Global Energy Prospects and Scenarios. *Sustainability* **2017**, *9*, 833. [CrossRef]
2. IAEA TRS 472. In *Handbook of Parameter Values for the Prediction of Radionuclide Transfer in Terrestrial and Freshwater Environments*; Technical Reports Series; International Atomic Energy Agency: Vienna, Austria, 2010; ISBN 978-92-0-113009-9.
3. IAEA TRS 486. In *Guidelines on Soil and Vegetation Sampling for Radiological Monitoring*; Technical Reports Series; International Atomic Energy Agency: Vienna, Austria, 2019; ISBN 978-92-0-102218-9.
4. Hameed, M.; Moradkhani, H.; Ahmadalipour, A.; Moftakhari, H.; Abbaszadeh, P.; Alipour, A. A Review of the 21st Century Challenges in the Food-Energy-Water Security in the Middle East. *Water* **2019**, *11*, 682. [CrossRef]
5. Semioshkina, N.; Voigt, G. Soil—Plant Transfer of Radionuclides in Arid Environments. *J. Environ. Radioact.* **2021**, *237*, 106692. [CrossRef]

6. Raj, P.; Padiyath, N.; Semioshkina, N.; Addad, Y.; Foulon, F.; Francis, D.; Voigt, G. Conceptualization of Arid Region Radioecology Strategies for Agricultural Ecosystems of the United Arab Emirates (UAE). *Sci. Total Environ.* **2022**, *832*, 154965. [CrossRef]
7. IAEA TECDOC 1979. In *Soil-Plant Transfer of Radionuclides in Non-Temperate Environments*; Technical Document; International Atomic Energy Agency: Vienna, Austria, 2021; ISBN 978-92-0-129121-9.
8. ENEC Emirates Nuclear Energy Corporation. Available online: https://www.enec.gov.ae/ (accessed on 25 May 2022).
9. Conatser, R.L.; Abdouli, A.A.; Sallam, A.; Bosc, E.; Nuaimi, F.A.; Ameri, H.A.; Zaabi, M.A.; Al-Hasani, S.; Mansour, T. Radiological Environmental Monitoring in the United Arab Emirates, First Report, 1 January—31 December 2015; Federal Authority for Nuclear Regulations (FANR). Available online: https://bit.ly/2DhwvVQ (accessed on 25 August 2022).
10. UAE MOCCAE Open Data | UAE Ministry of Climate Change and Environment. Available online: https://www.moccae.gov.ae/en/open-data.aspx#page=1 (accessed on 25 May 2022).
11. FAOSTAT, U.N.O. Food and Agriculture Data; Food and Agriculture Organization (FAO) of the United Nations. Available online: https://www.fao.org/faostat/en/#data (accessed on 25 May 2022).
12. Naushad, M.; Lichtfouse, E. (Eds.) *Sustainable Agriculture Reviews 34: Date Palm for Food, Medicine and the Environment*; Sustainable Agriculture Reviews; Springer International Publishing: Cham, Switzerland, 2019; Volume 34, ISBN 978-3-030-11344-5.
13. Alrefae, T. Long-Lived Gamma Emitting Radionuclides in Palm Dates and Estimates of Annual Effective Doses. *Health Phys.* **2015**, *108*, 547–550. [CrossRef]
14. Rahman, R.; Solodov, A. Analysis of Natural and Anthropogenic Radionuclide Content in Palm Date Fruit of the United Arab Emirates: A Baseline Study. *Health Phys.* **2016**, *111*, 465–470. [CrossRef]
15. El-Taher, A. A Study on Transfer Factors of Radionuclides from Soil to Plant. *Life Sci. J.* **2013**, *10*, 532–539.
16. Shayeb, M.A.; Alharbi, T.; Baloch, M.A.; Rahman Alsamhan, O.A. Transfer Factors for Natural Radioactivity into Date Palm Pits. *J. Environ. Radioact.* **2017**, *167*, 75–79. [CrossRef]
17. Al-Alawi, R.A.; Al-Mashiqri, J.H.; Al-Nadabi, J.S.M.; Al-Shihi, B.I.; Baqi, Y. Date Palm Tree (*Phoenix dactylifera* L.): Natural Products and Therapeutic Options. *Front. Plant Sci.* **2017**, *8*, 845. [CrossRef]
18. Emirates Soil Museum, U.A.E. International Center for Biosaline Agriculture (ICBA), Dubai. Available online: https://www.emiratessoilmuseum.org/ (accessed on 25 May 2022).
19. Napier, B.A.; Fellows, R.J.; Krupka, K.M. *Soil-to-Plant Concentration Ratios for Assessing Food Chain Pathways in Biosphere Models*; NUREG/CR-6941, 926965; Pacific Northwest National Laboratory (PNNL): Richland, WA, USA, 2007; p. PNNL-16741.
20. Duong Van, H.; Nguyen, T.D.; Peka, A.; Hegedus, M.; Csordas, A.; Kovacs, T. Study of Soil to Plant Transfer Factors of 226Ra, 232Th, 40K and 137Cs in Vietnamese Crops. *J. Environ. Radioact.* **2020**, *223–224*, 106416. [CrossRef]
21. Karam, M.C.; Petit, J.; Zimmer, D.; Baudelaire Djantou, E.; Scher, J. Effects of Drying and Grinding in Production of Fruit and Vegetable Powders: A Review. *J. Food Eng.* **2016**, *188*, 32–49. [CrossRef]
22. Gilmore, G. *Practical Gamma-Ray Spectrometry*, 2nd ed.; Wiley: Chichester, UK; Hoboken, NJ, USA, 2008; ISBN 978-0-470-86196-7.
23. Shirazi, M.A.; Boersma, L. A Unifying Quantitative Analysis of Soil Texture. *Soil Sci. Soc. Am. J.* **1984**, *48*, 142–147. [CrossRef]
24. USDA Handbook No. 60 Handbook No. 60: Diagnosis and Improvement of Saline and Alkali Soils. Available online: https://www.ars.usda.gov/pacific-west-area/riverside-ca/agricultural-water-efficiency-and-salinity-research-unit/docs/publications/handbook-no-60/ (accessed on 25 May 2022).
25. Day, P.R. Particle Fractionation and Particle-Size Analysis. In *Methods of Soil Analysis*; John Wiley & Sons, Ltd.: Hoboken, NJ, USA, 1965; pp. 545–567, ISBN 978-0-89118-203-0.
26. Genie 2000 GenieTM 2000 Gamma Analysis Software. Available online: https://www.mirion.com/products/genie-2000-gamma-analysis-software (accessed on 25 May 2022).
27. Kendall, M.G. A New Measure of Rank Correlation. *Biometrika* **1938**, *30*, 81–93. [CrossRef]
28. Chandrasekaran, A.; Ravisankar, R.; Rajalakshmi, A.; Eswaran, P.; Vijayagopal, P.; Venkatraman, B. Assessment of Natural Radioactivity and Function of Minerals in Soils of Yelagiri Hills, Tamilnadu, India by Gamma Ray Spectroscopic and Fourier Transform Infrared (FTIR) Techniques with Statistical Approach. *Spectrochim. Acta Part A Mol. Biomol. Spectrosc.* **2015**, *136*, 1734–1744. [CrossRef] [PubMed]
29. Nisbet, A.F.; Woodman, R.F.M. Soil-to-Plant Transfer Factors for Radiocesium and Radiostrontium in Agricultural Systems. *Health Phys.* **2000**, *78*, 279–288. [CrossRef] [PubMed]
30. Ajaj, R.M.B. Determination of Primordial and Anthropogenic Radionuclide Concentrations in Agricultural Soil of the United Arab Emirates Using Gamma-Ray Spectrometry. Ph.D. Thesis, United Arab Emirates University, Al Ain, United Arab Emirates, 2017.
31. Al-Ali, Y.A.M. Natural Radioactivity of UAE Coastal Areas. Master's Thesis, United Arab Emirates University, Al Ain, United Arab Emirates, 2003.
32. Ahmad, A.Y.; Al-Ghouti, M.A.; AlSadig, I.; Abu-Dieyeh, M. Vertical Distribution and Radiological Risk Assessment of 137Cs and Natural Radionuclides in Soil Samples. *Sci. Rep.* **2019**, *9*, 12196. [CrossRef] [PubMed]
33. *Sources and Effects of Ionizing Radiation: United Nations Scientific Committee on the Effects of Atomic Radiation: UNSCEAR 2000 Report to the General Assembly, with Scientific Annexes*; United Nations: New York, NY, USA, 2000; ISBN 978-92-1-142238-2.
34. Ibrahim, M.M.; El-Beshbeshy, R.T.; Kamh, N.R.; Abou-Amer, A.I. Effect of NPK and Biofertilizer on Date Palm Trees Grown in Siwa Oasis, Egypt. *Soil Use Manag.* **2013**, *29*, 315–321. [CrossRef]
35. Vandenhove, H.; Smolders, E.; Cremers, A. Potassium Bentonites Reduce Radiocaesium Availability to Plants. *Eur. J. Soil Sci.* **2003**, *54*, 91–102. [CrossRef]

36. Khandaker, M.U.; Shuaibu, H.K.; Alklabi, F.A.A.; Alzimami, K.S.; Bradley, D.A. Study of Primordial 226Ra, 228Ra, and 40K Concentrations in Dietary Palm Dates and Concomitant Radiological Risk. *Health Phys.* **2019**, *116*, 789–798. [CrossRef] [PubMed]
37. Prasad, N.G.S.; Nagaiah, N.; Ashok, G.V.; Karunakara, N. Concentrations of 226Ra, 232Th, AND 40K in the Soils of Bangalore Region, India. *Health Phys.* **2008**, *94*, 264–271. [CrossRef] [PubMed]
38. IAEA CRP K41022 New CRP: Transfer of Radionuclides in Arid and Semi-Arid Environments for Radiological Environmental Impact Assessment (K41022). Available online: https://www.iaea.org/newscenter/news/new-crp-transfer-of-radionuclides-in-arid-and-semi-arid-environments-for-radiological-environmental-impact-assessment-k41022 (accessed on 25 May 2022).

sustainability MDPI

Article

Elicitation Promoability with Gamma Irradiation, Chitosan and Yeast to Perform Sustainable and Inclusive Development for Marjoram under Organic Agriculture

Tarek E. Sayed * and El-Sayed S. Ahmed

Radioisotope Department, Egyptian Atomic Energy Authority, Cairo 11787, Egypt
* Correspondence: dr.tarekelsayed64@gmail.com

Abstract: Sweet marjoram (*Majorana hortensis*) is an important aromatic herbal plant that has long been used and well managed in the traditional and general medical, pharmaceutical, food, cosmetic, and perfume industries. Thus, the increase in its productivity appears to be of great value since there is a large number of bioactive secondary metabolites as well as an increase in the demand in domestic or foreign markets. The purpose of this study is the possibility of promoting the sustainable development of marjoram in the framework of organic farming through gamma irradiation, chitosan and yeast. Field experiments were conducted in a factorial split-plot design with three iterations over two consecutive seasons (2019 and 2020). The main plot is an abiotic elicitor (15 Gy gamma irradiation), two biotic elicitors 500 ppm chitosan, 0.5% yeast, and a non-elicitor (as control), while in the sub-main plot, there were two organic fertilizers, water extract of moringa 20 g/m^2 dry leaves, 20 g/m^2 fulvic acid, and 20 g/m^2 (NPK); the latter is a traditional agrochemical. Statistical analysis of all characteristics of production and quality of biomass and biologically active secondary metabolites revealed that the use of organic fertilizers helped in increasing the yield of marjoram, both qualitatively and quantitatively, and significantly outperformed the chemical fertilizer. The experiment enhances the comprehensive and integrated development of marjoram under organic cultivation and achieves a promising alternative to traditional cultivation without the use of microbicides and/or agrochemical pesticides.

Keywords: elicitation; gamma irradiation; organic agriculture; secondary metabolites; yeast

Citation: Sayed, T.E.; Ahmed, E.-S.S. Elicitation Promoability with Gamma Irradiation, Chitosan and Yeast to Perform Sustainable and Inclusive Development for Marjoram under Organic Agriculture. *Sustainability* **2022**, *14*, 9608. https://doi.org/10.3390/su14159608

Academic Editors: Hosam M. Saleh and Mohammad Mahmoud Dawoud

Received: 3 June 2022
Accepted: 1 August 2022
Published: 4 August 2022

1. Introduction

Sustainability is currently a global issue that has provoked significant challenges for modern economic technologies and researchers, especially in using plants to solve major environmental problems worldwide [1,2]. With great application of nuclear technologies, the risk for humans and the environment has been considered of significant importance. In concert with sustainability and nuclear applications, green and dry plants [3,4] have an important role in the remediation of wastewater contaminated with stable and radioisotopes then followed by sustainable stabilization in cement [5,6]. Moreover, plants could be used in treatment such as aiding the recovery of the workers at the Chernobyl accident site using a daily oral dose of G. Biloba [7]. Medicinal plants and their constituents can alter radiolabeling and biodistribution via several mechanisms [8]. On the other hand, nuclear radiation can be used to improve the efficiency and performance of medicinal plants used in modern and traditional medicinal applications [9].

Sweet marjoram (*Majorana hortensis*) is a precious herbaceous aromatic plant native to the Mediterranean region that belongs to the *Lamiaceae* family [10]. In addition, it is a medicinal plant that is widely applied in domestic medical systems. Sweet marjoram is grown in numerous Asian, North African, and European nations, including Tunisia, Egypt, India, France, Hungary, Germany, Spain, Hungary, Portugal, Poland, and France [11].

Egypt has been internationally known for the cultivation and export of quality marjoram fresh or dry tender stems and the leaves are used for condiments and spices [12].

The leaves and flowers of the plant contain delicate fragrant essential oils that are widely used in traditional medicinal including, but not limited to, lotions, perfumes, creams, and soap [13]. The use of the plant has been documented to treat many ailments such as headaches, asthma, ear problems, and others. Additionally, cramps, depression, dizziness, gastrointestinal disorders, migraine, headache, paroxysmal cough, and a diuretic are all treated with *O. majorana*. Its essential oil is used in perfumery, the pharmaceutical industry, and cosmetics.

Recent studies have shown the vital role of marjoram in the treatment of cancer and as an active anti-fungal, antioxidant and cytotoxic agent. The antioxidant properties of marjoram have been studied extensively [14]. However, it is used mainly as a spice and natural preservative for foods, especially meat.

Elicitors are described as herbal or synthetic (organic or non-biological) substances, and while they are permeable to plants at low levels, they stimulate stress responses in plants and aid in the synthesis of secondary metabolites [15]. Elicitors serve as signals, and elicitation begins with signal perception by eliciting specific receptors on the plant cell membrane, which is followed by the initiation of a signal transduction cascade, which subsequently changes the expression levels of different governance transcription genes of secondary metabolic pathways [16,17]. Elicitation is a successful and commonly used biotechnological method for inducing new secondary metabolites in plants [18].

Elicitors' innate plant resistance mechanisms are induced using biotic and abiotic factors. Elicitation can be an important strategy for producing bioactive secondary metabolites (BSMs). It has been suggested that foliar application of these compounds' (abiotic & biotic elicitors) under normal and stressful conditions improved phytochemicals and biosynthetic pathways of secondary metabolite (BSM) content in plants, including medicinal and aromatic plants [18,19].

Elicitors improve the quality and quantity of BSMs that promote health [17,19,20]. Changes in elicitor growth and development can affect morphological, physiological, and biochemical properties and improve biomass production and quality [21,22].

Attention to agriculture has become imperative over time and implementation of sustainability in this field is crucial. To move forward towards sustainability involves maintaining modern technology and rationing the use of pesticides of various kinds to increase production. These pesticides have been selectively used on medicinal and aromatic plants as well as cereals and horticulture. Pesticides have a negative impact on the environment and hinder long-term growth [23], and several studies have linked the increased use of these pesticides to health issues [24,25]. Organic farming has grown in popularity in recent years, as has pharmaceutical production of organic medicinal and aromatic plants as an alternative to regular agriculture [26]. Considering the above facts according to recent trends and future prospects of various strategies to direct increased productivity of biomass and bioactive secondary metabolites in medicinal and aromatic plants are highlighted. It is notwithstanding works dealing with these aspects are scarce. Hence, the present article highlights elicitation coupled with organic fertilizer mediating biomass and bioactive secondary metabolites production and quality to promote the achievement of sustainable development for marjoram under organic agriculture.

2. Materials and Methods

2.1. Practical Field Experiment

Two consecutive seasons of field experiment trials, 2019 and 2020, were constructed as split plot design with three replications based on randomized full block design.

The primary plot had three elicitors: 15 G gamma irradiation (GI), 500 ppm chitosan (CH), and 0.5 g/L (YS) solution in water with 0.1 percent (*v*/*v*) Tween 20 as a surfactant. The sub-main plot contained three fertilizers, two organic fertilizers, Moringa dry leaves water- extract 20 g/m^2 (MO), fulvic acid 20 g/m^2 (FA), and traditional chemical fertilizer NPK, 20 g/m^2.

Gamma irradiated and non-irradiated marjoram seeds on 1st February for both seasons were planted in trays containing soil, sand, beat mixed (1:1:1) ratio (*v*/*v*) subsequently established in a greenhouse. Seedlings 4 weeks of age were transplanted to the field in plots 4 × 4 m in row 50 cm interspacing that contain 64 plants/plot. At such seasons, plants' leaves were sprayed with water, (Zero elicitor), (NA)CH, and YS a non-elicitor or solution two times 1st April, 1st May before 1st harvesting on 1st June. They were also sprayed on 1st July and 1st August before harvesting on 1st September.

2.2. *Biomass Yield Production*

The fresh weight of leaves was determined pre-plot and per m^2 for the 2019 and 2020 seasons.

2.3. *Bioactive Secondary Metabolites Production (BSMs)*

2.3.1. Phenolics (TPC)

The Folin–Ciocalteu reagent was used to determine total phenol content, as described in [27]. A diluted sample extract (1 mg/mL) aliquot was mixed with 125 L of Folin-Clocaiteu reagent and 500 L of water. The mixture was stirred and left for 5 min before adding 125 g of 7% of Na_2CO_3. A 1 mL quantity of distilled water was added and well mixed. After 90 min in the dark, the absorbance at 760 nm was compared to a blank. Using a gallic acid solution, a calibration curve was constructed. Equations drawn from the standard gallic acid diagram are used to create content.

2.3.2. Flavonoids (TFC)

Flavonoid content was determined according to [28]. Sechium edule (jacq.) antioxidant activity of Swartz extract. Food Chemistry, 97:452:458. Each extract was dissolved in 1 mL of ethanol. One mL of the extract was combined with 1 mL of a 2% methanolic $AlCl_3$ solution. After 15 min of incubation at room temperature, the absorbance at 430 nm was measured, and the total flavonoids in the sample extract were quantified using a calibration curve (quercetin).

2.3.3. Flavonols (TFL)

The total flavonoids in *O. majorana* ethanolic extracts were calculated using the method described by [29]. Two mL $AlCl_3$ (2%) ethanol and 3 mL of sodium acetate solution (50 g/L) were mixed together. The blend was mixed before being incubated at 20 °C for 2 h. At 440 nm, the absorbance was measured. The amount of total flavonoids was measured in mg of quercetin equivalents per gram of dry weight (mg EQ/g DW).

2.3.4. Tannins (TAE)

Total tannins were calculated using a procedure devised by [30] based on the protein bovine serum albumin precipitation. The approach works by measuring the absorption of colored compound Fe^{2+} phenols at 510 nm using spectrophotometry. The results were calculated in milligrams of tannic acid equivalent per gram of dry weight (mg TA/g DW).

2.3.5. Essential Oils (EO)

Fresh leaf samples from the plots were subjected to hydrodistillation for three hours to obtain content. The extracted essential oil was concentrated using sodium sulphate (anhydrous Na_2SO_4) and reserved in an amber glass sealed with Teflon kept at 4 °C even analysis as the following:

1-EO% = Extracted EO, g/ground leaves sample X100-moisture

2-EOY, Kg/m^2 = EO% X FLY, Kg/m^2 EO

Essential oil contents (EOC), analyzed by GC/MS using Shimadzu-HPLC System (Shim-adze Co., Kyoto, Japan) prepared with a CBM–20 AH controller, and LC20 AP Pump and an SPD–M_2OA Photo Diode Array (FDA) detector.

3. Statistical Analysis

Using a statistical analysis method, the data sets were first checked for normality using the Anderson and Darling normality tests (SAS, 2003). There are no statistically significant differences in the data from the two seasons. As a result, the pooled mean values of two seasons for all attributes examined were statistically analyzed. LSD at the 1% level was used to compare the significant means.

4. Results and Discussion

Statistical analysis revealed that solitary application with elicitors and fertilizers achieved a significant positive impact for such lasted trait as, chitosan (CH) > yeast (YS) > γ- irradiation (GI) under Moringa (MO) > fulvic acid (FA) > NPK for both two seasons However, the drenches between the two seasons were insignificant.

4.1. Biomass Yield Production

Pooled fresh leaves yield (FLY), 6.060 KG/m^2 as control (NE.NPK), was increased significantly, expressed as % of control (0) as the flowing:

CH.MO, CH.FA, and CH.NPK were 27, 22, and 19 respectively. In addition, YS.MO, YS.FA, YS.NPK were 24, 21, and 17 respectively and GI.MO, GI.FA, GI.NPK were 20, 17, and 11 respectively. Additionally, NE.MO, NE.FA, NE.NPK were 11, 8, and 0 (control 6.060 Kg/m^2) respectively as listed in Table 1 and represented in Figure 1.

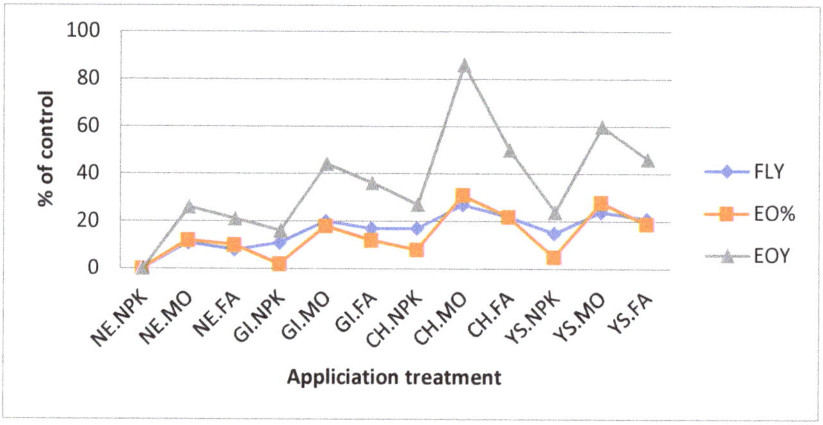

Figure 1. FLY, EO%, EOY.

4.2. Essential oils % (EO%)

PM EO %, 0.532% (control NE Significant, as % of control according to the following: CH.MO, CH.FA, CH.NPK were 3122, and 8, respectively. As well as, YS.MO, YS.FA, YS.NPK were 28, 19, and 5 respectively, and GI.MO, GI.FA, GI.NPK were 18, 12, and 2 respectively. In addition, NE.MO, NE.FA, and NE.NPK were 12, 10, and 0 (control 0.532 Kg/m^2) respectively as represented in Table 1 and Figure 1.

Table 1. Marjoram; fresh leaves yield, Kg/m^2, essential oil % and essential yield oil, Kg/m^2, in response to elicitation with gamma irradiation (GI), chitosan (CH) and yeast (YS) under the use of chemical fertilizer (NPK) and organic fertilizer; Moringa (MO) and fulvic acid (FA) over 2 subsequent seasons (2019 and 2020).

Application Treatment	Fresh Leaves Yield, Kg/m^2			Essential Oil %			Essential Oil Yield, Kg/m^2		
	2019	2020	PM	2019	2020	PM	2019	2020	PM
NE.NPK	5.920	6.200	6.060(0)	0.522	0.524	0.532(0)	3.090	3.294	3.192(0)
NE.MO	6.630	6.820	6.725(11)	0.600	0.592	0.596(12)	3.979	4.038	4.009(26)
NE.FA	6.453	6.696	6.574(8)	0.590	0.582	0.586(10)	3.807	3.897	3.852(21)
GI.NPK	6.690	6.882	6.786(11)	0.548	0.540	0.544(2)	3.666	3.716	3.691(16)
GI.MO	7.163	7.440	7.301(20)	0.632	0.629	0.630(18)	4.527	4.680	4.604(44)
GI.FA	7.045	7.254	7.149(17)	0.616	0.603	0.609(12)	4.340	4.374	4.357(36)
CH.NPK	6.926	7. 228	7.228(19)	0.580	0.571	0.575(8)	4.017	4.071	4.044(27)
CH.MO	7.637	7.750	7.693(27)	0.705	0.692	0.699(31)	5.385	5.363	5.374(68)
CH.FA	7.282	7.502	7.392(22)	0.653	0.645	0.649(22)	4.744	4.839	4.792(50)
YS.NPK	6.808	7.316	7.062(17)	0.564	0.556	0.560(5)	3.840	4.068	3.954(24)
YS.MO	7.518	7.564	7.541(24)	0.679	0.676	0.678(28)	5.105	5.113	5.109(60)
YS.FA	7.400	7.378	7.389(21)	0.642	0. 624	0.633(19)	4.751	4.604	4.678(46)
LSD1%	0.035	0.031	0.027	0.003	0.005	0.004	0.022	0.025	0.026

Values between parentheses ware % over control. PM: pooled mean for two seasons (2019–2020).

4.3. Essential Oils Yield (EOY, Kg/m^2)

EOY, 3.192 Kg/m^2 (control) increased significantly as % of control as the following: CH.MO, CH.FA, CH.NPK, 68, 50, 27, respectively, and YS.MO, YS.FA, and YS.NPK, 60, 46 and 24, respectively.

Furthermore, GI.MO, GI.FA and GI.NPK were 4436,and 16, respectively. In addition, NE.MO, NE.FA, and NE.NPK were 26, 21, and 0 (control 3.192 Kg/m^2), respectively as represented in Table 1 and Figure 1. Biomass, FLY, Kg/m^2, EO%, EOC and Kg/m^2 were significantly increased in response to elicitor, CH > YS > GI integrated with MO > FA > NPK, despite the fact that no pest infestation and microbial disease incidences in the field experiment were detected in either season without using any agrochemical pesticides and/or micro-biocides. The disappearance of any pests and microbial diseases was in agreement with [31,32].) In addition, many researchers supported our results and declared a positive, significant impact on biomass yield production [33,34].

4.4. Quali–Quantitative Bioactive Secondary Metabolites (BSMs)

4.4.1. Essential Oil Contents (EOC)

EOC, 11 tirpenoids; &-terpnene, cineol, y-Terpinene, p-cymene, Terpineolene, D-linnalol, Terpineol-4ol, Bcaryphellene, &-Terpineol, Tymol, carvacol (Table 2) were measured. The total percentages of these 11 terpenoids were increased significantly as the following:

Values of CH.MO, CH.FA, and CH.NPK were 79.81, 76.48, and 69.66, respectively. Furthermore, YS.MO, YS.FA, and YS.NPK were 70.94, 73.80, and 71.32 respectively, and GI.MO, GI.FA, and GI.NPK were 71.60, 71.04, and 69.58 respectively. In addition, NE.MO, NE.FA, and NE.NPK were 69.03, 68.75, and 67.43 (control).

Table 2. Marjoram essential oil contents; in response elicitation with gamma irradiation (GI), chitosan (CH) and yeast (YS) under the use of chemical fertilizer (NPK) and bioorganic fertilizer; Moringa (MO)and fulvic acid (FA) in 2 subsequent seasons (2019 and 2020).

Application Treatmen	Essential Oil Content from Terpinoids											
	∞-Ter-Pinene	Cineol	Y-Ter Pinene	P–Cy–Mene	Terp-Ineolene	D-Lin-Alool	Terpin-Eol 4ol	Bcar-Yophe Llene	∞-Ter-Pineol	Tymol	Carvacol	Total%
NE.NPK	7.15	1.44	12.51	2.62	2.24	1.23	23.41	1.92	3.64	9.95	9.32	67.43
NE.MO	7.75	1.62	12.81	2.75	2.62	1.71	23.99	1.97	3.92	9.89	1.52	69.03
NE.FA	7.53	1.48	12.60	2.68	2.42	1.53	23.66	1.95	3.71	9.75	1.44	68.75
GI.NPK	7.75	1.10	12.79	2.72	2.64	1.52	23.89	1.85	3.75	9.87	1.70	69.58
GI.MO	7.92	1.18	12.91	2.82	2.86	1.76	23.93	1.92	3.89	9.91	1.87	71.60
GI.FA	7.89	1.77	12.99	2.83	2.71	1.60	23.95	1.90	3.81	9.97	1.62	71.04
CH.NPK	7.69	1.52	12.75	2.66	2.63	1.05	23.76	1.78	3.72	9.75	1.75	69.66
CH.MO	8.77	2.72	13.80	3.51	3.45	2.32	24.81	2.75	3.63	10.80	2.25	79.81
CH.FA	8.42	2.35	13.31	3.25	3.31	2.25	24.35	2.27	4.35	10.42	2.20	76.48
YS.NPK	7.80	1.72	12.82	2.71	2.60	2.55	23.85	1.88	3.81	9.86	1.72	71.32
YS.MO	8.35	2.18	13.32	3.15	3.21	3.32	24.35	2.27	4.19	10.25	2.35	70.94
YS.FA	8.25	2.11	13.25	3.12	3.18	3.27	24.11	2.15	4.12	10.17	2.23	75.86
LSD1%												0.25

4.4.2. Total Phenolic (TPC)

TPC, 7.83 mg GAE/g. DLW. (the control was increased significantly as % of control (Table 3 and Figure 2) as the following:

CH.MO, CH.FA, and CH.NPK were 171, 159, and 97, respectively, and YS.MO, YS.FA, and YS.NPK were 156, 135, and 85, respectively. In addition, GI.MO, GI.FA, GI, and NPK were 122, 109, and 44, respectively. Furthermore, NE.MO, NE.FA, and NE NPK were 63, 51, and 0 (control = 7.83 mg GAE/g DLW).

4.4.3. Total Flavonoids Content (TFC)

TFC, 0.15 mg QE/g DLW (control) was increased significantly, as % of control (Table 3 and Figure 2) as the following:

Table 3. Marjoram total phenolic (TPC), flavonoid (TFC), flavonols (TFL), tannin (TAN) in response elicitation with gamma irradiation (GI), chitosan (CH) and yeast (YS) under chemical fertilizer (NPK) and organic fertilizer; Moringa (MO)and fulvic acid (FA) at 2 subsequent seasons (2019 and 2020).

Application Treatment	TPC, mg GAE, g/DW			TFC, mg QE/DW			TFL, mg QE/gDW			TAN, mg TAE/gDW		
	2019	2020	PM	2019	2020	PM	2019	2020	PM	2019	2020	PM
NE.NPK	7.85	7.81	7.83(0)	0.16	0.14	0.15(0)	2.58	2.53	2.55(0)	1.26	1.23	1.25(0)
NE.MO	12.64	12.89	12.77(63)	0.29	0.24	0.26(73)	3.90	3.97	3.93(54)	1.70	1.69	1.70(36)
NE.FA	11.78	11.87	11.83(51)	0.22	0.20	0.21(41)	3.61	3.59	3.60(41)	1.59	1.58	1.58(26)
GI.NPK	11.15	11.32	11.24(44)	0.24	0.22	0.23(53)	4.26	4.28	4.27(67)	1.74	1.75	1.75(40)
GI.MO	17.27	17.57	17.42(122)	0.33	0.29	0.31(106)	4.82	4.83	4.82(89)	2.03	2.03	2.03(62)
GI.FA	16.25	16.48	16.37(109)	0.28	0.25	0.26(73)	4.59	4.60	4.59(80)	1.85	1.86	1.86(49)
CH.NPK	15.31	15.46	15.39(97)	0.34	0.30	0.32(113)	4.67	4.73	4.70(84)	2.03	2.07	2.05(64)
CH.MO	21.20	21.24	21.22(71)	0.43	0.38	0.41(173)	5.94	6.55	6.25(144)	2.33	2.32	2.33(86)
CH.FA	20.18	20.38	20.28(59)	0.38	0.33	0.35(133)	5.03	5.01	5.02(97)	2.16	2.09	2.12(69)
YS.NPK	14.37	14.60	14.49(85)	0.29	0.26	0.28(87)	4.44	4.43	4.44(74)	1.92	1.89	1.91(53)
YS.MO	19.70	20.15	19.93(125)	0.35	0.32	0.34(127)	5.39	5.87	5.63(121)	2.17	2.14	2.16(73)
YS.FA	18.13	18.67	18.40(135)	0.32	0.29	0.30(100)	4.02	4.70	4.00(00)	2.03	1.99	2.01(61)
LSD1%	0.07	0.08	-	0.04	0.03	-	0.05	0.06	-	0.06	0.07	-

PM: pooled mean for two Seasons (2019–2020) parentheses were % of control (NE > NPK) Values between parentheses are % of control.

Values of CH.MO, CH.FA, CH.NPK were 173, 133, and 113, respectively, and YS.MO, YS.FA, and YS.NPK were 127, 100, and 87 respectively. In addition, values of GI.MO, GI.FA, GI, NPK were 106, 73, and 53 respectively, and NE.MO, NE.FA, NE NPK were 73, 41, and 0 (control = 0.15 mg QE/g DLW).

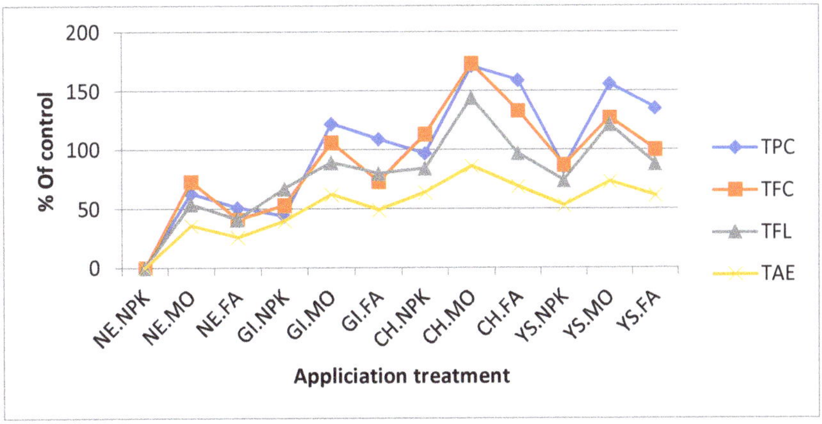

Figure 2. TPC, TFC, TFL, TAE, respectively.

4.4.4. Total Flavonols (TFL)

TFL, 2.55 mg QE/g. DLW, increased significantly, as % of control (Table 3 and Figure 2) as the following: values of CH.MO, CH.FA, and CH.NPK, 144, 97, and 84, respectively. Furthermore, YS.MO, YS.FA, YS. and NPK, 121, 88, and 74, respectively. In addition, values of GI.MO, GI.FA, GI, and NPK were 89, 80, and 67, respectively, and NE.MO, NE.FA, and NE NPK were 54, 41, and 0 (control = 2.55 mg QE/g DLW).

4.4.5. Tannin (TAN):

TAN;1.25 mgTAE/g. DLW, was increased significantly, as % of control, Table 3 and Figure 2 summarize the result, and show the following: CH.MO, CH.FA, and CH.NPK were 86, 69, and 64, respectively. In addition, values of YS.MO, YS.FA, and YS.NPK were 73, 61, and 53, respectively.

In addition, values of GI.MO, GI.FA, GI.NPK were 62, 49, and 40, respectively, and NE.MO, NE.FA, and NE.NPK were 36, 26, and 0 (control = 1.55 mg TAE/g DLW)

Elicitor application resulted in significant positive impacts for marjoram BSM production and quality (EO, TPC, TFC, TFL and TAN) as CH >YS > GI integrated with MO > FA > NPK.

Several researchers have supported our results [17,33] under agrochemical agriculture which impacts the environment preventing sustainable development [23].

Furthermore, under bio-fertilizer bio or organic fertilizers, organic, BSMs were increased [35,36].

The overall results manifested strong evidence that CH > YS > GI integrated with MO > FA > NPK could be considered as a reliable technological strategy to improve biomass and BSM (EO, TPC, TFC, TFL, TAN) production and quality of marjoram plants. The disappearance of any microbial diseases or pest infestation from the plant without using any pesticides such as micro-biocides is attributed to the following: (1) enhancing tolerance of plants to biotic and abiotic stresses [37] and leading to strong resistance to pests, microbial agents and nematodes [34]; (2) induced systemic resistance (ISR) prior to infection by regulating systemic resistance (ISR) prior to infection by regulating the expression genes involved in the production and accumulation of bioactive secondary metabolites (phytoalexins) which specific toxins characterized a broad spectrum of biomachrobiocide and biopesticides, making them less susceptible against microbial diseases and pest infecting [38]. Furthermore, (3) chitosan (CH); exhibited strong antifungal and antibacterial [39] nematocidal, virucidal [40,41] and biopesticide attributes [42]. Moringa exhibited anti-fungal traits [43], and improved resistance to pests and microbial diseases [44,45].

5. Conclusions

The overall results show strong evidence for the potent of biotic elicitors chitosan > yeast > abiotic, gamma irradiation coupled with organic fertilizers moringa > fulvic acid > traditional agrochemical fertilizer NPK, as reliable CO-friendly solutions, significantly improving marjoram biomass, secondary metabolites production and quality without using agrochemical pesticides and/or microbicides. In addition, chitosan > yeast > gamma irradiation coupled with organic fertilizers exceeded their integration with chemical NPK fertilizer. This highlights the sustainable and reclusive development for marjoram under organic agriculture and could be an alternative to conventional agrochemical agriculture. This section is not mandatory but can be added to the manuscript if the discussion is unusually long or complex.

Author Contributions: All authors have equal contributions. All authors have read and agreed to the published version of the manuscript.

Funding: This research received no external funding.

Institutional Review Board Statement: Not applicable.

Informed Consent Statement: Not applicable.

Data Availability Statement: Not applicable.

Conflicts of Interest: The authors declare no conflict of interest.

References

1. Saleh, H.M.; Mahmoud, H.H.; Abdou, M.I.; Eskander, S.B. Health Risk Assessment Based on Metal Analysis of Soil and Crops in Al-Dakhla Oasis. *Arab. J. Geosci.* **2021**, *14*, 260. [CrossRef]
2. Saleh, H.M.; Eskander, S.B.; Mahmoud, H.H.; Abdou, M.I. Groundwater Quality and Health Assessments Based on Heavy Metals and Trace Elementscontent in DakhlaOasis, NewValley Governorate, Egypt. *Water Sci.* **2022**, *36*, 1–12. [CrossRef]
3. Saleh, H.M.; Mahmoud, H.H.; Aglan, R.F.; Bayoumi, T.A. Biological Treatment of Wastewater Contaminated with Cu(II), Fe(II) and Mn(II) Using Ludwigia Stolonifera Aquatic Plant. *Environ. Eng. Manag. J.* **2019**, *18*, 1327–1336. [CrossRef]
4. Saleh, H.M.; Moussa, H.R.; El-Saied, F.A.; Dawoud, M.; Nouh, E.S.A.; Abdel Wahed, R.S. Adsorption of Cesium and Cobalt onto Dried Myriophyllum Spicatum L. from Radio-Contaminated Water: Experimental and Theoretical Study. *Prog. Nucl. Energy* **2020**, *125*, 103393. [CrossRef]
5. Saleh, H.M.; Aglan, R.F.; Mahmoud, H.H. Qualification of Corroborated Real Phytoremediated Radioactive Wastes under Leaching and Other Weathering Parameters. *Prog. Nucl. Energy* **2020**, *119*, 103178. [CrossRef]
6. Saleh, H.M.; Moussa, H.R.; El-Saied, F.A.; Dawod, M.; Bayoumi, T.A.; Abdel Wahed, R.S. Mechanical and Physicochemical Evaluation of Solidifed Dried Submerged Plants Subjected to Extreme Climatic Conditions to Achieve an Optimum Waste Containment. *Prog. Nucl. Energy* **2020**, *122*, 103285. [CrossRef]
7. Jagetia, G.C. Radioprotective Potential of Plants and Herbs against the Effects of Ionizing Radiation. *J. Clin. Biochem. Nutr.* **2007**, *40*, 74–81. [CrossRef]
8. Khosravian, P.; Heidari-Soureshjani, S.; Yang, Q. Effects of Medicinal Plants on Radiolabeling and Biodistribution of Diagnostic Radiopharmaceuticals: A Systematic Review. *Plant Sci. Today* **2019**, *6*, 123–131. [CrossRef]
9. Shala, A.Y. Effect of Different Doses of Gamma Irradiation on Vegetative Growth and Oil Yield of *Ocimum basilicum* L. *J. Plant Prod.* **2019**, *10*, 1–6. [CrossRef]
10. Uritu, C.M.; Mihai, C.T.; Stanciu, G.-D.; Dodi, G.; Alexa-Stratulat, T.; Luca, A.; Leon-Constantin, M.-M.; Stefanescu, R.; Bild, V.; Melnic, S. Medicinal Plants of the Family Lamiaceae in Pain Therapy: A Review. *Pain Res. Manag.* **2018**, *2018*, 7801543. [CrossRef]
11. Shylaja, M.R.; Peter, K.V. The Functional Role of Herbal Spices. *Handb. Herbs Spices* **2004**, *2*, 26–45.
12. Peter, K.V.; Shylaja, M.R. Introduction to Herbs and Spices: Definitions, Trade and Applications. In *Handbook of Herbs and Spices*; Elsevier: Amsterdam, The Netherlands, 2012; pp. 1–24.
13. Khan, I.A.; Abourashed, E.A. *Leung's Encyclopedia of Common Natural Ingredients: Used in Food, Drugs and Cosmetics*; John Wiley & Sons: Hoboken, NJ, USA, 2011; ISBN 1118213068.
14. Hossain, M.B.; Brunton, N.P.; Barry-Ryan, C.; Martin-Diana, A.B.; Wilkinson, M. Antioxidant Activity of Spice Extracts and Phenolics in Comparison to Synthetic Antioxidants. *Rasayan J. Chem.* **2008**, *1*, 751–756.

15. Namlı, S.; Işıkalan, Ç.; Akbaş, F.; Toker, Z.; Tilkat, E.A. Effects of UV-B Radiation on Total Phenolic, Flavonoid and Hypericin Contents in Hypericum Retusum Aucher Grown under in Vitro Conditions. *Nat. Prod. Res.* **2014**, *28*, 2286–2292. [CrossRef] [PubMed]
16. Mohammadparast, B.; Rasouli, M.; Rustaiee, A.R.; Zardari, S.; Agrawal, V. Quantification of Asiatic Acid from Plant Parts of Centella Asiatica L. and Enhancement of Its Synthesis through Organic Elicitors in in Vitro. *Hortic. Environ. Biotechnol.* **2014**, *55*, 578–582. [CrossRef]
17. Mejdoub-Trabelsi, B.; Touihri, S.; Ammar, N.; Riahi, A.; Daami-Remadi, M. Effect of Chitosan for the Control of Potato Diseases Caused by Fusarium Species. *J. Phytopathol.* **2020**, *168*, 18–27. [CrossRef]
18. Andrys, D.; Adaszyńska-Skwirzyńska, M.; Kulpa, D. Jasmonic Acid Changes the Composition of Essential Oil Isolated from Narrow-Leaved Lavender Propagated in in Vitro Cultures. *Nat. Prod. Res.* **2018**, *32*, 834–839. [CrossRef]
19. Ahamed, T.E.S.; Ahamed, E.S.S. Synergy Prospect Low Gamma Irradiation Doses Incorporating Elicitation with Iron Nanoparticles to Hyper Production Biomass Yield and Bioactive Secondary Metabolites for Cress, Medicinal Plant. *J. Plant Sci.* **2018**, *6*, 157–163.
20. Ahamed, T.E.S. Bioprospecting Elicitation with Gamma Irradiation Combine with Chitosan to Enhance, Yield Production, Bioactive Secondary Metabolites and Antioxidant Activity for Saffron. *J. Plant Sci.* **2019**, *7*, 137–143.
21. Sheshbahreh, M.J.; Dehnavi, M.M.; Salehi, A.; Bahreininejad, B. Effect of Irrigation Regimes and Nitrogen Sources on Biomass Production, Water and Nitrogen Use Efficiency and Nutrients Uptake in Coneflower (*Echinacea purpurea* L.). *Agric. Water Manag.* **2019**, *213*, 358–367. [CrossRef]
22. Yadav, K.K.; Sarkar, S. Biofertilizers, Impact on Soil Fertility and Crop Productivity under Sustainable Agriculture. *Environ. Ecol.* **2019**, *37*, 89–93.
23. Bombardi, L.M. *Geografia Do Uso de Agrotóxicos No Brasil e Conexões Com a União Europeia*; FFLCH- USP: São Paulo, Brazil, 2017; ISBN 8575063103.
24. Pignati, W.A.; de Souza e Lima, F.A.N.; de Lara, S.S.; Correa, M.L.M.; Barbosa, J.R.; da Costa Leão, L.H.; Pignatti, M.G. Distribuição Espacial Do Uso de Agrotóxicos No Brasil: Uma Ferramenta Para a Vigilância Em Saúde. *Cienc. Saúde Colet.* **2017**, *22*, 3281–3293. [CrossRef] [PubMed]
25. de Abreu, P.H.B.; Alonzo, H.G.A. Rural Work and Health Risks: A Review into de"Safe Use" of Pesticides in Brazil. *Cienc. Saúde Colet.* **2014**, *19*, 4197.
26. Pereira, M.M.A.; Morais, L.C.; Marques, E.A.; Martins, A.D.; Cavalcanti, V.P.; Rodrigues, F.A.; Gonçalves, W.M.; Blank, A.F.; Pasqual, M.; Dória, J. Humic Substances and Efficient Microorganisms: Elicitation of Medicinal Plants—A Review. *J. Agric. Sci.* **2019**, *11*, 268–280. [CrossRef]
27. Arabshahi-Delouee, S.; Urooj, A. Antioxidant Properties of Various Solvent Extracts of Mulberry (*Morus indica* L.) Leaves. *Food Chem.* **2007**, *102*, 1233–1240. [CrossRef]
28. Ordonez, A.A.L.; Gomez, J.D.; Vattuone, M.A. Antioxidant Activities of Sechium Edule (Jacq.) Swartz Extracts. *Food Chem.* **2006**, *97*, 452–458. [CrossRef]
29. Adedapo, A.A.; Jimoh, F.O.; Koduru, S.; Afolayan, A.J.; Masika, P.J. Antibacterial and Antioxidant Properties of the Methanol Extracts of the Leaves and Stems of Calpurnia Aurea. *BMC Complement. Altern. Med.* **2008**, *8*, 53. [CrossRef] [PubMed]
30. Hagerman, A.E.; Butler, L.G. Protein Precipitation Method for the Quantitative Determination of Tannins. *J. Agric. Food Chem.* **1978**, *26*, 809–812. [CrossRef]
31. Towler, M.J.; Weathers, P.J. Variations in Key Artemisinic and Other Metabolites throughout Plant Development in Artemisia Annua L. for Potential Therapeutic Use. *Ind. Crops Prod.* **2015**, *67*, 185–191. [CrossRef]
32. Boghozian, A.; Amjad, L.; Shahanipour, K. Chemical Constituents and Identification of the Essential Oil of Artemisia Aucheri Boiss. in Iran. *Adv. Environ. Biol.* **2014**, *8*, 2339–2344.
33. Adekiya, A.O.; Agbede, T.M.; Aboyeji, C.M.; Dunsin, O.; Ugbe, J.O. Green Manures and NPK Fertilizer Effects on Soil Properties, Growth, Yield, Mineral and Vitamin C Composition of Okra (*Abelmoschus esculentus* (L.) Moench). *J. Saudi Soc. Agric. Sci.* **2019**, *18*, 218–223. [CrossRef]
34. Kaleri, A.A.; Baloch, A.W.; Baloch, M.; Wahocho, N.A.; Abro, T.F.; Jogi, Q.; Soomro, A.A.; Marri, A.; Bhutto, L.A. 30. Heritability and Correlation Analysis in Bt and Non-Bt Cotton (*Gossypium hirsutum* L.) Genotypes. *Pure Appl. Biol.* **2021**, *5*, 906–912.
35. El-Mohamedy, R.S.R.; Mohamed, S.K. Effect of Moringa Oleifera Seed Oil, Root and Leave Extracts on Growth of Major Pathogenic Fungi of Tomato, Green Bean and Potato in Vitro. *Int. J. Agric. Technol.* **2018**, *14*, 505–520.
36. Tarek Elsayed, S.A.; El Sayed, S.A. The Potentiality Biotic-Elicitation with Chitosan or Vitamin C to Achieve Integrated and Sustainable Development for Sage Salvia Officiealis Under Sustainable Agriculture Systems. *J. Plant Sci.* **2021**, *9*, 151–162.
37. Valletta, A.; De Angelis, G.; Badiali, C.; Brasili, E.; Miccheli, A.; Di Cocco, M.E.; Pasqua, G. Acetic Acid Acts as an Elicitor Exerting a Chitosan-like Effect on Xanthone Biosynthesis in *Hypericum perforatum* L. Root Cultures. *Plant Cell Rep.* **2016**, *35*, 1009–1020. [CrossRef]
38. Gabaston, J.; El Khawand, T.; Waffo-Teguo, P.; Decendit, A.; Richard, T.; Mérillon, J.-M.; Pavela, R. Stilbenes from Grapevine Root: A Promising Natural Insecticide against Leptinotarsa Decemlineata. *J. Pest Sci.* **2018**, *91*, 897–906. [CrossRef]
39. Sathiyabama, M.; Akila, G.; Charles, R.E. Chitosan-Induced Defence Responses in Tomato Plants against Early Blight Disease Caused by Alternaria Solani (Ellis and Martin) Sorauer. *Arch. Phytopathol. Plant Prot.* **2014**, *47*, 1963–1973. [CrossRef]
40. Bondok, A.M. Response of Tomato Plants to Salicyli c Acid and Chitosan under Infection with Tomato Mosaic Virus. *Am. J. Agric. Environ. Sci.* **2015**, *15*, 1520–1529. [CrossRef]

41. Firmansyah, D.; Widodo; Hidayat, S.H. Chitosan and Plant Growth Promoting Rhizobacteria Application to Control Squash Mosaic Virus on Cucumber Plants. *Asian J. Plant Pathol.* **2017**, *11*, 148–155. [CrossRef]
42. Li, Y.; Qin, Y.; Liu, S.; Xing, R.; Yu, H.; Li, K.; Li, P. Preparation, Characterization, and Insecticidal Activity of Avermectin-Grafted-Carboxymethyl Chitosan. *BioMed Res. Int.* **2016**, *2016*, 9805675. [CrossRef]
43. Zaker, M. Natural Plant Products as Eco-Friendly Fungicides for Plant Diseases Control—A Review. *Agriculturists* **2016**, *14*, 134–141. [CrossRef]
44. Phiri, C. Influence of Moringa Oleifera Leaf Extracts on Germination and Early Seedling Development of Major Cereals. *Agric. Biol. J. N. Am.* **2010**, *1*, 774–777. [CrossRef]
45. Ho, T.-T.; Lee, J.-D.; Jeong, C.-S.; Paek, K.-Y.; Park, S.-Y. Improvement of Biosynthesis and Accumulation of Bioactive Compounds by Elicitation in Adventitious Root Cultures of Polygonum Multiflorum. *Appl. Microbiol. Biotechnol.* **2018**, *102*, 199–209. [CrossRef] [PubMed]

sustainability MDPI

Article

Rhyolite as a Naturally Sustainable Thermoluminescence Material for Dose Assessment Applications

Elsayed Salama [1,*], Dalal A. Aloraini [2], Sara A. El-Khateeb [3] and Mohamed Moustafa [4]

1 Basic Science Department, Faculty of Engineering, The British University in Egypt (BUE), El-Shorouk City 11837, Egypt
2 Physics Department, College of Science, Princess Nourah bint Abdulrahman University (PNU), Riyadh 11671, Saudi Arabia; daalorainy@pnu.edu.sa
3 Nuclear Medicine Technology Department, Inaya Medical Colleges, Riyadh 11352, Saudi Arabia; smohamed@inaya.edu.sa
4 Department of Physics, School of Science and Engineering, The American University in Cairo (AUC), New Cairo 11835, Egypt; mohamed.orabi@aucegypt.edu
* Correspondence: elsayed.salama@bue.edu.eg

Abstract: Thermoluminescence characteristics of natural rhyolite have been studied. Dose response at a wide dose range of 0.5–2000 Gy has been determined. Minimum detectable dose and thermal fading rate are evaluated. Glow curve deconvolution is conducted after determining the best read-out conditions. The repeated initial rise (RIR) method is used to detect the overlapping peaks, and a glow curve deconvolution procedure is used to extract the thermoluminescence parameters of rhyolite. According to the findings, rhyolite glow curves show five interfering peaks corresponding to five electron trap levels at 142, 176, 221, 298, and 355 °C, respectively, at a heating rate of 3 °C/s. The obtained kinetic order for the deconvoluted peaks showed mixed-order kinetic. The reported results might be useful to introduce rhyolite as a natural sustainable material for radiation dosimetry applications.

Keywords: rhyolite; sustainable natural material; thermoluminescence; kinetic parameters

Citation: Salama, E.; Aloraini, D.A.; El-Khateeb, S.A.; Moustafa, M. Rhyolite as a Naturally Sustainable Thermoluminescence Material for Dose Assessment Applications. *Sustainability* **2022**, *14*, 6918. https://doi.org/10.3390/su14116918

Academic Editors: Hosam M. Saleh and Mohammad Mahmoud Dawoud

Received: 6 May 2022
Accepted: 2 June 2022
Published: 6 June 2022

Publisher's Note: MDPI stays neutral with regard to jurisdictional claims in published maps and institutional affiliations.

1. Introduction

Thermoluminescence dosimetry (TLD) is a proper method for assessing ionizing radiation dosage. TLD materials come in a broad range of physical shapes, allowing for evaluating various radiation characteristics at dosage levels ranging from micro to kilo Gray. The tiny physical size of TL dosimeters and the point that no cables or additional equipment are required during dose measurement are significant benefits. As a result, they are well-suited to many dosimetric applications [1–3].

Sustainable materials are those that can be obtained in enough amounts without exhausting the non-renewable resources or altering the environment natural resources.

These materials can range from bio-based polymers derived from polysaccharides to highly recyclable materials, such as glass, which can be reused indefinitely without the need for additional mineral resources. Recyclable materials, such as waste composites, are used in many industrial applications, such as in stabilization and solidification of radioactive waste [4]. Reuse of cement kiln dust and polystyrene waste could contribute to environmental protection [5]. Cement can be used as a shielding material with modification by using various additives, such as cellulosic waste [6,7], bitumen [8,9], glass [10], polymers [4,11,12], nanomaterials [13,14], and cement wastes [5,15].

Natural minerals are used for a large variety of purposes in modern science and technology, which enhances human life. Quartz is a typical natural mineral with several benefits in a variety of TL dosimetry applications, such as for determining a substance's radiological history [16,17], monitoring of nuclear accidence [18], and food irradiation control [19]. Thermoluminescence characteristics of gamma irradiated sandstone showed that its linear dose-response covers the range of 1–50 Gy followed by sub-linearity at a high dose level [20]. Based on radiation induced free radicals, Egyptian limestone was used as an effective and low-cost electron paramagnetic resonance (EPR) gamma dosimeter [21]. Thermoluminescence characteristics of sedimentary natural calcite in Gebel El Galala, Egypt, revealed an average TL sensitivity of 322.356 Gy^{-1} mg^{-1} with minimum detectable dosage of 976.221 μGy and a wide range of linearity, from 1.1 to 330 Gy in response to beta particle irradiation [22]. For the sake of retrospective dosimetry, thermoluminescence properties of natural tuff were investigated in the dose range 0.5–5 Gy, where linear dose-response is obtained [23].

Rhyolite is an igneous rock that extrudes and has a high silica content. It is generally pink or grey in appearance, with tiny grains that are difficult to notice without a hand lens. Rhyolite is made up of quartz, plagioclase, and sanidine, with some hornblende and biotite thrown in [24]. Thermoluminescence emission of rhyolite was recently studied without in-depth investigation of the its full dosimetric properties and kinetic parameters involved [25].

A thorough understanding of the kinetic parameters is required for reliable research of thermoluminescent materials for usage in diverse nuclear applications.

The kinetic parameters, on the other hand, are a crucial aspect in describing the physical characteristics of TL materials. This may be done by looking at the thermoluminescence curve of the TL materials after they have been exposed to ionizing radiation. When electrons are liberated from the traps during heating, they recombine with trapped holes, causing light to be emitted. A collection of glow peaks is formed based on the amount of trapping bands in the material, which serves as a fingerprint for each substance. The kind of incident radiation and its energy, as well as the geological origin of minerals, chemical forms, kinds of impurities, and flaws, all influence the glow peaks [26].

The current study aims to investigate the capability of using rhyolite as a natural thermoluminescent dosimeter material. Thermoluminescence characterization, as well as its structural and compositional analysis, will also be investigated.

2. Material and Methods

Rhyolite rock was obtained from the Makkah area, Kingdom of Saudi Arabia, around 190 km southeast of Taif, in the mountainous region, at 20.64° N longitude and 42.12° E latitude. This rock sample was crushed into 100 μm granules and then washed with distilled water to remove organic contaminants. Magnetic separation was used to remove magnetic particles from the sample after it was dried. The powder was annealed at 400 °C in a muffle furnace for 2 h to eliminate any past radiation exposure, then quickly cooled to room temperature.

At room temperature, a Bruker S8 TIGER XRF spectrometer was used to identify the elemental composition of rhyolite samples while a Rigaku Ultima 4 XRD was used for crystallographic phase identification. An Optica FT-IR Spectrometer was used to measure Fourier transformation infra-red (FTIR) absorption spectra of prepared samples in the spectral range 400–4000 cm^{-1}. According to a qualitative XRF examination of the natural rhyolite sample, silicon was the predominant ingredient (53%), with lesser amounts of aluminum (15.8%), potassium (15.2%), calcium (2.5%), and a variety of other elements as impurities.

For the sake of shaping and forming small rigid disks, 25 mg average weight suitable for the TLD reader samples are compressed at 3 tons using a hydraulic press. The obtained small rhyolite disks of 2 mm diameter and 1 mm thickness are shown in Figure 1.

A Harshaw TL-reader model 3500 (Harshaw, WI, USA) was used for the thermoluminescence measurements. The temperature range for the recorded TL glow curves was 50–400 °C, with the heating rate of 3–25 °C/s.

The ^{60}Co irradiation source (GC220), factory-made by the Canadian Atomic Energy Authority, was utilized to gamma irradiate the samples at 0.3 Gy/s. The source is provided by the Egyptian National Center for Radiation Research and Technology.

Figure 1. Rhyolite compressed disks at 3 tons.

3. Results and Discussions

3.1. Structural Analysis (XRD and FTIR)

The X-ray diffraction pattern (XRD) for natural rhyolite is shown in Figure 2. Crystallographic phase identification and line profile analysis were performed using the Maud computer program by Luca Lutterotti [27]. Sodium aluminum silicate (42.193 ± 0 wt. %), potassium aluminum silicate (32.826 ± 1.079 wt. %) and silicon oxide (24.982 ± 0.671 wt. %) are three identified phases corresponding to three minerals: albite—low, microcline—intermediate, and quartz—low, respectively. The obtained peaks are matched to the assigned peaks regarding the forementioned three phases, which were obtained from the crystal structure database of the *American Mineralogist* [28–30].

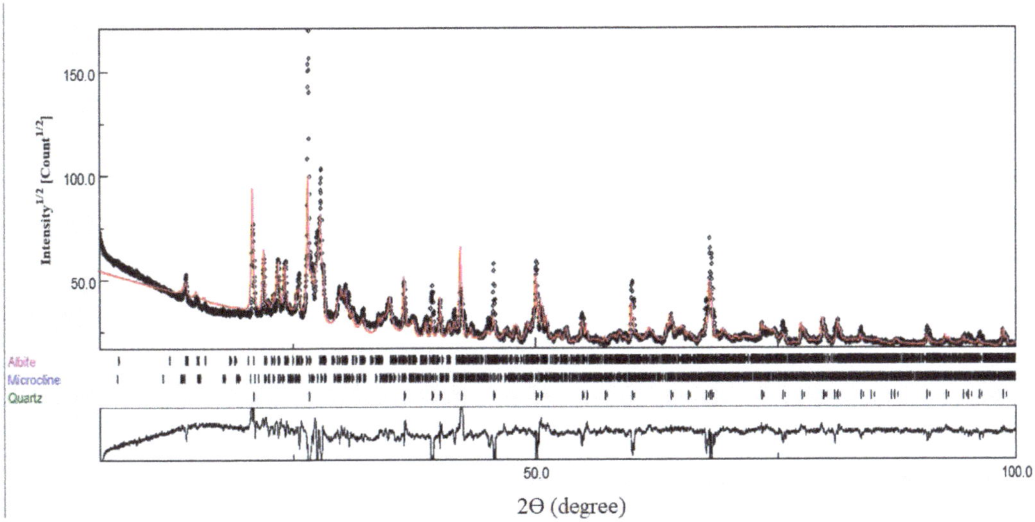

Figure 2. The *x*-ray diffraction pattern (XRD) for natural rhyolite.

The line profile fitting estimates that the sodium aluminum silicate phase has triclinic structure with crystallite size of 73 nm and the lattice parameters, a, b, c, α, β, and γ of 8.14, 12.79, 7.16 Å, 94.26, 116.62, and 87.80°, respectively. The potassium aluminum silicate phase also has triclinic structure with crystallite size of 73 nm and the lattice parameters, a, b, c, α, β, and γ to be 8.60, 13.00, 7.20 Å, 89.98, 115.93, and 89.43°, respectively. Silicon oxide phase has trigonal structure with crystallite size of 73 nm and the lattice parameters, a and b of 4.92 and 5.41 Å, respectively.

Figure 3 shows the FTIR transmission spectrum of the natural rhyolite under investigation. The observed absorption peak at 3435.91 cm^{-1} can be attributed to the OH group's stretching and bending vibrations [31,32], while the those observed at 1093 cm^{-1} and 661 cm^{-1} are attributed to the stretching and bending vibrations of the Si-O in quartz, respectively. The most intense bands recorded at around 1100 cm^{-1} and 460 cm^{-1}, are associated with the asymmetric stretching vibrations Si-O(Si) and bending vibrations O-Si-O present in silicate tetrahedra, respectively. The doublet 800–780 cm^{-1} is related to the symmetric stretching vibrations of Si-O-Si bridges [33]. This doublet is found in low-temperature quartz and is utilized as the analytical band for determining the phase's quantitative properties [34]. The Al-O-Si vibration in aluminum silicate is shown by the absorption band at 1384.16 cm^{-1} [35].

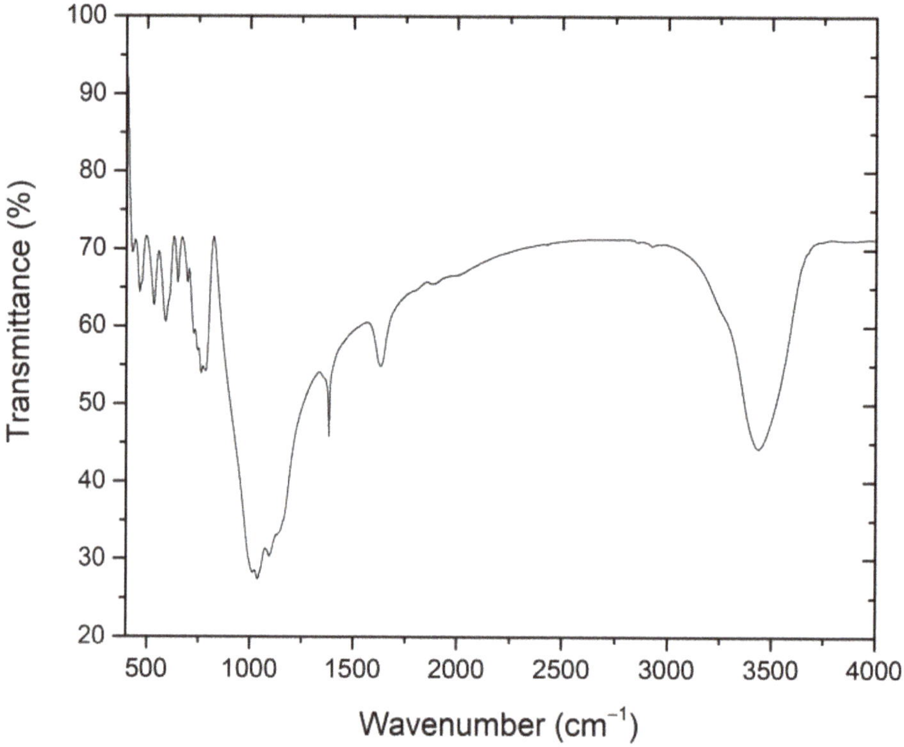

Figure 3. The FTIR transmission spectrum of the investigated natural rhyolite.

Metal cation vibrations, such as Na^+ and Al^+, are responsible for the bands detected at 588.41, 532.27, 463.94, and 428.08 cm^{-1} [36–38]. These findings are matched with those obtained with XRD study.

3.2. Glow Curves

The recorded glow curves of 50 Gy gamma irradiated natural rhyolite disks at 3 °C/s heating rate display one intensified TL peak around 135 °C, as shown in Figure 4.

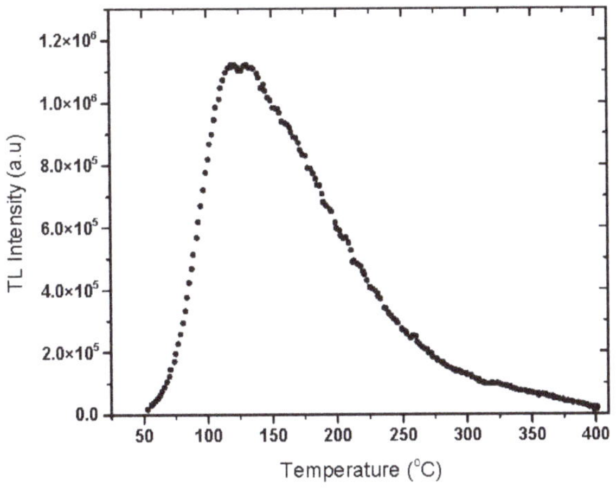

Figure 4. Thermoluminescence glow curve of 50 Gy gamma irradiated natural rhyolite disk heated at 3 °C/s heating rate.

The dosimetric properties, such as the effect of heating rate, dose-response, minimum detectable dose, reproducibility of the obtained response, and thermal fading, will be studied as follows.

3.3. The Effect of Heating Rate

The impact of various heating rates on natural rhyolite disks irradiated with 50 Gy gamma radiation is shown in Figure 5. The acquired findings verified that as the heating rate was increased, the dosimetric peak was pushed to a higher temperature.

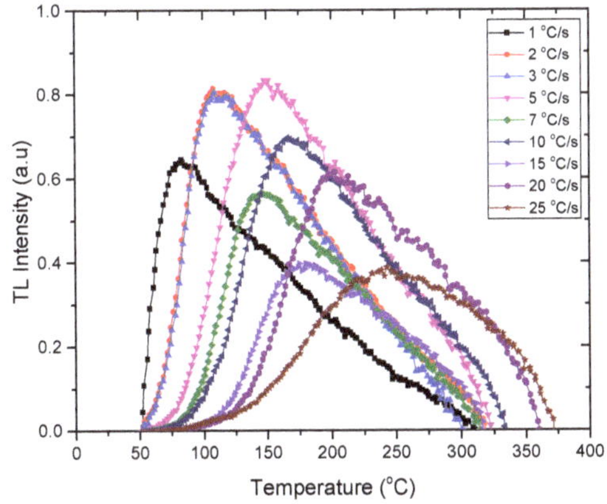

Figure 5. The effect of heating rate on the thermoluminescence glow curve of 50 Gy gamma irradiated natural rhyolite.

Due to increased interfering of sub-peaks at higher heating rates, the width of all dosimetric peaks at its half maximum increases [39]. This is due to a temperature discrepancy between the TL reading tray (planchet) and the sample's real recorded temperature at high heating rates. As a result, the peak seems to be at a greater temperature than it actually is. Slow heating rates, on the other hand, do not cause this because heat is more homogenously distributed in the sample, as a result of which there is no change in the peak temperature [40]. Figure 6 shows the results of TL response at various heating rates from 1 to 25 °C/s, where the maximum value was recorded at the 3 °C/s heating rate.

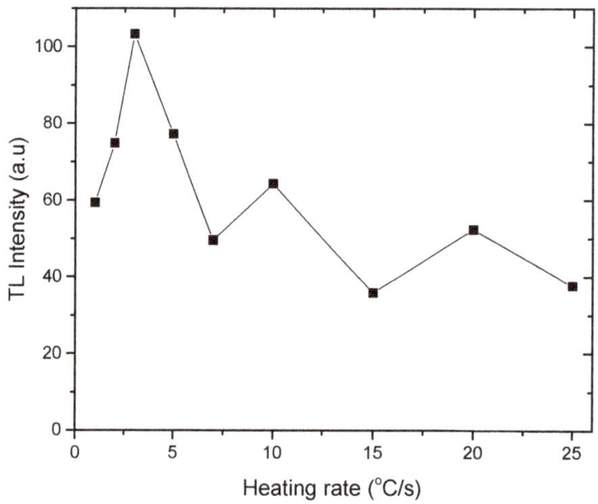

Figure 6. Gamma dose response of natural rhyolite at various heating rates.

3.4. Dose Response

As shown in Figure 7, the variation of the induced TL with gamma dose over the range 1–20 Gy displays a linear fitting (R^2 = 0.999) described by the formula I = (1.011 ± 0.026) D; followed by a supralinearity behavior up to 2000 Gy described by the formula I = (202.35 ± 20.98) $D^{1.28\pm0.02}$, where I refers to the intensity of the TL signal and D is the irradiation dose. The strong value of R^2 suggested that the studied material possesses a homogeneous delivery of deep electron traps, providing a linear dose-response across the irradiation dosage range studied. The following formula can be used to compute the linearity index f(D) at a given dosage D [41]:

$$f(D) = \frac{[S(D) - S_0]/D}{[S(D_1) - S_0]/D_1} \quad (1)$$

where S_0, S(D), D_1, and $S(D_1)$ denote the TL response at zero dose, S(D), dosage in the linear area; and $S(D_1)$ denotes the TL response corresponding to dose D_1; respectively. The value of f(D) equals one for linear behavior, higher than 1 for supralinearity, and lower than 1 for sublinearity.

The calculated f(D) versus irradiation dose from 0.5 to 2000 Gy confirmed the linear-supralinearity criteria within this range because f(D) equals approximately one for the range of 0.5–25 Gy, while supralinearity behavior for the residual dose range up to 2 kGy where the values of f(D) are greater than unity are shown in Figure 8.

The unified interaction model has been used extensively to model the behavior of f(D). Many experimental measurements have also shown that as the glow peak temperature rises, the supralinearity of the various glow peaks rises. This can be explained based on the increased charge carrier migration distances during the recombination (heating) stage resulting in increased effects of track interaction and, in turn, increased luminescence recombination efficiency [42,43].

3.5. Thermal Fading

The TL-response of Dosimeters is affected by a variety of environmental conditions. The temperature of the surrounding medium is maybe the most critical. After irradiation, heat can liberate electrons from its traps, a process known as thermal fading. The trapping parameters affecting this process are the frequency factor (s) and activation energy (E) [44]. Figure 9 shows the thermal fading behavior of 50 Gy gamma irradiated rhyolite disks during a 28-day period. These samples were annealed before the irradiation process and stored at room temperature (20–25 °C) with a humidity of 40–50%.

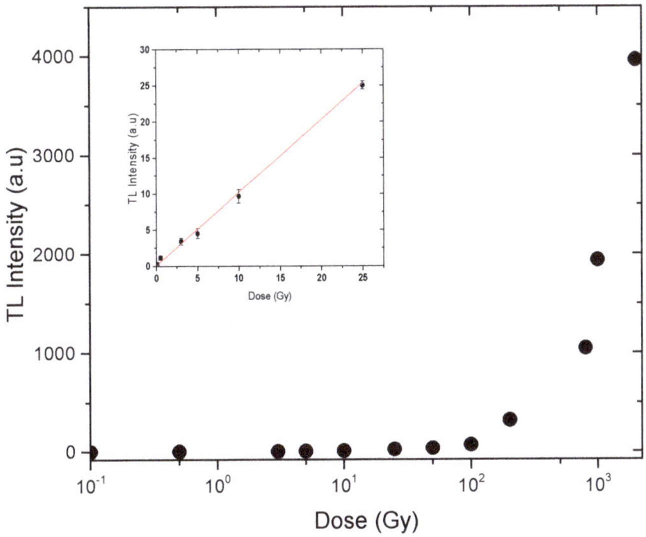

Figure 7. Dose response of natural rhyolite disks in the range of 0.5–2000 Gy.

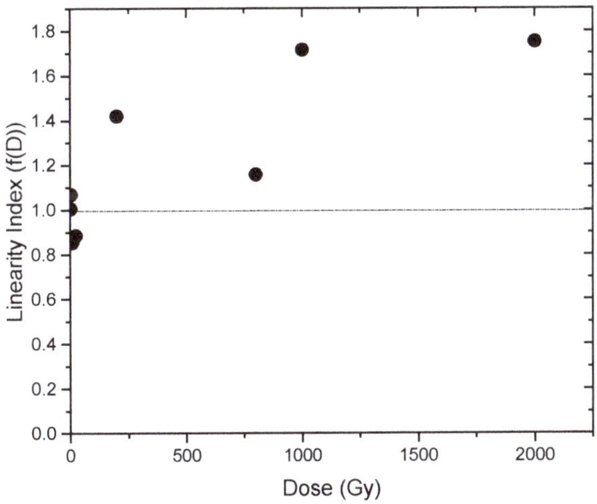

Figure 8. Linearity index of rhyolite at dose range 0.5–2000 Gy.

After 24 h, 52 h, one week, and two weeks, the results showed a 32 percent decrease, 52 percent reduction, and 57 percent reduction. During the remaining research time of two weeks, no additional thermal fading was noticed. In the case where the TL readout is delayed, these data will aid in estimating the precise absorbed dosages.

3.6. Reproducibility

Another key criterion in determining reusability of the recommended TL material is its reproducibility for dosimetry applications. The average TL intensity of a collection of annealed rhyolite disks was measured after numerous cycles of 50 Gy gamma irradiation. Based on a standard deviation of six measurements, the obtained data in Figure 10 show that the rhyolite dosimeter has acceptable reusability, with roughly 4% variance in successive measurements.

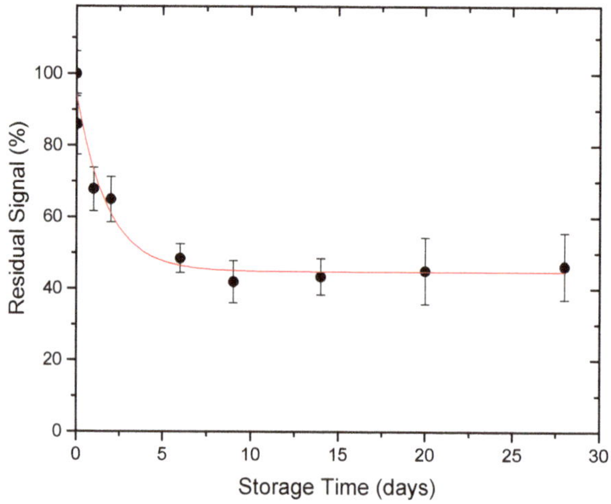

Figure 9. Thermal fading of 50 Gy gamma irradiated natural rhyolite.

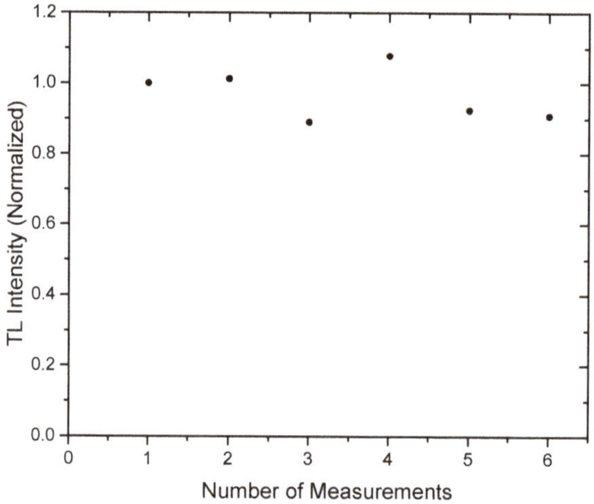

Figure 10. Reproducibility use of natural rhyolite for dosimetry applications.

3.7. Minimum Detectable Dose (MDD)

By using the dose calibration factor (F), the average background reading (B^*), and its standard deviation (σ_B), the following formula provided by Furetta et al. can be used to compute the rhyolite disks' minimal detectable dose (MDD) (2000) [45].

$$\mathrm{MDD} = (B^* + 2\sigma_B)F \quad (2)$$

Rhyolite has dose calibration factor F = 0.98 (0.75) Gy g nC^{-1}, the average background signal B^* = 0.40 nC g^{-1}, and its standard deviation σ_B = 0.06 nC g^{-1}. The calculated values of the MDD based on the above equation are about 0.5 Gy. Thus, natural rhyolite can measure low gamma doses up to 0.5 Gy.

3.8. Kinetic Parameters Determination

3.8.1. Repeated Initial Rise (RIR) Method

This method was developed by Garlick and Gibson in 1948 and is wildly used for determining the activation energy (E) [46]. The RIR method is based on the fact that at low temperatures, where TL intensities are less than 10–15% of the maximum intensity corresponding to a specific cut-off temperature, the change in charge carrier concentration in the traps is minimal and independent of the kinetic parameters [47]. The TL-intensity I(T) can be formulated as:

$$I(T) = \text{const.}e^{-E/kT} \quad (3)$$

The plot of ln (I) against 1/kT where k is the Boltzmann constant, gives a straight line where the activation energy (E) can be determined from the slope. This approach may recur with various stopping temperatures T_{stop} to acquire a large number of activation energies, and then the number of peaks in the glow curve can be determined if there are many peaks. The irradiation sample was heated at a continuous pace until it reached a certain cut-off temperature T_{stop}, at which point a thermoluminescence decay was recorded. Several heating and cooling cycles have produced a set of data I(T) spanning the temperature range of 333–673 K.

For a section of the data acquired at stopping temperatures of 100, 110, 130, 140, and 150 °C, Figure 11 depicts the relationship between ln (I) vs. 1/kT. The activation energy value is anticipated by the slope for each stopping temperature. The graph of the acquired values of the activation energy E vs. T_{stop} is shown in Figure 12. The obtained results displayed that the activation energies are nearly similar in five places (plateau region), indicating that the glow curve has five overlapping peaks with average activation energies of 0.77 ± 0.01, 0.86 ± 0.01, 1.01 ± 0.01, 0.99 ± 0.00, and 1.78 ± 0.01 eV, respectively.

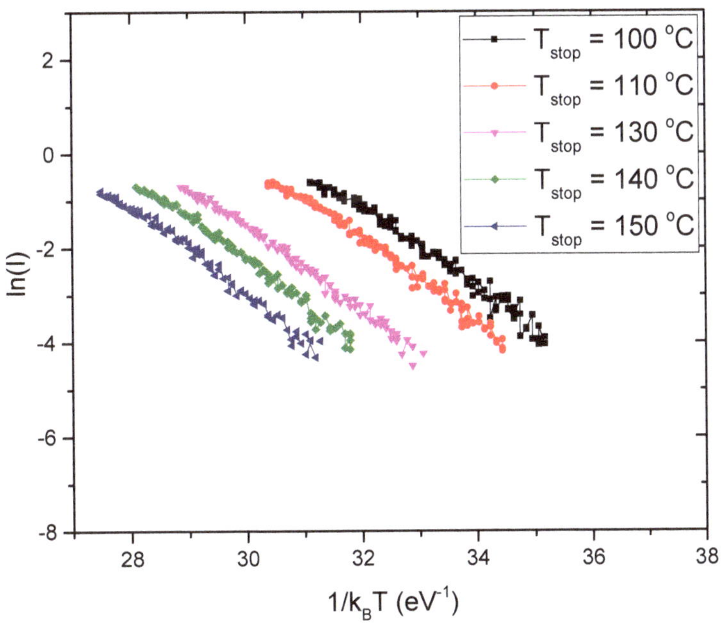

Figure 11. Initial raise plots of natural rhyolite sample at some values of T_{stop}.

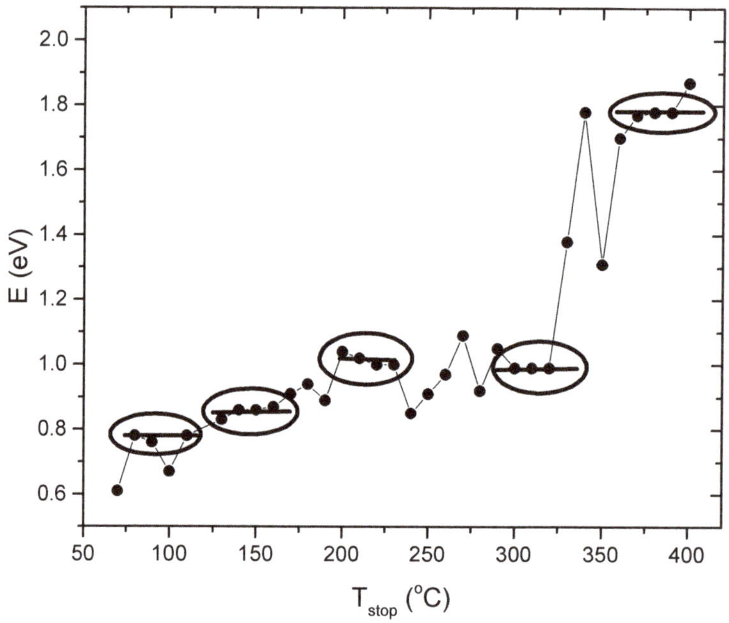

Figure 12. The activation energies (E) of natural rhyolite given by the RIR method.

3.8.2. Computerized Glow Curve Deconvolution (CGCD) Method

The glow curves of natural rhyolite have been deconvoluted into five interfering peaks using the general order kinetics equation given by Kitis et al. [48].

$$I(T) = n_o s \ \exp\left(-\frac{E}{kT}\right)\left[\frac{(b-1)skT^2}{\beta E} \times \exp\left(-\frac{E}{kT}\right)(1-\Delta)+1\right]^{-\frac{b}{b-1}} \quad (4)$$

where I is the glow-peak intensity, s (s^{-1}) is the frequency factor, E (eV) is the activation energy, n_0 is the initial concentration of trapped carriers, T (K) is the absolute temperature, k (eV K^{-1}) is the Boltzmann constant, $\Delta = 2\ kT/E$, and β is the heating rate. The deconvolution process was carried out using the Korean atomic energy institute's TL-ANAL tool [49]. As initial approximations, the number of peaks (five) and the corresponding values of their activation energies (obtained by the RIR method) were entered into the algorithm. The calculated value of the figure of merit (FOM) for all TL glow peaks determines the accuracy of the study [50]. The fit is satisfactory if the FOM values are between 0.0 and 2.5 percent, 2.5 and 3.5 percent is minor fit, and >3.5% is bad fit.

Figure 13 shows the glow curve deconvolution of a natural rhyolite sample irradiated with 50 Gy and recorded at a 3 °C/s heating rate. The glow curve deconvolution findings are summarized in Table 1. The computed standard deviation of the mean owing to many measurements or fitting peaks during the deconvolution process is incorporated in the considered activation energy data, which is the uncertainty type A. The acquired findings indicated that the glow curve of rhyolite material has five overlapping peaks created at 142, 176, 221, 298, and 355 °C, respectively, at a heating rate of 3 °C/s, and that the activation energies are compatible with the RIR method's results. The obtained kinetic order for the deconvoluted peaks showed mixed-order kinetic.

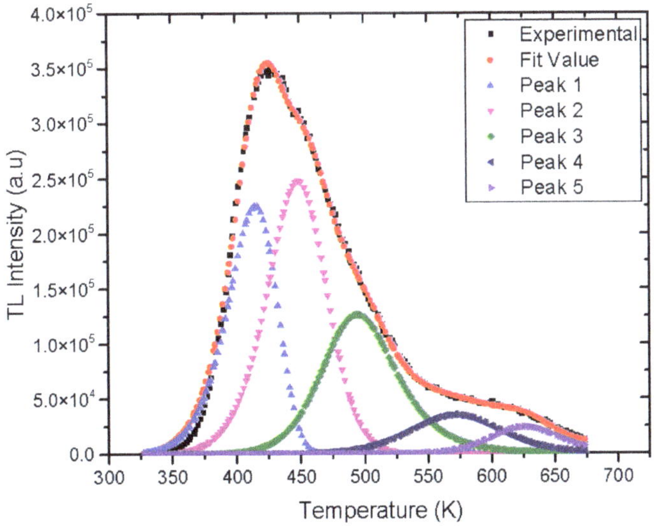

Figure 13. Glow curve deconvolution of natural rhyolite at 3 °C/s heating rate and irradiated by 50 Gy.

Table 1. The kinetic parameters for natural rhyolite.

Peak Number	Peak Temperature	RIR Method	CGCD Method		
	(°C)	E (eV)	E (eV)	s (s^{-1})	b
1	142	0.77 ± 0.01	0.78 ± 0.02	4.60×10^8	1.07
2	176	0.86 ± 0.01	0.87 ± 0.02	8.29×10^8	1.41
3	221	1.01 ± 0.01	1.02 ± 0.05	3.02×10^9	2.04
4	298	0.99 ± 0.00	0.98 ± 0.01	4.46×10^7	1.66
5	355	1.78 ± 0.01	1.76 ± 0.02	2.00×10^{13}	2.02

4. Conclusions

In this paper, natural rhyolite material is shaped into small disks to be used as a thermoluminescence dosimeter. Chemical composition and crystal structure of its composite phases showed that sodium aluminum silicate (42.193 ± 0 wt. %), potassium aluminum silicate (32.826 ± 1.079 wt. %), and silicon oxide (24.982 ± 0.671 wt. %) are the three identified phases of rhyolite composition. Linear dose response up to 25 Gy followed by supralinearity up to 2000 Gy are obtained. The obtained TL characteristics of rhyolite showed that this material has linear dose response up to 25 Gy, followed by supralinearity up to 2000 Gy, as well as relatively high fading rate of 57% after two weeks with no more thermal fading noted. Low detection limit of roughly 0.5 Gy and reasonable repeatability of about 4% variation in successive measurements are noted. These characteristics may qualify rhyolite as the sustainable natural material used in this study for a variety of applications involved radiation dose assessment.

Author Contributions: E.S.: methodology, validation, data curation, writing—original draft, writing—review and editing. D.A.A.: conceptualization, methodology, resources. S.A.E.-K.: methodology, writing—original draft. M.M.: methodology. All authors have read and agreed to the published version of the manuscript.

Funding: This research was funded by Princess Nourah bint Abdulrahman University, Riyadh, Saudi Arabia, Researchers Supporting Project number [PNURSP2022R57].

Institutional Review Board Statement: Not applicable.

Informed Consent Statement: Not applicable.

Data Availability Statement: Not applicable.

Acknowledgments: The authors express their gratitude to Princess Nourah bint Abdulrahman University Researchers Supporting Project number (PNURSP2022R57), Princess Nourah bint Abdulrahman University, Riyadh, Saudi Arabia.

Conflicts of Interest: The authors declare no conflict of interest.

References

1. Kron, T. Thermoluminescence dosimetry and its applications in medicine-Part 1: Physics, materials and equipment. *Australas. Phys. Eng. Sci. Med.* **1994**, *17*, 175–199. [PubMed]
2. Moscovitch, M.; John, T.J.; Cassata, J.R.; Blake, P.K.; Rotunda, J.E.; Ramlo, M.; Velbeck, K.J.; Luo, L.Z. The application of LiF:Mg,Cu,P to large scale personnel dosimetry: Current status and future directions. *Radiat. Prot. Dosim.* **2006**, *119*, 248–254. [CrossRef]
3. Chand, S.; Mehra, R.; Chopra, V. Recent developments in phosphate materials for their thermoluminescence dosimeter (TLD) applications. *Luminescence* **2021**, *36*, 1808–1817. [CrossRef]
4. Saleh, H.M.; Salman, A.A.; Faheim, A.A.; El-Sayed, A.M. Polymer and polymer waste composites in nuclear and industrial applications. *J. Nucl. Energy Sci. Power Gener. Technol.* **2020**, *9*, 1000199.
5. Saleh, H.M.; Salman, A.A.; Faheim, A.A.; El-Sayed, A.M. Sustainable composite of improved lightweight concrete from cement kiln dust with grated poly (styrene). *J. Clean. Prod.* **2020**, *277*, 123491. [CrossRef]
6. Dawoud, M.M.A.; Hegazi, M.M.; Saleh, H.M.; El Helew, W.K. Removal of stable and radio isotopes from wastewater by using modified microcrystalline cellulose based on Taguchi L16. *Int. J. Environ. Sci. Technol.* **2022**, 1–12. [CrossRef]
7. Dawoud, M.M.A.; Hegazy, M.M.; Helew, W.K.; Saleh, H.M. Overview of Environmental Pollution and Clean Management of Heavy Metals and Radionuclides by using Microcrystalline Cellulose. *J. Nucl. Ene. Sci. Power Gener. Technol.* **2021**, *3*, 2.
8. Saleh, H.M.; Bondouk, I.I.; Salama, E.; Esawii, H.A. Consistency and shielding efficiency of cement-bitumen composite for use as gamma-radiation shielding material. *Prog. Nucl. Energy* **2021**, *137*, 103764. [CrossRef]
9. Reda, S.M.; Saleh, H.M. Calculation of the gamma radiation shielding efficiency of cement-bitumen portable container using MCNPX code. *Prog. Nucl. Energy* **2021**, *142*, 104012. [CrossRef]
10. Eid, M.S.; Bondouk, I.I.; Saleh, H.M.; Omar, K.M.; Sayyed, M.I.; El-Khatib, A.M.; Elsafi, M. Implementation of waste silicate glass into composition of ordinary cement for radiation shielding applications. *Nucl. Eng. Technol.* **2021**, *54*, 1456–1463. [CrossRef]
11. Eskander, S.B.; Saleh, H.M.; Tawfik, M.E.; Bayoumi, T.A. Towards potential applications of cement-polymer composites based on recycled polystyrene foam wastes on construction fields: Impact of exposure to water ecologies. *Case Stud. Constr. Mater.* **2021**, *15*, e00664. [CrossRef]
12. Saleh, H.M.; Eskander, S.B. Impact of water flooding on hard cement-recycled polystyrene composite immobilizing radioactive sulfate waste simulate. *Constr. Build. Mater.* **2019**, *222*, 522–530. [CrossRef]
13. Saleh, H.M.; El-Saied, F.A.; Salaheldin, T.A.; Hezo, A.A. Influence of severe climatic variability on the structural, mechanical and chemical stability of cement kiln dust-slag-nanosilica composite used for radwaste solidification. *Constr. Build. Mater.* **2019**, *218*, 556–567. [CrossRef]
14. Saleh, H.M.; El-Sheikh, S.M.; Elshereafy, E.E.; Essa, A.K. Performance of cement-slag-titanate nanofibers composite immobilized radioactive waste solution through frost and flooding events. *Constr. Build. Mater.* **2019**, *223*, 221–232. [CrossRef]
15. Saleh, H.M.; Salman, A.A.; Faheim, A.A.; El-Sayed, A.M. Influence of aggressive environmental impacts on clean, lightweight bricks made from cement kiln dust and grated polystyrene. *Case Stud. Constr. Mater.* **2021**, *15*, e00759. [CrossRef]
16. Preusser, F.; Chithambo, M.L.; Götte, T.; Martini, M.; Ramseyer, K.; Sendezera, E.J.; Susino, G.J.; Wintle, A.G. Quartz as a natural luminescence dosimeter. *Earth Sci. Rev.* **2009**, *97*, 184–214. [CrossRef]
17. Thamóné Bozsó, E.; Füri, J.; Kovács István, J.; Biró, T.; Király, E.; Nagy, A.; Törökné Sinka, M.; Kónya, P.; Mészárosné Turi, J.; Vígh, C. Characteristics of quartz separates of different formations in Hungary from the aspect of osl dating. *Foldt. Kozlony* **2020**, *150*, 61. [CrossRef]
18. Bailiff, I.K.; Stepanenko, V.F.; Göksu, H.Y.; Bøtter-Jensen, L.; Brodski, L.; Chumak, V.; Correcher, V.; Delgado, A.; Golikov, V.; Jungner, H.; et al. Comparison of retrospective luminescence dosimetry with computational modeling in two highly contaminated settlements downwind of the chernobyl NPP. *Health Phys.* **2004**, *86*, 25–41. [CrossRef]
19. Beneitez, P.; Correcher, V.; Millán, A.; Calderón, T. Thermoluminescence analysis for testing the irradiation of spices. *J. Radioanal. Nucl. Chem. Artic.* **1994**, *185*, 401–410. [CrossRef]
20. Soliman, C.; Salama, E. Investigation on the suitability of natural sandstone as a gamma dosimeter. *Nucl. Instrum. Methods Phys. Res. Sect. B Beam Interact. Mater. At.* **2009**, *267*, 3323–3327. [CrossRef]
21. Salama, E. Egyptian limestone for gamma dosimetry: An electron paramagnetic resonance study. *Radiat. Eff. Defects Solids* **2014**, *169*, 325–333. [CrossRef]
22. El-Faramawy, N.; Alazab, H.A.; Gad, A.; Sabry, M. Study of the thermoluminescence kinetic parameters of a β-irradiated natural calcite. *Radiat. Phys. Chem.* **2022**, *190*, 109793. [CrossRef]

23. Almuqrin, A.; Soliman, C.; ELShokrof, K.; Aloraini, D. Tuff rock as a new thermoluminescent material for gamma dosimetry. *Asian J. Sci. Res.* **2018**, *11*, 126–133. [CrossRef]
24. DeMouthe, J.F. Natural materials: Sources, properties, and uses. *Rocks Miner.* **2006**, *81*, 3972006.
25. Aloraini, D.A. Sensitization of natural rhyolite rock using high gamma doses for thermoluminescence dosimetry. *Luminescence* **2020**, *35*, 1043–1047. [CrossRef]
26. Azorín, J.; Furetta, C.; Scacco, A. Preparation and properties of thermoluminescent materials. *Phys. Stat. Solidi* **1993**, *138*, 9–46. [CrossRef]
27. Lutterotti, L. Total pattern fitting for the combined size-strain-stress-texture determination in thin film diffraction. *Nucl. Instrum. Methods Phys. Res. Sect. B Beam Interact. Mater. At.* **2010**, *268*, 334–340. [CrossRef]
28. Harlow, G.E. The anorthoclase structures: The effects of temperature and composition. *Am. Mineral.* **1982**, *67*, 975–996.
29. Bailey, S.W. Refinement of an intermediate microcline structure. *Am. Mineral.* **1969**, *54*, 1540–1545.
30. Levien, L.; Prewitt, C.T.; Weidner, D.J.; Prewir, C.T.; Weidner, D.J. Structure and Elastic Properties of Quartz at Pressure. *Am. Mineral.* **1980**, *65*, 920–930.
31. El Makhloufy, S.; Tridane, M.; Majdi, E.M.; Marouani, H.; Zerraf, S.; Belhabra, M.; Cherqaoui, A.; Belaaouad, S. Chemical preparation, thermal behavior and infrared studies of the new cyclotriphosphate tetrahydrate of manganese and distrontium, MnSr2(P3O9)2.4H2O. *Mediterr. J. Chem.* **2019**, *9*, 280–289. [CrossRef]
32. Liang, Y.; Ouyang, J.; Wang, H.; Wang, W.; Chui, P.; Sun, K. Synthesis and characterization of core-shell structured SiO_2@YVO_4 :Yb^{3+},Er^{3+} microspheres. *Appl. Surf. Sci.* **2012**, *58*, 3689–3694. [CrossRef]
33. Mozgawa, W.; Król, M.; Dyczek, J.; Deja, J. Investigation of the coal fly ashes using IR spectroscopy. *Spectrochim. Acta Part A Mol. Biomol. Spectrosc.* **2014**, *132*, 889–894. [CrossRef]
34. Moenke, H.H.W. Silica, the Three-dimensional Silicates, Borosilicates and Beryllium Silicates. In *The Infrared Spectra of Minerals*; Mineralogical Society of Great Britain and Ireland: Oxford, UK, 2015.
35. Treto-Suárez, M.A.; Prieto-García, J.O.; Mollineda-Trujillo, Á.; Lamazares, E.; Hidalgo-Rosa, Y.; Mena-Ulecia, K. Kinetic study of removal heavy metal from aqueous solution using the synthetic aluminum silicate. *Sci. Rep.* **2020**, *10*, 10836. [CrossRef]
36. Mandal, A.K.; Agrawal, D.; Sen, R. Preparation of homogeneous barium borosilicate glass using microwave energy. *J. Non. Cryst. Solids* **2013**, *371*, 41–46. [CrossRef]
37. Study, S.; Consumption, L.E.; Part, D.U.; Utilization, D.; Yoshizawa, N.; Harimoto, K.; Ichihara, M.; Miki, Y.; Takase, K.; Inoue, T. Synthesis, Characterization and Bioactive Study of Borosilicate Sol-Gel Glass Khairy. *Nat. Sci.* **2015**, *13*, 475–476.
38. Rada, S.; Dehelean, A.; Culea, E. FTIR and UV-VIS spectroscopy investigations on the structure of the europium-lead-tellurate glasses. *J. Non. Cryst. Solids* **2011**, *357*, 3070–3073. [CrossRef]
39. Bos, A.J.J.; Vijverberg, R.N.M.; Piters, T.M.; McKeever, S.W.S. Effects of Cooling and Heating Rate on Trapping Parameters in LiF: Mg, Ti Crystals. *J. Phys. D Appl. Phys.* **1992**, *25*, 1249–1257. [CrossRef]
40. Amer, H.A.; Elashmawy, M.M.; Alazab, H.A.; Ezz El-Din, M.R. Suitability of pure nano crystalline LiF as a TLD dosimeter for high dose gamma radiation. *Nucl. Technol. Radiat. Prot.* **2018**, *33*, 93–99. [CrossRef]
41. Chen, R.; McKeever, S.W.S. Characterization of nonlinearities in the dose dependence of thermoluminescence. *Radiat. Meas.* **1994**, *23*, 667–673. [CrossRef]
42. Horowitz, Y.S.; Oster, L.; Datz, H. The thermoluminescence dose-response and other characteristics of the high-temperature TL in LiF:Mg,Ti (TLD-100). *Radiat. Prot. Dosim.* **2007**, *124*, 191–205. [CrossRef]
43. Attix, F.H. Further consideration of the track-interaction model for thermoluminescence in LiF(TLD-100). *J. Appl. Phys.* **1975**, *46*, 81–88. [CrossRef]
44. Chen, R.; Pagonis, V.; Lawless, J.L. Evaluated thermoluminescence trapping parameters-What do they really mean? *Radiat. Meas.* **2016**, *91*, 21–27. [CrossRef]
45. Furetta, C.; Prokic, M.; Salamon, R.; Prokic, V.; Kitis, G. Dosimetric characteristics of tissue equivalent thermoluminescent solid TL detectors based on lithium borate. *Nucl. Instrum. Methods Phys. Res. Sect. A Accel. Spectrometers Detect. Assoc. Equip.* **2001**, *456*, 411–417. [CrossRef]
46. Jose, M.T.; Anishia, S.R.; Annalakshmi, O.; Ramasamy, V. Determination of thermoluminescence kinetic parameters of thulium doped lithium calcium borate. *Radiat. Meas.* **2011**, *46*, 1026–1032. [CrossRef]
47. Garlick, G.F.J.; Gibson, A.F. The electron trap mechanism of luminescence in sulphide and silicate phosphors. *Proc. Phys. Soc.* **1948**, *60*, 574–590. [CrossRef]
48. Kitis, G.; Gomez-Ros, J.M.; Tuyn, J.W.N. Thermoluminescence glow-curve deconvolution functions for first, second and general orders of kinetics. *J. Phys. D. Appl. Phys.* **1998**, *31*, 2636–2641. [CrossRef]
49. Chung, K.S.; Choe, H.S.; Lee, J.I.; Kim, J.L.; Chang, S.Y. A computer program for the deconvolution of thermoluminescence glow curves. *Radiat. Prot. Dosim.* **2005**, *115*, 345–349. [CrossRef]
50. Balian, H.G.; Eddy, N.W. Figure-of-merit (FOM), an improved criterion over the normalized chi-squared test for assessing goodness-of-fit of gamma-ray spectral peaks. *Nucl. Instrum. Methods* **1977**, *145*, 389–395. [CrossRef]

sustainability

MDPI

Article

Fluorine-18 Fluorodeoxyglucose Isolation Using Graphene Oxide for Alternative Radiopharmaceutical Spillage Decontamination in PET Scan

Mohammad Khairul Azhar Abdul Razab [1,*], Norazlina Mat Nawi [2,*], Fara Hana Mohd Hadzuan [1], Nor Hakimin Abdullah [3], Maimanah Muhamad [4], Rosidah Sunaiwi [1], Fathirah Ibrahim [1], Farah Amanina Mohd Zin [3] and An'amt Mohamed Noor [3]

1 Medical Radiation Programme, School of Health Sciences, Universiti Sains Malaysia, Health Campus, Kubang Kerian 16150, Kelantan, Malaysia; farahadzuan@student.usm.my (F.H.M.H.); rosidah98@student.usm.my (R.S.); fathirah_ibrahim94@yahoo.com (F.I.)
2 Department of Nuclear Medicine, Radiotherapy & Oncology, School of Medical Sciences, Universiti Sains Malaysia, Health Campus, Kubang Kerian 16150, Kelantan, Malaysia
3 Advanced Materials Research Cluster, Faculty of Bioengineering and Technology, Universiti Malaysia Kelantan, Jeli Campus, Jeli 17600, Kelantan, Malaysia; norhakimin@umk.edu.my (N.H.A.); farahamanina2751@gmail.com (F.A.M.Z.); anamt@umk.edu.my (A.M.N.)
4 Nuclear Medicine & Molecular Imaging Department, Chancellor Tuanku Muhriz Hospital, Universiti Kebangsaan Malaysia, Jalan Yaacob Latif, Bandar Tun Razak, Cheras 56000, Kuala Lumpur, Malaysia; maimanah@ppukm.ukm.edu.my
* Correspondence: khairul.azhar@usm.my (M.K.A.A.R.); norazlina@usm.my (N.M.N.)

Abstract: Radiopharmaceuticals (RPC) used for diagnostic and therapeutic purposes in nuclear medicine may contaminate surface areas due to spillage during its preparation or accident during RPC transfer from laboratory to the treatment room. Fluorine-18 Fluorodeoxyglucose (^{18}F-FDG) is the most common RPC for positron emission tomography (PET) scan in nuclear medicine due to its ideal annihilation converted energy at 511 keV and short half-life at 109.8 min. Ineffective medical waste management of ^{18}F-FDG may pose a risk to the environment or cause unnecessary radiation doses to the personnel and public. Depending on the incident rate of these events, simple decontamination methods such as the use of chemicals and swabs might not be cost-effective and sustainable in the environment. This study aims to propose an alternative method to decontaminate ^{18}F-FDG by using graphene oxide (GO). GO was synthesised using the Hummers method while the physical morphology was analysed using a field emission scanning electron microscope (FESEM). ^{18}F-FDG adsorption efficiency rate using GO nanolayers was analysed based on the kinetic study of the GO:^{18}F-FDG mixtures. The chemical adsorbability of the material was analysed via UV–vis spectrophotometer to interlink the microstructures of GO with the sorption affinity interaction. Resultantly, the adsorption rate was effective at a slow decay rate and the optical adsorption of GO with ^{18}F-FDG was dominated by the $\pi \rightarrow \pi^*$ plasmon peak, which was near 230 nm. By elucidating the underlining GO special features, an alternative technique to isolate ^{18}F-FDG for the decontamination process was successfully proven.

Keywords: radiopharmaceutical; radioactive spillage; decontamination; radioactivity; adsorption; nuclear medicine

Citation: Razab, M.K.A.A.; Nawi, N.M.; Hadzuan, F.H.M.; Abdullah, N.H.; Muhamad, M.; Sunaiwi, R.; Ibrahim, F.; Zin, F.A.M.; Noor, A.M. Fluorine-18 Fluorodeoxyglucose Isolation Using Graphene Oxide for Alternative Radiopharmaceutical Spillage Decontamination in PET Scan. *Sustainability* **2022**, *14*, 4492. https://doi.org/10.3390/su14084492

Academic Editors: Hosam M. Saleh and Mohammad Mahmoud Dawoud

Received: 16 March 2022
Accepted: 7 April 2022
Published: 9 April 2022

Publisher's Note: MDPI stays neutral with regard to jurisdictional claims in published maps and institutional affiliations.

1. Introduction

Radiopharmaceutical (RPC) resources are essential for nuclear imaging and therapeutic purposes in nuclear medicine. Fluorine-18 Fluorodeoxyglucose (^{18}F-FDG) is the most ideal RPC for positron emission tomography (PET) scans due to its ideal half-life and versatile molecular structure [1,2]. Fluorine is considered a favourable atom in drug development due to its physical properties, including a small van der Waals radius (1.47 Å), high electronegativity, and ability to form a strong bond with carbon (C-F energy bond of 112 kcal/mol) [3]. In comparison with a carbon-hydrogen bond (C-H energy bond of 98 kcal/mol), ^{18}F-FDG is more thermally stable and oxidation resistant [3]. Unexpected radioactive contamination might occur in a nuclear medicine department either from source handling, such as breakage of a radionuclide vial, or may originate from the patient either by sneezing, vomiting, or urinating [4]. The hazard associated with radioactive contamination and waste depends on the concentration and nature of the activity of the radionuclide [5]. ^{18}F-FDG, an unstable radioisotope, will decay in the human metabolic process and produce positrons, thereby leading to an annihilation process with electrons and producing pairs of 511 keV gamma-ray photons in 180° directions [6]. This energy is quite high and exposes the personnel, patient, and public to unnecessary radiation doses.

The high demand for PET scans requires mass production of ^{18}F-FDG using cyclotron, which indirectly leads to an increase in the incident rate of radionuclide spillage contaminations. Meanwhile, the usage of a large volume of simple decontamination chemical agents might not be sustainable in the environment [7], and is not cost-effective as its periodic purchase is met with limited production. Thus, alternative decontamination agents must be considered by ensuring the application of a competent material for radionuclide decontamination, which is more efficient, environmentally friendly, and economical in terms of long-term usage of the resources [8].

Graphene oxide (GO) is a nanolayer material that has a carbon compound and is rich in oxygenated functional groups, which makes it a fitting adsorbent for various molecules either in aqueous or dispersed in a polymeric solution [9]. GO also has several desirable properties, such as high surface area, high mechanical strength, electrical conductivity, large pore volume structure, and high solubility due to abundant oxygen-based functional groups [9]. GO is a two-dimensional (2D), carbon-based material that has a single-atom-thin full sp^2 hybridised carbon structure with minimal defects, which makes it a well-known adsorbent [10]. Due to its uniquely tuneable physicochemical characteristics [11] and biocompatibility [12], GO has been use in many applications such as the removal of metal ions [13], biomedical [14], electrochemical energy [15], catalysis [16], sensing [17], polymers composite [18], removal of heavy metals [19], wastewater treatment [20] and antibacterial agents [21].

The availability of various functional groups such as carboxyl, hydroxyl, epoxy and keto groups on its nano surface [22] give advantages to GO as a super adsorbent carbon material to efficiently interact with any hazardous and toxic liquid, gas or solid contaminants [23]. Practically, radioactive waste has been decontaminated using special chemical agents which might chemically produce other chemical by-products [24] due to chemical interactions with unstable energetic radionuclide.

Application of GO nanosheets in trapping radionuclides has been conducted on ^{238}U [25], ^{137}Cs [26], ^{131}I [27], ^{232}Th, ^{95}Am, ^{94}Pu, ^{63}Eu, ^{38}Sr and ^{99m}Tc [28]. However, adsorption and chemical adsorption characterization of short-live radionuclides such as the ^{18}F radionuclide by using GO has not been comprehensively reported. Herewith, we conduct experiments to prove the efficiency of ^{18}F-FDG decontamination by using GO, which is a stable carbon-based material with minimal production of chemical by-products. The interactions between these compound mixtures with different concentration of GO and activities for energy variations were characterize by using UV–vis spectrometer for complex adsorption capabilities.

2. Materials and Methods

Graphite powder, a naturally occurring form of crystalline carbon along with inorganic compound potassium permanganate ($KMnO_4$) and corrosive substances, sulphuric acid (H_2SO_4) and phosphoric acid (H_3PO_4), were used to make up the GO mixture. Hydrogen peroxide (H_2O_2, 30%) and ice cubes were used to initiate and phase out the oxidation process. Consequently, hydrochloric acid (HCl) and distilled water were used for the centrifuge process. ^{18}F-FDG was obtained at the Nuclear Medicine & Molecular Imaging

Department, Chancellor Tuanku Muhriz Hospital, Universiti Kebangsaan Malaysia. A dose calibrator (Capintec CRC-25R, Capintec Inc., Florham Park, NJ, USA) was used to monitor and record the activity of the radionuclides. The characterisation of the synthesised GO was performed using FESEM (FEI QuantaTM 450 FEG, FEI Company, Hillsboro, OR, USA) at SEM Laboratory, School of Health Sciences, Universiti Sains Malaysia. Furthermore, a morphological analysis of the nanolayers and their porosity was conducted. Meanwhile, the chemical adsorbability affinity of the GO:^{18}F-FDG was analysed using a UV–vis spectrometer (CARY 100 Bio, Agilent Technologies Inc., Santa Clara, CA, USA) at the Analytical Laboratory, School of Health Sciences, Universiti Sains Malaysia.

2.1. Synthesis of Graphene Oxide

The Graphene oxide was synthesised by the widely used Hummer's method [29–31]. About 3 g of graphite and 18 g of $KMnO_4$ were added to the 320 mL of H_2SO_4 and 80 mL of H_3PO_4 in a ratio of 4:1, respectively. The mouth of the beaker was closed with an aluminium foil and stirred for three days continuously before transferring the reagent into a 500 mL ice cube along with 27 mL of H_2O_2 (30%) to stop the oxidation and cooling process. The dark brown mixture formed was known as graphite oxide, which was then washed with HCl to separate the GO from the chemical by-products and unexfoliated graphite. About 15 mL of the mixture and 15 mL of HCL were mixed and centrifuged at 5000 g/rpm for 10 min. This washing process was repeated thrice before washing with distilled water several times and centrifuged with the same setting until a pH of 5 was achieved.

2.2. Isolation of ^{18}F-FDG

The GO concentrations were varied into 1 mg/mL, 2 mg/mL, and 3 mg/mL, which were obtained by applying Equation (1), where M_1 is the initial concentration of the solution, V_1 is the volume of the solution, M_2 is the concentration of the diluted solution after adding more solvent, and V_2 is the volume of the diluted solution.

$$M_1V_1 = M_2V_2 \quad (1)$$

The FDG dispenser (NUCLEMED 317R3, NUCLEMED BV, Roeselare, Belgium) was used to accurately prepare the 37 MBq, 74 MBq, and 111 MBq of ^{18}F-FDG, and three different concentrations of GO were added into each glass serum vial as shown in Figure 1a. The mixture of GO:^{18}F-FDG was then poured on a filter paper, and the sediment was obtained as depicted in Figure 1b. The radioactivity of the sediment along with the filter paper and the filtrate (if available) were obtained using the dose calibrator ionisation chamber over the function of time (A_o = initial activity, $A_o/4$ = 54.9 min, $A_o/2$ = 109.8 min, $A_o3/4$ = 164.7 min, and $A_o/16$ = 219.6 min). The radioactivity of each activity with different concentrations was compared to the natural decay of ^{18}F-FDG by plotting the decay graph. A Geiger–Muller survey meter (Model Fluke 451B, Fluke Corporation, Everett, WA, USA) was used during the experiment to measure the radiation exposure, ensuring there was no leakage or spillage of the radionuclide source on any surfaces in the laboratory.

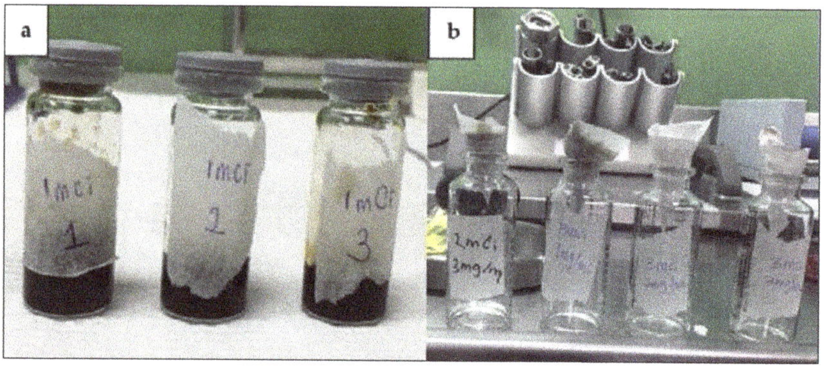

Figure 1. (**a**) GO solution with 1 mg/mL, 2 mg/mL and 3 mg/mL mixed with ^{18}F-FDG of 1mCi (~37 MBq) in glass serum vials. (**b**) GO:^{18}F-FDG mixture poured into filter papers.

3. Results and Discussions

3.1. Characterisation of GO Nanolayers

Graphene oxide (GO) images obtained from the FESEM were well defined and interlinked two-dimensional (2D) graphene sheets. Figure 2a,b show random sophisticated 2D structures with different magnifications. The surface appears as a wrinkly 'wax tissue layer' with hollow structures (red arrows in Figure 2a) and a continuous vast layer-by-layer surface area with random size of GO flakes (numbering layers in Figure 2b). This feature is advantageous to the GO in terms of 'wrapping' or entrapping other materials with various chemical bonds by the numerous carbon functional groups existing at the molecular structure level [32]. It also shows that the 'wax tissue layer' [33,34], which resembles the GO sheet, reflects that the deformation of the graphene layers was due to the oxidation process in the synthesised GO. Hence, the carbon lattice became distorted following the addition of oxygen lattice into the structures [35]. The generated hollow pores as shown in Figure 2a could be formed via the hydrothermal method by crumpling graphene sheets together, which results in the formation of voids between the sheets [32]. These features in the GO appear to be advantageous in terms of the hydrophilicity and dispersion into various media, including aqueous and organic solvents.

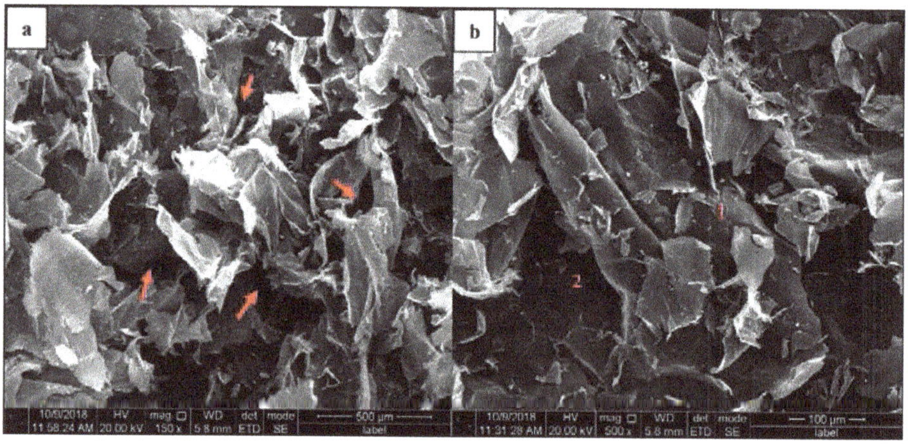

Figure 2. GO flake 2D structures with random 'wax-tissue' layers and complex hollow structures. (**a**) Complex hollow structures with wrinkles and exfoliated tissue layers, (**b**) wrinkles and exfoliated stacked structures with random small flakes.

Stacked GO structures can be clearly seen in Figure 3a,b for 250×, and 500× magnification. The stacking structures enhance the continuous interaction probabilities with other compounds, where the large substrate surface area will participate in the molecular interactions [33]. The interactions probabilities are continuously increased with stacking layers by layers of GO and might be trapped within the compound with increasing depth via complex interactions.

Figure 3. GO substrate surface stacked nanostructures. (**a**) Random stacks of small-flake GO nano-layers, (**b**) flat surface of large-flake GO nano-layers.

Meanwhile, a single substrate of GO nanolayers can be seen in a close-up micrograph in Figure 4a,b. The micrograph enhances the size of the nano substrate surface for molecular interactions to trap the ^{18}F-FDG and other compounds. Some contaminants on the GO surface can be seen in Figure 4a, which might be due to the incomplete washing of some foreign particles (red arrows). However, Figure 4b shows a significant structure of single wax tissue layers of GO with wrinkles and exfoliated networking lines on the surface (arrows), ready for the adsorption process via the molecular interaction of many available functional groups.

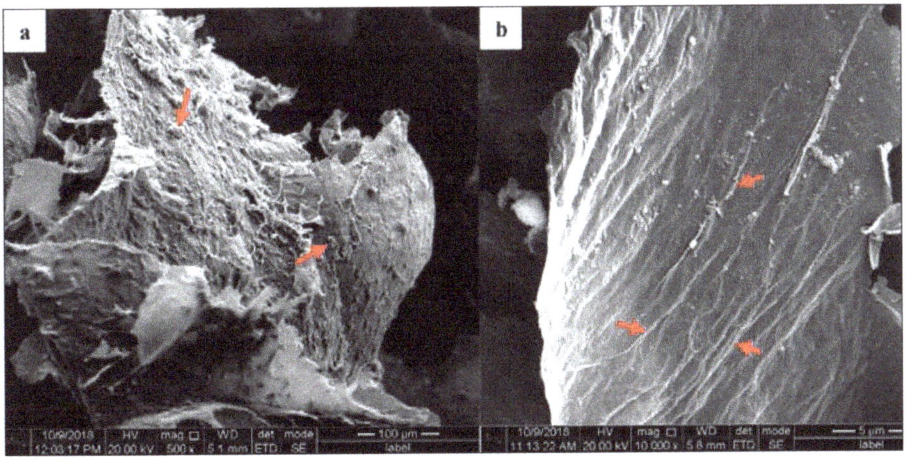

Figure 4. Close-up of single substrate GO nanolayers. (**a**) Contaminants existed on the GO nano-layers, (**b**) continuous wrinkle and exfoliated networking lines ready for adsorption interactions.

In additions, the complex structures of GO nanolayers have been focused within smaller scales as those shown in Figure 5a,b. The agglomerated 'wax tissue' nanolayers with non-uniform structures are shown in Figure 5a. The complex wrinkles and exfoliated structures can be seen via the red arrows, whereas the agglomerated long strand 'wax-tissue' nanolayers can be seen on another site. On the other hand, Figure 5b shows the multiple stacking layers that can still be identified from the agglomerated wax tissue layers. These structures demonstrate that the continuous multiplex nanolayers randomly existed for fine focused micrograph and widely spread on the GO substrates.

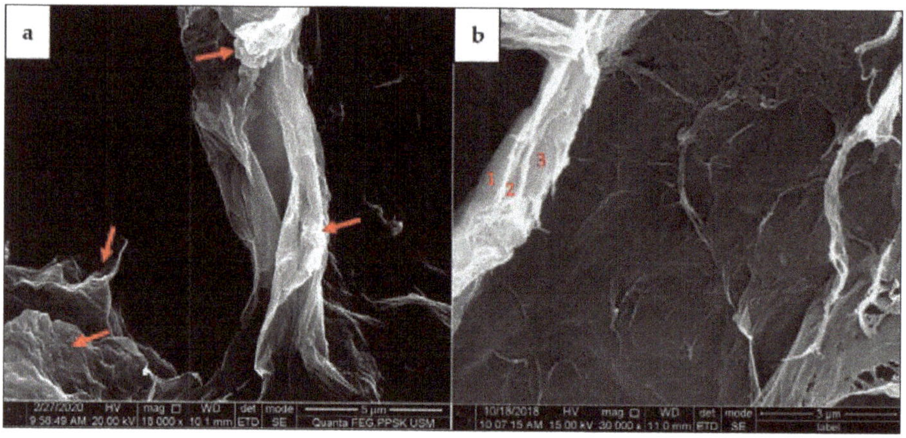

Figure 5. Agglomerated 'wax-tissues' GO with stacking layers. (**a**) Random and long strand agglomerated 'wax-tissue' nanolayers, (**b**) identification of stacking layers on agglomerated GO structures.

Moreover, agglomerated wrinkle and exfoliated structures have been further characterized with smaller scales to show a close up image of the 2D single nanolayers of the 'wax-tissue' GO structures as shown in Figure 6a,b. It can be seen that the small spherical agglomerated GO nanostructures are randomly attached on the upper site as shown in Figure 6a and the lower site as shown in Figure 6b of the single nanolayers. This phenomenon proves the complex random structures of agglomerated GO nanolayers existed along the stacking surfaces and promote high interactions with any materials via molecular interactions [36]. However, agglomerated unattached spherical nano structures are also randomly seen in Figure 6a, which leads to random complex interactions efficiency during the adsorption process [37].

In Figure 7a, a stacking wrinkles and exfoliated GO nanolayers has been selected for characterization by using X-ray diffraction technique (XRD) (Bruker D2, Bruker Corporation, Billerica, MA, USA) and UV–vis spectrometer. XRD spectrum of the GO is similar to the GO reported in other previous published works. It can be seen in Figure 7b that the characteristic peak of GO centred at 10.0°, which was assigned to the (0,0,1) reflection of GO. The UV–visible adsorption spectra of GO are shown in the Figure 7c. As revealed by the spectrum, two characteristic peaks of GO were observed at 227 and 229 nm, which were $\pi \rightarrow \pi^*$ transition of aromatic C-C bond and the $n \rightarrow \pi^*$ of C=O group, respectively. This is in agreement with the results published elsewhere [38].

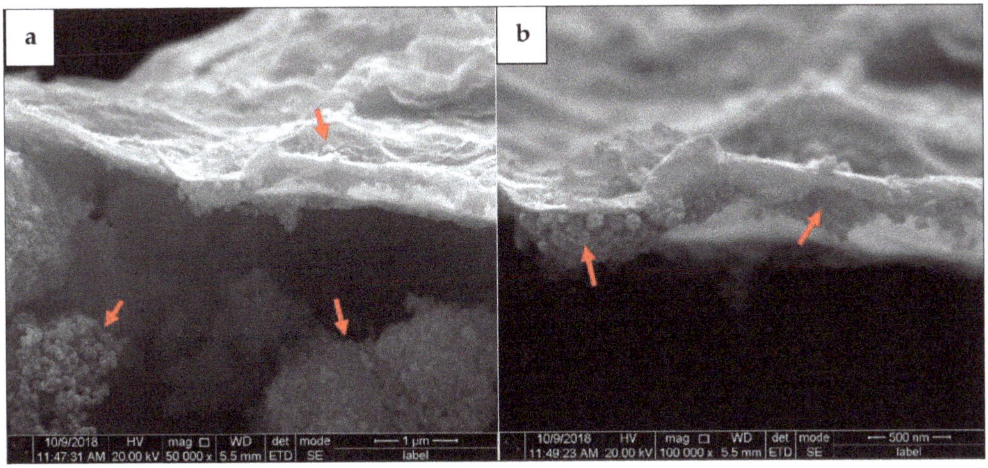

Figure 6. (**a**) Attached and free spherical agglomerated GO structures and (**b**) small spherical agglomerated GO nano structures have been randomly attached on the stacked wrinkles and exfoliated single GO nanolayer.

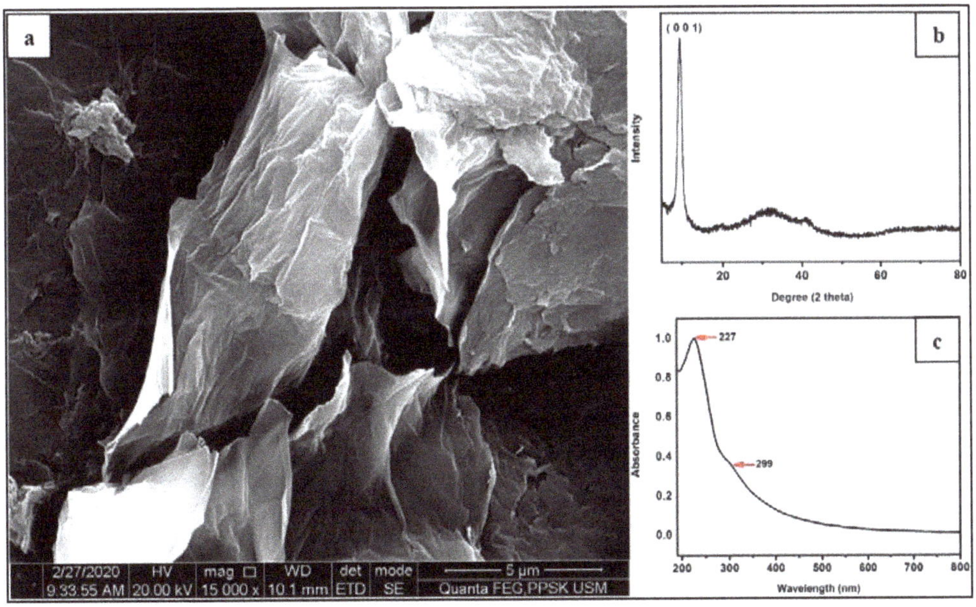

Figure 7. (**a**) Stacking wrinkles and exfoliated GO nanolayers [34], (**b**) XRD characterization of GO centred at 10.0°, and (**c**) UV–vis spectrum characteristic peaks of GO observed at 227 and 229 nm.

3.2. Kinetic Study

The kinetic study of the isolation of GO:^{18}F-FDG mixtures for different concentrations of GO concentrations and RPC activities was extrapolated using the radioactive decay equation as shown in Equation (2), where A is the radionuclide activity, A_o is the initial radionuclide activity, λ is the decay constant, and t is the decay time. Each of the activities (37 MBq, 74 MBq and 111 MBq) that were mixed with different GO concentrations (1, 2, 3 mg/mL) and trapped by using a filter paper were measured using a dose calibrator within specific periods (54.9 min,

109.8 min, 164.7 min, and 219.6 min). The activities were differentiated in order to investigate the influence of the positrons and gamma-ray photons' production rate on the molecular adsorption process. Three graphs to study the kinetics of the ^{18}F-FDG isolated by GO based on activity over time were plotted for different values of A_0 as shown in Figures 8–10.

$$\ln (A/A_o) = -\lambda t \quad (2)$$

Figures 8–10 depict the exponentially plotted graph of GO:^{18}F-FDG activities versus time for three different concentrations (1, 2, and 3 mg/ mL). From these graphs, the activity of GO:^{18}F-FDG was exponentially decreased to half from their approximate initial activities within 109.8 min, which is the natural half-life of ^{18}F [39]. From the graph, the first term half-life for three different activities (37 MBq, 74 MBq and 111 MBq) are sharply reduced due to their active decay processes with respect to its initial activities (high activity), becoming stagnant for the second term of its half-life and approximately close to zero for total decay (219.6 min). The decay process depends on the activities of the radionuclide, where higher activities promote active interactions with random decay processes in the nucleus. In this study, the existence of different concentrations of GO with RPC influenced the nuclear decay process within the unstable 1^{8}F nucleus, where there are different activities has been measured for different concentrations in a specific natural half-life. This phenomenon might be due to the adsorption of GO at the molecular level which is mostly via van der walls and ionic interactions [40], causing some inter-molecular changes of weak nuclear and electromagnetic forces in ^{18}F during its decay process. Influence of the weak nuclear forces that control the nuclear activities for any radioisotope decay [41], might cause some changes in the decay rate of the materials during complex molecular interactions.

Generally, 1 mg/mL concentration of GO can trap the ^{18}F-FDG effectively, where the activity is always higher in each measured time as shown in Figures 8–10. This might be due to random nuclear interaction activities in the ^{18}F nucleus, where vast numbers of electrons for higher concentrations might fluctuate or slow the molecular interactions. Fluctuation at the initial activity due to a high decay rate may be attributed to the reactivity of the orbital electrons of fluorine, which actively tend to be stabilised via complex interaction [42].

The adsorption interactions might also be influenced by the physical volume of the RPC that coagulate with certain GO concentrations, where the insoluble process might occur due to saturated solutions. It also shows that a higher concentration of GO could reduce the activity of ^{18}F-FDG due to the interactions occurring at molecular levels. High activities of ^{18}F-FDG (111 MBq and 74 Mbq) actively produce positron, which originates from the nucleus and leads to an annihilation process rather than adsorption interaction with the GO. Figures 9 and 10 show the undifferentiation for low and high concentrations of GO and their influence on the activity decay of ^{18}F-FDG, where the line is almost redundant with each other. However, the slower decay process provides a likelihood for adsorption interactions, where the line is seen to be identical at 50 min onwards in Figure 6, as well as the whole measurable activity in Figure 8. This might be due to the availability of free electron clouds in ^{18}F-FDG, which have more chance to be interacted with and form functional groups in the GO chemical structures at low activity.

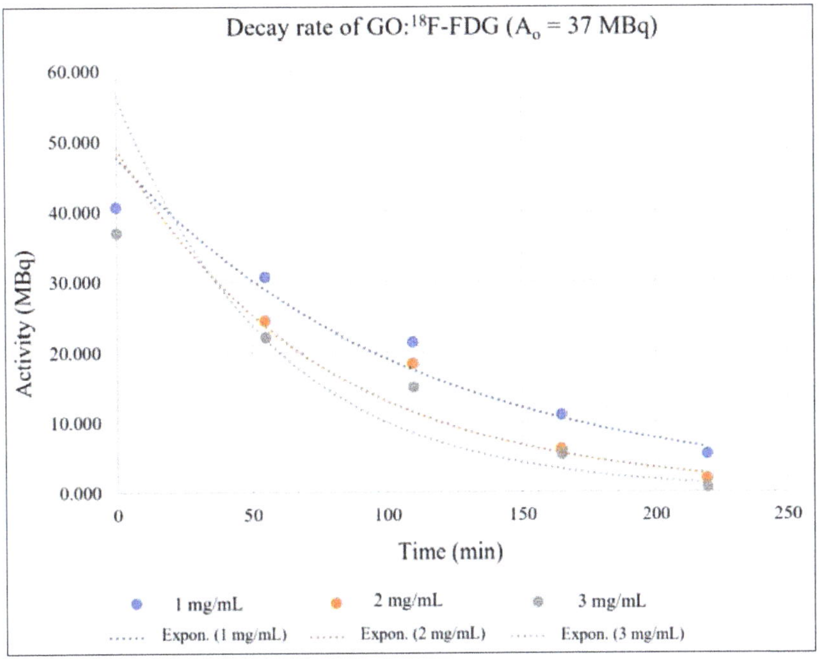

Figure 8. GO:^{18}F-FDG mixtures with an initial activity of 37 MBq and different concentrations of GO.

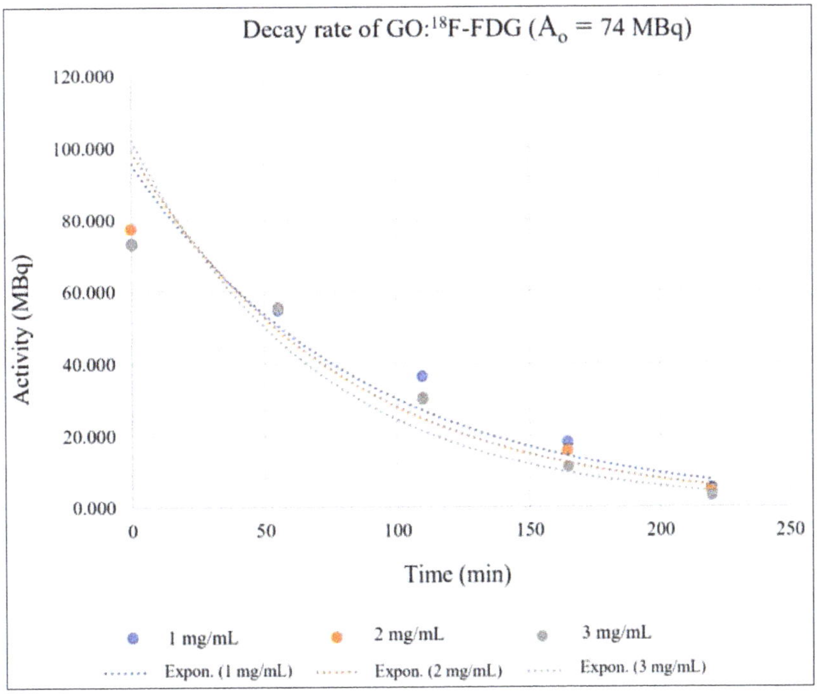

Figure 9. GO:^{18}F-FDG mixtures with an initial activity of 74 MBq and different concentrations of GO.

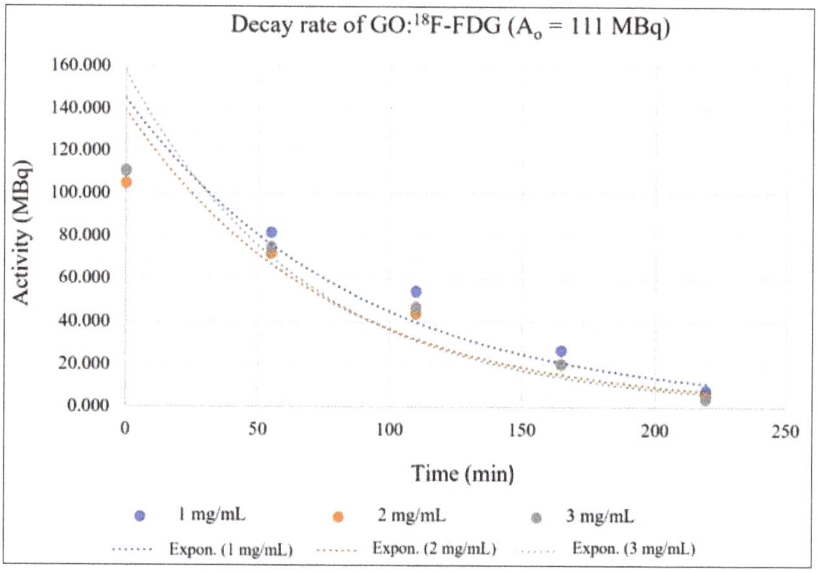

Figure 10. GO:^{18}F-FDG mixtures with an initial activity of 111 MBq and different concentrations of GO.

3.3. *UV–Vis Spectrometry and Chemical Analysis*

The study of absorbability of the graphene oxide and the radionuclide using the UV–vis spectrophotometer provided detailed information about the peak of the wavelength based on the Beer–Lambert law [43]. The GO and ^{18}F-FDG UV–visible adsorption spectra were obtained as shown in Figure 11. A single band GO adsorption peak can be observed, which was centred at 220 nm as depicted by the dashed red line. This band value indicates the $\pi \rightarrow \pi^*$, the transition of aromatic C-C bonds, which have numerous sp^2 bonds into the molecular mixtures [44]. The molecule absorbs the UV–vis light starting at the wavelength of 200 nm and begins to decline before attaining the peak at 220 nm. Thereafter, the molecule continues to decline to show that the functional group of GO has partially decreased as it passes a higher value of wavelength. The optical adsorption of GO is dominated by the $\pi \rightarrow \pi^*$ plasmon peak, which is near 230 nm [45]. This occurrence depends on two conjugative effects: clusters of nanometer-scale sp^2 and the chromophore unit, such as C=C, C=O and C-O bonds [45].

The chemical structure of ^{18}F is $1s^2 2s^2 2p^5$ and has a single free orbital electron. It is able to produce high energetic photons with 511 keV which activate the orbital electrons of the functional groups and delocalize on to the GO nano sheets via continuous $\pi \rightarrow \pi^*$ stacking interactions due to the stacked nanolayers of GO structures. The 511 keV photons energy might eject the orbital electrons at available functional groups to provide free molecular vacancy to be filled in by unstable 1^{8}F atoms. The electrons might interact with epoxy bridges, hydroxyl groups, and pairwise carboxyl groups in GO nanolayers [45] and might be influenced by excess energies received during the internal nuclear decay, leading to the annihilation process as shown in Figure 12.

The peak obtained at 220 nm was probably due to the aggregation of chromophore effects that were affected by the mixture of GO:^{18}F-FDG for different GO concentrations [46]. The 1 mg/mL showed the highest peak and the highest adsorbability compared to 2 mg/mL and 3 mg/mL of GO concentrations. This result is in line with the kinetic study illustrated in Figure 8, where a slow decay rate of low ^{18}F-FDG activity provided the chances for the molecular adsorption rate to occur. Higher concentrations might yield a complex saturated phase between the molecular mixtures to interact within a short time, whereas higher activities will generate excess energy to the particles for the annihilation process or eject the electrons from orbitals without interactions. The mixture probably modified the oxygen-containing functional groups, thereby causing the mixture to absorb a slightly lower range of UV–vis wavelength. The wavelength of adsorption also depends on the energy difference between the bonding or antibonding and non-bonding orbital concerned [47].

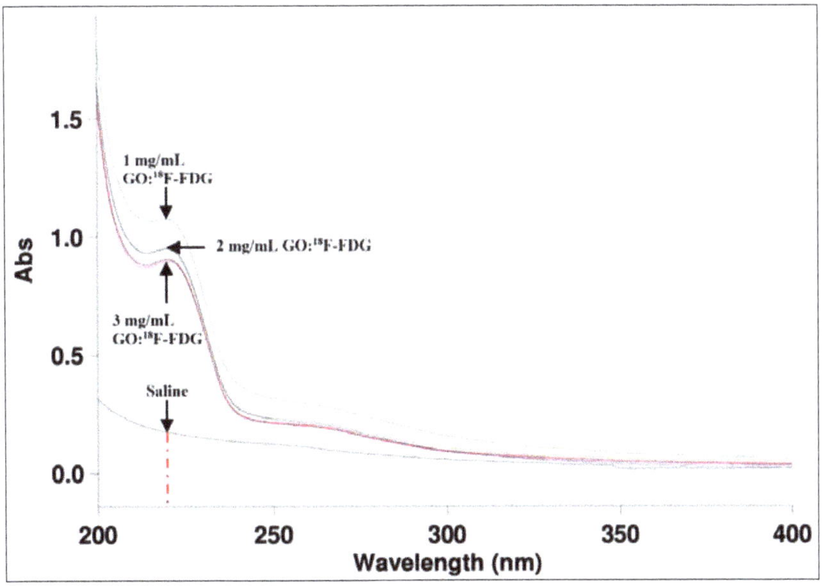

Figure 11. The UV–vis adsorption of GO:^{18}F-FDG with different concentrations has been seen on at the peak of 220 nm.

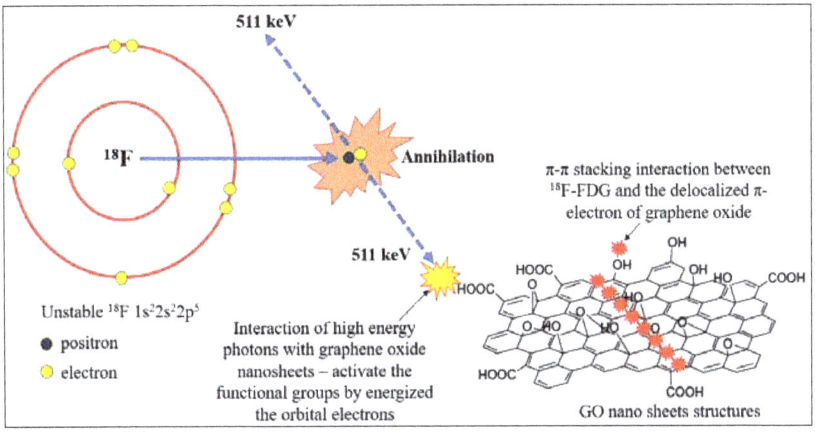

Figure 12. Illustration of energetic gamma rays produced by unstable ^{18}F when activating the adsorption process with functional groups and GO nanosheet structures.

4. Conclusions

Conclusively, GO has demonstrated high sorption affinity towards the radionuclide of a pure radiopharmaceutical. The high sorption capacity and the ability to coagulate with any reactive elements at molecular structures, such as ^{18}F-FDG, makes it a prime option for alternative radionuclides decontamination. The 'wax tissue' nanolayers and vast surface area have been shown to help GO wrap and adsorb radionuclides effectively. The adsorption rate was effective at a slow decay rate of ^{18}F-FDG, where more available free electrons are ready for the adsorption interaction with GO functional groups. Fluctuation of electron interactions and the active annihilation process occurred at high activity, thereby reducing the coagulation process within a short time due to the energetic nuclear process. The optical adsorption of GO with ^{18}F-FDG is dominated by the $\pi \rightarrow \pi^*$ plasmon peak, which is near 230 nm.

Author Contributions: Conceptualization, M.K.A.A.R. and A.M.N.; methodology, M.K.A.A.R., A.M.N. and N.H.A.; software, N.H.A. and F.A.M.Z.; validation, M.K.A.A.R., A.M.N. and N.H.A.; formal analysis, M.K.A.A.R., F.H.M.H., A.M.N. and F.A.M.Z.; investigation, M.K.A.A.R., N.M.N.; resources, M.K.A.A.R., N.M.N., A.M.N., M.M. and F.A.M.Z.; data curation, F.H.M.H., R.S. and F.I.; writing—original draft preparation, M.K.A.A.R. and F.H.M.H.; writing—review and editing, M.K.A.A.R., N.M.N., A.M.N.; visualization, M.K.A.A.R., A.M.N.; supervision, M.K.A.A.R., N.M.N., N.H.A. and M.M.; project administration, F.H.M.H., R.S. and F.I., M.M.; funding acquisition, M.K.A.A.R. and N.M.N. All authors have read and agreed to the published version of the manuscript.

Funding: This work was supported by a Short-Term Research Grant from Universiti Sains Malaysia [304/PPSK/6315499]. The APC was funded by Universiti Sains Malaysia.

Institutional Review Board Statement: Not applicable.

Informed Consent Statement: Not applicable.

Data Availability Statement: Not applicable.

Acknowledgments: We express our appreciation to Nik Fakurudin Nik Ali and Wan Norhasikin Wan Marizam of the School of Health Sciences, Universiti Sains Malaysia, for their help in obtaining the FESEM micrographs. Our appreciation goes to the Department of Nuclear Medicine, Radiotherapy & Oncology, Hospital Universiti Sains Malaysia and Nuclear Medicine & Molecular Imaging Department, Chancellor Tuanku Muhriz Hospital, Universiti Kebangsaan Malaysia for the use of their facilities when conducting the research project.

Conflicts of Interest: The authors declare no conflict of interest.

References

1. Payolla, B.; Massabni, A.F.; Orvig, C. Radiopharmaceuticals for Diagnosis in Nuclear Medicine: A short review. *Eclet. Quim.* **2019**, *44*, 11–19. [CrossRef]
2. Jacobson, O.; Kiesewetter, D.; Chen, X. Fluorine-18 Radiochemistry, Labeling Strategies and Synthetic Routes. *Bioconjugate Chem.* **2014**, *26*, 1–18. [CrossRef] [PubMed]
3. Park, B.; Kitteringham, N. Effects of Fluorine Substitution on Drug Metabolism: Pharmacological and Toxicological Implications. *Drug Metab. Rev.* **1994**, *26*, 605–643. [CrossRef] [PubMed]
4. Nawi, N.M.; Ahmad, N.S.; Abdullah, R.; Zainon, W.M.N.W.; Razab, M.K.A.A. Correlation of external dose rate with whole body clearance estimation in radioiodine therapy for rhTSH and THW patients. *J. Radiat. Res. Appl. Sci.* **2020**, *13*, 240–245. [CrossRef]
5. Ramesha, G.; Vijaya Kumara, A.; Muralidhara, H.; Sampath, S. Graphene and graphene oxide as effective adsorbents toward anionic and cationic dyes. *J. Colloid Interface Sci.* **2011**, *361*, 270–277. [CrossRef]
6. Jiang, W.; Chalich, Y.; Deen, M.J. Sensors for Positron Emission Tomography Applications. *Sensors* **2019**, *19*, 5019. [CrossRef]
7. Shivanand, B.; Akchata, S.; Lavanya, K. Influence of decontaminating agents and swipe materials on laboratory simulated working surfaces wet spilled with sodium pertechnetate. *Indian J. Nucl. Med.* **2017**, *32*, 173–176. [CrossRef]
8. Lee, W.; Ojovan, M.; Jantzen, C. *Radioactive Waste Management and Contaminated Site Clean-Up*; Woodhead Publishing: Cambridge, UK, 2013; pp. 301–326.
9. Fu, S.; Sun, Z.; Huang, P.; Li, Y.; Hu, N. Some Basic Aspects of Polymer Nanocomposites: A critical review. *Nano Mater. Sci.* **2019**, *1*, 2–30. [CrossRef]
10. Smith, A.; LaChance, A.; Zeng, S.; Liu, B.; Sun, L. Synthesis, properties, and applications of graphene oxide/reduced graphene oxide and their nanocomposites. *Nano Mater. Sci.* **2019**, *1*, 31–47. [CrossRef]
11. Sajjad, S.; Leghari, S.A.K.; Iqbal, A. Study of Graphene Oxide Structural Features for Catalytic, Antibacterial, Gas Sensing and Metals Decontamination Environmental Applications. *ACS Appl. Mater. Interfaces* **2017**, *9*, 43393–43414. [CrossRef]
12. Zhao, Y.; Zhang, Z.; Pan, Z.; Liu, Y. Advanced Bioactive Nanomaterials for Biomedical Applications. *Exploration* **2021**, *1*, 20210089. [CrossRef]
13. Novacek, M.; Jankovsky, O.; Luxa, J.; Sedmidubsky, D.; Pumera, M.; Fila, V.; Lhotka, M.; Klimova, K.; Matejkova, S.; Sofer, Z. Tuning of Graphene Oxide Composition by Multiple Oxidations for Carbon Dioxide Storage and Capture of Toxic Metals. *J. Mater. Chem. A* **2017**, *5*, 2739–2748. [CrossRef]

14. Prasadh, S.; Suresh, S.; Wong, R. Osteogenic Potential of Graphene in Bone Tissue Engineering Scaffolds. *Materials* **2018**, *11*, 1430. [CrossRef]
15. Raccichini, R.; Varzi, A.; Passerini, S.; Scrosati, B. The Role of Graphene for Electrochemical Energy Storage. *Nat. Mater.* **2015**, *14*, 271–279. [CrossRef] [PubMed]
16. Ng, Y.H.; Ikeda, S.; Matsumura, M.; Amal, R. A Perspective on Fabricating Carbon-based Nanomaterials by Photocatalysis and their Applications. *Energy Environ. Sci.* **2012**, *5*, 9307–9318. [CrossRef]
17. Wen, J.; Xu, Y.; Li, H.; Lu, A.; Sun, S. Recent Applications of Carbon Nanomaterials in Fluorescence Biosensing and Bioimaging. *Chem. Commun.* **2015**, *51*, 11346–11358. [CrossRef]
18. Shtein, M.; Nadiv, R.; Buzaglo, M.; Kahil, K.; Regev, O. Thermally Conductive Graphene Polymer Composites: Size, Percolation and Synergy Effects. *Chem. Mater.* **2015**, *27*, 2100–2106. [CrossRef]
19. Zunita, M.; Irawanti, R.; Koesmawati, T.A.; Lugitoa, G.; Wenten, I.G. Graphene Oxide (GO) Membrane in Removing Heavy Metals from Wastewater: A Review. *Chem. Eng. Trans.* **2020**, *82*, 415–420.
20. Ng, L.Y.; Chua, H.Y.; Ng, C.Y. Incorporation of Graphene Oxide-based Nanocomposite in the Polymeric Membrane for Water and Wastewater Treatment: A review on Recent Development. *J. Environ. Chem. Eng.* **2021**, *9*, 105994. [CrossRef]
21. Aunkor, M.T.H.; Raihan, T.; Prodhan, S.H.; Metselaar, H.S.C.; Malik, S.U.F.; Azad, A.K. Antibacterial Activity of Graphene Oxide Nanosheet Against Multidrug Resistant Superbugs Isolated from Infected Patients. *R. Soc. Open Sci.* **2020**, *7*, 200640. [CrossRef]
22. Eigler, S.; Hirsch, A. Chemistry with Graphene and Graphene Oxide—Challenges for Synthetic Chemists. *Angew. Chem. Int. Ed.* **2014**, *53*, 7720–7738. [CrossRef] [PubMed]
23. Sabzehmeidani, M.M.; Mahnaee, S.; Ghaedi, M.; Heidari, H.; Roy, V.A.L. Carbon Based Materials: A review of Adsorbents for Inorganic and Organic Compounds. *Mater. Adv.* **2021**, *2*, 598–627. [CrossRef]
24. Feltcorn, E. *Technology Reference Guide for Radiologically Contaminated Surfaces*; U.S. Environmental Protection Agency: Washington, DC, USA, 2006; pp. 11–18.
25. Zhao, G.; Wen, T.; Yang, X.; Yang, S.; Liao, L.; Hu, J.; Shao, D.; Wang, X. Preconcentration of U(VI) Ions on Few-layered Graphene Oxide Nanosheets from Aqueous Solutions. *Dalton Trans.* **2012**, *41*, 6182–6188. [CrossRef]
26. Rauwel, P.; Rauwel, E. Towards the Extraction of Radioactive Cesium-137 from Water via Graphene/CNT and Nanostructured Prussian Blue Hybrid Nanocomposites: A Review. *Nanomaterials* **2019**, *9*, 682. [CrossRef] [PubMed]
27. Suksompong, T.; Thongmee, S.; Sudprasert, W. Efficacy of a Graphene Oxide/Chitosan Sponge for Removal of Radioactive Iodine-131 from Aqueous Solutions. *Life* **2021**, *11*, 721. [CrossRef] [PubMed]
28. Romanchuk, A.Y.; Slesarev, A.S.; Kalmykov, S.N.; Kosynkin, D.V.; Tour, T.M. Graphene Oxide for Effective Radionuclide Removal. *Phys. Chem. Chem. Phys.* **2013**, *15*, 2321–2327. [CrossRef]
29. Noor, A.M.; Yusoff, N.F.A.; Huang, N.M.; Kari, Z.A.; Razab, M.K.A.A.; Bakar, M.H.A.; Lee, S.W.; Abdullah, N.H. Facile Preparation of Graphene Oxide Silver Aerogel for Antibacterial. *J. Trop. Resour. Sustain. Sci.* **2018**, *6*, 41–44.
30. Zin FA, M.; Noor, A.M.; Razab, M.K.A.A.; Abdullah, N.H.; Lee, S.W. Synthesis of Silver Graphene Oxide Nanocomposite Reinforced with Kenaf Cellulose Nanofibril Aerogel. *AIP Conf. Proc.* **2019**, *2068*, 020045-1–020045-6.
31. Noor, A.M.; Zin, F.M.; Wei, L.S.; Huang, N.M.; Bakar, M.H.A.; Sajab, M.S.; Razab, M.K.A.A.; Abdullah, N.H.; Kari, Z.A. Laser Scribe Silver-reduced Graphene Oxide as Novel Bactericidal Filter. *AIP Conf. Proc.* **2019**, *2068*, 020026-1–020026-6.
32. Zobir, S.A.M.; Rashid, S.A.; Tan, T. Recent Development on the Synthesis Techniques and Properties of Graphene Derivatives. In *Synthesis, Technology and Applications of Carbon Nanomaterials*; Rashid, S.A., Othman, R.N.I.R., Hussein, M.Z., Eds.; Matthew Deans: Oxford, UK, 2019; pp. 77–107.
33. Razab, M.K.A.A.; Mansor, M.S.; Noor, A.M.; Latif, N.F.F.A.; Rozi, S.M.; Jaafar, K.N.; Jamaludin, F. Characterization of GO:I-131 for Radioactive Clinical Waste Water Management in Nuclear Medicine. *Mater. Sci. Forum.* **2020**, *1010*, 561–566. [CrossRef]
34. Razab, M.K.A.A.; Mansor, M.S.; Noor, A.M.; Rozi, S.M.; Latif, N.F.F.A.; Jaafar, K.N.; Jamaludin, F. Preliminary Study of the Potential Graphene Oxide as Radioactive Clinical Wastewater Adsorbability in Nuclear Medicine. *IOP Conf. Ser. Earth Environ. Sci.* **2020**, *96*, 012037. [CrossRef]
35. Díez-Pascual, A.; Sainz-Urruela, C.; Vallés, C.; Vera-López, S.; San Andrés, M. Tailorable Synthesis of Highly Oxidized Graphene Oxides via an Environmentally-Friendly Electrochemical Process. *Nanomaterials* **2020**, *10*, 239. [CrossRef] [PubMed]
36. Yang, N.; Yang, D.; Zhang, G.; Chen, L.; Liu, D.; Cai, M.; Fan, X. The Effects of Graphene Stacking on the Performance of Methane Sensor: A First-Principles Study on the Adsorption, Band Gap and Doping of Graphene. *Sensors* **2018**, *18*, 422. [CrossRef] [PubMed]
37. Kamali, N.; Ghasemi, J.B.; Ziarani, G.M.; Moradian, S.; Badiei, A. Design, Synthesis, and Nanoengineered Modification of Spherical Graphene Surface by LDH for Removal of As(III) from Aqueous Solutions. *Chin. J. Chem. Eng.* 2022, in press. [CrossRef]
38. Aliyev EFiliz, V.; Khan, M.M.; Lee, Y.J.; Abetz, C.; Abetz, V. Structural Characterization of Graphene Oxide: Surface Functional Groups and Fractionated Oxidative Debris. *Nanomaterials* **2019**, *9*, 1180. [CrossRef]
39. Smajlagic, I.; Rowshanpour, R.; Milkin, L.; Dudding, T. Organofluorine Compounds in Fluorine-18 Positron Emission Tomography Imaging. *Res. Rev. Health Care Open Acc. J.* **2020**, *5*, 488–491.
40. Kuilla, T.; Bhadra, S.; Yao, D.; Kim, N.H.; Bose, S.; Lee, J.H. Recent Advances in Graphene Based Polymer Composites. *Prog. Polym. Sci.* **2010**, *35*, 1350–1375. [CrossRef]
41. Jenkins, D. Nuclear Structure and Radioactive Decay. In *Radiation Detection for Nuclear Physics*; IOP Publishing: Bristol, UK, 2020; pp. 1–32.

42. Alauddin, M.M. Positron Emission Tomography (PET) Imaging with ^{18}F-based Radiotracers. *Am. J. Nucl. Med. Mol. Imaging* **2012**, *2*, 55–76.
43. Yang, S.; Chen, Q.; Shi, M.; Zhang, Q.; Lan, S.; Maimaiti, T.; Li, Q.; Ouyang, P.; Tang, K.; Yang, S.T. Fast Identification and Quantification of Graphene Oxide in Aqueous Environment by Raman Spectroscopy. *Nanomaterials* **2020**, *10*, 770. [CrossRef]
44. Zin, F.A.M.; Noor, A.M.; Lee, S.W.; Sajab, M.S.; Razab, M.K.A.A.; Abdullah, N.H.; Ishak, W.M.F.W.; Wong, K.N.S.S.; Zaudin, N.A.C. Graphene Oxide Silver Cellulose Alginate for Antibacterial. *Mater. Sci. Forum.* **2020**, *1010*, 590–595.
45. Maslekar, N.; Zetterlund, P.B.; Kumar, P.V.; Agarwal, V. Mechanistic Aspects of the Functionalization of Graphene Oxide with Ethylene Diamine: Implications for Energy Storage Applications. *ACS Appl. Nano Mater.* **2021**, *4*, 3232–3240. [CrossRef]
46. Lai, Q.; Zhu, S.; Luo, X.; Zou, M.; Huang, S. Ultraviolet-visible Spectroscopy of Graphene Oxides. *AIP Adv.* **2012**, *2*, 032146-1–032146-5. [CrossRef]
47. Yu, L.; Yan, Q.; Ruzsinszky, A. Key Role of Antibonding Electron Transfer in Bonding on Solid Surfaces. *Phys. Rev. Mater.* **2019**, *3*, 092801. [CrossRef]

sustainability MDPI

Article

Asphaltene or Polyvinylchloride Waste Blended with Cement to Produce a Sustainable Material Used in Nuclear Safety

Hosam M. Saleh [1,*], Ibrahim I. Bondouk [2], Elsayed Salama [3], Hazem H. Mahmoud [1], Khalid Omar [2] and Heba A. Esawii [3,4]

1 Radioisotope Department, Nuclear Research Center, Egyptian Atomic Energy Authority (EAEA), Cairo 11787, Egypt; hazem_h_mansour@yahoo.com
2 Physics Department, Faculty of Science, University of Tanta, Tanta 31111, Egypt; ibrahim.bondok@science.tanta.edu.eg (I.I.B.); khalid.omar@science.tanta.edu.eg (K.O.)
3 Basic Science Department, Faculty of Engineering, The British University in Egypt (BUE), El Shorouk City 11837, Egypt; elsayed.salama@bue.edu.eg (E.S.); hyba.esawii@bue.edu.eg or heba.esawii@aucegypt.edu (H.A.E.)
4 School of Science and Engineering, The American University in Cairo (AUC), Cairo 11835, Egypt
* Correspondence: hosam.saleh@eaea.org.eg

Abstract: The current research uses sustainable methods to preserve the environment, such as exploiting municipal or industrial waste that may harm the environment. The wreckage of polyvinyl chloride (PVC) pipes and asphaltene are used as additives to cement to improve its mechanical properties, while stabilizing the radioactive waste resulting from the peaceful uses of nuclear materials, or enhancing its radiation shielding efficiency. New composites of Portland cement with ground PVC or asphaltene up to 50% are investigated. Fast neutron removal cross-section (Σ_R) and gamma shielding parameters, such as mass attenuation coefficient (MAC), half-value layer (HVL), effective atomic number (Z_{eff}), and exposure build-up factor (EBF) at wide energy range and thickness, are determined. The compressive strength and apparent porosity of the examined composites are examined to test the durability of the prepared composites as stabilizers for radioactive waste. The obtained results show that the bulk density of hardened cementitious composites was slightly increased by increasing the additive amount of PVC or asphaltene. The compressive strength of cement composites reached more than 4.5 MP at 50 wt.% PVC and 8.8 MPa at 50 wt.% asphaltene. These values are significantly higher than those recommended by the US Nuclear Regulatory Commission (3.4 MPa). Additionally, the obtained results demonstrate that although the gamma MAC is slightly decreased by adding asphaltene or PVC, the neutron removal cross-section was highly increased, reaching 171% in the case of 50 wt.% asphaltene and 304% in the case of 50 wt.% PVC. We can conclude that cement composites with PVC or asphaltene have optimized radiation shielding properties and can stabilize radioactive waste.

Keywords: asphaltene; cement; polyvinylchloride; radiation shielding; waste immobilization

Citation: Saleh, H.M.; Bondouk, I.I.; Salama, E.; Mahmoud, H.H.; Omar, K.; Esawii, H.A. Asphaltene or Polyvinylchloride Waste Blended with Cement to Produce a Sustainable Material Used in Nuclear Safety. *Sustainability* **2022**, *14*, 3525. https://doi.org/10.3390/su14063525

Academic Editor: Castorina Silva Vieira

Received: 28 February 2022
Accepted: 15 March 2022
Published: 17 March 2022

1. Introduction

Radiation shielding and radioactive waste stabilization are two critical requirements for safe peaceful nuclear applications. The various applications of radioactive materials have adhere to general radiation protection principles, such as justification, optimization, and dose and risk limits [1]. The reduction in personnel dose is a fundamental principle of radiation protection. To reduce the occupational radiation hazard, the level of radiation must be assessed, and three parameters are typically emphasized in controlling the radiation hazard: time, distance, and shielding [1].

The stabilization of radioactive waste reduces the possibility of radioactive contamination, migration, or the dispersing of radionuclides. Stabilization is defined as the transformation of waste into another form through solidification, embedding, or encapsulation for safe handling, transportation, storage, and disposal in designated landfills. Radioactive waste is encapsulated by physically surrounding it in materials such as bitumen or cement, which isolates it and retains radionuclides. Packaging or immobilization protects the radioactive substances, and prevents them from being accessed by the environment or escaping into the environment [2]. Furthermore, radioactive materials constantly irradiate through the immobilizing medium, sometimes at high levels of radiation. Radionuclides with long half-lives have extended irradiation periods and make the immobilization process a problem with no simple solution, necessitating the use of another material with high radiation shielding properties.

In the solidification/stabilization process, radioactive waste is converted into a stable solid form that is insoluble by cementation or vitrification, so as to prevent the radio contamination of the surrounding environment [3].

The excellent compressive strength and cost-effectiveness of cement allow it to be widely used as a stabilizer for hazardous and radioactive waste produced initially or as a by-product from treatment operations [4,5]. Cementation technology is used for the immobilization of radioactive waste. The practice of encapsulating radioactive waste in Portland cement began during the early years of the nuclear industry due to its low cost, availability, and compatibility with aqueous waste. The properties of cement vary depending on the type of cement used and the additives used, as well as the desired characteristics that the slurry design should have in order to maintain the durability of the cement. To achieve this goal, several studies and experiments have been carried out with the target of increasing the cement's stability, lifetime, and quality, while decreasing the cement's deterioration. Portland cement can be modified using a range of additives. These may include cellulosic waste [6,7], bitumen [8,9], glass [10], polymers [11–14], nanomaterials [15,16] and cement wastes [17,18].

With the growing problem of the high levels of problematic plastic waste produced today, polyvinylchloride (PVC) is the third most-produced synthetic polymer on a global scale. Annually, approximately 39.3 million tons are consumed worldwide, with a 3.2% increase in demand per year [19]. The recycling of PVC waste, including mechanical, chemical, re-extrusion, and burning, has been a leading cause of the presence of dioxin in incinerators when they are burned, and poses a harmful waste problem [20]. PVC wastes require recycling to overcome the many troubles of waste disposal and attain a sustainable solution for various industries, including nuclear applications. PVC wastes can be mixed with cement as a sustainable additive to produce cementitious products suitable for the immobilization of radioactive waste or as a shielding construction material. This technique is an environmentally friendly and economic method for developing economies such as Egypt, due to the great need to dispose of PVC and radioactive waste at the same time.

Another type of problematic material is asphaltene, since it can form dense flocculation and deposits in reservoirs, wellbores, and transportation pipelines, and thus can cause severe operational and production problems. Fortunately, asphaltene is a low-cost hydrocarbon byproduct produced during deep oil refining, with no significant uses in the pavement industry. Recently, asphaltenes were chemically modified for use as a novel thermal conductivity enhancer for liquid paraffin [21], and have been used with cement as additives to asphalt emulsion-stabilized layers [22]. Presently, asphaltene is characterized by its proper ability to mix with cement and produce a new composite of reasonable radiation shielding properties.

This research offers a comparative study to evaluate the mechanical properties, gamma attenuation and fast neutron removal cross-section of two cement composites, namely, a cement–PVC composite and a cement–asphaltene composite. This research reports on a systematic study demonstrating the effects of the incorporation of various ratios of grated PVC waste or asphaltene emulsion into Portland cement to improve the durability and mechanical properties of the produced composite, which will be used as a solidifying agent for radioactive waste, to enhance the attenuation coefficient required to achieve radiation safety in a shielding construction material, or to achieve the two purposes at the same time. This study aimed to investigate the suitability of some cement additives of definite compositions for producing proper composites with good performances and high densities, using different types of aggregates (asphaltene or PVC) that could enhance the shielding efficiency or achieve the safe stabilization of radioactive waste. The most significant issue in the present study is the elimination of two problematic wastes (asphaltene and PVC), and the production of an innovative product with reasonable efficiency that can be used in the immobilization of radioactive waste or radiation shielding construction, with eco-sustainability, low energy demands, and cost minimization.

2. Methodology of Research

2.1. Materials

2.1.1. Portland Cement

A commercially available Portland cement supplied by El Sewedy cement company, CEM1(42.5 N), was used in this study as the binding material, developed according to Egyptian Standard Specifications, ES 4756-1/2005 [23], and the British Standard Institution (BSI), EN 197-1/2011 [24].

2.1.2. Polyvinylchloride (PVC)

Scraps of polyvinylchloride pipes were collected as municipal solid waste and cleaned with tap water. The large pieces were crushed and sieved to obtain fine particles of polyvinyl waste appropriate for the subsequent blending with cement [25].

2.1.3. Asphaltene

The asphaltene used in this study was of commercial grade. Asphaltene is a fraction of a hydrocarbon fuel, consisting of condensed polynuclear aromatic ring systems at the center bearing alkyl side chains with hetero elements, the FT-IR spectrum has been presented previously in the literature [26].

2.2. Preparation of Cementitious Samples

Portland cement, the major component used in this study, was hydrated with 35% tap water without any additives to prepare the blank samples, or was blended with various additives, such as asphaltene or PVC, to prepare the different composites at various additive ratios.

Each group of compositions contained more than six specimens, three of which were subjected to compressive strength tests, three to porosity and water absorption measurements, and the remaining were subjected to attenuation testing. The pastes of various groups were poured into cylindrical bottles that were internally coated with Vaseline, manually compacted, tightly closed, and safely stored in the laboratory at room temperature (25.5 °C in wet conditions) for 28 days to allow complete reaction and proper solidification. After curing, the solidified samples were demolded and subjected to mechanical integrity tests, permeability spectroscopic analysis, and attenuation measurements.

2.3. Assessment of Mechanical Properties

The mechanical integrity is here expressed as compressive strength in MPa for cementitious samples—either cement mixed with asphaltene or PVC, or cement hydrated without any additives. In the Italian testing machine, Ma test E 159 SP, cylindrical samples were loaded axially between the loading plates to determine the maximum load before sample cracking, determined via the ASTM C109 standard test method [27]. Furthermore, the boiling water technique was used to measure porosity, bulk density, and water absorption following ASTM C 20-00 [28]. Using the water displacement method, which is based on Archimedes' saturation technique, the masses of porous specimens, both dry and in water, were recorded. The following two equations can be used to calculate apparent porosity.

$$P = [(W - D)/V] \times 100 \quad (1)$$

$$V = W - S \quad (2)$$

P—apparent porosity (%), W—saturated mass of the specimen (g), D—dry mass of the specimen (g), S—suspended weight of the specimen (g), V—exterior volume of the specimen (cm^3).

Then, the bulk density (r) can be calculated from the volume and saturated mass of the sample, as:

$$r = D/V \quad (3)$$

2.4. Determination of Gamma-Ray Shielding by Theoretical Aspects

The attenuation of γ-rays by a medium can be described by the well-known Lambert–Beer formula [29]:

$$I = B \times I_0 \times e^{-\mu_m x \rho} \quad (4)$$

where I and I_0 are, respectively, the transmitted and initial intensities of photons; B (X, E) is the build-up factor, depending on the penetration depth x of the material and energy E of the incident photon; μ_m is the mass attenuation coefficient (cm^2/g) and ρ is the mass density (g/cm^3). The mass attenuation coefficients (MAC) of the prepared samples at an energy range of 0.015–15 MeV were calculated using Phy-X/PSD, a user-friendly online software. This program, published by the National Institute of Standards and Technology, includes the coefficients of attenuation of all elements in the periodic table at various energies (NIS) [30].

The radiation protection efficiency (RPE), which is provided by the following equation, may be used to study the shielding efficiency of an absorber sample.

$$I = (1 - \frac{I}{I_0}) \times 100 \quad (5)$$

The effective atomic number (Z_{eff}) of a sample containing a mixture of elements is the ratio of its effective atomic cross-section (σ_a) to its electronic cross-section (σ_e), and can be calculated using the following formula [31];

$$Z_{eff} = \frac{\sigma_a}{\sigma_e} = \frac{\sum_i f_i A_i (\mu_m)_i}{\sum_i f_i \frac{A_i}{Z_i} (\mu_m)_i} \quad (6)$$

where A_i and Z_i are the atomic weight and the atomic number, respectively, of the i-element inside the sample. The term f_i refers to the fractional abundance. The Auto-Z_{eff} program is a user-friendly software presented recently by Taylor that can be used for the quick computation of Z_{eff} [32].

The equivalent atomic number (Z_{eq}) of the prepared sample at a specific energy may be calculated using the fractional mass attenuation coefficient owing to Compton scattering $R = (\mu_m)_{comp}/(\mu_m)_{total}$. This may be accomplished by matching R_1 and R_2 for two pure elements with two consecutive atomic numbers Z_1 and Z_2 at the same energy, resulting in R being positioned between R_1 and R_2 (near to R). The logarithmic interpolation formula of Z_{eq} is as follows [33]:

$$Z_{eq} = \frac{Z_1(\log R_2 - \log R) + Z_2(\log R - \log R_1)}{\log R_2 - \log R_1} \quad (7)$$

Harima et al. created the Geometric Progression (G-P) approximation, which is used to calculate the exposure build-up factors (EBF) of the prepared samples (1993). Using the interpolation formulas below, one can calculate the G-P fitting parameters and the related EBF [34,35].

$$B(E, X) = 1 + \frac{b-1}{K-1}\left(K^X - 1\right) \text{ for } K \neq 1 \quad (8)$$

$$B(E, X) = 1 + (b-1)X \text{ for } K = 1 \quad (9)$$

where X is the penetration depth ($X \leq 40$ mfp), b is the EBF value at 1 mfp, $K(E, X)$ is the multiplicative dose factor, and b, c, a, X_K and d are the computed G-P fitting parameters, which depend on the attenuating medium and the source energy. The G-P fitting parameters (b, c, a, X_K, and d) of the prepared samples throughout the 0.015–15 MeV gamma-ray energy range up to 40 mfp may be interpolated using a comparable logarithmic formula, such as Formula (5) [36,37]. The G-P fit requirements for the elements were derived from a research study conducted by the American Nuclear Society [38].

The fast neutron removal cross-section (Σ_R) of this material is the proportion of fast neutrons removed from a beam of neutrons during its penetration and after the first contact with the target material. For neutron energies of 2–12 MeV, the effective removal cross-section is nearly constant [36]. In the case of a combination of components, the removal cross-section may be calculated using the following formula:

$$\Sigma_R = \sum_i w_i (\Sigma_R)_i \quad (10)$$

where w_i and $(\Sigma_R)_i$ are the elemental weight fraction and removal cross-section, respectively. Dividing by the mass density (ρ), the fast neutron mass removal cross-section Σ_R/ρ (cm^2 g^{-1}) can be calculated.

3. Results and Discussion

3.1. Investigation of Mechanical Integrity and Porosity

3.1.1. Cement-Based PVC

The mechanical stability, reflected in the compressive strength of the reference sample of cement without any additives, had the highest value of 24.4 MPa; this is compared with the steady decrease in compressive strength, to 4.5 MPa, with an increasing PVC ratio from 10 to 50 wt.%, as shown in Figure 1. This fall in compressive strength is attributed to the heterogeneity between the two components, and their pores. On the other hand, the porosity showed the opposite trend; a gradual increase in porosity was correlated with an increasing PVC content in cement. The behavior of porosity explains the decreasing compressive strength.

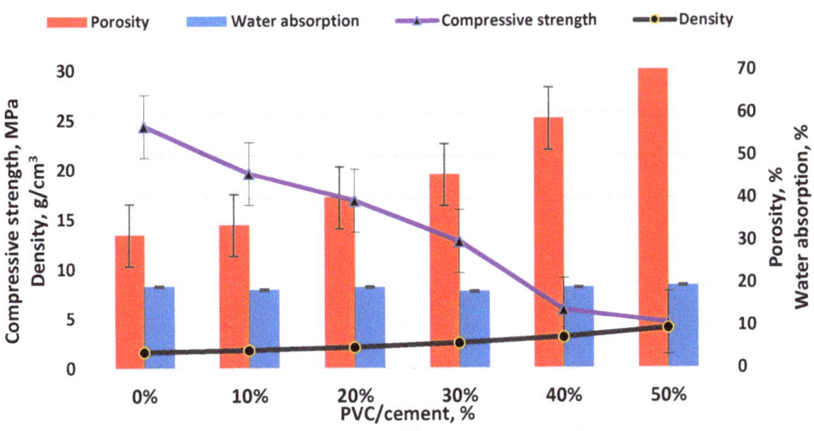

Figure 1. Compressive strength, density, water absorption, and porosity of cement blended with extended ratios of PVC.

However, the bulk density of the hardened cementitious composites was slightly increased by increasing the additive amount of PVC granules, due to their relatively high density, while the water absorption showed a constant value while increasing the addition of recycled PVC. According to previous research and in agreement with the current experiment, absorption affinity indirectly reflects the porosity due to permeable pore volume and connectivity [39]. As per the US Nuclear Regulatory Commission, the standard value relative to the compressive strength of solidified radioactive waste has to be more than 3.4 MPa [40], and the compressive strength of cement mixed with 50 wt.% PVC has reached more than 4.5 MPa; this value is significantly higher than recommended, and displays the standard parameters required for the immobilization of low and intermediate levels of radioactive waste.

3.1.2. Cement-Based Asphaltene

To examine the effect of increasing the addition of asphaltene into cement paste on the mechanical properties of the produced composite, the compressive strength, bulk density, porosity and water absorption are reported in Figure 2. In this condition, asphaltene will produce a homogeneous mixture with cement compared with PVC. A significant improvement in porosity to 24.24 can be observed with the gradual decrease in water absorption by increasing the amount of asphaltene, due to the potential filling of voids and pores inside the cementitious paste with asphaltene granules. A reasonable decrease in compressive strength was achieved with higher additions of the asphaltene, to reach 8.8 MPa at 50 wt.% asphaltene. It can be concluded from the results presented in Figures 1 and 2 that the heterogeneity and low density of the PVC with cement are higher than in the case of using asphaltene with cement, thus the porosity and water absorption increase while the compressive strength decreases [41,42].

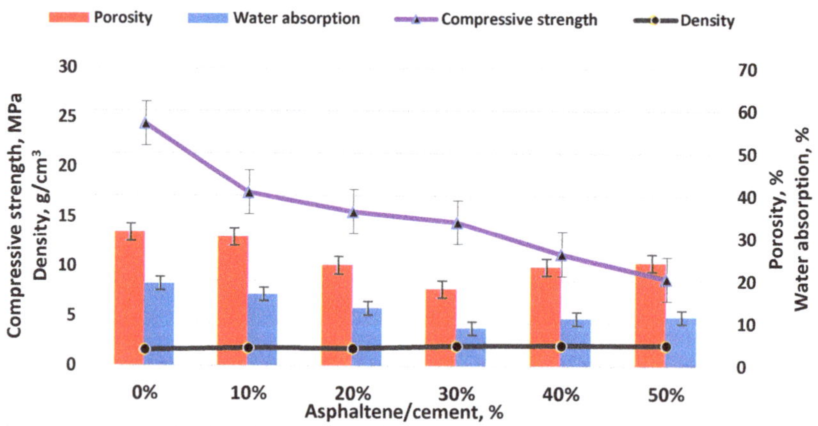

Figure 2. Compressive strength, density, water absorption, and porosity of cement blended with extended ratios of asphaltene.

A slight increase in the bulk density of the cementitious samples including asphaltene was detected when increasing the ratio of mixed asphaltene. Despite the presence of asphaltene not causing improvements in the mechanical integrity, the compressive strength of cement-based asphaltene at 50 wt.% has a value more than twice that required of materials used in radioactive waste stabilization, according to the standard specifications previously mentioned (3.4 MPa) [40], and this value is higher than in the case of PVC with cement (4.5 MPa). It is possible to conclude that asphaltene is more advantageous than PVC when mixed with cement up to 50 wt.%, and that it could be considered as a good additive to Portland cement for use in the solidification/stabilization of radioactive waste. Table 1 shows a descriptive comparison between the results here obtained and those reported previously in the recent literature.

Table 1. Comparing the mechanical properties of the present materials and others previously studied.

System	Compressive Strength, MPa	Porosity, %	Reference
Standard value	3.4	-	[40]
Cement without additives	32.0–36.5	27.5–30.0	[43,44]
Cement mixed with bitumen (30 wt.%)	7.62	60.14	[8]
Cement mixed with PVC (50 wt.%)	4.5	78.3	Present study
Cement mixed with asphaltene (50 wt.%)	8.8	24.24	Present study

3.2. Radiation Shielding Performance

3.2.1. Gamma-Rays Shielding Properties

The obtained results of the mass attenuation coefficients (MAC) of the prepared samples are shown in Figure 3. A regular trend in the MAC for all the prepared samples is observed. The MAC values are dramatically reduced at low photon energies of less than 0.1 MeV, where photoelectric interactions are prominent. The controlling of the photoelectric effect in such a region is perceived as a dramatic drop in the MAC. Compton scattering eventually became the dominant interaction as the energy range increased from 0.1 to 1 MeV, and the likelihood of interaction was inversely related to the energy. Therefore, the values of the MAC for all samples were also gradually decreasing. The MAC values for all samples increased somewhat in the energy range 1.02–15 MeV, where the pair formation interaction became the most prominent and was directly proportional to the energy. The obtained results indicate that adding asphaltene or PVC to cement samples slightly decreases the mass attenuation coefficient, and in turn increases the half-value layer of the cement, as shown in Figures 3 and 4. This was expected due to the low Z_{eff} of asphaltene and PVC compared with cement.

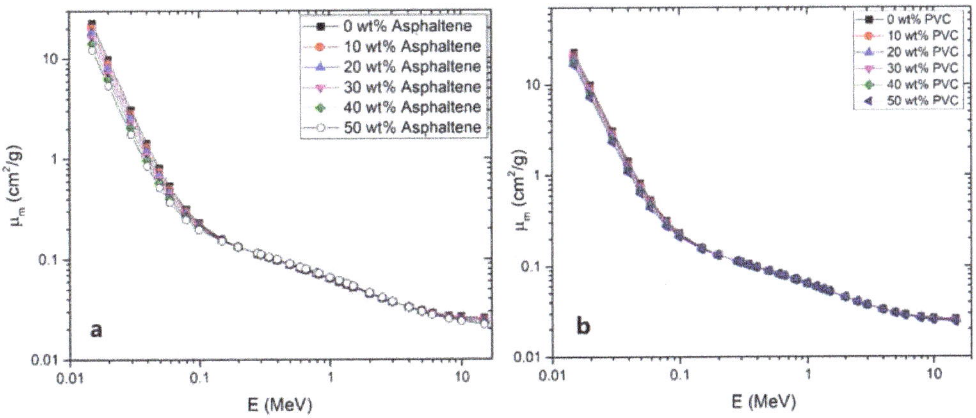

Figure 3. Mass attenuation coefficients of cement samples containing different concentrations of (**a**) asphaltene and (**b**) PVC.

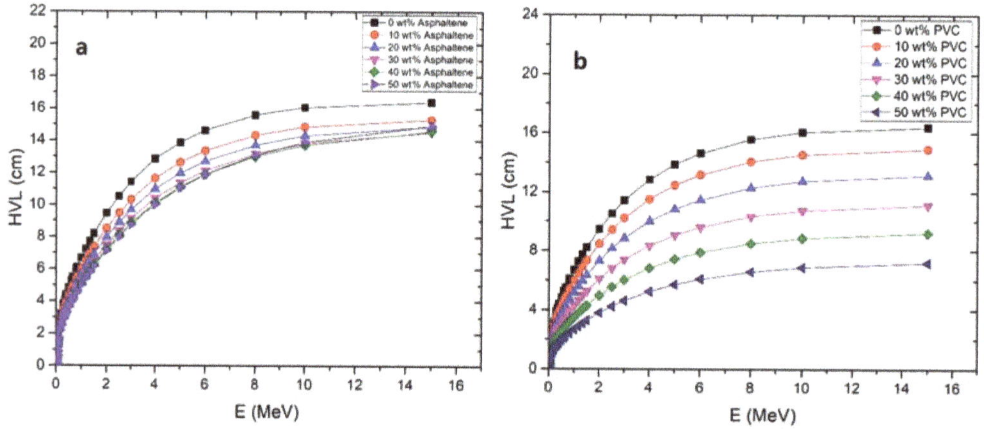

Figure 4. Half value layer (HVL) results of the cement samples containing different concentrations of (**a**) asphaltene and (**b**) PVC.

3.2.2. Radiation Protection Efficiency (RPE)

The prepared samples' radiation protection efficiencies compared with those of known materials, such as barite concrete at 0.662 MeV and 2 cm thickness, are shown in Figure 5. The obtained results show that cement with asphaltene and cement with PVC have larger RPEs than pure cement. Additionally, cement with 50% PVC has the largest RPE (79.4%) compared with pure cement (46.8%) and barite concrete (63.2%). The enhanced RPE of cement with the addition of asphaltene or PVC indicates these composites for use in radioactive waste stabilization. Moreover, the highest RPE of cement with 50% PVC recommends its use for gamma radiation shielding applications, more strenuously than concrete.

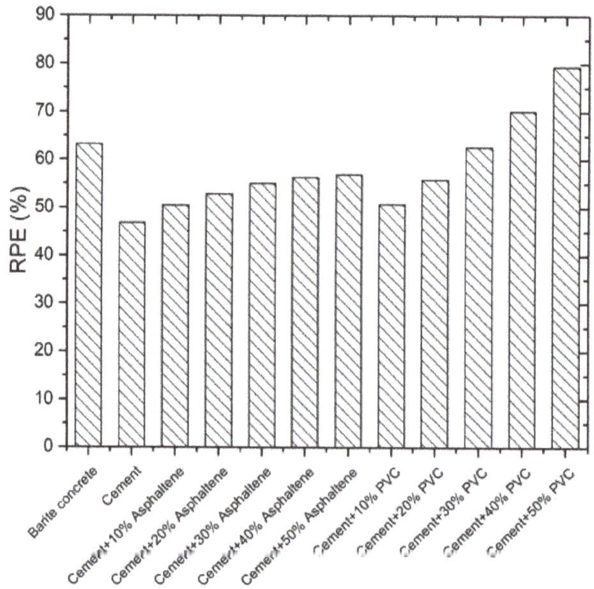

Figure 5. Radiation protection efficiency (RPE) results of the cement samples containing different concentrations of asphaltene and PVC.

3.2.3. Effective Atomic Number (Z_{eff})

Figure 6 shows the predicted effective atomic number Z_{eff} findings for the samples including asphaltene or PVC at energy ranges of 0.015–15 MeV. The Z_{eff} varies depending on the dominant interaction in each energy range, since the attenuation cross-section is proportional to Z^{4-5} for photoelectric attenuation, to Z for a Compton interaction, and to Z^2 for pair creation [45].

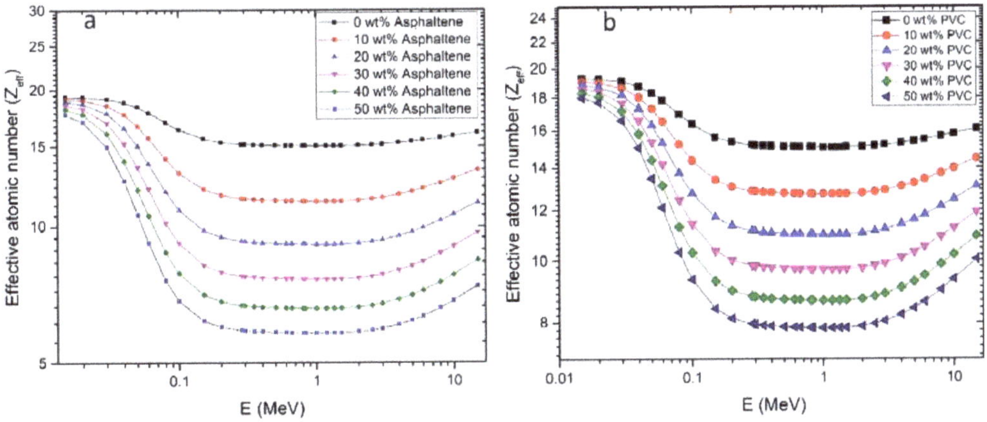

Figure 6. Effective atomic numbers of the cement samples containing different concentrations of (**a**) asphaltene and (**b**) PVC.

3.2.4. Exposure Build-Up Factors (EBF)

The computed gamma-ray energy exposure buildup factors (EBF), as a function of the incoming energy at 1, 2, 5, 10, 20, 30 and 40 MFP, are presented in Figures 7 and 8 for the cement samples with asphaltene and PVC, respectively, based on the acquired G-P fitting parameters (b, c, a, X_k, and d) for each sample.

Using the photon energy, additive concentrations (PVC or asphaltene), and penetration depth dependencies of the EBF, the data may be understood. The lowest EBF values were found in low- and high-energy areas, whereas values in the intermediate energy zone were greater for all asphaltene and PVC concentrations, as shown in Figures 7a–f and 8a–f, respectively. In the low- and high-energy areas, total absorption was most likely caused by the photoelectric effect and pair creation. Multiple Compton scattering, and hence larger EBF values, were seen in the intermediate area [45].

3.2.5. Fast Neutron Removal Cross-Section

Furthermore, the calculated fast neutron removal cross-sections Σ_R of the examined samples are shown in Figure 9a,b. Although the gamma MAC is slightly decreased with the adding of asphaltene or PVC, the neutron removal cross-section was highly increased, reaching 171% in the case of 50 wt.% asphaltene and 304% in the case of 50 wt.% PVC.

This can be attributed to the higher fast neutron removal cross-sections of the constituent elements of both PVC and asphaltene. The effective fast neutrons mass removal cross-sections Σ_R/ρ (cm^2 g^{-1}) of cement and PVC are 0.03069 and 0.06252 cm^2 g^{-1}, respectively [46], while for asphaltene this value is about 0.08820 cm^2 g^{-1}. Finally, based on the obtained radiation shielding results, we have found that adding PVC or asphaltene does not significantly change the gamma shielding qualities, but it does improve the neutron removal cross-section and the mechanical properties of the cement.

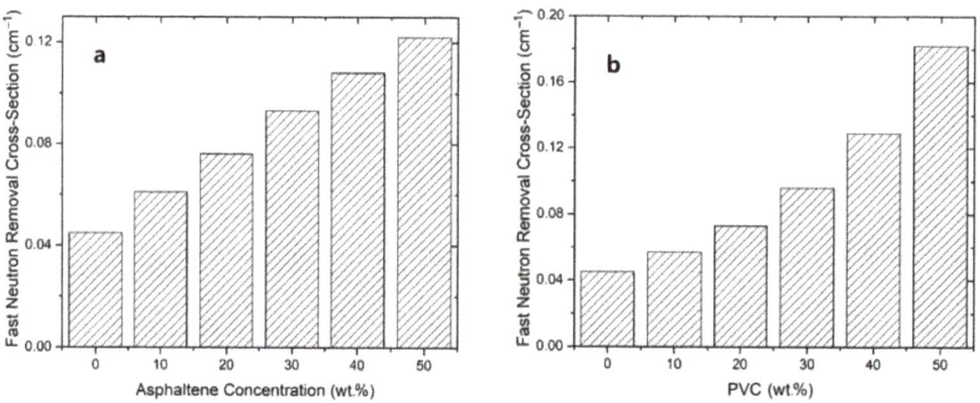

Figure 9. Fast neutron removal cross-sections of the cement samples containing different concentrations of (**a**) asphaltene and (**b**) PVC.

Comparisons between the results obtained here for the prepared composites and those for the well-known ones such as Barite concrete and Portland cement are shown in Table 2. The tabulated results indicate that cement mixed with bitumen has a higher mass attenuation coefficient than the other composites, while the fast neutron removal cross section of cement mixed with PVC or asphaltene showed the highest values.

Table 2. Comparison between similar shielding materials.

System	Rang of μ_m (cm^2/g) $\times 10^{-2}$ (at 0.662 MeV)	Fast Neutron Mass Removal Cross-Sections, Σ_R/ρ (cm^2 g^{-1})	Reference
Barite concrete	6.7–7.8	0.041–0.027	[47,48]
Portland cement	7.76	0.031	[30]
Cement mixed with bitumen (10–50 wt.%)	9.94–14.11	-	[8]
Cement mixed with PVC (10–50 wt.%)	7.78–7.84	0.045–0.182	Present study
Cement mixed with asphaltene (10–50 wt.%)	7.80–8.01	0.045–0.122	Present study

4. Conclusions

In this study, the elimination of two problematic waste products (asphaltene and PVC), in order to produce an innovative product with a reasonable efficiency that can be used in the immobilization of radioactive waste or radiation shielding construction, has been conducted, while the achieving of environmental sustainability, low energy demand, and cost minimization have been considered in parallel.

Asphaltene or PVC were mixed with Portland cement in different proportions up to 50%. The performances of the resulting compounds were evaluated using several parameters, such as compressive strength, porosity, bulk density, water absorption, and attenuation coefficient. The physical and mechanical stabilities of the modified cement composites gradually decreased with the increases in the amounts of both additives, while the attenuation and isolation performance, as well as γ-radiation shielding, were slightly affected. The effective cross-sections for removing neutron mass from cement with PVC or asphaltene were significantly increased. The results of this study show that the new compound prepared by mixing cement with asphaltene or PVC can be used to safely encapsulate hazardous toxic and radioactive wastes, while providing adequate radiation shielding from gamma-rays for the environment at the same time.

Author Contributions: H.M.S.: methodology, validation, data curation, writing—original draft, writing—review and editing. I.I.B.: conceptualization, methodology, resources. E.S.: methodology, writing—original draft. H.H.M.: methodology. K.O.: visualization, H.A.E.: methodology. All authors have read and agreed to the published version of the manuscript.

Funding: This research received no external funding.

Institutional Review Board Statement: Not applicable.

Informed Consent Statement: Not applicable.

Data Availability Statement: Not applicable.

Conflicts of Interest: The authors declare that they have no known competing financial interests or personal relationships that could have appeared to influence the work reported in this paper.

References

1. Do, K.-H. General Principles of Radiation Protection in Fields of Diagnostic Medical Exposure. *J. Korean Med. Sci.* **2016**, *31*, S6–S9. [CrossRef] [PubMed]
2. Bayoumi, T.A.; Reda, S.M.; Saleh, H.M. Assessment Study for Multi-Barrier System Used in Radioactive Borate Waste Isolation Based on Monte Carlo Simulations. *Appl. Radiat. Isot.* **2012**, *70*, 99–102. [CrossRef] [PubMed]
3. Saleh, H.M.; Eskander, S.B. Long-Term Effect on the Solidified Degraded Cellulose-Based Waste Slurry in Cement Matrix. *Acta Montan. Slovaca* **2009**, *14*, 291–297.
4. Saleh, H.M.; Mahmoud, H.H.; Aglan, R.F.; Bayoumi, T.A. Biological Treatment of Wastewater Contaminated with Cu(Ii), Fe(Ii) and Mn(Ii) Using Ludwigia Stolonifera Aquatic Plant. *Environ. Eng. Manag. J.* **2019**, *18*, 1327–1336. [CrossRef]
5. Saleh, H.M.; Moussa, H.R.; El-Saied, F.A.; Dawoud, M.; Nouh, E.S.A.; Abdel Wahed, R.S. Adsorption of Cesium and Cobalt onto Dried *Myriophyllum spicatum* L. from Radio-Contaminated Water: Experimental and Theoretical Study. *Prog. Nucl. Energy* **2020**, *125*, 103393. [CrossRef]
6. Saleh, H.M.; Eskander, S.B.; Fahmy, H.M. Mortar Composite Based on Wet Oxidative Degraded Cellulosic Spinney Waste Fibers. *Int. J. Environ. Sci. Technol.* **2014**, *11*, 1297–1304. [CrossRef]
7. Dawoud, M.M.A.; Hegazy, M.M.; Helew, W.K.; Saleh, H.M. Overview of Environmental Pollution and Clean Management of Heavy Metals and Radionuclides by Using Microcrystalline Cellulose. *J. Nucl. Energy Sci. Power Gener. Technol.* **2021**, *10*, 100201.
8. Saleh, H.M.; Bondouk, I.I.; Salama, E.; Esawii, H.A. Consistency and Shielding Efficiency of Cement-Bitumen Composite for Use as Gamma-Radiation Shielding Material. *Prog. Nucl. Energy* **2021**, *137*, 103764. [CrossRef]
9. Reda, S.M.; Saleh, H.M. Calculation of the Gamma Radiation Shielding Efficiency of Cement-Bitumen Portable Container Using MCNPX Code. *Prog. Nucl. Energy* **2021**, *142*, 104012. [CrossRef]
10. Eid, M.S.; Bondouk, I.I.; Saleh, H.M.; Omar, K.M.; Sayyed, M.I.; El-Khatib, A.M.; Elsafi, M. Implementation of Waste Silicate Glass into Composition of Ordinary Cement for Radiation Shielding Applications. *Nucl. Eng. Technol.* **2021**. [CrossRef]
11. Eskander, S.B.; Saleh, H.M.; Tawfik, M.E.; Bayoumi, T.A. Towards Potential Applications of Cement-Polymer Composites Based on Recycled Polystyrene Foam Wastes on Construction Fields: Impact of Exposure to Water Ecologies. *Case Stud. Constr. Mater.* **2021**, *15*, e00664. [CrossRef]
12. Saleh, H.; Salman, A.; Faheim, A.; El-Sayed, A. Polymer and Polymer Waste Composites in Nuclear and Industrial Applications. *J. Nucl. Energy Sci. Power Gener. Technol.* **2020**, *9*, 1000199.
13. Saleh, H.M.; Eskander, S.B. Impact of Water Flooding on Hard Cement-Recycled Polystyrene Composite Immobilizing Radioactive Sulfate Waste Simulate. *Constr. Build. Mater.* **2019**, *222*, 522–530. [CrossRef]
14. Smirnova, O.M.; Menéndez Pidal de Navascués, I.; Mikhailevskii, V.R.; Kolosov, O.I.; Skolota, N.S. Sound-Absorbing Composites with Rubber Crumb from Used Tires. *Appl. Sci.* **2021**, *11*, 7347. [CrossRef]
15. Saleh, H.M.; El-Saied, F.A.; Salaheldin, T.A.; Hezo, A.A. Influence of Severe Climatic Variability on the Structural, Mechanical and Chemical Stability of Cement Kiln Dust-Slag-Nanosilica Composite Used for Radwaste Solidification. *Constr. Build. Mater.* **2019**, *218*, 556–567. [CrossRef]
16. Saleh, H.M.; El-Sheikh, S.M.; Elshereafy, E.E.; Essa, A.K. Performance of Cement-Slag-Titanate Nanofibers Composite Immobilized Radioactive Waste Solution through Frost and Flooding Events. *Constr. Build. Mater.* **2019**, *223*, 221–232. [CrossRef]
17. Saleh, H.M.; Salman, A.A.; Faheim, A.A.; El-Sayed, A.M. Sustainable Composite of Improved Lightweight Concrete from Cement Kiln Dust with Grated Poly (Styrene). *J. Clean. Prod.* **2020**, *277*, 123491. [CrossRef]
18. Saleh, H.M.; Salman, A.A.; Faheim, A.A.; El-Sayed, A.M. Influence of Aggressive Environmental Impacts on Clean, Lightweight Bricks Made from Cement Kiln Dust and Grated Polystyrene. *Case Stud. Constr. Mater.* **2021**, *15*, e00759. [CrossRef]
19. Ragab, A.A.; Mohammedy, M.M.; El-Shafie, M. Using Waste Flexible Polyvinyl Chloride Treated with DOP/Calcium Hydroxide for Enriching the Performance of Oxidizing Bitumen. *J. Therm. Anal. Calorim.* **2019**, *136*, 1079–1091. [CrossRef]

20. Kou, S.C.; Lee, G.; Poon, C.S.; Lai, W.L. Properties of Lightweight Aggregate Concrete Prepared with PVC Granules Derived from Scraped PVC Pipes. *Waste Manag.* **2009**, *29*, 621–628. [CrossRef] [PubMed]
21. Glova, A.D.; Nazarychev, V.M.; Larin, S.V.; Lyulin, A.V.; Lyulin, S.V.; Gurtovenko, A.A. Asphaltenes as Novel Thermal Conductivity Enhancers for Liquid Paraffin: Insight from in Silico Modeling. *J. Mol. Liq.* **2021**, *346*, 117112. [CrossRef]
22. Uddin, M.M.; Kamran, F.; Hashemian, L. Performance Comparison of Asphalt Emulsion Stabilized Granular Base Modified with Cement or Asphaltenes. *Can. J. Civ. Eng.* **2021**. [CrossRef]
23. *ESS-4756-1/2005*; Ordinary Portland Cement (OPC) CEM1 (42.5 N). Egyptian Standard Specifications: Cairo, Egypt, 2005.
24. *BSI-197-1/2011*; Cement, Composition, Specifications and Conformity Criteria for Common Cements. British Standards Institution: London, UK, 2011; Volume 1.
25. Wang, J.; Wang, H.; Yue, D. Separation of Waste Polymethyl Methacrylate and Polyvinyl Chloride Mixtures by Flotation after Fenton Oxidation. *J. Clean. Prod.* **2019**, *228*, 1218–1228. [CrossRef]
26. Speight, J.G.; Moschopedis, S.E. On the Molecular Nature of Petroleum Asphaltenes. In *Chemistry of Asphaltenes*; ACS Publications: Washington, DC, USA, 1981.
27. *ASTM C109/C109M-16a*; ASTM Standard Test Method for Compressive Strength of Hydraulic Cement Mortars (Using 2-in. or [50-Mm] Cube Specimens). ASTM International: West Conshohocken, PA, USA, 2016.
28. *ASTM C20-00*; ASTM Standard Test Methods for Apparent Porosity, Water Absorption, Apparent Specific Gravity, and Bulk Density of Burned Refractory Brick and Shapes by Boiling Water. ASTM International: West Conshohocken, PA, USA, 2015.
29. Singh, V.P.; Badiger, N.M.; Kaewkhao, J. Radiation Shielding Competence of Silicate and Borate Heavy Metal Oxide Glasses: Comparative Study. *J. Non-Cryst. Solids* **2014**, *404*, 167–173. [CrossRef]
30. Berger, M.J. Photon Cross-Sections Database. NIST Standard Reference Database 8 (XGAM). 1998. Available online: http://physics.nist.gov/PhysRefData/Xcom/Text/XCOM.html (accessed on 1 February 2022).
31. Tijani, S.A.; Kamal, S.M.; Al-Hadeethi, Y.; Arib, M.; Hussein, M.A.; Wageh, S.; Dim, L.A. Radiation Shielding Properties of Transparent Erbium Zinc Tellurite Glass System Determined at Medical Diagnostic Energies. *J. Alloys Compd.* **2018**, *741*, 293–299. [CrossRef]
32. Taylor, M.L.; Smith, R.L.; Dossing, F.; Franich, R.D. Robust Calculation of Effective Atomic Numbers: The Auto-Zeff Software. *Med. Phys.* **2012**, *39*, 1769–1778. [CrossRef]
33. Sathiyaraj, P.; Samuel, E.J.J.; Valeriano, C.C.S.; Kurudirek, M. Effective Atomic Number and Buildup Factor Calculations for Metal Nano Particle Doped Polymer Gel. *Vacuum* **2017**, *143*, 138–149. [CrossRef]
34. Harima, Y. An Historical Review and Current Status of Buildup Factor Calculations and Applications. *Radiat. Phys. Chem.* **1993**, *41*, 631–672. [CrossRef]
35. Kavaz, E.; Yorgun, N.Y. Gamma Ray Buildup Factors of Lithium Borate Glasses Doped with Minerals. *J. Alloys Compd.* **2018**, *752*, 61–67. [CrossRef]
36. Kaplan, M.F. *Concrete Radiation Shielding: Nuclear Physics, Concrete Properties, Design and Construction*; Longman Scientific & Technical: Harlow, UK, 1989; ISBN 0470213388.
37. Singh, V.P.; Badiger, N.M. Gamma Ray and Neutron Shielding Properties of Some Alloy Materials. *Ann. Nucl. Energy* **2014**, *64*, 301–310. [CrossRef]
38. *ANSI/ANS-6.4.3*; Gamma-Ray Attenuation Coefficients and Buildup Factors for Engineering Materials. American Nuclear Society: La Grange Park, IL, USA, 1991.
39. Babu, K.G.; Babu, D.S. Behaviour of Lightweight Expanded Polystyrene Concrete Containing Silica Fume. *Cem. Concr. Res.* **2003**, *33*, 755–762. [CrossRef]
40. NRC. *Waste Form Technical Position, Revision 1*; US Nuclear Regulatory Commission: Washington, DC, USA, 1991.
41. Zingg, L.; Briffaut, M.; Baroth, J.; Malecot, Y. Influence of Cement Matrix Porosity on the Triaxial Behaviour of Concrete. *Cem. Concr. Res.* **2016**, *80*, 52–59. [CrossRef]
42. Prêt, D.; Sardini, P.; Beaufort, D.; Zellagui, R.; Sammartino, S. Porosity Distribution in a Clay Gouge by Image Processing of 14C-PolyMethylMethAcrylate (14C-PMMA) Autoradiographs: Case Study of the Fault of St. Julien (Basin of Lodève, France). *Appl. Clay Sci.* **2004**, *27*, 107–118. [CrossRef]
43. Saleh, H.M.; Aglan, R.F.; Mahmoud, H.H. Qualification of Corroborated Real Phytoremediated Radioactive Wastes under Leaching and Other Weathering Parameters. *Prog. Nucl. Energy* **2020**, *119*, 103178. [CrossRef]
44. Saleh, H.M.; Moussa, H.R.; El-Saied, F.A.; Dawod, M.; Bayoumi, T.A.; Abdel Wahed, R.S. Mechanical and Physicochemical Evaluation of Solidifed Dried Submerged Plants Subjected to Extreme Climatic Conditions to Achieve an Optimum Waste Containment. *Prog. Nucl. Energy* **2020**, *122*, 103285. [CrossRef]
45. Salama, E.; Maher, A.; Youssef, G.M. Gamma Radiation and Neutron Shielding Properties of Transparent Alkali Borosilicate Glass Containing Lead. *J. Phys. Chem. Solids* **2019**, *131*, 139–147. [CrossRef]
46. El-Khayatt, A.M. Calculation of Fast Neutron Removal Cross-Sections for Some Compounds and Materials. *Ann. Nucl. Energy* **2010**, *37*, 218–222. [CrossRef]
47. Picha, R.; Channuie, J.; Khaweerat, S.; Liamsuwan, T.; Promping, J.; Ratanatongchai, W.; Silva, K.; Wonglee, S. Gamma and Neutron Attenuation Properties of Barite-Cement Mixture. *J. Phys. Conf. Ser.* **2015**, *611*, 012002. [CrossRef]
48. Akkurt, I.; Akyildirim, H.; Karipçin, F.; Mavi, B. Chemical Corrosion on Gamma-Ray Attenuation Properties of Barite Concrete. *J. Saudi Chem. Soc.* **2012**, *16*, 199–202. [CrossRef]

sustainability

Review

Chelating Agents in Assisting Phytoremediation of Uranium-Contaminated Soils: A Review

Yue You [1], Junfeng Dou [1], Yu Xue [2], Naifu Jin [1] and Kai Yang [1,*]

[1] College of Water Sciences, Beijing Normal University, Beijing 100875, China; 202021470042@mail.bnu.edu.cn (Y.Y.); doujf@bnu.edu.cn (J.D.); n.jin@bnu.edu.cn (N.J.)
[2] School of Chemical and Environmental Engineering, China University of Mining and Technology, Beijing 100083, China; 1810380315@student.cumtb.edu.cn
* Correspondence: yangk@bnu.edu.cn; Tel.: +86-138-1175-3081

Abstract: Massive stockpiles of uranium (U) mine tailings have resulted in soil contamination with U. Plants for soil remediation have low extraction efficiency of U. Chelating agents can mobilize U in soils and, hence, enhance phytoextraction of U from the soil. However, the rapid mobilization rate of soil U by chelating agents in a short period than plant uptake rate could increase the risk of groundwater contamination with soluble U leaching down the soil profile. This review summarizes recent progresses in synthesis and application of chelating agents for assisting phytoremediation of U-contaminated soils. In detail, the interactions between chelating agents and U ions are initially elucidated. Subsequently, the mechanisms of phytoextraction and effectiveness of different chelating agents for phytoremediation of U-contaminated soils are given. Moreover, the potential risks associated with chelating agents are discussed. Finally, the synthesis and application of slow-release chelating agents for slowing down metal mobilization in soils are presented. The application of slow-release chelating agents for enhancing phytoextraction of soil U is still scarce. Hence, we propose the preparation of slow-release biodegradable chelating agents, which can control the release speed of chelating agent into the soil in order to match the mobilization rate of soil U with plant uptake rate, while diminishing the risk of residual chelating agent leaching to groundwater.

Keywords: chelator; soil uranium; mobilization; phytoextraction; slow release

Citation: You, Y.; Dou, J.; Xue, Y.; Jin, N.; Yang, K. Chelating Agents in Assisting Phytoremediation of Uranium-Contaminated Soils: A Review. *Sustainability* **2022**, *14*, 6379. https://doi.org/10.3390/su14106379

Academic Editors: Hosam M. Saleh and Mohammad Mahmoud Dawoud

Received: 13 April 2022
Accepted: 18 May 2022
Published: 23 May 2022

1. Introduction

Nuclear energy is a major source of low-carbon electricity and plays an important role in the achievement of carbon neutrality. According to the World Nuclear Performance Report 2021, there are 443 operational reactors with a total capacity of 3.94×10^5 MWe, and 57 reactors with a total capacity of 5.88×10^4 MWe under construction worldwide [1]. The global nuclear power generation reached 2.55×10^3 TWh in 2020, meanwhile the annual global demand for uranium (U), a primary element used in nuclear energy, has increased to 6.25×10^4 t [1,2]. At the same time, soil contamination with U has become a serious environmental issue.

There are two main sources of U in soils, i.e., natural source and anthropogenic source. U naturally occurs in soil parent materials. Due to different accumulation levels of U in different soil parent materials, the concentration of U in soils can vary greatly. For example, the average background concentration of soil U in the United States (US) is 3.5 mg/kg, and that in China is 3.13 mg/kg, whereas Portugal reaches 25.1 mg/kg [3]. The background concentration of U in the environment is usually low. However, anthropogenic activities (e.g., industrial and agricultural production and military activities) have emitted a large amount of U into the environment, resulting in distinct accumulation of U in soils [4–9]. U tailings are the main source of U in contaminated soils. For example, U contamination in the soil and groundwater within the catchment of a U mill tailings pond can be caused by the hydrological cycle around the tailings pond [10].

Although U has a long half-life and relatively weak radioactivity, it may induce chemical and radiological toxicity in organisms after long-term exposure to U-contaminated soils or U accumulation in the organism through the food chain [11–17]. On the one hand, U can induce ionic toxicity similar to that of cadmium (Cd), copper (Cu), zinc (Zn), and other heavy metals. The ionic toxicity of U increases with U concentration in the contaminated environmental substrate. On the other hand, the radiation from the decay of U can cause oral necrosis, anemia or chronic diseases of lung and kidney in mild cases, and central nervous system damage or death in severe cases [17,18]. In addition, chronic U exposure can cause chromosome aberrations and increase cancer risk [19,20]. Therefore, it is of great importance to clean up U-contaminated soils.

Among common remediation technologies for soil contamination with heavy metals and radionuclides, phytoextraction is a popular in situ technology for large-scale soil remediation due to its low cost and environmental friendliness. Although several plants (e.g., broad bean (*Vicia faba* L.), willow (*Salix smithiana* Willd.), radish (*Raphanus sativus* L.), Indian mustard (*Brassica juncea* L.), and sunflower (*Helianthus annuus* L.)) have the ability to absorb U from soils [21–25], hyperaccumulator plant species of U have not yet been discovered. In addition, plants for remediation of U-contaminated soils usually have low extraction efficiency of U, which results in a long-term remediation period and consequently hinders large-scale phytoremediation of U-contaminated soils. Apart from searching for U hyperaccumulators, numerous studies have showed that chelating agents can mobilize heavy metals and radionuclides in soils and therefore enhance plant uptake of these contaminants [26–28]. Due to the lack of U hyperaccumulators, chelating agent-assisted phytoremediation can be an effective approach for U-contaminated soil remediation.

It should be noted several previous studies found that the mobilization rate of soil U by chelating agents exceeded the plant U uptake rate, consequently causing groundwater contamination with soluble U through leaching processes as well as transient phytotoxicity [26,28]. In fact, U contamination in groundwater is also a global environmental issue. It is reported that the groundwater in India, South Korea, the US, and Zambia has been contaminated with U to a varying extent [29]. In Nambe, New Mexico, US, the maximum groundwater U concentration reached 1200 ng/mL, which is conspicuously greater than the drinking water standard (20 ng/mL) recommended by the US Environmental Protection Agency (US EPA) [30]. U in contaminated groundwater can enter into the human body via groundwater drinking pathway and consequently pose a risk to human health. Epidemiological studies found that U mainly accumulates in kidney. Long-term consumption of U-contaminated groundwater can induce kidney tubular cell death, resulting in kidney diseases [30,31]. In addition, chronic drinking of U-contaminated groundwater can cause other detrimental health effects, such as hindering bone growth [32], disrupting metabolic processes in liver [33], affecting germ cell growth [34,35], and triggering inflammatory responses in nervous system [36]. Therefore, it is necessary to tackle the secondary pollution issue due to the short-term rapid mobilization of soil U so as to achieve effective phytoremediation and ensure groundwater safety.

This review elucidates the interactions between chelating agents and U ions, and then explains the phytoextraction mechanisms and summarizes recent progresses in synthesis and application of chelating agents in assisting phytoremediation of U-contaminated soils. The major factors influencing phytoextraction efficiency are given. The environmental risks of existing chelating agents are also discussed. Finally, recent advances in slow-release technology are reviewed, with synthesis and application of slow-release chelating agents for slowing down metal mobilization in soils in order to match plant uptake rate presented. This review aims at shedding light on the promise of slow-release biodegradable chelating agents for enhancing phytoextraction of soil U. This novel approach is considered as a green and sustainable remediation technology for U-contaminated soils.

2. Interactions between Chelating Agents and U Ions

2.1. Mechanisms of Chelation

The atomic number of U is 92. There are 92 protons and 146 neutrons in the nucleus, forming four electron-filled layers (K, L, M, N) and three under-filled electron layers (O, P, Q). The six electrons ($5f^3 6d^1 7s^2$) on the three underfilled electron sublayers are valence electrons, which determine the chemical properties of U [11]. After losing six valence electrons, the outermost energy level of U is filled (S^2P^6), forming noble gas structure, which has a strong affinity with oxygen. Therefore, U usually exists in the forms of oxides and oxysalts in the crust and soil. Specifically, U is mainly present in the form of uranyl ion (UO_2^{2+}) in soils, which has a strong complexing ability to form stable complexes with phosphate (PO_4^{3-}), carbonate (CO_3^{2-}), hydroxyl (OH^-), sulfate (SO_4^{2-}), etc., and consequently reduces the mobility and bioavailability of U in soils [37,38].

Chelating agent refers to ligand which can form a cyclic structure complex with central ions (usually metal ions). Once in the soil, it can chelate with heavy metals and radionuclides to form water-soluble and exchangeable metal-chelating agent complexes, which can increase their mobility and bioavailability, and consequently promote plant uptake of these contaminants [39]. Chelating agents can be divided into two main categories including aminopolycarboxylate chelating agents (APCAs) (e.g., aminotriacetic acid (NTA), ethylenediaminetetraacetic acid (EDTA), ethylenediaminedisuccinic acid (EDDS), and N, N-Bis(carboxymethyl)-glutamic acid (GLDA)) [40,41] and low molecular weight organic acids (LMWOAs) (e.g., citric acid (CA), oxalic acid (OA), and succinic acid (SA)) [42,43]. In addition, APCAs can be classified as biodegradable and non-biodegradable [44], whereas LMWOAs are biodegradable [26,45]. The chemical structures and biodegradability of commonly used chelating agents are shown in Table 1.

Table 1. Structures and biodegradability of commonly used chelating agents, and structures and stability of chelates formed by UO_2^{2+} with these chelating agents.

Category	Chelator	Chelator Structure	Biodegradability	U Chelate Structure	$\log K_U$ *
Aminopolycarboxylate chelating agents (APCAs)	Ethylenediaminetetraacetic acid (EDTA)	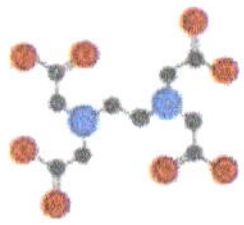	Half-life > 40 d [46]		7.4 (UO_2^{2+}:EDTA = 1:1) 17.87 (UO_2^{2+}:EDTA = 2:1)
	Ethylenediaminedisuccinic acid (EDDS)	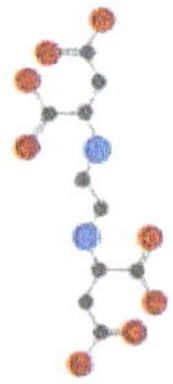	Completely degraded after 54 d [47]	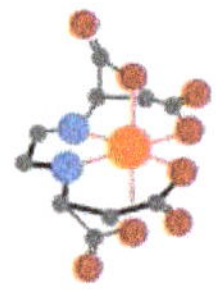	10.7
	Nitrilotriacetic acid (NTA)	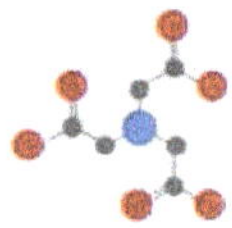	75% degraded after 21 d [48]	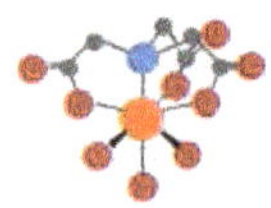	9.5
	Diethylenetriaminepentaacetic acid (DTPA)	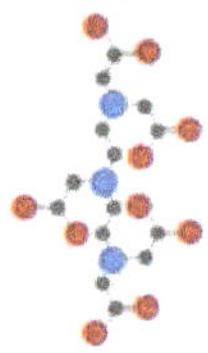	Similar to EDTA [49]	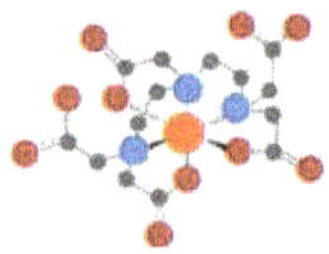	11
Low molecular weight organic acids (LMWOAs)	Citric acid (CA)	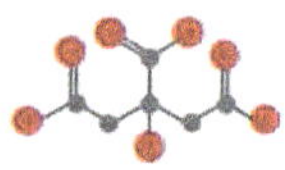	69% degraded after 20 d [50]	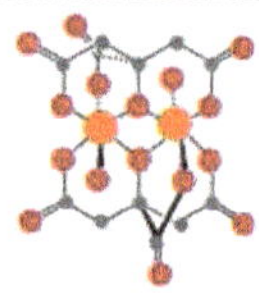	8.96
	Oxalic acid (OA)	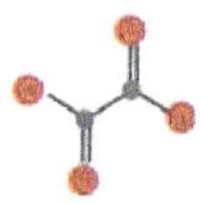	Hard to degrade in nature due to the formation of Ca-oxalate [21]	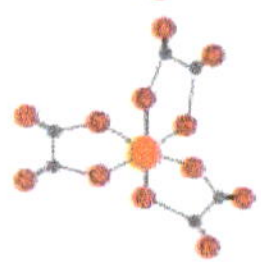	6.36
	Succinic acid (SA)	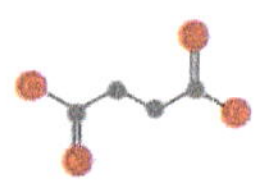	Half-life < 2 min and 200 h in soil solution and for microbial mineralization, respectively [51]		3.87

Notes: N C O UO_2^{2+}; * Data was sourced from [26,52,53].

Cyclic chelate formed by metal ion with ligand through coordination bonds has strong stability. Table 1 shows the structures of chelates formed by UO_2^{2+} with commonly used chelating agents. The stability constant (K) is used to indicate the stability of metal chelate, which increases with K value. K value is calculated using the following equation [54–57]:

$$K_{ML}^{T} = \frac{C_{ML}}{C_M C_L}$$

where C_{ML}, C_M, and C_L are concentrations of metal chelate, metal ion, and ligand, respectively; K_{ML}^{T} is the stability constant.

It is worth noting that EDTA has the highest K values for various heavy metal chelates compared to other chelating agents (Table 1) and therefore has been widely used in mobilization of metal contaminants in soils and soil washing over past decades [58]. However, the tertiary amine in EDTA molecule makes it difficult to be biodegraded. As a result, long-term retention of non-degradable EDTA in the environment has posed a critical risk to the ecology and human health [59–63]. As shown in Table 1, biodegradable EDDS has a high *K* value for U chelate. Moreover, its chemical structure is similar to that of EDTA. Thus, EDDS has become an ideal substitute for EDTA [64–67]. The *K* value of NTA for U chelate is slightly lower than that of EDDS. However, the use of NTA should be restricted due to its carcinogenic effect [44,68]. For example, iron (Fe)-NTA can induce renal cell carcinoma, which is widely used to perform oxidative tissue damage and carcinogenesis experiments [69–71]. CA, which is a natural LMWOA, also has a high *K* value for U chelate. Previous studies showed that CA increased the bioavailability of U in soils more effectively compared to other chelating agents (i.e., OA, EDDS, and NTA) [26,72].

2.2. Mobilization of Soil U by Chelating Agents

Soil texture, pH, and redox potential (Eh) are dominant factors influencing the effect of chelating agents on mobilizing U in soils [26,73–75]. Table 2 presents the mobilization efficiency of soil U by different chelating agents. The dosages of chelating agents ranged from 0.5 to 50 mmol/kg. The experimental soil pH values ranged from 5.6 to 7.3. The soil U concentration reached nearly 1000 mg/kg. Among the chelating agents listed in Table 2, the effect of CA on mobilizing U in soils, in particular, in sandy soil, was the most significant, with the exception of NH_4-citrate > CA in the loamy sand and heavy clay in the study by Duquène, et al. [26]. A possible explanation for the different effects of CA on mobilizing U in soils with contrasting textures is that the soil containing higher silt and clay contents has a larger specific surface area, which can result in stronger adsorption of contaminants and consequently reduce the mobilization efficiency of soil U by chelating agents [76,77]. In addition, there are a large number of functional groups (e.g., amino and carboxyl groups) in soils with high organic matter (OM) content, which has a strong adsorption capacity for U ions [78–80]. This can also affect the effectiveness of chelating agents in mobilization of soil U. A similar trend was observed in Jiang, et al., which showed that the mobilization efficiency of chromium (Cr) in contaminated soils by CA was in the following decreasing order: sandy soil > loam > clay soil [81]. The mobilization efficiency of soil heavy metals by chelating agents tends to increase with increasing soil acidity [82]. For example, Yang, et al., found that the mobilization efficiency of heavy metals in the soil solution with low pH by EDDS was significantly higher than that with high pH [83]. Likewise, Wang, et al., showed that the mobilization efficiency of heavy metals by chelating agents (i.e., GLDA and NTA) in the strongly acidic soil was 4-times higher than that in the alkaline soil, and the chelating agents were more effective in mobilizing heavy metals in the sandy soil compared to the clay soil at a wide soil pH range of 4 to 10 [82]. The leaching experiment of Kantar and Honeyman showed that the lower pH of eluent facilitated the mobilization of soil U by CA [84]. In addition, Eh can significantly influence the chemical forms of U present in soils and therefore control the solubility of U in the soil [73]. Specifically, at low Eh, U is present in solid phases, whereas at high Eh, U carbonates are transformed into soluble phases, which tend to migrate with water. To sum up, chelating agents can transform soil U from solid phase to soluble phase by forming high *K* value complex compounds with UO_2^{2+}. In particular, sandy soil with low pH, high Eh, and low OM content is favorable for the mobilization of U by chelating agents.

Table 2. Effects of different chelating agents on mobilizing U in soils with different textures.

Item	Reference						
	[85]	[72]	[86]	[26]			[27]
Soil characteristics							
Texture	Silty loam	Loam	Sandy loam	Sand	Loamy sand	Heavy clay	Medium clay
OM (%)	3.4	4.2	3.47	4.9	7.8	9.2	12.1
pH	6.8	7.3	5.58	6.9	7.2	6.7	6.67
Total U (mg/kg)	909	280	272	14	13	41	18
Source of U contamination	Industrial	Industrial	U mine site	Industrial	Industrial	Uraniferous shale	$UO_2(NO_3)_3 \cdot 6H_2O$
Experimental design							
Chelator	CA and OA	CA, EDTA, and MA	CA, EDTA, and EDDS	CA, EDDS, OA, NTA, and NH_4-citrate			CA and EDDS
Chelator dosage (mmol/kg)	2, 10, and 20	0.5, 1, 5, 10, and 20	0.5, 2, 2.5, 5, 10, 15, 25, and 50	5			5, 5 + 5, 10, and 10 + 10
Test conditions	Soil incubation	Soil incubation and pot experiment	Column leaching test	Soil incubation	Soil incubation	Soil incubation	Soil incubation and pot experiment
Result of most effective mobilization of soil U							
Chelator and dosage (mmol/kg)	CA 20	CA 20	CA 50	CA 5	NH_4-citrate 5	NH_4-citrate 5	CA 10 + 10
Days taken to reach the maximum U concentration in soil solution	1	1	6	1	1	1	8
Maximum U concentration in soil solution and increment compared to control	775 mg/kg, increased by 140 times	240 mg/L, increased by 200 times	2000–2400 Bq/kg, increased by 356 times	5019 µg/L, increased by 479 times	1106 µg/L, increased by 368 times	733 µg/L, increased by 366 times	1463.6 µg/L, increased by 215 times

3. Chelating Agent-Assisted Phytoremediation of U-Contaminated Soils

3.1. Mechanisms of Phytoextraction of U from Soils

Phytoextraction is a common technology for phytoremediation of U-contaminated soils, which transfers U from the soil to plant roots, stems, and leaves, and therefore cleans up the soil. Specifically, plant uptakes U^{4+} or U^{6+} ions through epidermal cells [3,87]. U cations (e.g., UO_2^{2+}) can be absorbed by roots through the same carrier or ion channel as calcium (Ca^{2+}), Fe^{3+}, magnesium (Mg^{2+}), and other necessary elements. U anions (e.g., $UO_2(CO_3)_2^{2-}$) can pass through plant cell membranes in a way similar to CO_3^{2-} [88]. For most plants, the majority of U is enriched in roots [89,90], with a small proportion of plants, such as sesbania (*Sesbania rostrata*) and water lily (*Nymphaea tetragona* Georgi) can further transfer the absorbed U upward to stems, leaves, and fruits [3].

The dominant factors affecting root uptake of soil U include chemical forms of U, soil environmental conditions (e.g., pH and Eh), and plant species [3,72,89–92]. U is mainly present in the form of UO_2^{2+} in soils. PO_4^{3-} and CO_3^{2-} are common ligands for U ions and can form uranyl phosphate, uranyl carbonate or triple uranyl-calcium carbonate complexes. The formation of uranyl phosphate and uranyl carbonate in soils usually depends on the contents of PO_4^{3-} and CO_3^{2-} in the soil [90]. Moreover, soil pH can influence U solubility and sorption or desorption in soils, and consequently affect the bioavailability of soil U. Soil OM is rich in functional groups (e.g., carboxyl, hydroxyl, and aromatic groups) containing lone pairs of electrons, which can cause the formation of different U compounds by altering soil Eh [93]. Moreover, when U is taken up by root cells, it is translocated to the mid-column before being released into the xylem [3]. The transpiration and expression of transporter protein genes can influence the translocation of U to the aboveground parts. For example, three genes including IRT1, FRO2, and FIT1 were found to affect U translocation in *Arabidopsis thaliana* L. [94,95].

3.2. Mechanisms of Assisted Phytoextraction of U from Soils

The principle of chelating agent-assisted phytoremediation of U-contaminated soils is to use chelating agents to desorb U ions from soil particle surface into soil solution, and therefore increase the solubility of soil U. The chelating agent-mobilized U can be readily available for plant uptake and facilitate phytoextraction of U from the soil. Table 3 lists the effects of commonly used chelating agents on phytoextraction of U from soils. It can be seen that although CA has a relatively lower *K* value for U chelate than EDDS (Table 1), CA was the most effective in enhancing the phytoextraction of soil U in all phytoremediation trials presented in Table 3. This can be related to the following three reasons: (1) CA can reduce soil pH value, which may be favorable for the mobilization of soil U [72,96]. Lozano, et al., found that CA had the most significant effect on the dissolution of U with acidic pH, whereas the maximum dissolution of U with the addition of EDDS, EDTA or no chelating agent was under alkaline conditions [86]. (2) CA can form U-CA complexes, which increase the solubility and bioavailability of U in soils [91]. (3) CA can improve plant tolerance to the bioavailable U in soils [97]. Rong, et al., found that 5 mmol/kg CA addition could alleviate cell damage and improve the tolerance of perennial ryegrass (*Lolium perenne* L.) to soil U [98]. They claimed that the optimal dosage of CA can avoid adverse effects (e.g., etiolation, withering, and even death) of CA in excess amounts on plant growth as well as improve the buffer capacity of cytosol, enhance the photosynthesis of perennial ryegrass, and decrease the electrical conductivity (EC) and malondialdehyde (MDA) content while increasing the contents of soluble proteins and enhancing the activities of antioxidant enzymes in the shoots and roots. Likewise, a significant increase in antioxidant enzyme activities in the leaves of *Macleaya cordata* grown in the CA-treated soils has been noted, thereby mitigating the oxidative stress induced by U and chelating agent and facilitating the phytoextraction of soil U [27]. As shown in Table 3, the application rates of chelating agents in assisting phytoremediation of U-contaminated soils ranged from 0.5 to 25 mmol/kg. At a high application rate, the toxic effect of EDDS on plant growth was more pronounced than that of CA and OA [99]. In addition, compared to the single application at a high rate, multiple consecutive applications of chelating agents at a relatively low rate can alleviate the toxicity of mobilized U to plants and increase phytoextraction efficiency of soil U. It is well known that hyperaccumulator plant species play an important role in phytoremediation. At present, the Global Hyperaccumulator Database records 759 hyperaccumulators. However, there is still a lack of U-related hyperaccumulator plant species [100]. Therefore, the enhancement of phytoextraction of soil U as assisted by chelating agents can be a possible approach for filling the current gap of U hyperaccumulator plant species.

Table 3. Effects of different chelating agents on assisting phytoextraction of U from soils.

Chelator	Plant	Total U (mg/kg)	Source of U Contamination	Incubation Period	Dosage (mmol/kg)	Application Frequency	Result of Most Effective Mobilization of Soil U	Ref.
AA [1], CA, and MA	10 species [6]	280 and 750	Industrial	28 d of growth prior to chelator addition and harvested 7 d after addition	20	Single	CA was the most effective in enhancing U accumulation in plants. After 20 mmol/kg CA addition, shoot U concentrations in four plant species (*B. juncea*, *B. chinensis*, *B. narinosa*, and amaranth) increased by more than 1000 times compared to control within a few days.	[72]
CA, CDTA [2], DTPA, EDTA, HEDTA [3], and OA	Sunflower (*Helianthus annuus* L.) and Indian mustard (*Brassica juncea* L.)	300	Additional $UO_2(NO_3)_3 \cdot 6H_2O$	28 d of growth prior to chelator addition and harvested 42–56 d after addition	1–25	Single	CA was the most effective in desorption of U ions and enhancing plant accumulation of U. 20 mmol/kg CA addition to loamy acid soil resulted in the highest U concentration in sunflower shoots, being 150 times greater than control.	[89]

Table 3. *Cont.*

Chelator	Plant	Total U (mg/kg)	Source of U Contamination	Incubation Period	Dosage (mmol/kg)	Application Frequency	Result of Most Effective Mobilization of Soil U	Ref.
AC [4], CA, EDDS, NTA, and OA	Indian mustard (*Brassica juncea* cv. Vitasso) and ryegrass (*Lolium perenne* cv. Melvina)	14 41	Industrial U-contaminated soil (sand, labelled as BK1) Natural U-enriched soil (heavy clay, labelled as BI)	28 d of growth prior to chelator addition and harvested 4–14 d after addition	5	Single	The Indian mustard shoot U concentration was in the following decreasing order: EDDS > AC > CA > OA > NTA in BK1; and CA > AC > EDDS > OA = NTA in BI. The ryegrass shoot U concentration was in the following decreasing order: CA > AC > OA > EDDS > NTA in BK1; and CA = AC > OA = EDDS > NTA in BI.	[21]
AC and CA	Kochia (*Kochia scoparia* L. Schrad.), sunflower (*Helianthus annuus* L.), and sweet corn (*Zea mays* L.)	78.7	Depleted U-contaminated soil	45 d of growth prior to chelator addition and harvested 6 d after addition	20	Single	AC was as effective in enhancing plant uptake of U as CA. The kochia leaves U concentration was in the following decreasing order: CA > AC, whereas that in the stems: AC > CA. The sunflower leaves U concentration was in the following decreasing order: AC > CA, whereas that in the stems: AC = CA. Corn had the lowest U uptake capacity and was considered as a negative control.	[101]
CA and MA	Indian mustard (*Brassica juncea* L.)	100	Additional $UO_2(NO_3)_3 \cdot 6H_2O$	66 d of growth prior to chelator addition and harvested 7 d after addition	5, 10, and 20	Single	A total of two restorations were carried out. In the first round, the maximum plant accumulation of U was noted in 10 mmol/kg CA treatment after 45 d and in 20 mmol/kg CA treatment after 55 d. In the second round, the maximum plant accumulation of U was noted in 20 mmol/kg MA treatment after both 55 and 65 d.	[102]
CA and mixture of CA, MA, OA, and LA [5] (labelled as Mix)	Mustard (*Brassica juncea var. tumida*)	47.75	Additional $UO_2(NO_3)_3 \cdot 6H_2O$	60 d of growth prior to chelator addition and harvested 7 d after addition	5 for CA and molar ratio of CA: MA: OA: LA in Mix = 2.5:2.31:1.1 5:0.044	Single	Mix was more effective in enhancing U accumulation in mustard.	[103]
CA, EDDS, and OA	Red wetter grass (*Zebrina pendula* Schnizl)	150	Additional $UO_2(CH_3CO_2)_2 \cdot 2H_2O$	90 d of growth prior to chelator addition and harvested 7 d after the last addition	2.5, 5, and 7.5	Three consecutive applications every three days	CA was the most effective in enhancing U accumulation in *Zebrina pendula* Schnizl. Addition of 5 mmol/kg CA resulted in the highest U concentration in the plant, being 5.7 times greater than control.	[99]
CA, EDDS, and OA	*Macleaya cordata*	18	Additional $UO_2(NO_3)_3 \cdot 6H_2O$	Chelator addition at the beginning of flowering and harvested 14 d after the last addition	5 and 10	Except for 10 mmol/kg EDDS, all other chelators were applied both in a single application and two consecutive applications.	The enhanced U uptake was in the following decreasing order: CA > EDDS > OA. Two consecutive applications of 10 mmol/kg CA resulted in the most significantly promoted solubilization of soil U, being 215 times greater than control.	[27]
CA, EDDS, and OA	Sunflower (*Helianthus annuus* L.)	15	Additional $UO_2(CH_3CO_2)_2 \cdot 2H_2O$	60 d of growth prior to chelator addition and harvested 7 d after the last addition	2.5, 5, and 7.5	Three consecutive applications every three days	5 mmol/kg CA addition resulted in the highest U phytoextraction efficiency, being 1.78 times greater than control.	[28]

Notes: [1] AA: Acetic acid; [2] CDTA: Trans-1,2-diaminocyclohexane-N,N,N′,N′-tetraacetic acid; [3] HEDTA: N-hydroxyethlenediaminetriacetic acid; [4] AC: Ammonium citrate; [5] LA: Lactic acid; [6] 10 species: Amaranth (*Amaranth cruentus* L.), Indian mustard (*Brassica juncea* L.), bush bean (*Phaseolus vulgaris* L.), Chinese cabbage (*Brassica chinensis* L.), Chinese mustard (*Brassica narinosa* L.), corn (*Zea mays*), cow pea (*Pisum sativum* L.), field pea (*Pisum sativum* L.), sunflower (*Helianthus annuus* L.), and winter wheat (*Triticum aestivum* L.).

3.3. Environmental Risks

Chelating agents can rapidly increase the mobility of soil heavy metals and radionuclides through chelation once applied to the soil. However, during the short period of rapid mobilization of soil heavy metals and radionuclides, plants may only be able to absorb a small proportion of the mobilized metal ions in the soil, whereas the remaining large amount of mobilized metal ions exists in the soil and may cause contamination to subsurface soil and groundwater through leaching [60,62,63,104]. In addition, a sudden short-term release of bioavailable and substantial heavy metals and radionuclides in the soil may cause transient phytotoxicity and consequently inhibit plant growth [97,105–108]. For example, Chen, et al., compared the growth of sunflowers in soils treated with EDDS, CA, and OA at high (7.5 mmol/kg), medium (5 mmol/kg), and low (2.5 mmol/kg) application rates [28]. They found that all chelating agents reduced sunflower biomass, and a greater reduction was noted in soils treated with higher dosage of chelating agents. Whereas Römkens, et al., found that although EDTA increased the concentrations of bioavailable heavy metals in the soil, the plant uptake of heavy metals did not exhibit an increasing trend after EDTA application [109]. This could be attributed to the limitations in plant uptake capacity for and tolerance to U, beyond which plant growth and uptake of soil bioavailable U will be affected. Similar results were obtained by Hou, et al., which showed that the germination of tomato, cabbage, and radish seeds was significantly inhibited at a soil U concentration greater than 320 mg/kg [24]. Therefore, plants have a limited uptake capacity for soil U, and excess amounts of mobilized U in soils in a short period can be toxic to plants.

In addition, chelating agents can influence the soil ecosystem by affecting soil microbial metabolism. For example, additional chelating agents (e.g., EDTA and EDDS) can affect the activity of soil microorganisms, especially dehydrogenase activity and basal respiration [110]. Lee and Sung found that EDTA inhibited the microbial activity in heavy metal-contaminated soils [111]. Soil microorganisms depend directly or indirectly on soil solutions to absorb food and water, and increased concentrations of bioavailable metals in soils may poison microorganisms [109]. Nevertheless, Cao, et al., found that biodegradable chelating agents (e.g., EDDS and methylglycinediacetic acid (MGDA)) could alleviate soil heavy metal stress and therefore benefit the bacterial community in the soil [112]. Thus, varying chelating agents may bring different effects on the soil ecosystem.

Barona, et al., found that EDTA increased the leachability of heavy metals in the soil and led to a weak adsorption of metal ions by soil compounds, which contributed to phytoextraction [113]. The study of Udovic and Lestan showed that the mobility of residual lead (Pb) in the EDTA-washed soil increased during the soil aging process [114]. The column leaching experiment of Wu, et al., also demonstrated that Cu, Zn, and Pb migrated into the leachate with rainwater after EDTA application to the soil, and the heavy metal concentrations in the leachate increased linearly with increasing rainfall [60]. This indicated that the migration of soil heavy metals increased by EDTA, which was conducive to phytoextraction. However, if the mobilized heavy metals in soils were not absorbed by plants in time, they may migrate to groundwater with rainwater, and consequently cause more serious environmental problems. Huang, et al., found that DTPA contributed more significantly to the leakage of soil Pb than acid rain under planting conditions, posing a great risk of groundwater pollution [115]. Nowack, et al., elucidated the unavoidability of chelated metal leaching by preferential flow processes during the chelating agent-enhanced phytoextraction [104]. They pointed out that the application of chelating agents to soils should be limited to areas where the connection to groundwater has been broken, or where groundwater contamination is not an issue. Chang, et al., found that CA enhanced the accumulation of U in crop plants (i.e., edible rape, Indian mustard, canola, and sunflower), whereas additional CA could lead to groundwater pollution with downward U migration [97]. Additionally, CA promoted the phytoextraction of U from the soil in the short term, whereas the additional U-CA complexes could be converted into less available forms (e.g., U trioxide) after dissociation in the long term, which could consequently reduce the phytoextraction efficiency after the first few croppings [97]. Thayalakumaran, et al., found that Cu leaching was much greater than herbaceous uptake after EDTA application [116]. Therefore, when using chelating agents to assist phytoremediation of U-contaminated soils, it is necessary to consider the reasonable application rate and precipitation during the application.

4. Slow-Release Chelating Agents

4.1. Slow-Release Technology

Slow-release technology, which aims at extending the release time of a substance, was first used in the pharmaceutical industry and has since been applied to synthesize a wide range of slow-release materials such as slow-release fertilizers, slow-release pesticides, and food packaging [117–119]. Controlled-release technology is also a way to extend the release time, which allows the core material to be released at a specific rate or concentration level [120,121]. There is a slight difference between the two technologies, but essentially both extend the release time of core material, so this paper does not make a detailed distinction between the two. Slow-release fertilizers are designed to control and slow down the release speed of nutrients so as to reduce fertilization as well as nutrient loss and increase crop yield [119]. In the food packaging industry, the quality of food products for long-term storage is maintained or improved by slow release of substances such as antimicrobials, antioxidants, enzymes, and spices [122–124]. Although used in different industries, slow-release materials usually consist of three parts, i.e., core material which is the active substance to be released, wall/carrier material which can regulate the release speed of core material, and binding agent which forms the core material and wall/carrier material into a cohesive whole.

There are four common approaches which can be used to achieve slow-release effect, including blending, multilayer composite-controlled release, multi-hole adsorption carrier, and microencapsulation [125,126]. To be specific, the blending approach uses the principle of chaotic convection in fluid mechanics to prepare different forms of polymers and films [127]. The multi-layer composite controlled-release approach originated from the layer-by-layer coating (LBL) approach in the pharmaceutical field. Han and Floros first proposed the concept of multi-layer composite controlled-release approach, which is mainly composed of internal controlled-release layer, intermediate active film layer, and external barrier layer [128]. The inner layer is used to control the diffusion rate of active substance to food surface, the intermediate matrix layer contains the active substance, and the outer layer serves to prevent the extravasation loss of active substance. Multi-hole adsorption carrier controls the drug release speed by modulating and modifying the porous material to allow selective adsorption of target molecules. Microencapsulation uses coating material to encapsulate the core material to form particles less than 1000 μm in diameter [129,130]. Microencapsulation is popular due to its simple operation and inexpensive equipment. The common methods for preparing slow-release microcapsules include emulsion solvent evaporation [131], spray drying [105], emulsion dispersion [132], chemical crosslinking, ion exchange, and complex coalescence [133].

Slow-release technology has recently received growing attention in environmental pollution treatment. In in situ chemical remediation, slow-release technology can control the release of active compounds, reduce the non-selective consumption of oxidants, and maintain the treatment effect for a longer period [134]. For example, Christenson, et al., prepared a slow-release oxidant by heating the mixed paraffin and potassium permanganate and cooling them in the mold [135]. Their five-year field-scale test showed that the slow-release oxidant gradually reduced trichloroethene (TCE) in a low permeable aquifer by 89% and was only refurbished yearly. Tang, et al., used chitosan and urea as carriers to prepare sodium persulfate slow-release material, which can prolong the generation time of free radicals and improve the degradation of OM [136]. In bioremediation, slow-release technology can continuously provide substrate, improve the removal efficiency of pollutants, and maintain a longer period of anaerobic dechlorination [134]. For example, Tsai, et al., reported that the slow-release material prepared by vegetable oil, cane molasses, and surfactants, provided nutrients for microorganisms and changed the subsurface environment to anaerobic conditions, which were conducive to reductive dechlorination. The removal efficiency of TCE in the contaminated groundwater reached 99% after 50 d [137]. In the study of Yeum, et al., slow-release precipitating tablets and slow-release floating tablets, both prepared from hydroxypropyl methylcellulose, were applied to in situ biological denitrification systems to provide a continuous carbon source for microorganisms [138].

4.2. Slow-Release Chelating Agent-Assisted Phytoremediation

To address the current bottleneck problem that the rapid mobilization rate of soil heavy metals and radionuclides by chelating agents does not match the plant uptake rate of bioavailable metal ions in the soil [28,105], slow-release chelating agents have been developed to control the release speed of chelating agent so as to avoid a sudden increase in soil bioavailable heavy metals and radionuclides which exceed plant tolerance and consequently cause phytotoxicity. The mechanisms of slow-release chelating agent-assisted phytoremediation of heavy metal- or radionuclide-contaminated soils are illustrated in Figure 1. As shown in Figure 1, the chelating agent as core material is wrapped by degradable wall material. After being applied to the soil, the wall material will gradually degrade meanwhile the internal chelating agent can be released to form chelates with metal ions in the soil, which can increase their solubility and therefore facilitate plant uptake of the contaminants.

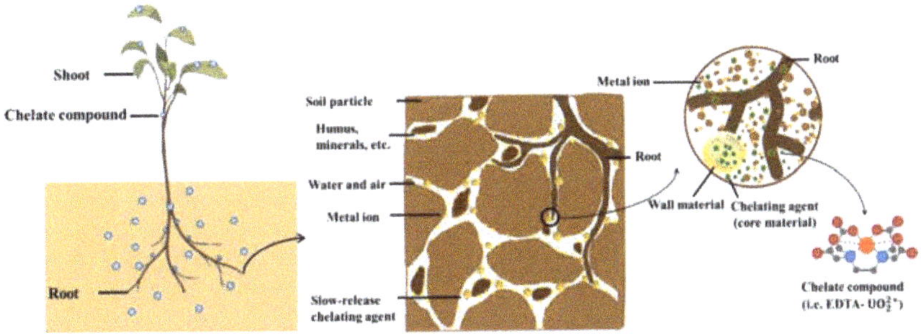

Figure 1. Mechanisms of slow-release chelating agent-assisted phytoremediation of heavy metal- and radionuclide-contaminated soils.

Slow-release chelating agents and slow-release fertilizers have conspicuous similarities in their slow-releasing behaviors. Based on the comparatively mature preparation methods for slow-release fertilizers, slow-release chelating agents can also be prepared in two main ways, i.e., one method is to synthesize slow-release carrier and then use diffusion mechanism to import chelating agent into the carrier in liquid form (e.g., [139–141]). The other method is to use granular solid chelating agent as core material, and polymer organic or inorganic materials is selected as wall material to encapsulate chelating agent particles inside the slow-release chelating agent (e.g., [105,131,142,143]).

Until now, most studies have used synthetic organic polymers (e.g., polyethylene, polycaprolactone, and ethylene-vinyl acetate copolymer) as wall/carrier material to prepare slow-release chelating agents. However, these synthetic materials can be harmful to the environment. In contrast, biodegradable natural organic polymers (e.g., starch, cellulose, and chitosan) and organic-inorganic hybrid materials (e.g., acrylamide + montmorillonite and cellulose + zeolite) are ideal wall/carrier materials, which have been used to synthesize slow-release fertilizers. Current research on slow-release chelating agents with biodegradable materials as wall/carrier material is relatively limited. Very recently, Wang, et al., prepared a slow-release chelating agent with chitosan and cyclodextrin as carrier materials and CA as core material [141]. In addition, considering the similarity between slow-release chelating agents and slow-release fertilizers, some wall/carrier materials applied to slow-release fertilizers (e.g., starch, cellulose, and chitosan) may also be ideal for the synthesis of slow-release chelating agents. Furthermore, slow-release materials can cause a burst-release effect after application in the water and soil environment [144,145]. The burst-release effect can be induced when a large amount of core material is released after the shell of slow-release material with a shell-core structure degrades or the shell breaks down due to the increased internal pressure. To mitigate the above disadvantage, polymers, such as polyvinyl alcohol (PVA) and gelatine, can be added to increase the toughness of outer wall and enable it to withstand a certain amount of internal pressure. Zhang, et al., prepared a biodegradable carboxymethylcellulose and PVA blended film. They found that the water permeability of the film could be adjusted by changing the ratios of PVA and formaldehyde [146]. If this film was applied to slow-release chelating agents, the slow-release period could be prolonged. Han, et al., prepared a blended film from starch and PVA that is biodegradable and has good compatibility [147], which may also be used as a slow-release carrier for slow-release chelating agents.

It is worth noting that the aforementioned organic polymers and inorganic materials have been used in preparation of adsorbents [148–152], suggesting that these materials may promote the adsorption of some metal ions onto the wall/carrier material, leading to a temporary passivation effect. Kos, et al., used EDDS and EDTA in conjunction with acrylamide hydrogel to assist the phytoextraction of soil Pb [153]. They found that 10 mmol/kg EDDS addition significantly increased the leachable Pb, whereas 5 mmol/kg EDDS addition not only increased the phytoextraction efficiency of soil Pb, but also effectively controlled the leaching of soil Pb. Hydrogel can be a promising carrier for the synthesis of slow-release chelating agents as the chelating agent can be released into the soil through water loss from the hydrogel. Moreover, hydrogel can maintain the soil nutrient balance by retaining water in the soil. Hydrogel-based slow-release fertilizers have also been reported by other researchers [154–158]. If the functions of nutrient supply, water retention, and controlled release of chelating agent were incorporated into hydrogel-based slow-release chelating agents, chelating agent-assisted phytoextraction technology could be significantly advanced.

Until now, only a few studies have examined the effectiveness of slow-release chelating agents in assisting phytoremediation of heavy metal-contaminated soils, whereas its application in phytoremediation of U-contaminated soils has been scarce. Table 4 lists slow-release chelating agents which have been used to remediate heavy metal- and radionuclide-contaminated soils in recent years. As shown in Table 4, the particle sizes of slow-release chelating agents ranged from 5–20 μm to 3–5 mm. EDTA is the most commonly used chelating agent in these studies. The slow-release period of these chelating agents ranged from several days to months, which is mainly related to the particle size of chelating agent and properties of wall/carrier material. Among the chelating agents listed in Table 4, those with higher molecular weight organic compounds as the wall/carrier material possessed a longer slow-release period. Recently, the first biodegradable slow-release chelating agent with CA as core material has been prepared by Wang, et al., who used carboxymethyl-β-cyclodextrin (CMCD) and hydroxypropyl chitosan as carriers [141]. They found that the novel slow-release chelating agent had a noticeable slow-release effect on CA, which could reduce the impact of soil bioavailable U on Indian mustard (*Brassica juncea* L.) growth and promote the absorption and accumulation of U in the plant. However, compared with the slow-release chelating agent (i.e., microencapsulated EDTA) in Xie, et al., the release rate of the aforementioned slow-release CA in water is faster [131]. In addition, due to the lack of long-term soil incubation experiment, the dynamic interactions between the slow-release CA and U as well as other metals in the soil remain unclear. Therefore, the slow-release period of slow-release CA prepared by Wang, et al., in soils is hard to be estimated [141].

Table 4. Components, synthesis methods, particle size, and slow-release performance of slow-release chelating agents applied in remediation of heavy metal- and radionuclide-contaminated soils.

Wall/Carrier Material	Core Material	Synthesis Method	Particle Size	Slow-Release Period	Contaminant	Dosage (mmol/kg)	Test Conditions	Ref.
Silicate	EDTA-Na_2	Spray drying	3–5 mm	Release of EDTA from slow-release EDTA was still significantly lower than from uncoated solid EDTA in soil after 18 d.	Pb and Zn	13	Pot experiment	[105]
Talc, polyethylene, ethylene-vinyl acetate copolymer, ethylene-octene-1 copolymer, and polyoxyethylene monomethyl ether	EDTA-Na_2	Coating	-	Release of 75% EDTA ranged from 3 to 210 d in water	Pb	4	Pot experiment	[142]
Diethylenetriamine and hexamethylene diisocyanate	EDTA-Na_4	Interfacial polymerization	5.78 μm in average	Release of 85% EDTA in water after 5 d	-	-	Pot experiment	[143]
Polycaprolactone	EDTA	Solvent evaporation	65 ± 15 μm	Release of 93% EDTA in water after 30 d	Cu and Pb	3 and 6	Pot experiment	[131]
Chitosan and its derivatives	EDTA-Na_2	Spray drying	5–20 μm	Release of EDTA from slow-release EDTA was still significantly lower than from non-microencapsulated EDTA in water after 3 d.	Cd, Cu, and Pb	4 and 8	Pot experiment	[159]
Hydroxypropyl chitosan-graft-carboxymethyl-β-cyclodextrin	CA	Spray drying	-	Cumulative release of 80% CA in water after 2 d	U	5	Pot experiment	[141]

Application rate of slow-release chelating agents is one of the most important factors influencing mobilization of soil U. As summarized from Table 3, the application rates of chelating agents alone in assisting phytoremediation of U-contaminated soils ranged from 0.5 to 25 mmol/kg. In addition, previous studies reported that chelating agents (e.g., EDDS, EDTA, and DIPA) reduced plant biomass and even caused plant mortality at application rates greater than 5–7.5 mmol/kg [27,28,160,161]. The impact of chelating agents on plant growth is also related to the types of chelating agent and plant. For example, Chen, et al., found that 7.5 mmol/kg EDDS addition significantly inhibited the growth of *Zebrina pendula* Schnizl and sunflower (*Helianthus annuus* L.) [28,160]. Hu, et al., also noted some toxic symptoms such as chlorosis and necrotic spots occurred during *Macleaya cordata* growth after the addition of 5 and 10 mmol/kg EDDS [27]. In contrast, under the same experimental conditions, the aforementioned plants exhibited much higher tolerance to CA treatments. As shown in Table 4, the application rates of slow-release chelating agents in assisting phytoremediation of heavy metal and radionuclide-contaminated soils ranged from 3 to 13 mmol/kg, which is close to the above application rate range of chelating agents alone. Moreover, when comparing the effectiveness of chelating agents and slow-release chelating agents in enhancing phytoextraction of soil contaminants, the amount of core material in the slow-release chelating agent should be equal to that of chelating agent alone. At present, research on the use of slow-release chelating agents in assisting phytoremediation of U-contaminated soils is still at juvenile stage. The evaluation of their effectiveness in enhancing phytoextraction of soil heavy metals and radionuclides reported in scientific literature are exclusively obtained from laboratory pot experiments with no field applications (Table 4). Hence, it is recommended that pilot-scale tests should be conducted to investigate the effect of slow-release chelating agents on enhancing phytoextraction of soil U under field conditions. This effort can provide a better evaluation of slow-release performance. In addition, bench-scale accelerated aging tests (e.g., wet–dry cycling and freeze–thaw cycling) can be carried out to investigate the effect of slow-release chelating agents on controlling the release speed of core material under simulated natural aging conditions, which can make a contribution to elucidation of long-term slow-release mechanisms.

4.3. Evaluation of Slow-Release Performance

The slow-release performance of slow-release chelating agents can be indicated by the increasing rate of core material (i.e., chelating agent) content in the soil with time after application. The common methods for evaluating slow-release performance of slow-release materials include hydrostatic dissolution, soil incubation, and soil column leaching [162–166]. In the hydrostatic dissolution method, slow-release material is added to deionized water. The mixture is then maintained at a constant temperature. The amount of core material released into the water is measured regularly to calculate its release speed with time. The method is simple, fast, and well reproducible. However, it ignores the effect of soil environmental conditions on the slow-release performance. The soil incubation method simulates the release of core material in the soil and is closer to natural environmental conditions compared to the hydrostatic dissolution test. In the soil column leaching method, slow-release material is subjected to leaching test, and the concentration of core material in the leachate collected with time is measured. Among the above methods for evaluating slow-release performance, hydrostatic dissolution method has been widely used (e.g., [141,167–169]) due to its simple operating procedure, controllable environmental conditions, and convenient comparison between different slow-release materials. Considering the complexity of actual contaminated site environment, it is necessary to evaluate the slow-release performance of slow-release chelating agents in soils. Li, et al., first synthesized a slow-release chelating agent with EDTA as core material and silicate as coating material [105]. They demonstrated that the change of DOC concentration in the leachate could reflect the slow-release performance of coated EDTA. In contrast to the inorganic wall material in Li, et al., Xie, et al., used polyethylene as the wall material of microcapsule, and the concentration of EDTA in aqueous solution released from the microcapsule was determined by an ion chromatography (IC) [143]. Likewise, Shibata, et al., prepared polymer-coated EDTA and also characterized the slow-release property based on the variation of EDTA concentration in distilled water using a high-performance liquid chromatography (HPLC) [142]. The EDTA concentration in the solution can be determined by HPLC after pretreatment with an extractant containing Cu or Fe ions [170], and Shibata, et al., used a $CuSO_4$ solution as extractant. However, due to the lack of details about the operating procedures of the above methods, it is difficult to compare the slow-release results obtained from different studies. The longest release period of slow-release chelating agent recorded in literature is a polymer-coated EDTA synthesized by Shibata, et al., who reported that 75% of the EDTA in the slow-release chelating agent was released into the water after 210 d [142]. The pot experiment showed that among the five slow-release chelating agents with different wall materials, the EDTA coated with 40% talc, 45% polyethylene, 10% ethylene-octene-1 copolymer, and 5% polyoxoethylene monomethyl ether (0.72 mmol/g) were the most effective in reducing Pb and EDTA concentrations in the soil solution and increasing Pb enrichment in Sorghum bicolor *L.* seedlings with a release period of 80 d [142].

5. Conclusions and Future Perspectives

With the rapid development of nuclear energy and on-going U mining, cleaning up of U-contaminated soils has become a major global challenge. At present, no U hyperaccumulator plant species have been found, whereas plants for soil remediation have low extraction efficiency and long-term remediation period for U-contaminated soils. Both APCAs (e.g., EDTA and EDDS) and LMWOAs (e.g., CA and OA) can enhance phytoextraction of soil U by increasing bioavailable U in the soil via forming soluble chelate compounds. CA, which can be generated by organisms and is biodegradable, has a pronounced effect on mobilizing U in soils, and therefore assists phytoextraction of U from the soil. It should be noted that after chelating agents are applied to the soil, the mobility and bioavailability of soil U can increase rapidly in a short period, resulting in the problem that the increasing rate of soil bioavailable U does not match the plant uptake rate of and tolerance to U. The majority of mobilized U may not be absorbed by roots and remained in the soil, posing a non-negligible contamination risk to groundwater through leaching processes. Slow-release chelating agents synthesized by slow-release technology can control the release speed of chelating agent so that the mobilization rate of soil U matches plant uptake rate of U.

Currently, research on synthesis and application of slow-release chelating agents in assisting phytoremediation of U-contaminated soils is still in its infancy. The slow-release CA synthesized by Wang, et al., is the only slow-release chelating agent which was used to enhance phytoextraction of soil U [141]. Meanwhile most of the existing slow-release chelating agents designed for phytoremediation of heavy metal-contaminated soils contain non-biodegradable EDTA as core material, which would cause secondary pollution. In addition, the effective performance of the aforementioned slow-release CA was noted in laboratory pot experiments, whereas its effectiveness in assisting phytoremediation of U-contaminated soils in the field remains unclear. Future work can mainly include: (1) selecting and developing environmentally friendly wall/carrier materials and cost-effective preparation methods. Natural materials (e.g., starch, cellulose, and chitosan) and simple preparation methods with low cost (e.g., interfacial polymerization and spray drying) should be screened for synthesis of slow-release chelating agents prior to large-scale application. (2) Some existing wall/carrier materials (e.g., chitosan and its derivatives) of slow-release chelating agents have high adsorption capacity for heavy metals and radionuclides. They can have a passivation effect on these contaminants in soils. Therefore, it is necessary to eliminate the above negative impact of wall/carrier materials of slow-release chelating agents on mobilizing soil U. This can be achieved by optimizing the ratio of wall/carrier material to core material and modifying the surface porous structure of slow-release chelating agent. (3) Existing slow-release chelating agents have been mainly added to soils in the form of solutions. Synthesis of hydrogel-based slow-release chelating agent can be a promising research direction because hydrogel as a carrier for chelating agents has the potential to combine controlled release of chelating agent into the soil with soil and water conservation, thus improving the soil environment. (4) The present systematic review showed that CA is the most effective in enhancing phytoextraction of U from contaminated soils compared to other commonly used chelating agents. However, CA is easily decomposed in soils. Thus, it is encouraged to evaluate the long-term effect of CA on mobilization of soil U as well as explore methods for maintaining the effectiveness of CA during natural aging processes. (5) Soil microorganisms play an important role in the formation of soil fertility, transformation of contaminants as well as nutrient elements [171]. Therefore, it is also necessary to investigate the effect of microbial metabolic activities in rhizosphere soils on the slow-release performance of slow-release chelating agents and explore the transport mechanisms of complex compound formed by chelating agents with U ions at the plant-soil interface. This endeavor can contribute to the combination of chelating agent- and microorganisms-assisted phytoremediation of U-contaminated soils. (6) The studies of Saleh, et al.,e. demonstrated that aquatic plants including *Eichhornia crassipes*, *Ludwigia stolonifera*, and *Myriophyllum spicatum* have an enormous accumulation capacity of heavy metals (e.g., Cd, Cr, Cu, and Pb) and radionuclides (e.g., cesium (Cs) and cobalt (Co)) [172–177]. Therefore, apart from the above on-going research relevant to slow-release chelating agents, the effect of chelating agents on enhancing the accumulation of U and other radionuclides in these aquatic plants in contaminated paddy soils and riparian zones can be investigated.

Author Contributions: Conceptualization, K.Y. and Y.Y.; Literature Search and Data Analysis, Y.Y., K.Y. and Y.X.; Writing—Original Draft Preparation, Y.Y. and K.Y.; Writing—Review and Editing, K.Y., J.D. and N.J.; Funding Acquisition, K.Y.; Supervision, K.Y. and J.D. All authors have read and agreed to the published version of the manuscript.

Funding: This research was funded by the National Key Research and Development Program of China, grant number 2020YFC1806603, the National Natural Science Foundation of China, grant number 41907095, and the Fundamental Research Funds for the Central Universities, grant number 2020NTST24. The APC was funded by the National Key Research and Development Program of China, grant number 2020YFC1806603.

Institutional Review Board Statement: Not applicable.

Informed Consent Statement: Not applicable.

Data Availability Statement: Data is contained within the article.

Conflicts of Interest: The authors declare no conflict of interest.

References

1. World Nuclear Association. *World Nuclear Performance Report 2021—COP26 Edition*; World Nuclear Performance Report; World Nuclear Association: London, UK, 2021; pp. 3–13.
2. World Nuclear Association. *The Nuclear Fuel Report: Global Scenarios for Demand and Supply Availability 2021–2040*; World Nuclear Association: London, UK, 2021.
3. Chen, L.; Liu, J.; Zhang, W.; Zhou, J.; Luo, D.; Li, Z. Uranium (U) Source, Speciation, Uptake, Toxicity and Bioremediation Strategies in Soil-Plant System: A Review. *J. Hazard. Mater.* **2021**, *413*, 125319. [CrossRef] [PubMed]
4. Winde, F. Uranium Pollution of the Wonderfonteinspruit, 1997–2008 Part 1: Uranium Toxicity, Regional Background and Mining-Related Sources of Uranium Pollution. *Water SA* **2010**, *36*, 239–256.
5. Yan, X.; Luo, X. Radionuclides Distribution, Properties, and Microbial Diversity of Soils in Uranium Mill Tailings from Southeastern China. *J. Environ. Radioact.* **2015**, *139*, 85–90. [CrossRef] [PubMed]
6. Lu, Z.; Liu, Z. Pollution Characteristics and Risk Assessment of Uranium and Heavy Metals of Agricultural Soil around the Uranium Tailing Reservoir in Southern China. *J. Radioanal. Nucl. Chem.* **2018**, *318*, 923–933. [CrossRef]
7. Sabol, J. Uranium in the Beginning of the Nuclear Age: Reflections on the Historical Role of Jáchymov and an Overview of Early and Present Epidemiological Studies. In *Uranium in Plants and the Environment*; Gupta, D.K., Walther, C., Eds.; Springer International Publishing: Cham, Switzerland, 2020; ISBN 978-3-030-14961-1.
8. Chen, L.; Wang, J.; Beiyuan, J.; Guo, X.; Wu, H.; Fang, L. Environmental and Health Risk Assessment of Potentially Toxic Trace Elements in Soils near Uranium (U) Mines: A Global Meta-Analysis. *Sci. Total Environ.* **2021**, *816*, 151556. [CrossRef]
9. Gongalsky, K.B. Impact of Pollution Caused by Uranium Production on Soil Macrofauna. *Environ. Monit. Assess.* **2003**, *89*, 197–219. [CrossRef]
10. Ma, W.; Gao, B.; Guo, Y.; Sun, Z.; Zhang, Y.; Chen, G.; Zhu, X.; Zhang, C. Occurrence and Distribution of Uranium in a Hydrological Cycle around a Uranium Mill Tailings Pond, Southern China. *Int. J. Environ. Res. Public Health* **2020**, *17*, 773. [CrossRef]
11. Bourdon, B.; Turner, S.; Henderson, G.M.; Lundstrom, C.C. Introduction to U-Series Geochemistry. *Rev. Mineral. Geochem.* **2003**, *52*, 1–21. [CrossRef]
12. Brugge, D.; de Lemos, J.L.; Oldmixon, B. Exposure Pathways and Health Effects Associated with Chemical and Radiological Toxicity of Natural Uranium: A Review. *Rev. Environ. Health* **2005**, *20*, 177–194. [CrossRef]
13. Zhu, G.; Xiang, X.; Chen, X.; Wang, L.; Hu, H.; Weng, S. Renal Dysfunction Induced by Long-Term Exposure to Depleted Uranium in Rats. *Arch. Toxicol.* **2009**, *83*, 37–46. [CrossRef]
14. Brugge, D.; Buchner, V. Health Effects of Uranium: New Research Findings. *Doug Brugge Va Buchner* **2011**, *26*, 231–249. [CrossRef] [PubMed]
15. Cheng, H.; Lawrence Edwards, R.; Shen, C.-C.; Polyak, V.J.; Asmerom, Y.; Woodhead, J.; Hellstrom, J.; Wang, Y.; Kong, X.; Spötl, C.; et al. Improvements in ^{230}Th Dating, ^{230}Th and ^{234}U Half-Life Values, and U–Th Isotopic Measurements by Multi-Collector Inductively Coupled Plasma Mass Spectrometry. *Earth Planet. Sci. Lett.* **2013**, *371–372*, 82–91. [CrossRef]
16. Dewar, D. Uranium Mining: Environmental and Human Health Effects. In *Nuclear Non-Proliferation in International Law—Volume IV: Human Perspectives on the Development and Use of Nuclear Energy*; Black-Branch, J.L., Fleck, D., Eds.; T.M.C. Asser Press: Hague, The Netherlands, 2019; ISBN 978-94-6265-267-5.
17. Ma, M.; Wang, R.; Xu, L.; Xu, M.; Liu, S. Emerging Health Risks and Underlying Toxicological Mechanisms of Uranium Contamination: Lessons from the Past Two Decades. *Environ. Int.* **2020**, *145*, 106107. [CrossRef] [PubMed]
18. Meinrath, A.; Schneider, P.; Meinrath, G. Uranium Ores and Depleted Uranium in the Environment, with a Reference to Uranium in the Biosphere from the Erzgebirge/Sachsen, Germany. *J. Environ. Radioact.* **2003**, *64*, 175–193. [CrossRef]
19. Mészáros, G.; Bognár, G.; Köteles, G.J. Long-Term Persistence of Chromosome Aberrations in Uranium Miners. *J. Occup. Health* **2004**, *46*, 310–315. [CrossRef]
20. Wolf, G.; Arndt, D.; Kotschy-Lang, N.; Obe, G. Chromosomal Aberrations in Uranium and Coal Miners. *Int. J. Radiat. Biol.* **2009**, *80*, 147–153. [CrossRef]

21. Duquène, L.; Vandenhove, H.; Tack, F.; Meers, E.; Baeten, J.; Wannijn, J. Enhanced Phytoextraction of Uranium and Selected Heavy Metals by Indian Mustard and Ryegrass Using Biodegradable Soil Amendments. *Sci. Total Environ.* **2009**, *407*, 1496–1505. [CrossRef]
22. Mihalík, J.; Tlustoš, P.; Szaková, J. Comparison of Willow and Sunflower for Uranium Phytoextraction Induced by Citric Acid. *J. Radioanal. Nucl. Chem.* **2010**, *285*, 279–285. [CrossRef]
23. Stojanović, M.D.; Stevanović, D.R.; Milojković, J.V.; Grubišić, M.S.; Ileš, D.A. Phytotoxic Effect of the Uranium on the Growing Up and Development the Plant of Corn. *Water Air Soil Pollut.* **2010**, *209*, 401–410. [CrossRef]
24. Hou, J.; Wang, C.; Zhou, Y.; Li, S.; Hayat, T.; Alsaedi, A.; Wang, X. Effects of Uranium Stress on Physiological and Biochemical Characteristics in Seedlings of Six Common Edible Vegetables. *J. Radioanal. Nucl. Chem.* **2018**, *316*, 1001–1010. [CrossRef]
25. Lai, J.; Liu, Z.; Luo, X. A Metabolomic, Transcriptomic Profiling, and Mineral Nutrient Metabolism Study of the Phytotoxicity Mechanism of Uranium. *J. Hazard. Mater.* **2020**, *386*, 121437. [CrossRef] [PubMed]
26. Duquène, L.; Tack, F.; Meers, E.; Baeten, J.; Wannijn, J.; Vandenhove, H. Effect of Biodegradable Amendments on Uranium Solubility in Contaminated Soils. *Sci. Total Environ.* **2008**, *391*, 26–33. [CrossRef] [PubMed]
27. Hu, N.; Lang, T.; Ding, D.; Hu, J.; Li, C.; Zhang, H.; Li, G. Enhancement of Repeated Applications of Chelates on Phytoremediation of Uranium Contaminated Soil by *Macleaya cordata*. *J. Environ. Radioact.* **2019**, *199–200*, 58–65. [CrossRef] [PubMed]
28. Chen, L.; Yang, J.; Wang, D. Phytoremediation of Uranium and Cadmium Contaminated Soils by Sunflower (*Helianthus Annuus* L.) Enhanced with Biodegradable Chelating Agents. *J. Clean Prod.* **2020**, *263*, 121491. [CrossRef]
29. Balaram, V.; Rani, A.; Rathore, D.P.S. Uranium in Groundwater in Parts of India and World: A Comprehensive Review of Sources, Impact to the Environment and Human Health, Analytical Techniques, and Mitigation Technologies. *Geosyst. Geoenviron.* **2022**, *1*, 100043. [CrossRef]
30. Kurttio, P.; Harmoinen, A.; Saha, H.; Salonen, L.; Karpas, Z.; Komulainen, H.; Auvinen, A. Kidney Toxicity of Ingested Uranium From Drinking Water. *Am. J. Kidney Dis.* **2006**, *47*, 972–982. [CrossRef]
31. Bjørklund, G.; Semenova, Y.; Pivina, L.; Dadar, M.; Rahman, M.M.; Aaseth, J.; Chirumbolo, S. Uranium in Drinking Water: A Public Health Threat. *Arch. Toxicol.* **2020**, *94*, 1551–1560. [CrossRef]
32. Arzuaga, X.; Gehlhaus, M.; Strong, J. Modes of Action Associated with Uranium Induced Adverse Effects in Bone Function and Development. *Toxicol. Lett.* **2015**, *236*, 123–130. [CrossRef]
33. Guéguen, Y.; Souidi, M.; Baudelin, C.; Dudoignon, N.; Grison, S.; Dublineau, I.; Marquette, C.; Voisin, P.; Gourmelon, P.; Aigueperse, J. Short-Term Hepatic Effects of Depleted Uranium on Xenobiotic and Bile Acid Metabolizing Cytochrome P450 Enzymes in the Rat. *Arch. Toxicol.* **2006**, *80*, 187–195. [CrossRef]
34. Wang, S.; Ran, Y.; Lu, B.; Li, J.; Kuang, H.; Gong, L.; Hao, Y. A Review of Uranium-Induced Reproductive Toxicity. *Biol. Trace Elem. Res.* **2020**, *196*, 204–213. [CrossRef]
35. Angenard, G.; Muczynski, V.; Coffigny, H.; Duquenne, C.; Frydman, R.; Habert, R.; Livera, G.; Rouiller-Fabre, V. In Vitro Effects of Uranium on Human Fetal Germ Cells. *Reprod. Toxicol.* **2011**, *31*, 470–476. [CrossRef]
36. Saint-Marc, B.; Elie, C.; Manens, L.; Tack, K.; Benderitter, M.; Gueguen, Y.; Ibanez, C. Chronic Uranium Contamination Alters Spinal Motor Neuron Integrity via Modulation of SMN1 Expression and Microglia Recruitment. *Toxicol. Lett.* **2016**, *254*, 37–44. [CrossRef] [PubMed]
37. Aide, M.T. Chemical Thermodynamics of Uranium in the Soil Environment. In *Uranium—Safety, Resources, Separation and Thermodynamic Calculation*; IntechOpen: London, UK, 2017; ISBN 978-1-78923-119-9.
38. Wang, X.; Shi, Z.; Kinniburgh, D.G.; Zhao, L.; Ni, S.; Wang, R.; Hou, Y.; Cheng, K.; Zhu, B. Effect of Thermodynamic Database Selection on the Estimated Aqueous Uranium Speciation. *J. Geochem. Explor.* **2019**, *204*, 33–42. [CrossRef]
39. Sarkar, D.; Andra, S.S.; Saminathan, S.K.M.; Datta, R. Chelant-Aided Enhancement of Lead Mobilization in Residential Soils. *Environ. Pollut.* **2008**, *156*, 1139–1148. [CrossRef] [PubMed]
40. Bucheli-Witschel, M.; Egli, T. Environmental Fate and Microbial Degradation of Aminopolycarboxylic Acids. *Fems Microbiol. Rev.* **2001**, *25*, 69–106. [CrossRef]
41. Egli, T. Biodegradation of Metal-Complexing Aminopolycarboxylic Acids. *J. Biosci. Bioeng.* **2001**, *92*, 89–97. [CrossRef]
42. Hu, Y.-H.; Wei, S.-H.; Zhou, Q.-X.; Zhan, J.; Ma, L.-H.; Niu, R.-C.; Li, Y.-M.; Wang, S.-S. Application of Chelator in Phytoremediation of Heavy Metals Contaminated Soils: A Review. *J. Agro-Environ. Sci.* **2010**, *29*, 2055–2063.
43. Bian, X.; Cui, J.; Tang, B.; Yang, L. Chelant-Induced Phytoextraction of Heavy Metals from Contaminated Soils: A Review. *Pol. J. Environ. Stud.* **2018**, *27*, 2417–2424. [CrossRef]
44. Pinto, I.S.S.; Neto, I.F.F.; Soares, H.M.V.M. Biodegradable Chelating Agents for Industrial, Domestic, and Agricultural Applications—a Review. *Environ. Sci. Pollut. Res.* **2014**, *21*, 11893–11906. [CrossRef]
45. Huang, G.; You, J.; Zhou, X.; Ren, C.; Islam, M.S.; Hu, H. Effects of Low Molecular Weight Organic Acids on Cu Accumulation by Castor Bean and Soil Enzyme Activities. *Ecotox. Environ. Safe* **2020**, *203*, 110983. [CrossRef]
46. Lai, H.-Y. Negative Effects of Chelants on Soil Qualities of Five Soil Series. *Int. J. Phytoremediat.* **2014**, *17*, 228–234. [CrossRef] [PubMed]
47. Meers, E.; Tack, F.M.G.; Verloo, M.G. Degradability of Ethylenediaminedisuccinic Acid (EDDS) in Metal Contaminated Soils: Implications for Its Use Soil Remediation. *Chemosphere* **2008**, *70*, 358–363. [CrossRef] [PubMed]
48. Baek, N.H.; Clesceri, N.L. NTA Biodegradation and Removal in Subsurface Sandy Soil. *Water Res.* **1986**, *20*, 345–349. [CrossRef]

49. Means, J.L.; Kucak, T.; Crerar, D.A. Relative Degradation Rates of NTA, EDTA and DTPA and Environmental Implications. *Environ. Pollut.* **1980**, *1*, 45–60. [CrossRef]
50. Wen, J.; Stacey, S.P.; McLaughlin, M.J.; Kirby, J.K. Biodegradation of Rhamnolipid, EDTA and Citric Acid in Cadmium and Zinc Contaminated Soils. *Soil Biol. Biochem.* **2009**, *41*, 2214–2221. [CrossRef]
51. Gunina, A.; Smith, A.R.; Kuzyakov, Y.; Jones, D.L. Microbial Uptake and Utilization of Low Molecular Weight Organic Substrates in Soil Depend on Carbon Oxidation State. *Biogeochemistry* **2017**, *133*, 89–100. [CrossRef]
52. Smith, R.M.; Martell, A.E. *Critical Stability Constants Volume 6: Second Supplement*; Springer: New York, NY, USA, 1989; Volume 6, ISBN 978-1-4615-6766-0.
53. Leydier, A.; Lin, Y.; Arrachart, G.; Turgis, R.; Lecerclé, D.; Favre-Reguillon, A.; Taran, F.; Lemaire, M.; Pellet-Rostaing, S. EDTA and DTPA Modified Ligands as Sequestering Agents for Uranyl Decorporation. *Tetrahedron* **2012**, *68*, 1163–1170. [CrossRef]
54. Smith, R.M.; Martell, A.E. Critical Stability Constants, Enthalpies and Entropies for the Formation of Metal Complexes of Aminopolycarboxylic Acids and Carboxylic Acids. *Sci. Total Environ.* **1987**, *64*, 125–147. [CrossRef]
55. Chen, C.-Y.; Chen, C.-Y. Stability Constants of Polymer-Bound Iminodiacetate-Type Chelating Agents with Some Transition-Metal Ions. *J. Appl. Polym. Sci.* **2002**, *86*, 1986–1994. [CrossRef]
56. Vukovic, S.; Hay, B.P.; Bryantsev, V.S. Predicting Stability Constants for Uranyl Complexes Using Density Functional Theory. *Inorg. Chem.* **2015**, *54*, 3995–4001. [CrossRef]
57. Singh, J.; Srivastav, A.N.; Singh, N.; Singh, A. Stability Constants of Metal Complexes in Solution. In *Stability and Applications of Coordination Compounds*; IntechOpen: London, UK, 2019; ISBN 978-1-83880-058-1.
58. Tandy, S.; Bossart, K.; Mueller, R.; Ritschel, J.; Hauser, L.; Schulin, R.; Nowack, B. Extraction of Heavy Metals from Soils Using Biodegradable Chelating Agents. *Environ. Sci. Technol.* **2004**, *38*, 937–944. [CrossRef] [PubMed]
59. Jiang, X.J.; Luo, Y.M.; Zhao, Q.G.; Baker, A.J.M.; Christie, P.; Wong, M.H. Soil Cd Availability to Indian Mustard and Environmental Risk Following EDTA Addition to Cd-Contaminated Soil. *Chemosphere* **2003**, *50*, 813–818. [CrossRef]
60. Wu, L.H.; Luo, Y.M.; Xing, X.R.; Christie, P. EDTA-Enhanced Phytoremediation of Heavy Metal Contaminated Soil with Indian Mustard and Associated Potential Leaching Risk. *Agric. Ecosyst. Environ.* **2004**, *102*, 307–318. [CrossRef]
61. Liphadzi, M.S.; Kirkham, M.B. Availability and Plant Uptake of Heavy Metals in EDTA-Assisted Phytoremediation of Soil and Composted Biosolids. *S. Afr. J. Bot.* **2006**, *72*, 391–397. [CrossRef]
62. Na, Z.; Yanshan, C.; Yu, F. Leaching of Cd and Pb from a Contaminated Soil with EDTA and EDDS: Effectiveness and Health Risk Assessment. *Environ. Chem.* **2011**, *30*, 958–963.
63. Udovic, M.; Lestan, D. EDTA and HCl Leaching of Calcareous and Acidic Soils Polluted with Potentially Toxic Metals: Remediation Efficiency and Soil Impact. *Chemosphere* **2012**, *88*, 718–724. [CrossRef]
64. Luo, C.; Shen, Z.; Li, X. Enhanced Phytoextraction of Cu, Pb, Zn and Cd with EDTA and EDDS. *Chemosphere* **2005**, *59*, 1–11. [CrossRef]
65. Beiyuan, J.; Tsang, D.C.W.; Valix, M.; Baek, K.; Ok, Y.S.; Zhang, W.; Bolan, N.S.; Rinklebe, J.; Li, X.-D. Combined Application of EDDS and EDTA for Removal of Potentially Toxic Elements under Multiple Soil Washing Schemes. *Chemosphere* **2018**, *205*, 178–187. [CrossRef]
66. Tao, Y.; Brigante, M.; Zhang, H.; Mailhot, G. Phenanthrene Degradation Using Fe(III)-EDDS Photoactivation under Simulated Solar Light: A Model for Soil Washing Effluent Treatment. *Chemosphere* **2019**, *236*, 124366. [CrossRef]
67. Zhao, Y.; Li, H.; Li, B.; Lai, Y.; Zang, L.; Tang, X. Process Design and Validation of a New Mixed Eluent for Leaching Cd, Cr, Pb, Cu, Ni, and Zn from Heavy Metal-Polluted Soil. *Anal. Methods* **2021**, *13*, 1269–1277. [CrossRef]
68. Leibold, E.; Deckardt, K.; Mellert, W.; Potthoff-Karl, B.; Grundler, O.; Jäckh, R. NTA and Fe(III)NTA: Differential Patterns of Renal Toxicity in Subchronic Studies. *Hum. Exp. Toxicol.* **2002**, *21*, 445–452. [CrossRef] [PubMed]
69. Toyokuni, S. Reactive Oxygen Species-Induced Molecular Damage and Its Application in Pathology. *Pathol. Int.* **1999**, *49*, 91–102. [CrossRef] [PubMed]
70. Iqbal, M.; Okazaki, Y.; Okada, S. Curcumin Attenuates Oxidative Damage in Animals Treated with a Renal Carcinogen, Ferric Nitrilotriacetate (Fe-NTA): Implications for Cancer Prevention. *Mol. Cell. Biochem.* **2009**, *324*, 157–164. [CrossRef] [PubMed]
71. Hirayama, T. Development of Chemical Tools for Imaging of Fe(II) Ions in Living Cells: A Review. *Acta Histochem. Cytochem.* **2018**, *51*, 137–143. [CrossRef]
72. Huang, J.W.; Blaylock, M.J.; Kapulnik, Y.; Ensley, B.D. Phytoremediation of Uranium-Contaminated Soils: Role of Organic Acids in Triggering Uranium Hyperaccumulation in Plants. *Environ. Sci. Technol.* **1998**, *32*, 2004–2008. [CrossRef]
73. Gavrilescu, M.; Pavel, L.V.; Cretescu, I. Characterization and Remediation of Soils Contaminated with Uranium. *J. Hazard. Mater.* **2009**, *163*, 475–510. [CrossRef]
74. Prieto, C.; Lozano, J.C.; Rodríguez, P.B.; Tomé, F.V. Enhancing Radium Solubilization in Soils by Citrate, EDTA, and EDDS Chelating Amendments. *J. Hazard. Mater.* **2013**, *250–251*, 439–446. [CrossRef]
75. Orr, R.; Hocking, R.K.; Pattison, A.; Nelson, P.N. Extraction of Metals from Mildly Acidic Tropical Soils: Interactions between Chelating Ligand, pH and Soil Type. *Chemosphere* **2020**, *248*, 126060. [CrossRef]
76. Pepper, I.L.; Brusseau, M.L. Chapter 2—Physical-Chemical Characteristics of Soils and the Subsurface. In *Environmental and Pollution Science*, 3rd ed.; Brusseau, M.L., Pepper, I.L., Gerba, C.P., Eds.; Academic Press: Cambridge, MA, USA, 2019; ISBN 978-0-12-814719-1.

77. Rakhsh, F.; Golchin, A.; Beheshti Al Agha, A.; Nelson, P.N. Mineralization of Organic Carbon and Formation of Microbial Biomass in Soil: Effects of Clay Content and Composition and the Mechanisms Involved. *Soil Biol. Biochem.* **2020**, *151*, 108036. [CrossRef]
78. Crançon, P.; van der Lee, J. Speciation and Mobility of Uranium(VI) in Humic-Containing Soils. *Radiochim. Acta* **2003**, *91*, 673–679. [CrossRef]
79. Kantar, C. Heterogeneous Processes Affecting Metal Ion Transport in the Presence of Organic Ligands: Reactive Transport Modeling. *Earth-Sci. Rev.* **2007**, *81*, 175–198. [CrossRef]
80. Ratnikov, A.N.; Sviridenko, D.G.; Popova, G.I.; Sanzharova, N.I.; Mikailova, R.A. The Behaviour of Uranium in Soils and the Mechanisms of Its Accumulation by Agricultural Plants. In *Uranium in Plants and the Environment*; Gupta, D.K., Walther, C., Eds.; Springer International Publishing: Cham, Switzerland, 2020; ISBN 978-3-030-14961-1.
81. Jiang, X.; Li, X.; Zhang, J.; Lv, J.; Chang, J.; Cui, D. Study of Citric Acid on Extraction and Washing from Different Soil Polluted by Cr. *Chin. Agric. Sci. Bull.* **2012**, *28*, 278–281.
82. Wang, K.; Liu, Y.; Song, Z.; Khan, Z.H.; Qiu, W. Effects of Biodegradable Chelator Combination on Potentially Toxic Metals Leaching Efficiency in Agricultural Soils. *Ecotox. Environ. Safe* **2019**, *182*, 109399. [CrossRef] [PubMed]
83. Yang, R.; Luo, C.; Zhang, G.; Li, X.; Shen, Z. Extraction of Heavy Metals from E-Waste Contaminated Soils Using EDDS. *J. Environ. Sci.* **2012**, *24*, 1985–1994. [CrossRef]
84. Kantar, C.; Honeyman, B.D. Citric Acid Enhanced Remediation of Soils Contaminated with Uranium by Soil Flushing and Soil Washing. *J. Environ. Eng.* **2006**, *132*, 247–255. [CrossRef]
85. Ebbs, S.D.; Norvell, W.A.; Kochian, L.V. The Effect of Acidification and Chelating Agents on the Solubilization of Uranium from Contaminated Soil. *J. Environ. Qual.* **1998**, *27*, 1486–1494. [CrossRef]
86. Lozano, J.C.; Blanco Rodríguez, P.; Vera Tomé, F.; Calvo, C.P. Enhancing Uranium Solubilization in Soils by Citrate, EDTA, and EDDS Chelating Amendments. *J. Hazard. Mater.* **2011**, *198*, 224–231. [CrossRef]
87. Sarthou, M.C.M.; Devime, F.; Baggio, C.; Figuet, S.; Alban, C.; Bourguignon, J.; Ravanel, S. Calcium-Permeable Cation Channels Are Involved in Uranium Uptake in *Arabidopsis thaliana*. *J. Hazard. Mater.* **2022**, *424*, 127436. [CrossRef]
88. Croteau, M.-N.; Fuller, C.C.; Cain, D.J.; Campbell, K.M.; Aiken, G. Biogeochemical Controls of Uranium Bioavailability from the Dissolved Phase in Natural Freshwaters. Available online: https://pubs.acs.org/doi/pdf/10.1021/acs.est.6b02406 (accessed on 1 April 2022).
89. Shahandeh, H.; Hossner, L.R. Enhancement of Uranium Phytoaccumulation from Contaminated Soils. *Soil Sci.* **2002**, *167*, 269–280. [CrossRef]
90. Shahandeh, H.; Hossner, L. Role of Soil Properties in Phytoaccumulation of Uranium. *Water Air Soil Pollut.* **2002**, *141*, 165–180. [CrossRef]
91. Laurette, J.; Larue, C.; Llorens, I.; Jaillard, D.; Jouneau, P.-H.; Bourguignon, J.; Carrière, M. Speciation of Uranium in Plants upon Root Accumulation and Root-to-Shoot Translocation: A XAS and TEM Study. *Environ. Exp. Bot.* **2012**, *77*, 87–95. [CrossRef]
92. Shahid, M.; Austruy, A.; Echevarria, G.; Arshad, M.; Sanaullah, M.; Aslam, M.; Nadeem, M.; Nasim, W.; Dumat, C. EDTA-Enhanced Phytoremediation of Heavy Metals: A Review. *Soil. Sediment. Contam.* **2013**, *23*, 389–416. [CrossRef]
93. Zhou, P.; Gu, B. Extraction of Oxidized and Reduced Forms of Uranium from Contaminated Soils: Effects of Carbonate Concentration and pH. *Environ. Sci. Technol.* **2005**, *39*, 4435–4440. [CrossRef] [PubMed]
94. Doustaly, F.; Combes, F.; Fiévet, J.B.; Berthet, S.; Hugouvieux, V.; Bastien, O.; Aranjuelo, I.; Leonhardt, N.; Rivasseau, C.; Carrière, M.; et al. Uranium Perturbs Signaling and Iron Uptake Response in *Arabidopsis thaliana* Roots. *Metallomics* **2014**, *6*, 809–821. [CrossRef] [PubMed]
95. Berthet, S.; Villiers, F.; Alban, C.; Serre, N.B.C.; Martin-Laffon, J.; Figuet, S.; Boisson, A.-M.; Bligny, R.; Kuntz, M.; Finazzi, G.; et al. *Arabidopsis thaliana* Plants Challenged with Uranium Reveal New Insights into Iron and Phosphate Homeostasis. *New Phytol.* **2018**, *217*, 657–670. [CrossRef] [PubMed]
96. Malaviya, P.; Singh, A. Phytoremediation Strategies for Remediation of Uranium-Contaminated Environments: A Review. *Crit. Rev. Environ. Sci. Technol.* **2012**, *42*, 2575–2647. [CrossRef]
97. Chang, P.; Kim, K.-W.; Yoshida, S.; Kim, S.-Y. Uranium Accumulation of Crop Plants Enhanced by Citric Acid. *Environ. Geochem. Health* **2005**, *27*, 529–538. [CrossRef]
98. Rong, L.; Zhang, S.; Wang, J.; Li, S.; Xie, S.; Wang, G. Phytoremediation of Uranium-Contaminated Soil by Perennial Ryegrass (*Lolium perenne* L.) Enhanced with Citric Acid Application. *Environ. Sci. Pollut. Res.* **2022**, *29*, 33002–33012. [CrossRef]
99. Li, C.; Dan, W.; Chan, L.; Cui, Z. Effect of Three Kinds of Chelating Agents on Phytoremediation in U and Cd Contaminated Soil by *Zebrina pendula*. *Atomic Energy Sci. Technol.* **2018**, *52*, 1359–1366. [CrossRef]
100. Manara, A.; Fasani, E.; Furini, A.; DalCorso, G. Evolution of the Metal Hyperaccumulation and Hypertolerance Traits. *Plant Cell Environ.* **2020**, *43*, 2969–2986. [CrossRef]
101. Sevostianova, E.; Lindemann, W.C.; Ulery, A.L.; Remmenga, M.D. Plant Uptake of Depleted Uranium from Manure-Amended and Citrate Treated Soil. *Int. J. Phytoremediat.* **2010**, *12*, 550–561. [CrossRef] [PubMed]
102. Wan, Q.; Pan, N.; Jin, Y.; Xia, C. Study on remediation for uranium contaminated soils enhanced by chelator using brassica mustard. In *Progress Report on Nuclear Science and Technology in China*; Atomic Energy Press: Beijing, China, 2012; Volume 2, ISBN 978-7-5022-5602-9.

103. Qi, F.; Zha, Z.; Du, L.; Feng, X.; Wang, D.; Zhang, D.; Fang, Z.; Ma, L.; Jin, Y.; Xia, C. Impact of Mixed Low-Molecular-Weight Organic Acids on Uranium Accumulation and Distribution in a Variant of Mustard (*Brassica juncea* VarTumida). *J. Radioanal. Nucl. Chem.* **2014**, *302*, 149–159. [CrossRef]
104. Nowack, B.; Schulin, R.; Robinson, B.H. Critical Assessment of Chelant-Enhanced Metal Phytoextraction. *Environ. Sci. Technol.* **2006**, *40*, 5225–5232. [CrossRef] [PubMed]
105. Li, H.; Wang, Q.; Cui, Y.; Dong, Y.; Christie, P. Slow Release Chelate Enhancement of Lead Phytoextraction by Corn (*Zea mays* L.) from Contaminated Soil—A Preliminary Study. *Sci. Total Environ.* **2005**, *339*, 179–187. [CrossRef] [PubMed]
106. Yadav, S.K. Heavy Metals Toxicity in Plants: An Overview on the Role of Glutathione and Phytochelatins in Heavy Metal Stress Tolerance of Plants. *S. Afr. J. Bot.* **2010**, *76*, 167–179. [CrossRef]
107. Yan, A.; Wang, Y.; Tan, S.N.; Mohd Yusof, M.L.; Ghosh, S.; Chen, Z. Phytoremediation: A Promising Approach for Revegetation of Heavy Metal-Polluted Land. *Front. Plant Sci.* **2020**, *11*, 359. [CrossRef] [PubMed]
108. Pietrini, I.; Grifoni, M.; Franchi, E.; Cardaci, A.; Pedron, F.; Barbafieri, M.; Petruzzelli, G.; Vocciante, M. Enhanced Lead Phytoextraction by Endophytes from Indigenous Plants. *Soil Syst.* **2021**, *5*, 55. [CrossRef]
109. Römkens, P.; Bouwman, L.; Japenga, J.; Draaisma, C. Potentials and Drawbacks of Chelate-Enhanced Phytoremediation of Soils. *Environ. Pollut.* **2002**, *116*, 109–121. [CrossRef]
110. Epelde, L.; Ma Becerril, J.; Alkorta, I.; Garbisu, C. Heavy Metal Phytoremediation: Microbial Indicators of Soil Health for the Assessment of Remediation Efficiency. In *Advances in Applied Bioremediation*; Singh, A., Kuhad, R.C., Ward, O.P., Eds.; Springer: Berlin/Heidelberg, Germany, 2009; ISBN 978-3-540-89621-0.
111. Lee, J.; Sung, K. Effects of Chelates on Soil Microbial Properties, Plant Growth and Heavy Metal Accumulation in Plants. *Ecol. Eng.* **2014**, *73*, 386–394. [CrossRef]
112. Cao, A.; Carucci, A.; Lai, T.; La Colla, P.; Tamburini, E. Effect of Biodegradable Chelating Agents on Heavy Metals Phytoextraction with *Mirabilis jalapa* and on Its Associated Bacteria. *Eur. J. Soil Biol.* **2007**, *43*, 200–206. [CrossRef]
113. Barona, A.; Aranguiz, I.; Elías, A. Metal Associations in Soils before and after EDTA Extractive Decontamination: Implications for the Effectiveness of Further Clean-up Procedures. *Environ. Pollut.* **2001**, *113*, 79–85. [CrossRef]
114. Udovic, M.; Lestan, D. Pb, Zn and Cd Mobility, Availability and Fractionation in Aged Soil Remediated by EDTA Leaching. *Chemosphere* **2009**, *74*, 1367–1373. [CrossRef] [PubMed]
115. Huang, X.; Luo, D.; Chen, X.; Wei, L.; Liu, Y.; Wu, Q.; Xiao, T.; Mai, X.; Liu, G.; Liu, L. Insights into Heavy Metals Leakage in Chelator-Induced Phytoextraction of Pb- and Tl-Contaminated Soil. *Int. J. Environ. Res. Public Health* **2019**, *16*, 1328. [CrossRef] [PubMed]
116. Thayalakumaran, T.; Robinson, B.; Vogeler, I.; Scotter, D.; Clothier, B.; Percival, H. Plant Uptake and Leaching of Copper during EDTA-Enhanced Phytoremediation of Repacked and Undisturbed Soil. *Plant Soil* **2003**, *254*, 415–423. [CrossRef]
117. Mastromatteo, M.; Mastromatteo, M.; Conte, A.; del Nobile, M.A. Advances in Controlled Release Devices for Food Packaging Applications. *Trends Food Sci. Technol.* **2010**, *21*, 591–598. [CrossRef]
118. Yusoff, S.N.M.; Kamari, A.; Aljafree, N.F.A. A Review of Materials Used as Carrier Agents in Pesticide Formulations. *Int. J. Environ. Sci. Technol.* **2016**, *13*, 2977–2994. [CrossRef]
119. Al-Rawajfeh, A.E.; Alrbaihat, M.R.; AlShamaileh, E.M. Chapter 4—Characteristics and Types of Slow- and Controlled-Release Fertilizers. In *Controlled Release Fertilizers for Sustainable Agriculture*; Lewu, F.B., Volova, T., Thomas, S., Rakhimol, K.R., Eds.; Academic Press: Cambridge, MA, USA, 2021; ISBN 978-0-12-819555-0.
120. Lubkowski, K.; Grzmil, B. Controlled Release Fertilizers. *Pol. J. Chem. Technol.* **2007**, *9*, 83–84. [CrossRef]
121. Liu, G.; Zotarelli, L.; Li, Y.; Dinkins, D.; Wang, Q.; Ozores-Hampton, M. Controlled-Release and Slow-Release Fertilizers as Nutrient Management Tools. Available online: https://edis.ifas.ufl.edu/publication/HS1255 (accessed on 21 September 2021).
122. Almasi, H.; Oskouie, M.J.; Saleh, A. A Review on Techniques Utilized for Design of Controlled Release Food Active Packaging. *Crit. Rev. Food Sci. Nutr.* **2020**, *61*, 2601–2621. [CrossRef]
123. Kuai, L.; Liu, F.; Chiou, B.-S.; Avena-Bustillos, R.J.; McHugh, T.H.; Zhong, F. Controlled Release of Antioxidants from Active Food Packaging: A Review. *Food Hydrocoll.* **2021**, *120*, 106992. [CrossRef]
124. Vasile, C.; Baican, M. Progresses in Food Packaging, Food Quality, and Safety—Controlled-Release Antioxidant and/or Antimicrobial Packaging. *Molecules* **2021**, *26*, 1263. [CrossRef]
125. Heirlings, L.; Siró, I.; Devlieghere, F.; Bavel, E.V.; Cool, P.; Meulenaer, B.D.; Vansant, E.F.; Debevere, J. Influence of Polymer Matrix and Adsorption onto Silica Materials on the Migration of α-Tocopherol into 95% Ethanol from Active Packaging. *Food Addit. Contam.* **2007**, *21*, 1125–1136. [CrossRef] [PubMed]
126. Cerisuelo, J.P.; Bermúdez, J.M.; Aucejo, S.; Catalá, R.; Gavara, R.; Hernández-Muñoz, P. Describing and Modeling the Release of an Antimicrobial Agent from an Active PP/EVOH/PP Package for Salmon. *J. Food Eng.* **2013**, *116*, 352–361. [CrossRef]
127. LaCoste, A.; Schaich, K.M.; Zumbrunnen, D.; Yam, K.L. Advancing Controlled Release Packaging through Smart Blending. *Packag. Technol. Sci.* **2005**, *18*, 77–87. [CrossRef]
128. Han, J.H.; Floros, J.D. Simulating Diffusion Model and Determining Diffusivity of Potassium Sorbate Through Plastics to Develop Antimicrobial Packaging Films. *J. Food Process Preserv.* **1998**, *22*, 107–122. [CrossRef]
129. Whelehan, M.; Marison, I.W. Microencapsulation Using Vibrating Technology. *J. Microencapsul.* **2011**, *28*, 669–688. [CrossRef] [PubMed]

130. Rathore, S.; Desai, P.M.; Liew, C.V.; Chan, L.W.; Heng, P.W.S. Microencapsulation of Microbial Cells. *J. Food Eng.* **2013**, *116*, 369–381. [CrossRef]
131. Xie, Z.; Wu, L.; Chen, N.; Liu, C.; Zheng, Y.; Xu, S.; Li, F.; Xu, Y. Phytoextraction of Pb and Cu Contaminated Soil With Maize and Microencapsulated EDTA. *Int. J. Phytoremediat.* **2012**, *14*, 727–740. [CrossRef]
132. Ding, C.; Li, Y. A Slow-Release Chelating Agent for Heavy Metal Soil Remediation. Patent Registration Number 201910960626, 14 January 2020.
133. Ocak, B. Complex Coacervation of Collagen Hydrolysate Extracted from Leather Solid Wastes and Chitosan for Controlled Release of Lavender Oil. *J. Environ. Manag.* **2012**, *100*, 22–28. [CrossRef]
134. O'Connor, D.; Hou, D.; Ok, Y.S.; Song, Y.; Sarmah, A.K.; Li, X.; Tack, F.M.G. Sustainable in Situ Remediation of Recalcitrant Organic Pollutants in Groundwater with Controlled Release Materials: A Review. *J. Control Release* **2018**, *283*, 200–213. [CrossRef]
135. Christenson, M.; Kambhu, A.; Reece, J.; Comfort, S.; Brunner, L. A Five-Year Performance Review of Field-Scale, Slow-Release Permanganate Candles with Recommendations for Second-Generation Improvements. *Chemosphere* **2016**, *150*, 239–247. [CrossRef]
136. Tang, X.; Yu, C.; Lei, Y.; Wang, Z.; Wang, C.; Wang, J. A Novel Chitosan-Urea Encapsulated Material for Persulfate Slow-Release to Degrade Organic Pollutants. *J. Hazard. Mater.* **2022**, *426*, 128083. [CrossRef] [PubMed]
137. Tsai, T.T.; Liu, J.K.; Chang, Y.M.; Chen, K.F.; Kao, C.M. Application of Polycolloid-Releasing Substrate to Remediate Trichloroethylene-Contaminated Groundwater: A Pilot-Scale Study. *J. Hazard. Mater.* **2014**, *268*, 92–101. [CrossRef] [PubMed]
138. Yeum, Y.; Han, K.; Kang, J.-H.; Kim, D.-W.; Park, C.-W.; Kwon, S.; Kim, Y. Production, Characterization, and Evaluation of Two Types of Slow-Releasing Carbon Source Tablets for in-Situ Heterotrophic Nitrate Denitrification in Aquifers. *Chemosphere* **2020**, *260*, 127478. [CrossRef] [PubMed]
139. Essawy, H.A.; Ghazy, M.B.M.; El-Hai, F.A.; Mohamed, M.F. Superabsorbent Hydrogels via Graft Polymerization of Acrylic Acid from Chitosan-Cellulose Hybrid and Their Potential in Controlled Release of Soil Nutrients. *Int. J. Biol. Macromol.* **2016**, *89*, 144–151. [CrossRef] [PubMed]
140. Pang, L.; Gao, Z.; Feng, H.; Wang, S.; Wang, Q. Cellulose Based Materials for Controlled Release Formulations of Agrochemicals: A Review of Modifications and Applications. *J. Control Release* **2019**, *316*, 105–115. [CrossRef]
141. Wang, G.; Wang, B.; Fan, W.; Deng, N. Enhanced Phytoremediation of Uranium-Contaminated Soils by Indian Mustard (*Brassica juncea* L.) Using Slow Release Citric Acid. *Environ. Sci. Pollut. Res.* **2021**, *28*, 61061–61071. [CrossRef]
142. Shibata, M.; Konno, T.; Akaike, R.; Xu, Y.; Shen, R.; Ma, J.F. Phytoremediation of Pb Contaminated Soil with Polymer-Coated EDTA. *Plant Soil* **2007**, *290*, 201–208. [CrossRef]
143. Xie, Z.; Chen, N.; Liu, C.; Zhou, J.; Xu, S.; Zheng, Y.; Li, F.; Xu, Y. Synthesis and Characterization of Ethylenediamine Tetraacetic Acid Tetrasodium Salt Loaded in Microcapsules with Slow Release Properties. *Chin. J. Chem. Eng.* **2010**, *18*, 149–155. [CrossRef]
144. Yoo, J.; Won, Y.-Y. Phenomenology of the Initial Burst Release of Drugs from PLGA Microparticles. *ACS Biomater. Sci. Eng.* **2020**, *6*, 6053–6062. [CrossRef]
145. Chen, J.; Lü, S.; Zhang, Z.; Zhao, X.; Li, X.; Ning, P.; Liu, M. Environmentally Friendly Fertilizers: A Review of Materials Used and Their Effects on the Environment. *Sci. Total Environ.* **2018**, *613–614*, 829–839. [CrossRef]
146. Zhang, L.; Zhang, G.; Lu, J.; Liang, H. Preparation and Characterization of Carboxymethyl Cellulose/Polyvinyl Alcohol Blend Film as a Potential Coating Material. *Polym. Plast. Technol. Eng.* **2013**, *52*, 163–167. [CrossRef]
147. Han, X.; Chen, S.; Hu, X. Controlled-Release Fertilizer Encapsulated by Starch/Polyvinyl Alcohol Coating. *Desalination* **2009**, *240*, 21–26. [CrossRef]
148. Anghel, N.; Marius, N.; Spiridon, I. Heavy Metal Adsorption Ability of a New Composite Material Based on Starch Strengthened with Chemically Modified Cellulose. *Polym. Adv. Technol.* **2019**, *30*, 1453–1460. [CrossRef]
149. Wan, M.-W.; Petrisor, I.G.; Lai, H.-T.; Kim, D.; Yen, T.F. Copper Adsorption through Chitosan Immobilized on Sand to Demonstrate the Feasibility for in Situ Soil Decontamination. *Carbohydr. Polym.* **2004**, *55*, 249–254. [CrossRef]
150. Zhou, Y.; Gao, B.; Zimmerman, A.R.; Fang, J.; Sun, Y.; Cao, X. Sorption of Heavy Metals on Chitosan-Modified Biochars and Its Biological Effects. *Chem. Eng. J.* **2013**, *231*, 512–518. [CrossRef]
151. Li, B.; Li, M.; Zhang, P.; Pan, Y.; Huang, Z.; Xiao, H. Remediation of Cd (II) Ions in Aqueous and Soil Phases Using Novel Porous Cellulose/Chitosan Composite Spheres Loaded with Zero-Valent Iron Nanoparticles. *React. Funct. Polym.* **2022**, *173*, 105210. [CrossRef]
152. O'Connell, D.W.; Birkinshaw, C.; O'Dwyer, T.F. Heavy Metal Adsorbents Prepared from the Modification of Cellulose: A Review. *Bioresour. Technol.* **2008**, *99*, 6709–6724. [CrossRef]
153. Kos, B.; Le tan, D. Influence of a Biodegradable ([S,S]-EDDS) and Nondegradable (EDTA) Chelate and Hydrogel Modified Soil Water Sorption Capacity on Pb Phytoextraction and Leaching. *Plant Soil* **2003**, *253*, 403–411. [CrossRef]
154. Noppakundilograt, S.; Pheatcharat, N.; Kiatkamjornwong, S. Multilayer-Coated NPK Compound Fertilizer Hydrogel with Controlled Nutrient Release and Water Absorbency. *J. Appl. Polym. Sci.* **2015**, *132*, 41249. [CrossRef]
155. Ramli, R.A. Slow Release Fertilizer Hydrogels: A Review. *Polym. Chem.* **2019**, *10*, 6073–6090. [CrossRef]
156. Zhang, X.; Liu, Y.; Lu, P.; Zhang, M. Preparation and Properties of Hydrogel Based on Sawdust Cellulose for Environmentally Friendly Slow Release Fertilizers. *Green Process. Synth.* **2020**, *9*, 139–152. [CrossRef]
157. Kareem, S.A.; Dere, I.; Gungula, D.T.; Andrew, F.P.; Saddiq, A.M.; Adebayo, E.F.; Tame, V.T.; Kefas, H.M.; Joseph, J.; Patrick, D.O. Synthesis and Characterization of Slow-Release Fertilizer Hydrogel Based on Hydroxy Propyl Methyl Cellulose, Polyvinyl Alcohol, Glycerol and Blended Paper. *Gels* **2021**, *7*, 262. [CrossRef] [PubMed]

158. Rizwan, M.; Rubina Gilani, S.; Iqbal Durani, A.; Naseem, S. Materials Diversity of Hydrogel: Synthesis, Polymerization Process and Soil Conditioning Properties in Agricultural Field. *J. Adv. Res.* **2021**, *33*, 15–40. [CrossRef] [PubMed]
159. Wang, G.; Zhang, S.; Xu, L.; Li, Q. The Slow-Release Chelating Agent and Its Preparation Method for Phytoremediation of Soil. Patent Registration Number 201610567644.8, 7 December 2016.
160. Chen, L.; Wang, D.; Long, C.; Cui, Z. Effect of Biodegradable Chelators on Induced Phytoextraction of Uranium- and Cadmium-Contaminated Soil by *Zebrina Pendul.* Schnizl. *Sci. Rep.* **2019**, *9*, 19817. [CrossRef]
161. Arshad, M.; Naqvi, N.; Gul, I.; Yaqoob, K.; Bilal, M.; Kallerhoff, J. Lead Phytoextraction by *Pelargonium hortorum*: Comparative Assessment of EDTA and DIPA for Pb Mobility and Toxicity. *Sci. Total Environ.* **2020**, *748*, 141496. [CrossRef]
162. Broschat, T.K.; Moore, K.K. Release Rates of Ammonium-Nitrogen, Nitrate-Nitrogen, Phosphorus, Potassium, Magnesium, Iron, and Manganese from Seven Controlled-Release Fertilizers. *Commun. Soil Sci. Plant Anal.* **2007**, *38*, 843–850. [CrossRef]
163. Dai, J.; Fan, X.; Yu, J.; Liu, F.; Zhang, Q. Study on the Rapid Method to Predict Longevity of Controlled Release Fertilizer Coated by Water Soluble Resin. *Agric. Sci. China* **2008**, *7*, 1127–1132. [CrossRef]
164. Medina, L.C.; Sartain, J.B.; Obreza, T.A.; Hall, W.L.; Thiex, N.J. Evaluation of a Soil Incubation Method to Characterize Nitrogen Release Patterns of Slow- and Controlled-Release Fertilizers. *J. AOAC Int.* **2014**, *97*, 643–660. [CrossRef]
165. Jia, Y.; Hu, Z.; Mu, J.; Zhang, W.; Xie, Z.; Wang, G. Preparation of Biochar as a Coating Material for Biochar-Coated Urea. *Sci. Total Environ.* **2020**, *731*, 139063. [CrossRef]
166. Purnomo, C.W.; Saputra, H. Chapter 6—Manufacturing of Slow and Controlled Release Fertilizer. In *Controlled Release Fertilizers for Sustainable Agriculture*; Lewu, F.B., Volova, T., Thomas, S., Rakhimol, K.R., Eds.; Academic Press: Cambridge, MA, USA, 2021; ISBN 978-0-12-819555-0.
167. Kottegoda, N.; Sandaruwan, C.; Priyadarshana, G.; Siriwardhana, A.; Rathnayake, U.A.; Berugoda Arachchige, D.M.; Kumarasinghe, A.R.; Dahanayake, D.; Karunaratne, V.; Amaratunga, G.A.J. Urea-Hydroxyapatite Nanohybrids for Slow Release of Nitrogen. *ACS Nano* **2017**, *11*, 1214–1221. [CrossRef]
168. Ni, B.; Liu, M.; Lü, S. Multifunctional Slow-Release Urea Fertilizer from Ethylcellulose and Superabsorbent Coated Formulations. *Chem. Eng. J.* **2009**, *155*, 892–898. [CrossRef]
169. Xie, Z.; Chen, N.; Xu, S.; Xu, Y. A Complexing Agent Microcapsule for Phytoremediation of Soil and Its Preparation Method. Patent Registration Number 200810220712.9, 2 September 2009.
170. Cagnasso, C.E.; López, L.B.; Rodríguez, V.G.; Valencia, M.E. Development and Validation of a Method for the Determination of EDTA in Non-Alcoholic Drinks by HPLC. *J. Food Compos. Anal.* **2007**, *20*, 248–251. [CrossRef]
171. Khan, M.S.; Zaidi, A.; Wani, P.A.; Oves, M. Role of Plant Growth Promoting Rhizobacteria in the Remediation of Metal Contaminated Soils. *Environ. Chem. Lett.* **2009**, *7*, 1–19. [CrossRef]
172. Saleh, H.M. Water Hyacinth for Phytoremediation of Radioactive Waste Simulate Contaminated with Cesium and Cobalt Radionuclides. *Nucl. Eng. Des.* **2012**, *242*, 425–432. [CrossRef]
173. Saleh, H.M.; Bayoumi, T.A.; Mahmoud, H.H.; Aglan, R.F. Uptake of Cesium and Cobalt Radionuclides from Simulated Radioactive Wastewater by *Ludwigia stolonifera* Aquatic Plant. *Nucl. Eng. Des.* **2017**, *315*, 194–199. [CrossRef]
174. Saleh, H.M.; Aglan, R.F.; Mahmoud, H.H. *Ludwigia stolonifera* for Remediation of Toxic Metals from Simulated Wastewater. *Chem. Ecol.* **2018**, *35*, 164–178. [CrossRef]
175. Saleh, H.M.; Mahmoud, H.H.; Aglan, R.F.; Bayoumi, T.A. Biological Treatment of Wastewater Contaminated with Cu (II), Fe (II) and Mn (II) Using *Ludwigia stolonifera* Aquatic Plant. *Environ. Eng. Manag. J.* **2019**, *18*, 1327–1336. [CrossRef]
176. Saleh, H.M.; Moussa, H.R.; El-Saied, F.A.; Dawoud, M.; Nouh, E.S.A.; Abdel Wahed, R.S. Adsorption of Cesium and Cobalt onto Dried *Myriophyllum spicatum* L. from Radio-Contaminated Water: Experimental and Theoretical Study. *Prog. Nucl. Energy* **2020**, *125*, 103393. [CrossRef]
177. Saleh, H.M.; Moussa, H.R.; Mahmoud, H.H.; El-Saied, F.A.; Dawoud, M.; Abdel Wahed, R.S. Potential of the Submerged Plant *Myriophyllum spicatum* for Treatment of Aquatic Environments Contaminated with Stable or Radioactive Cobalt and Cesium. *Prog. Nucl. Energy* **2020**, *118*, 103147. [CrossRef]

MDPI
St. Alban-Anlage 66
4052 Basel
Switzerland
Tel. +41 61 683 77 34
Fax +41 61 302 89 18
www.mdpi.com

Sustainability Editorial Office
E-mail: sustainability@mdpi.com
www.mdpi.com/journal/sustainability

www.ingramcontent.com/pod-product-compliance
Lightning Source LLC
Chambersburg PA
CBHW061624210926
51850CB00009B/2428
9783036572482